Student Solutions Manual
CINDY TRIMBLE

Introductory Algebra
THIRD EDITION

Elayn Martin-Gay

PEARSON

Prentice
Hall

Upper Saddle River, NJ 07458

Editor-in-Chief: Chris Hoag
Executive Editor: Paul Murphy
Project Manager: Mary Beckwith
Editorial Assistant: Abigail Rethore
Executive Managing Editor: Kathleen Schiaparelli
Senior Managing Editor: Nicole M. Jackson
Assistant Managing Editor: Karen Bosch Petrov
Production Editor: Jessica Barna
Supplement Cover Manager: Paul Gourhan
Supplement Cover Designer: Christopher Kossa
Manufacturing Buyer: Ilene Kahn
Manufacturing Manager: Alexis Heydt-Long

© 2007 Pearson Education, Inc.
Pearson Prentice Hall
Pearson Education, Inc.
Upper Saddle River, NJ 07458

Printed in the United States of America

10 9 8 7 6 5 4 3 2 1

ISBN   0-13-227606-2   Standalone
       0-13-227607-0   Component

Pearson Education Ltd., *London*
Pearson Education Australia Pty. Ltd., *Sydney*
Pearson Education Singapore, Pte. Ltd.
Pearson Education North Asia Ltd., *Hong Kong*
Pearson Education Canada, Inc., *Toronto*
Pearson Educación de Mexico, S.A. de C.V.
Pearson Education—Japan, *Tokyo*
Pearson Education Malaysia, Pte. Ltd.

Contents

Chapter R

Section R.1 Practice Problems

1. $10 = 1 \cdot 10$, $10 = 2 \cdot 5$
   The factors of 10 are 1, 2, 5, and 10.

2. $18 = 1 \cdot 18$, $18 = 2 \cdot 9$, $18 = 3 \cdot 6$
   The factors of 18 are 1, 2, 3, 6, 9, and 18.

3. 5 is a prime number. Its factors are 1 and 5 only.
   16 is a composite number. Its factors are 1, 2, 4, 8, and 16.
   23 is a prime number. Its factors are 1 and 23 only.
   42 is a composite number. Its factors are 1, 2, 3, 6, 7, 14, 21, and 42.

4. $44 = 4 \cdot 11 = 2 \cdot 2 \cdot 11$
   The prime factorization of 44 is $2 \cdot 2 \cdot 11$.

5. $60 = 4 \cdot 15 = 2 \cdot 2 \cdot 3 \cdot 5$
   The prime factorization of 60 is $2 \cdot 2 \cdot 3 \cdot 5$.

6.
   $$\begin{array}{r} 11 \\ 3\overline{)33} \\ 3\overline{)99} \\ 3\overline{)297} \end{array}$$

   The prime factorization of 297 is $3 \cdot 3 \cdot 3 \cdot 11$.

7. $14 = 2 \cdot 7$
   $35 = 5 \cdot 7$
   $\text{LCM} = 2 \cdot 5 \cdot 7 = 70$

8. $5 = 5$
   $9 = 3 \cdot 3$
   $\text{LCM} = 3 \cdot 3 \cdot 5 = 45$

9. $4 = 2 \cdot 2$
   $15 = 3 \cdot 5$
   $10 = 2 \cdot 5$
   $\text{LCM} = 2 \cdot 2 \cdot 3 \cdot 5 = 60$

Exercise Set R.1

1. $9 = 1 \cdot 9$, $9 = 3 \cdot 3$
   The factors of 9 are 1, 3, and 9.

3. $24 = 1 \cdot 24$, $24 = 2 \cdot 12$, $24 = 3 \cdot 8$, $24 = 4 \cdot 6$
   The factors of 24 are 1, 2, 3, 4, 6, 8, 12, and 24.

5. $42 = 1 \cdot 42$, $42 = 2 \cdot 21$, $42 = 3 \cdot 14$, $42 = 6 \cdot 7$
   The factors of 42 are 1, 2, 3, 6, 7, 14, 21, and 42.

7. $80 = 1 \cdot 80$, $80 = 2 \cdot 40$, $80 = 4 \cdot 20$, $80 = 5 \cdot 16$, $80 = 8 \cdot 10$
   The factors of 80 are 1, 2, 4, 5, 8, 10, 16, 20, 40, and 80.

9. $19 = 1 \cdot 19$
   The factors of 19 are 1 and 19.

11. 13 is a prime number. Its factors are only 1 and 13.

13. 39 is a composite number. Its factors are 1, 3, 13, and 39.

15. 41 is a prime number. Its factors are only 1 and 41.

17. 201 is composite. Its factors are 1, 3, 67, and 201.

19. 2065 is a composite number. Its factors are 1, 5, 7, 35, 59, 295, 413, and 2065.

21. $18 = 2 \cdot 9 = 2 \cdot 3 \cdot 3$
    The prime factorization of 18 is $2 \cdot 3 \cdot 3$.

23. $20 = 4 \cdot 5 = 2 \cdot 2 \cdot 5$
    The prime factorization of 20 is $2 \cdot 2 \cdot 5$.

25. $56 = 2 \cdot 28 = 2 \cdot 2 \cdot 14 = 2 \cdot 2 \cdot 2 \cdot 7$
    The prime factorization of 56 is $2 \cdot 2 \cdot 2 \cdot 7$.

27. $81 = 3 \cdot 27 = 3 \cdot 3 \cdot 9 = 3 \cdot 3 \cdot 3 \cdot 3$
The prime factorization of 81 is $3 \cdot 3 \cdot 3 \cdot 3$.

29.
$$
\begin{array}{r}
5 \\
5\overline{)25} \\
3\overline{)75} \\
2\overline{)150} \\
2\overline{)300}
\end{array}
$$
The prime factorization of 300 is
$2 \cdot 2 \cdot 3 \cdot 5 \cdot 5$.

31.
$$
\begin{array}{r}
7 \\
7\overline{)49} \\
3\overline{)147} \\
2\overline{)294} \\
2\overline{)588}
\end{array}
$$
The prime factorization of 588 is
$2 \cdot 2 \cdot 3 \cdot 7 \cdot 7$.

33. $48 = 1 \cdot 48$, $48 = 2 \cdot 24$, $48 = 3 \cdot 16$,
$48 = 4 \cdot 12$, $48 = 6 \cdot 8$
The factors of 48 are 1, 2, 3, 4, 6, 8, 12, 16,
and 48, which is choice d.

35. $3 = 3$
$4 = 2 \cdot 2$
$LCM = 2 \cdot 2 \cdot 3 = 12$

37. $6 = 2 \cdot 3$
$14 = 2 \cdot 7$
$LCM = 2 \cdot 3 \cdot 7 = 42$

39. $20 = 2 \cdot 2 \cdot 5$
$30 = 2 \cdot 3 \cdot 5$
$LCM = 2 \cdot 2 \cdot 3 \cdot 5 = 60$

41. $5 = 5$
$7 = 7$
$LCM = 5 \cdot 7 = 35$

43. $9 = 3 \cdot 3$
$12 = 2 \cdot 2 \cdot 3$
$LCM = 2 \cdot 2 \cdot 3 \cdot 3 = 36$

45. $16 = 2 \cdot 2 \cdot 2 \cdot 2$
$20 = 2 \cdot 2 \cdot 5$
$LCM = 2 \cdot 2 \cdot 2 \cdot 2 \cdot 5 = 80$

47. $40 = 2 \cdot 2 \cdot 2 \cdot 5$
$90 = 2 \cdot 3 \cdot 3 \cdot 5$
$LCM = 2 \cdot 2 \cdot 2 \cdot 3 \cdot 3 \cdot 5 = 360$

49. $24 = 2 \cdot 2 \cdot 2 \cdot 3$
$36 = 2 \cdot 2 \cdot 3 \cdot 3$
$LCM = 2 \cdot 2 \cdot 2 \cdot 3 \cdot 3 = 72$

51. $2 = 2$
$8 = 2 \cdot 2 \cdot 2$
$15 = 3 \cdot 5$
$LCM = 2 \cdot 2 \cdot 2 \cdot 3 \cdot 5 = 120$

53. $2 = 2$
$3 = 3$
$7 = 7$
$LCM = 2 \cdot 3 \cdot 7 = 42$

55. $8 = 2 \cdot 2 \cdot 2$
$24 = 2 \cdot 2 \cdot 2 \cdot 3$
$48 = 2 \cdot 2 \cdot 2 \cdot 2 \cdot 3$
$LCM = 2 \cdot 2 \cdot 2 \cdot 2 \cdot 3 = 48$

57. $8 = 2 \cdot 2 \cdot 2$
$18 = 2 \cdot 3 \cdot 3$
$30 = 2 \cdot 3 \cdot 5$
$LCM = 2 \cdot 2 \cdot 2 \cdot 3 \cdot 3 \cdot 5 = 360$

59. a. $40 = 2 \cdot 20 = 2 \cdot 2 \cdot 10 = 2 \cdot 2 \cdot 2 \cdot 5$
The prime factorization of 40 is
$2 \cdot 2 \cdot 2 \cdot 5$.

    b. $40 = 4 \cdot 10 = 2 \cdot 2 \cdot 10 = 2 \cdot 2 \cdot 2 \cdot 5$
The prime factorization of 40 is
$2 \cdot 2 \cdot 2 \cdot 5$.

    c. answers may vary

61. True, the number 311 is a prime number.

63. $35 = 5 \cdot 7$
$20 = 2 \cdot 2 \cdot 5$
$LCM = 2 \cdot 2 \cdot 5 \cdot 7 = 140$
They are in New Orleans on the same day
every 140 days.

65. $1000 = 25 \cdot 40 = 25 \cdot 5 \cdot 8$
$1125 = 25 \cdot 45 = 25 \cdot 5 \cdot 9$
$LCM = 25 \cdot 5 \cdot 8 \cdot 9 = 9000$

Section R.2 Practice Problems

1. $\dfrac{4}{4} = 1$ since $4 \div 4 = 1$.

2. $\dfrac{9}{3} = 3$ since $9 \div 3 = 3$.

3. $\dfrac{10}{10} = 1$ since $10 \div 10 = 1$.

4. $\dfrac{5}{1} = 5$ since $5 \div 1 = 5$.

5. $\dfrac{0}{11} = 0$ since $0 \cdot 11 = 0$.

6. $\dfrac{11}{0}$ is undefined because there is no number
that when multiplied by 0 gives 11.

7. $\dfrac{1}{4} = \dfrac{1}{4} \cdot \dfrac{5}{5} = \dfrac{1 \cdot 5}{4 \cdot 5} = \dfrac{5}{20}$

8. $\dfrac{20}{35} = \dfrac{2 \cdot 2 \cdot 5}{5 \cdot 7} = \dfrac{4}{7}$

9. $\dfrac{7}{20}$ is already simplified.

10. $\dfrac{12}{40} = \dfrac{4 \cdot 3}{4 \cdot 10} = \dfrac{3}{10}$

11. $\dfrac{3}{4} \cdot \dfrac{8}{9} = \dfrac{3 \cdot 8}{4 \cdot 9} = \dfrac{3 \cdot 4 \cdot 2}{4 \cdot 3 \cdot 3} = \dfrac{2}{3}$

12. $\dfrac{2}{9} \div \dfrac{3}{4} = \dfrac{2}{9} \cdot \dfrac{4}{3} = \dfrac{2 \cdot 4}{9 \cdot 3} = \dfrac{8}{27}$

13. $\dfrac{8}{11} \div 24 = \dfrac{8}{11} \div \dfrac{24}{1} = \dfrac{8}{11} \cdot \dfrac{1}{24} = \dfrac{8 \cdot 1}{11 \cdot 8 \cdot 3} = \dfrac{1}{33}$

14. $\dfrac{5}{4} \div \dfrac{15}{8} = \dfrac{5}{4} \cdot \dfrac{8}{15} = \dfrac{5 \cdot 4 \cdot 2}{4 \cdot 5 \cdot 3} = \dfrac{2}{3}$

15. $\dfrac{2}{11} + \dfrac{5}{11} = \dfrac{2+5}{11} = \dfrac{7}{11}$

16. $\dfrac{1}{8} + \dfrac{3}{8} = \dfrac{1+3}{8} = \dfrac{4}{8} = \dfrac{4}{2 \cdot 4} = \dfrac{1}{2}$

17. $\dfrac{7}{6} - \dfrac{2}{6} = \dfrac{7-2}{6} = \dfrac{5}{6}$

18. $\dfrac{13}{10} - \dfrac{3}{10} = \dfrac{13-3}{10} = \dfrac{10}{10} = 1$

19. $\dfrac{3}{8} + \dfrac{1}{20} = \dfrac{3}{8} \cdot \dfrac{5}{5} + \dfrac{1}{20} \cdot \dfrac{2}{2} = \dfrac{15}{40} + \dfrac{2}{40} = \dfrac{17}{40}$

20. $\dfrac{8}{15} - \dfrac{1}{3} = \dfrac{8}{15} - \dfrac{1}{3} \cdot \dfrac{5}{5} = \dfrac{8}{15} - \dfrac{5}{15} = \dfrac{3}{15} = \dfrac{3}{3 \cdot 5} = \dfrac{1}{5}$

21. $5\dfrac{1}{6} = \dfrac{6 \cdot 5 + 1}{6} = \dfrac{31}{6}; \quad 4\dfrac{2}{5} = \dfrac{5 \cdot 4 + 2}{5} = \dfrac{22}{5}$

$5\dfrac{1}{6} \cdot 4\dfrac{2}{5} = \dfrac{31}{6} \cdot \dfrac{22}{5} = \dfrac{31 \cdot 2 \cdot 11}{2 \cdot 3 \cdot 5} = \dfrac{341}{15} = 22\dfrac{11}{15}$

22. $7\dfrac{3}{8} + 6\dfrac{3}{4} = \dfrac{59}{8} + \dfrac{27}{4} = \dfrac{59}{8} + \dfrac{54}{8} = \dfrac{113}{8} = 14\dfrac{1}{8}$

23.
$$76\dfrac{1}{12} \qquad 76\dfrac{1}{12} \qquad 75\dfrac{13}{12}$$
$$\underline{-35\dfrac{1}{4}} \qquad \underline{-35\dfrac{3}{12}} \qquad \underline{-35\dfrac{3}{12}}$$
$$\qquad\qquad\qquad\qquad\qquad 40\dfrac{10}{12} = 40\dfrac{5}{6}$$

Exercise Set R.2

1. $\dfrac{14}{14} = 1$ since $14 \cdot 1 = 14$.

3. $\dfrac{20}{2} = 10$ since $2 \cdot 10 = 20$.

5. $\dfrac{13}{1} = 13$ since $1 \cdot 13 = 13$.

7. $\dfrac{0}{9} = 0$ since $9 \cdot 0 = 0$.

9. $\dfrac{9}{0}$ is undefined.

11. $\dfrac{7}{10} = \dfrac{7 \cdot 3}{10 \cdot 3} = \dfrac{21}{30}$

13. $\dfrac{2}{9} = \dfrac{2 \cdot 2}{9 \cdot 2} = \dfrac{4}{18}$

15. $\dfrac{4}{5} = \dfrac{4 \cdot 4}{5 \cdot 4} = \dfrac{16}{20}$

17. $\dfrac{2}{4} = \dfrac{2 \cdot 1}{2 \cdot 2} = \dfrac{1}{2}$

19. $\dfrac{10}{15} = \dfrac{2 \cdot 5}{3 \cdot 5} = \dfrac{2}{3}$

21. $\dfrac{3}{7}$ cannot be simplified further.

23. $\dfrac{18}{30} = \dfrac{3 \cdot 6}{5 \cdot 6} = \dfrac{3}{5}$

25. $\dfrac{16}{20} = \dfrac{4 \cdot 4}{4 \cdot 5} = \dfrac{4}{5}$

27. $\dfrac{66}{48} = \dfrac{6 \cdot 11}{6 \cdot 8} = \dfrac{11}{8}$

29. $\dfrac{120}{244} = \dfrac{4 \cdot 30}{4 \cdot 61} = \dfrac{30}{61}$

31. $\dfrac{192}{264} = \dfrac{8 \cdot 24}{11 \cdot 24} = \dfrac{8}{11}$

33. $\dfrac{1}{2} \cdot \dfrac{3}{4} = \dfrac{1 \cdot 3}{2 \cdot 4} = \dfrac{3}{8}$

35. $\dfrac{2}{3} \cdot \dfrac{3}{4} = \dfrac{2 \cdot 3}{3 \cdot 4} = \dfrac{2 \cdot 3 \cdot 1}{2 \cdot 3 \cdot 2} = \dfrac{1}{2}$

37. $\dfrac{1}{2} \div \dfrac{7}{12} = \dfrac{1}{2} \cdot \dfrac{12}{7} = \dfrac{1 \cdot 12}{2 \cdot 7} = \dfrac{2 \cdot 6}{2 \cdot 7} = \dfrac{6}{7}$

39. $\dfrac{3}{4} \div \dfrac{1}{20} = \dfrac{3}{4} \cdot \dfrac{20}{1} = \dfrac{3 \cdot 20}{4 \cdot 1} = \dfrac{3 \cdot 4 \cdot 5}{4 \cdot 1} = \dfrac{15}{1} = 15$

41. $5\dfrac{1}{9} \cdot 3\dfrac{2}{3} = \dfrac{46}{9} \cdot \dfrac{11}{3} = \dfrac{506}{27} = 18\dfrac{20}{27}$

43. $8\dfrac{3}{5} \div 2\dfrac{9}{10} = \dfrac{43}{5} \div \dfrac{29}{10}$

$= \dfrac{43}{5} \cdot \dfrac{10}{29}$

$= \dfrac{43 \cdot 5 \cdot 2}{5 \cdot 29}$

$= \dfrac{86}{29}$

$= 2\dfrac{28}{29}$

45. $\dfrac{4}{5} + \dfrac{1}{5} = \dfrac{4+1}{5} = \dfrac{5}{5} = 1$

47. $\dfrac{4}{15} - \dfrac{1}{12} = \dfrac{4}{15} \cdot \dfrac{4}{4} - \dfrac{1}{12} \cdot \dfrac{5}{5}$

$= \dfrac{16}{60} - \dfrac{5}{60}$

$= \dfrac{16-5}{60}$

$= \dfrac{11}{60}$

49. $\dfrac{2}{3}+\dfrac{3}{7}=\dfrac{2\cdot 7}{3\cdot 7}+\dfrac{3\cdot 3}{7\cdot 3}=\dfrac{14}{21}+\dfrac{9}{21}=\dfrac{14+9}{21}=\dfrac{23}{21}$

51. $\dfrac{10}{3}-\dfrac{5}{21}=\dfrac{10\cdot 7}{3\cdot 7}-\dfrac{5}{21}$

$\qquad=\dfrac{70}{21}-\dfrac{5}{21}$

$\qquad=\dfrac{70-5}{21}$

$\qquad=\dfrac{65}{21}$

53. $\begin{array}{r} 8\dfrac{1}{8} \\[2mm] -\,6\dfrac{3}{8} \\[1mm]\hline \end{array}\qquad \begin{array}{r} 7\dfrac{9}{8} \\[2mm] -\,6\dfrac{3}{8} \\[1mm]\hline 1\dfrac{6}{8}=1\dfrac{3}{4} \end{array}$

55. $\begin{array}{r} 9\dfrac{7}{8} \\[2mm] +\,2\dfrac{3}{10} \\[1mm]\hline \end{array}\qquad \begin{array}{r} 9\dfrac{35}{40} \\[2mm] +\,2\dfrac{12}{40} \\[1mm]\hline 11\dfrac{47}{40}=11+1\dfrac{7}{40}=12\dfrac{7}{40} \end{array}$

57. $\dfrac{23}{105}+\dfrac{4}{105}=\dfrac{23+4}{105}=\dfrac{27}{105}=\dfrac{3\cdot 9}{3\cdot 35}=\dfrac{9}{35}$

59. $\dfrac{17}{21}-\dfrac{10}{21}=\dfrac{17-10}{21}=\dfrac{7}{21}=\dfrac{1\cdot 7}{3\cdot 7}=\dfrac{1}{3}$

61. $\dfrac{7}{10}\cdot\dfrac{5}{21}=\dfrac{7\cdot 5}{10\cdot 21}=\dfrac{7\cdot 5}{5\cdot 2\cdot 7\cdot 3}=\dfrac{1}{2\cdot 3}=\dfrac{1}{6}$

63. $\dfrac{9}{20}\div 12=\dfrac{9}{20}\div\dfrac{12}{1}=\dfrac{9}{20}\cdot\dfrac{1}{12}=\dfrac{3\cdot 3\cdot 1}{20\cdot 3\cdot 4}=\dfrac{3}{80}$

65. $\dfrac{5}{22}-\dfrac{5}{33}=\dfrac{5\cdot 3}{22\cdot 3}-\dfrac{5\cdot 2}{33\cdot 2}$

$\qquad=\dfrac{15}{66}-\dfrac{10}{66}$

$\qquad=\dfrac{15-10}{66}$

$\qquad=\dfrac{5}{66}$

67. $17\dfrac{2}{5}+30\dfrac{2}{3}=\dfrac{87}{5}+\dfrac{92}{3}$

$\qquad=\dfrac{87\cdot 3}{5\cdot 3}+\dfrac{92\cdot 5}{3\cdot 5}$

$\qquad=\dfrac{261}{15}+\dfrac{460}{15}$

$\qquad=\dfrac{721}{15}$

$\qquad=48\dfrac{1}{15}$

69. $7\dfrac{2}{5}\div\dfrac{1}{5}=\dfrac{37}{5}\div\dfrac{1}{5}=\dfrac{37}{5}\cdot\dfrac{5}{1}=\dfrac{37\cdot 5}{5\cdot 1}=\dfrac{37}{1}=37$

71. $4\dfrac{2}{11}\cdot 2\dfrac{1}{2}=\dfrac{46}{11}\cdot\dfrac{5}{2}=\dfrac{2\cdot 23\cdot 5}{11\cdot 2}=\dfrac{115}{11}=10\dfrac{5}{11}$

73. $\dfrac{12}{5}-1=\dfrac{12}{5}-\dfrac{5}{5}=\dfrac{12-5}{5}=\dfrac{7}{5}$

75. $\begin{array}{r} 8\dfrac{11}{12} \\[2mm] -\,1\dfrac{5}{6} \\[1mm]\hline \end{array}\qquad \begin{array}{r} 8\dfrac{11}{12} \\[2mm] -\,1\dfrac{10}{12} \\[1mm]\hline 7\dfrac{1}{12} \end{array}$

77. $\dfrac{2}{3}-\dfrac{5}{9}+\dfrac{5}{6}=\dfrac{2\cdot 6}{3\cdot 6}-\dfrac{5\cdot 2}{9\cdot 2}+\dfrac{5\cdot 3}{6\cdot 3}$

$\qquad=\dfrac{12}{18}-\dfrac{10}{18}+\dfrac{15}{18}$

$\qquad=\dfrac{12-10+15}{18}$

$\qquad=\dfrac{17}{18}$

79. a and c are incorrect.

b is correct.

$$\frac{12}{24} = \frac{1 \cdot 2 \cdot 2 \cdot 3}{2 \cdot 2 \cdot 2 \cdot 3} = \frac{1}{2}$$

81. answers may vary

83. $1 - \dfrac{3}{11} - \dfrac{2}{11} = \dfrac{11}{11} - \dfrac{3}{11} - \dfrac{2}{11} = \dfrac{11-3-2}{11} = \dfrac{6}{11}$

The unknown part is $\dfrac{6}{11}$.

85. $1 - \dfrac{1}{3} - \dfrac{5}{12} - \dfrac{1}{6} = \dfrac{12}{12} - \dfrac{4}{12} - \dfrac{5}{12} - \dfrac{2}{12}$

$$= \frac{12-4-5-2}{12}$$

$$= \frac{1}{12}$$

The unknown part is $\dfrac{1}{12}$.

87. $1 - \dfrac{41}{50} = \dfrac{50}{50} - \dfrac{41}{50} = \dfrac{50-41}{50} = \dfrac{9}{50}$

The fraction that does not agree is $\dfrac{9}{50}$.

89. a.   36 of the 459 Federated stores were Bloomingdale's.

$$\frac{36}{459} = \frac{9 \cdot 4}{9 \cdot 51} = \frac{4}{51}$$

  b.   95 + 144 = 239 of the 459 Federated stores were either Macy's East or Macy's West.

$\dfrac{239}{459}$ does not simplify further.

91. $A = lw = \dfrac{2}{5} \cdot \dfrac{3}{11} = \dfrac{2 \cdot 3}{5 \cdot 11} = \dfrac{6}{55}$

The area is $\dfrac{6}{55}$ square meter.

Section R.3 Practice Problems

1. $0.27 = \dfrac{27}{100}$

2. $5.1 = \dfrac{51}{10}$

3. $7.685 = \dfrac{7685}{1000}$

4. a.
$$\begin{array}{r} 7.19 \\ 19.782 \\ + 1.006 \\ \hline 27.978 \end{array}$$

  b.
$$\begin{array}{r} 12. \\ 0.79 \\ + 0.03 \\ \hline 12.82 \end{array}$$

5. a.
$$\begin{array}{r} 84.230 \\ - 26.982 \\ \hline 57.248 \end{array}$$

  b.
$$\begin{array}{r} 90.00 \\ - 0.19 \\ \hline 89.81 \end{array}$$

6. a.
$$\begin{array}{r} 0.31 \\ \times 4.6 \\ \hline 186 \\ 1\,24 \\ \hline 1.426 \end{array}$$

  b.
$$\begin{array}{r} 1.26 \\ \times 0.03 \\ \hline 0.0378 \end{array}$$

7. a.
$$\begin{array}{r} 43.5 \\ 0.5\overline{)\,21.75} \\ \underline{-20} \\ 1\,7 \\ \underline{-1\,5} \\ 25 \\ \underline{-25} \\ 0 \end{array}$$

b.
$$0.006\overline{)15.600}^{\;2600}$$
$$\begin{array}{r} -12 \\ \hline 3\;6 \\ -3\;6 \\ \hline 0 \end{array}$$

8. 12.9187 rounded to the nearest hundredth is 12.92.

9. 245.348 rounded to the nearest tenth is 245.3.

10.
$$5\overline{)2.0}^{\;0.4}$$
$$\begin{array}{r} -2\;0 \\ \hline 0 \end{array}$$

$$\frac{2}{5} = 0.4$$

11.
$$6\overline{)5.000}^{\;0.833}$$
$$\begin{array}{r} -4\;8 \\ \hline 20 \\ -18 \\ \hline 20 \\ -18 \\ \hline 2 \end{array}$$

$$\frac{5}{6} = 0.833... = 0.8\overline{3}$$

12.
$$9\overline{)1.0000}^{\;0.1111}$$
$$\begin{array}{r} -9 \\ \hline 10 \\ -9 \\ \hline 10 \\ -9 \\ \hline 10 \\ -9 \\ \hline 1 \end{array}$$

$$\frac{1}{9} = 0.1111... \approx 0.111$$

13. a. $20\% = 20.\% = 0.20$

b. $1.2\% = 01.2\% = 0.012$

c. $465\% = 465.\% = 4.65$

14. a. $0.42 = 0.42 = 42\%$

b. $0.003 = 0.003 = 0.3\%$

c. $2.36 = 2.36 = 236\%$

d. $0.7 = 0.70 = 70\%$

Exercise Set R.3

1. $0.6 = \dfrac{6}{10}$

3. $1.86 = \dfrac{186}{100}$

5. $0.114 = \dfrac{114}{1000}$

7. $123.1 = \dfrac{1231}{10}$

9.
$$\begin{array}{r} 5.7 \\ +\,1.13 \\ \hline 6.83 \end{array}$$

11.
$$\begin{array}{r} 24.6 \\ 2.39 \\ +\,0.0678 \\ \hline 27.0578 \end{array}$$

13.
$$\begin{array}{r} 8.8 \\ -\,2.3 \\ \hline 6.5 \end{array}$$

15.
$$\begin{array}{r} 18.00 \\ -\,2.78 \\ \hline 15.22 \end{array}$$

17.
$$\begin{array}{r} 0.2 \\ \times\,0.6 \\ \hline 0.12 \end{array}$$

19.
$$
\begin{array}{r}
0.063 \\
\times\quad 4.2 \\
\hline
126 \\
252\ \ \\
\hline
0.2646
\end{array}
$$

21.
$$
\begin{array}{r}
1.68 \\
5\overline{)\,8.40} \\
\underline{-5}\quad\ \\
3\ 4 \\
\underline{-3\ 0} \\
40 \\
\underline{-40} \\
0
\end{array}
$$

23.
$$
\begin{array}{r}
5.8 \\
0.82\overline{)\,4.756} \\
\underline{-4\ 10}\quad \\
656 \\
\underline{-656} \\
0
\end{array}
$$

25.
$$
\begin{array}{r}
45.02 \\
3.006 \\
+\ 8.405 \\
\hline
56.431
\end{array}
$$

27.
$$
\begin{array}{r}
6.75 \\
\times\ 10 \\
\hline
67.5
\end{array}
$$

29.　$0.6\overline{)\,42}$　　$\begin{array}{r}
70 \\
6\overline{)\,420.} \\
\underline{-42}\quad \\
00
\end{array}$

31.
$$
\begin{array}{r}
654.90 \\
-\ 56.67 \\
\hline
598.23
\end{array}
$$

33.
$$
\begin{array}{r}
5.62 \\
\times\ \ 7.7 \\
\hline
3\ 934 \\
39\ 34\ \ \\
\hline
43.274
\end{array}
$$

35.
$$
\begin{array}{r}
840. \\
0.063\overline{)\,52.920} \\
\underline{-50\ 4}\quad\ \ \\
2\ 52 \\
\underline{-2\ 52} \\
00
\end{array}
$$

37.
$$
\begin{array}{r}
16.003 \\
\times\quad 5.31 \\
\hline
16003 \\
4\ 8009\ \ \\
80\ 015\quad\ \ \\
\hline
84.97593
\end{array}
$$

39. 0.57 rounded to the nearest tenth is 0.6.

41. 0.234 rounded to the nearest hundredth is 0.23.

43. 0.5945 rounded to the nearest thousandth is 0.595.

45. 98,207.23 rounded to the nearest tenth is 98,207.2.

47. 12.347 rounded to the nearest hundredth is 12.35.

49. $\dfrac{3}{4} = \dfrac{3\cdot 25}{4\cdot 25} = \dfrac{75}{100} = 0.75$

51.
$$
\begin{array}{r}
0.333... \\
3\overline{)\,1.000} \\
\underline{-9}\quad\ \\
10 \\
\underline{-9} \\
10 \\
\underline{-9} \\
1
\end{array}
$$

$\dfrac{1}{3} = 0.\overline{3} \approx 0.33$

53. $\dfrac{7}{16} = \dfrac{7\cdot 625}{16\cdot 625} = \dfrac{4375}{10,000} = 0.4375$

55.
$$
\begin{array}{r}
0.5454... \\
11\overline{)6.0000} \\
-5\,5 \\
\hline
50 \\
-44 \\
\hline
60 \\
-55 \\
\hline
50 \\
-44 \\
\hline
6
\end{array}
$$

$$\frac{6}{11} = 0.\overline{54} \approx 0.55$$

57.
$$
\begin{array}{r}
4.833... \\
6\overline{)29.000} \\
-24 \\
\hline
5\,0 \\
-4\,8 \\
\hline
20 \\
-18 \\
\hline
20 \\
-18 \\
\hline
2
\end{array}
$$

$$\frac{29}{6} = 4.8\overline{3} \approx 4.83$$

59. $28\% = 0.28$

61. $3.1\% = 0.031$

63. $135\% = 1.35$

65. $200\% = 2.00$ or 2

67. $96.55\% = 0.9655$

69. $0.1\% = 0.001$

71. $51\% = 0.51$

73. $0.68 = 68\%$

75. $0.876 = 87.6\%$

77. $1 = 1.00 = 100\%$

79. $0.5 = 50\%$

81. $1.92 = 192\%$

83. $0.004 = 0.4\%$

85. a. In the number 3.659, the place value of the 6 is tenths.

   b. In the number 3.659, the place value of the 9 is thousandths.

   c. In the number 3.659, the place value of the 3 is ones.

87. 78.787878... rounded to the nearest thousandth is 78.788.

89. answers may vary

91. a.
$$
\begin{array}{r}
213.4 \\
-\ 30.8 \\
\hline
182.6
\end{array}
$$
The consumption of fluid milk products is 182.6 pounds more than cheese products.

   b.
$$
\begin{array}{r}
213.4 \\
30.8 \\
+\ \ 4.4 \\
\hline
248.6
\end{array}
$$
The total amount consumed is 248.6 pounds.

Chapter R Vocabulary Check

1. factor

2. multiple

3. composite number

4. percent

5. equivalent

6. improper fraction

7. prime number

8. simplified

9. proper fraction

10. mixed number

Chapter R Review

1. $42 = 2 \cdot 21 = 2 \cdot 3 \cdot 7$

2. $800 = 2 \cdot 400$
$= 2 \cdot 2 \cdot 200$
$= 2 \cdot 2 \cdot 2 \cdot 100$
$= 2 \cdot 2 \cdot 2 \cdot 2 \cdot 50$
$= 2 \cdot 2 \cdot 2 \cdot 2 \cdot 2 \cdot 25$
$= 2 \cdot 2 \cdot 2 \cdot 2 \cdot 2 \cdot 5 \cdot 5$

3. $12 = 2 \cdot 2 \cdot 3$
$30 = 2 \cdot 3 \cdot 5$
$\text{LCM} = 2 \cdot 2 \cdot 3 \cdot 5 = 60$

4. $7 = 7$
$42 = 2 \cdot 3 \cdot 7$
$\text{LCM} = 2 \cdot 3 \cdot 7 = 42$

5. $4 = 2 \cdot 2$
$6 = 2 \cdot 3$
$10 = 2 \cdot 5$
$\text{LCM} = 2 \cdot 2 \cdot 3 \cdot 5 = 60$

6. $2 = 2$
$5 = 5$
$7 = 7$
$\text{LCM} = 2 \cdot 5 \cdot 7 = 70$

7. $\dfrac{5}{8} = \dfrac{5 \cdot 3}{8 \cdot 3} = \dfrac{15}{24}$

8. $\dfrac{2}{3} = \dfrac{2 \cdot 20}{3 \cdot 20} = \dfrac{40}{60}$

9. $\dfrac{8}{20} = \dfrac{2 \cdot 2 \cdot 2}{2 \cdot 2 \cdot 5} = \dfrac{2}{5}$

10. $\dfrac{15}{100} = \dfrac{3 \cdot 5}{2 \cdot 2 \cdot 5 \cdot 5} = \dfrac{3}{20}$

11. $\dfrac{12}{6} = \dfrac{2 \cdot 6}{6} = 2$

12. $\dfrac{8}{8} = 1$

13. $\dfrac{1}{7} \cdot \dfrac{8}{11} = \dfrac{1 \cdot 8}{7 \cdot 11} = \dfrac{8}{77}$

14. $\dfrac{5}{12} + \dfrac{2}{15} = \dfrac{5 \cdot 5}{12 \cdot 5} + \dfrac{2 \cdot 4}{15 \cdot 4}$
$= \dfrac{25}{60} + \dfrac{8}{60}$
$= \dfrac{25 + 8}{60}$
$= \dfrac{33}{60}$
$= \dfrac{11 \cdot 3}{20 \cdot 3}$
$= \dfrac{11}{20}$

15. $\dfrac{3}{10} \div 6 = \dfrac{3}{10} \div \dfrac{6}{1}$
$= \dfrac{3}{10} \cdot \dfrac{1}{6}$
$= \dfrac{3 \cdot 1}{10 \cdot 6}$
$= \dfrac{3 \cdot 1}{10 \cdot 2 \cdot 3}$
$= \dfrac{1}{20}$

16. $\dfrac{7}{9} - \dfrac{1}{6} = \dfrac{7 \cdot 2}{9 \cdot 2} - \dfrac{1 \cdot 3}{6 \cdot 3} = \dfrac{14}{18} - \dfrac{3}{18} = \dfrac{14 - 3}{18} = \dfrac{11}{18}$

17. $3\dfrac{3}{8} \cdot 4\dfrac{1}{4} = \dfrac{27}{8} \cdot \dfrac{17}{4} = \dfrac{27 \cdot 17}{8 \cdot 4} = \dfrac{459}{32} = 14\dfrac{11}{32}$

18. $2\dfrac{1}{3} - 1\dfrac{5}{6} = \dfrac{7}{3} - \dfrac{11}{6}$

$\qquad = \dfrac{14}{6} - \dfrac{11}{6}$

$\qquad = \dfrac{14-11}{6}$

$\qquad = \dfrac{3}{6}$

$\qquad = \dfrac{1}{2}$

19. $\quad 16\dfrac{9}{10} \qquad 16\dfrac{27}{30}$

$\quad +3\dfrac{2}{3} \qquad +3\dfrac{20}{30}$

$\qquad\qquad\qquad 19\dfrac{47}{30} = 19 + 1\dfrac{17}{30} = 20\dfrac{17}{30}$

20. $6\dfrac{2}{7} \div 2\dfrac{1}{5} = \dfrac{44}{7} \div \dfrac{11}{5}$

$\qquad = \dfrac{44}{7} \cdot \dfrac{5}{11}$

$\qquad = \dfrac{44 \cdot 5}{7 \cdot 11}$

$\qquad = \dfrac{4 \cdot 11 \cdot 5}{7 \cdot 11}$

$\qquad = \dfrac{20}{7}$

$\qquad = 2\dfrac{6}{7}$

21. $A = lw = \dfrac{11}{12} \cdot \dfrac{3}{5} = \dfrac{11 \cdot 3}{12 \cdot 5} = \dfrac{11 \cdot 3}{3 \cdot 4 \cdot 5} = \dfrac{11}{20}$

The area is $\dfrac{11}{20}$ square mile.

22. $A = \dfrac{1}{2}bh = \dfrac{1}{2} \cdot \dfrac{5}{4} \cdot \dfrac{1}{2} = \dfrac{1 \cdot 5 \cdot 1}{2 \cdot 4 \cdot 2} = \dfrac{5}{16}$

The area is $\dfrac{5}{16}$ square meter.

23. $1.81 = \dfrac{181}{100}$

24. $0.035 = \dfrac{35}{1000}$

25. $\quad 76.358$

$\quad \underline{+\,18.76}$

$\quad 95.118$

26. $\quad 35$

$\quad\;\; 0.02$

$\quad \underline{+\,1.765}$

$\quad 36.785$

27. $\quad 18.00$

$\quad \underline{-\,4.62}$

$\quad 13.38$

28. $\quad 804.062$

$\quad \underline{-\,112.489}$

$\quad 691.573$

29. $\quad\;\; 7.6$

$\quad \underline{\times\, 12}$

$\quad\; 15\,2$

$\quad\;\; \underline{76}$

$\quad\; 91.2$

30. $\quad\; 14.63$

$\quad \underline{\times\;\; 3.2}$

$\quad\;\; 2\,926$

$\quad \underline{43\,89}$

$\quad 46.816$

31.
$$\begin{array}{r} 28.6 \\ 27\overline{)\,772.2} \\ \underline{-54} \\ 232 \\ \underline{-216} \\ 16\,2 \\ \underline{-16\,2} \\ 0 \end{array}$$

32.
$$\begin{array}{r} 230 \\ 0.06\overline{)\,13.80} \\ \underline{-12} \\ 1\,8 \\ \underline{-1\,8} \\ 00 \end{array}$$

33. 0.7652 rounded to the nearest hundredth is 0.77.

34. 25.6293 rounded to the nearest tenth is 25.6.

35. $\dfrac{1}{2} = \dfrac{1 \cdot 5}{2 \cdot 5} = \dfrac{5}{10} = 0.5$

36. $\dfrac{3}{8} = \dfrac{3 \cdot 125}{8 \cdot 125} = \dfrac{375}{1000} = 0.375$

37.
$$11\overline{)\,4.0000}$$

$$\begin{array}{r} 0.3636... \\ \hline 4.0000 \\ -3\,3 \\ \hline 70 \\ -66 \\ \hline 40 \\ -33 \\ \hline 70 \\ -66 \\ \hline 4 \end{array}$$

$\dfrac{4}{11} = 0.\overline{36} \approx 0.364$

38.
$$\begin{array}{r} 0.833... \\ \hline 5.000 \\ -4\,8 \\ \hline 20 \\ -18 \\ \hline 20 \\ -18 \\ \hline 2 \end{array}$$
$6\overline{)\,5.000}$

$\dfrac{5}{6} = 0.8\overline{3} \approx 0.833$

39. $29\% = 0.29$

40. $1.4\% = 0.014$

41. $0.39 = 39\%$

42. $1.2 = 120\%$

43. $68.3\% = 0.683$
    The decimal is 0.683.

44. $2.3\% = 0.023$
    $5 = 500\%$
    $40\% = 0.4$
    The true statement is b.

Chapter R Test

1.
$$\begin{array}{r} 3 \\ 3\overline{)\,9} \\ 2\overline{)\,18} \\ 2\overline{)\,36} \\ 2\overline{)\,72} \end{array}$$

The prime factorization of 72 is $2 \cdot 2 \cdot 2 \cdot 3 \cdot 3$.

2. $5 = 5$
    $18 = 2 \cdot 3 \cdot 3$
    $20 = 2 \cdot 2 \cdot 5$
    $\text{LCM} = 2 \cdot 2 \cdot 3 \cdot 3 \cdot 5 = 180$

3. $\dfrac{5}{12} = \dfrac{5 \cdot 5}{12 \cdot 5} = \dfrac{25}{60}$

4. $\dfrac{15}{20} = \dfrac{3 \cdot 5}{4 \cdot 5} = \dfrac{3}{4}$

5. $\dfrac{48}{100} = \dfrac{4 \cdot 12}{4 \cdot 25} = \dfrac{12}{25}$

6. $1.3 = 1\dfrac{3}{10} = \dfrac{13}{10}$

7. $\dfrac{5}{8} + \dfrac{7}{10} = \dfrac{5 \cdot 5}{8 \cdot 5} + \dfrac{7 \cdot 4}{10 \cdot 4}$
    $= \dfrac{25}{40} + \dfrac{28}{40}$
    $= \dfrac{25 + 28}{40}$
    $= \dfrac{53}{40}$

8. $\dfrac{2}{3} \cdot \dfrac{27}{49} = \dfrac{2 \cdot 27}{3 \cdot 49} = \dfrac{2 \cdot 3 \cdot 9}{3 \cdot 49} = \dfrac{18}{49}$

9. $\dfrac{9}{10} \div 18 = \dfrac{9}{10} \div \dfrac{18}{1}$

$\quad = \dfrac{9}{10} \cdot \dfrac{1}{18}$

$\quad = \dfrac{9 \cdot 1}{10 \cdot 18}$

$\quad = \dfrac{9 \cdot 1}{10 \cdot 9 \cdot 2}$

$\quad = \dfrac{1}{20}$

10. $\dfrac{8}{9} - \dfrac{1}{12} = \dfrac{8 \cdot 4}{9 \cdot 4} - \dfrac{1 \cdot 3}{12 \cdot 3}$

$\quad = \dfrac{32}{36} - \dfrac{3}{36}$

$\quad = \dfrac{32 - 3}{36}$

$\quad = \dfrac{29}{36}$

11. $1\dfrac{2}{9} + 3\dfrac{2}{3} = \dfrac{11}{9} + \dfrac{11}{3}$

$\quad = \dfrac{11}{9} + \dfrac{33}{9}$

$\quad = \dfrac{11 + 33}{9}$

$\quad = \dfrac{44}{9}$

$\quad = 4\dfrac{8}{9}$

12.
$$5\dfrac{6}{11} \qquad 5\dfrac{12}{22}$$
$$-3\dfrac{7}{22} \qquad -3\dfrac{7}{22}$$
$$\overline{\qquad\qquad} \quad \overline{\;\;2\dfrac{5}{22}\;}$$

13. $6\dfrac{7}{8} \div \dfrac{1}{8} = \dfrac{55}{8} \div \dfrac{1}{8} = \dfrac{55}{8} \cdot \dfrac{8}{1} = \dfrac{55 \cdot 8}{8 \cdot 1} = \dfrac{55}{1} = 55$

14. $2\dfrac{1}{10} \cdot 6\dfrac{1}{2} = \dfrac{21}{10} \cdot \dfrac{13}{2} = \dfrac{21 \cdot 13}{10 \cdot 2} = \dfrac{273}{20} = 13\dfrac{13}{20}$

15.
$$
\begin{array}{r}
43 \\
0.21 \\
+\,1.9 \\
\hline
45.11
\end{array}
$$

16.
$$
\begin{array}{r}
123.60 \\
-\,57.72 \\
\hline
65.88
\end{array}
$$

17.
$$
\begin{array}{r}
7.93 \\
\times \quad 1.6 \\
\hline
4758 \\
793 \\
\hline
12.688
\end{array}
$$

18.
$$
\begin{array}{r}
320 \\
0.25\overline{)\,80.00} \\
\underline{-75} \\
5\,0 \\
\underline{-5\,0} \\
00
\end{array}
$$

19. 23.7272 rounded to the nearest hundredth is 23.73.

20. $\dfrac{7}{8} = \dfrac{7 \cdot 125}{8 \cdot 125} = \dfrac{875}{1000} = 0.875$

21.
$$
\begin{array}{r}
0.166... \\
6\overline{)\,1.000} \\
\underline{-6} \\
40 \\
\underline{-36} \\
40 \\
\underline{-36} \\
4
\end{array}
$$

$\dfrac{1}{6} = 0.1\overline{6} \approx 0.167$

22. $63.2\% = 0.632$

23. $0.09 = 9\%$

24. $\dfrac{3}{4} = \dfrac{3 \cdot 25}{4 \cdot 25} = \dfrac{75}{100} = 0.75 = 75\%$

25. $\frac{3}{4}$ of the fresh water is icecaps and glaciers.

26. $\frac{1}{200}$ of the fresh water is active water.

27. $1 - \frac{3}{4} - \frac{1}{200} = \frac{200}{200} - \frac{150}{200} - \frac{1}{200}$

$= \frac{200 - 150 - 1}{200}$

$= \frac{49}{200}$

$\frac{49}{200}$ of the fresh water is groundwater.

28. $1 - \frac{1}{200} = \frac{200}{200} - \frac{1}{200} = \frac{200 - 1}{200} = \frac{199}{200}$

$\frac{199}{200}$ of the fresh water is groundwater or icecaps and glaciers.

29. Area $= \frac{1}{2}$(base)(height)

$= \frac{1}{2} \cdot \frac{3}{4} \cdot \frac{1}{3}$

$= \frac{1 \cdot 3 \cdot 1}{2 \cdot 4 \cdot 3}$

$= \frac{1}{8}$

The area is $\frac{1}{8}$ square foot.

30. $A = lw = \frac{9}{8} \cdot \frac{7}{8} = \frac{9 \cdot 7}{8 \cdot 8} = \frac{63}{64}$

The area is $\frac{63}{64}$ square centimeter.

Chapter 1

Section 1.2 Practice Problems

1. Since 8 is to the right of 6 on the number line, the statement $8 < 6$ is false.

2. Since 100 is to the right of 10 on the number line, the statement $100 > 10$ is true.

3. Since $21 = 21$, the statement $21 \le 21$ is true.

4. Since $21 = 21$, the statement $21 \ge 21$ is true.

5. Since neither $0 > 5$ nor $0 = 5$ is true, the statement $0 \ge 5$ is false.

6. Since $25 > 22$, the statement $25 \ge 22$ is true.

7. a. Fourteen is greater than or equal to fourteen is written as $14 \ge 14$.

   b. Zero is less than five is written as $0 < 5$.

   c. Nine is not equal to 10 is written as $9 \ne 10$.

8. The integer -8 represents 8 feet below sea level.

9. $\dfrac{5}{4} = 1\dfrac{1}{4}$

10. a. $-11 < -9$ since -11 is to the left of -9 on the number line.

    b. By comparing digits in the same places, we find that $4.511 > 4.151$, since $0.5 > 0.1$.

c. By dividing, we find that $\dfrac{7}{8} = 0.875$ and $\dfrac{2}{3} = 0.66....$ Since $0.875 > 0.66...,$ then $\dfrac{7}{8} > \dfrac{2}{3}.$

11. a. The natural numbers are 6 and 913.

    b. The whole numbers are 0, 6, and 913.

    c. The integers are -100, 0, 6, and 913.

    d. The rational numbers are -100, $-\dfrac{2}{5}$, 0, 6, and 913.

    e. The irrational number is π.

    f. All numbers in the given set are real numbers.

12. a. $|7| = 7$ since 7 is 7 units from 0 on the number line.

    b. $|-8| = 8$ since -8 is 8 units from 0 on the number line.

    c. $\left|\dfrac{2}{3}\right| = \dfrac{2}{3}$

    d. $|0| = 0$ since 0 is 0 units from 0 on the number line.

    e. $|-3.06| = 3.06$

13. a. $|-4| = 4$

    b. $-3 < |0|$ since $-3 < 0$.

    c. $|-2.7| > |-2|$ since $2.7 > 2$.

    d. $|-6| \le |-16|$ since $6 < 16$.

e.　$\left|10\right| < \left|-10\dfrac{1}{3}\right|$ since $10 < 10\dfrac{1}{3}$.

Exercise Set 1.2

1.　Since 4 is to the left of 10 on the number line, $4 < 10$.

3.　Since 7 is to the right of 3 on the number line, $7 > 3$.

5.　$6.26 = 6.26$

7.　Since 0 is to the left of 7 on the number line, $0 < 7$.

9.　Since 32 is to the left of 212 on the number line, $32 < 212$.

11.　Since 30 is to the left of 45 on the number line, $30 \le 45$.

13.　Since $11 = 11$, the statement $11 \le 11$ is true.

15.　Since -11 is to the left of -10 on the number line, $-11 > -10$ is false.

17.　Comparing digits with the same place value, we have $0.0 < 0.9$. Thus the statement $5.092 < 5.902$ is true.

19.　Rewrite the fractions with a common denominator and compare numerators.

$$\frac{9}{10} = \frac{81}{90}; \frac{8}{9} = \frac{80}{90}$$

Since $81 > 80$, then $\dfrac{9}{10} \le \dfrac{8}{9}$ is false.

21.　$25 \ge 20$ has the same meaning as $20 \le 25$.

23.　$0 < 6$ has the same meaning as $6 > 0$.

25.　$-10 > -12$ has the same meaning as $-12 < -10$.

27.　Seven is less than eleven is written as $7 < 11$.

29.　Five is greater than or equal to four is written as $5 \ge 4$.

31.　Fifteen is not equal to negative two is written as $15 \ne -2$.

33.　The integer 14,494 represents 14,494 feet above sea level. The integer -282 represents 282 feet below sea level.

35.　The integer $-43,413$ represents 43,413 fewer students.

37.　The integer 475 represents a $475 deposit. The integer -195 represents a $195 withdrawal.

39.　

41.　

43.　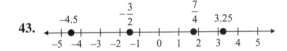

45.　0 is a whole number, an integer, a rational number, and a real number.

47.　-7 is an integer, a rational number, and a real number.

49.　265 is a natural number, a whole number, an integer, a rational number, and a real number.

51.　$\dfrac{2}{3}$ is a rational number and a real number.

53.　False; the rational number $\dfrac{2}{3}$ is not an integer.

55.　True; 0 is a real number.

57. False; the negative number $-\sqrt{2}$ is not a rational number.

59. False; the real number $\sqrt{7}$ is not a rational number.

61. $|8.9| = 8.9$ since 8.9 is 8.9 units from 0 on the number line.

63. $|-20| = 20$ since -20 is 20 units from 0 on the number line.

65. $\left|\frac{9}{2}\right| = \frac{9}{2}$ since $\frac{9}{2}$ is $\frac{9}{2}$ units from 0 on the number line.

67. $\left|-\frac{12}{13}\right| = \frac{12}{13}$ since $-\frac{12}{13}$ is $\frac{12}{13}$ unit from 0 on the number line.

69. $|-5| = 5$
$-4 = -4$
Since 5 is to the right of -4 on the number line, $|-5| > -4$.

71. $\left|-\frac{5}{8}\right| = \frac{5}{8}$
$\left|\frac{5}{8}\right| = \frac{5}{8}$
Since $\frac{5}{8} = \frac{5}{8}$, then $\left|-\frac{5}{8}\right| = \left|\frac{5}{8}\right|$.

73. $|-2| = 2$
$|-2.7| = 2.7$
Since 2 is to the left of 2.7 on the number line, $|-2| < |-2.7|$.

75. $|0| = 0$
$|-8| = 8$
Since 0 is to the left of 8 on the number line, $|0| < |-8|$.

77. The apple production in 1998 was 762 thousand bushels, while the apple production in 1999 was 1548 thousand bushels.
$762 < 1548$

79. The apple production in 2000 was 1190 thousand bushels, while the production in 2001 was 929 thousand bushels.
$1190 - 929 = 261$
The production went down by 261 thousand bushels, or -261.

81. Since -0.04 is to the right of -26.7 on the number line, $-0.04 > -26.7$.

83. Sun: -26.7
Arcturus: -0.04
Since $-26.7 < -0.04$, the sun is brighter than Arcturus.

85. Since the brightest star corresponds to the smallest apparent magnitude, which is -26.7, the brightest star is the sun.

87. answers may vary

Section 1.3 Practice Problems

1. a. $4^2 = 4 \cdot 4 = 16$

b. $2^2 = 2 \cdot 2 = 4$

c. $3^4 = 3 \cdot 3 \cdot 3 \cdot 3 = 81$

d. $9^1 = 9$

e. $\left(\frac{2}{5}\right)^3 = \left(\frac{2}{5}\right)\left(\frac{2}{5}\right)\left(\frac{2}{5}\right) = \frac{2 \cdot 2 \cdot 2}{5 \cdot 5 \cdot 5} = \frac{8}{125}$

2. $3 \cdot 2 + 4^2 = 3 \cdot 2 + 16 = 6 + 16 = 22$

3. $28 \div 7 \cdot 2 = 4 \cdot 2 = 8$

4. $\frac{9}{5} \cdot \frac{1}{3} - \frac{1}{3} = \frac{9}{15} - \frac{1}{3} = \frac{9}{15} - \frac{5}{15} = \frac{4}{15}$

5. $5 + 3[2(3 \cdot 4 + 1) - 20] = 5 + 3[2(12 + 1) - 20]$
$$= 5 + 3[2(13) - 20]$$
$$= 5 + 3[26 - 20]$$
$$= 5 + 3[6]$$
$$= 5 + 18$$
$$= 23$$

6. $\dfrac{1 + |7 - 4| + 3^2}{8 - 5} = \dfrac{1 + |3| + 3^2}{8 - 5}$
$$= \dfrac{1 + 3 + 3^2}{3}$$
$$= \dfrac{1 + 3 + 9}{3}$$
$$= \dfrac{13}{3}$$

7. a. Replace y with 4.
$$3y^2 = 3 \cdot (4)^2 = 3 \cdot 16 = 48$$

  b. Replace x with 1 and y with 4.
$$2y - x = 2(4) - 1 = 8 - 1 = 7$$

  c. Replace x with 1 and y with 4.
$$\dfrac{11x}{3y} = \dfrac{11 \cdot 1}{3 \cdot 4} = \dfrac{11}{12}$$

  d. Replace x with 1 and y with 4.
$$\dfrac{x}{y} + \dfrac{6}{y} = \dfrac{1}{4} + \dfrac{6}{4} = \dfrac{7}{4}$$

  e. Replace x with 1 and y with 4.
$$y^2 - x^2 = 4^2 - 1^2 = 16 - 1 = 15$$

8. $5x - 10 = x + 2$
$$5(3) - 10 \stackrel{?}{=} 3 + 2$$
$$15 - 10 \stackrel{?}{=} 5$$
$$5 = 5 \quad \text{True}$$
3 is a solution.

9. a. $5 \cdot x$ and $5x$ are both ways to denote the product of 5 and x.

  b. A number added to 7 is denoted by $7 + x$.

  c. A number divided by 11.2 is denoted by $x \div 11.2$ or $\dfrac{x}{11.2}$.

  d. A number subtracted from 8 is denoted by $8 - x$.

  e. Twice a number, plus 1 is denoted by $2x + 1$.

10. a. The ratio of a number and 6 is 24 is written as $\dfrac{x}{6} = 24$.

  b. The difference of 10 and a number is 18 is written as $10 - x = 18$.

  c. One less than twice a number is 99 is written as $2x - 1 = 99$.

Calculator Explorations

1. $5^3 = 125$

2. $7^4 = 2401$

3. $9^5 = 59,049$

4. $8^6 = 262,144$

5. $2(20 - 5) = 30$

6. $3(14 - 7) + 21 = 42$

7. $24(862 - 455) + 89 = 9857$

8. $99 + (401 + 962) = 1462$

9. $\dfrac{4623 + 129}{36 - 34} = 2376$

10. $\dfrac{956 - 452}{89 - 86} = 168$

Exercise Set 1.3

1. $3^5 = 3 \cdot 3 \cdot 3 \cdot 3 \cdot 3 = 243$

3. $3^3 = 3 \cdot 3 \cdot 3 = 27$

5. $1^5 = 1 \cdot 1 \cdot 1 \cdot 1 \cdot 1 = 1$

7. $5^1 = 5$

9. $7^2 = 7 \cdot 7 = 49$

11. $\left(\dfrac{2}{3}\right)^4 = \left(\dfrac{2}{3}\right)\left(\dfrac{2}{3}\right)\left(\dfrac{2}{3}\right)\left(\dfrac{2}{3}\right) = \dfrac{2 \cdot 2 \cdot 2 \cdot 2}{3 \cdot 3 \cdot 3 \cdot 3} = \dfrac{16}{81}$

13. $\left(\dfrac{1}{5}\right)^3 = \left(\dfrac{1}{5}\right)\left(\dfrac{1}{5}\right)\left(\dfrac{1}{5}\right) = \dfrac{1 \cdot 1 \cdot 1}{5 \cdot 5 \cdot 5} = \dfrac{1}{125}$

15. $(1.2)^2 = 1.2 \cdot 1.2 = 1.44$

17. $(0.7)^3 = 0.7 \cdot 0.7 \cdot 0.7 = 0.343$

19. $5 \cdot 5 = 5^2$ square meters

21. $5 + 6 \cdot 2 = 5 + 12 = 17$

23. $4 \cdot 8 - 6 \cdot 2 = 32 - 12 = 20$

25. $18 \div 3 \cdot 2 = 6 \cdot 2 = 12$

27. $2 + (5 - 2) + 4^2 = 2 + 3 + 4^2 = 2 + 3 + 16 = 21$

29. $5 \cdot 3^2 = 5 \cdot 9 = 45$

31. $\dfrac{1}{4} \cdot \dfrac{2}{3} - \dfrac{1}{6} = \dfrac{2}{12} - \dfrac{1}{6} = \dfrac{1}{6} - \dfrac{1}{6} = 0$

33. $\dfrac{6-4}{9-2} = \dfrac{2}{7}$

35. $\begin{aligned} 2[5 + 2(8 - 3)] &= 2[5 + 2(5)] \\ &= 2[5 + 10] \\ &= 2[15] \\ &= 30 \end{aligned}$

37. $\dfrac{19 - 3 \cdot 5}{6 - 4} = \dfrac{19 - 15}{2} = \dfrac{4}{2} = 2$

39. $\dfrac{|6 - 2| + 3}{8 + 2 \cdot 5} = \dfrac{4 + 3}{8 + 10} = \dfrac{7}{18}$

41. $\dfrac{3 + 3(5 + 3)}{3^2 + 1} = \dfrac{3 + 3(8)}{9 + 1} = \dfrac{3 + 24}{10} = \dfrac{27}{10}$

43. $\begin{aligned} \dfrac{6 + |8 - 2| + 3^2}{18 - 3} &= \dfrac{6 + |6| + 9}{15} \\ &= \dfrac{6 + 6 + 9}{15} \\ &= \dfrac{21}{15} \\ &= \dfrac{7}{5} \end{aligned}$

45. $\begin{aligned} &2 + 3[10(4 \cdot 5 - 16) - 30] \\ &= 2 + 3[10(20 - 16) - 30] \\ &= 2 + 3[10(4) - 30] \\ &= 2 + 3[40 - 30] \\ &= 2 + 3[10] \\ &= 2 + 30 \\ &= 32 \end{aligned}$

47. $\begin{aligned} \left(\dfrac{2}{3}\right)^3 + \dfrac{1}{9} + \dfrac{1}{3} \cdot \dfrac{4}{3} &= \dfrac{8}{27} + \dfrac{1}{9} + \dfrac{1}{3} \cdot \dfrac{4}{3} \\ &= \dfrac{8}{27} + \dfrac{1}{9} + \dfrac{4}{9} \\ &= \dfrac{8}{27} + \dfrac{3}{27} + \dfrac{12}{27} \\ &= \dfrac{23}{27} \end{aligned}$

49. Replace y with 3.
$3y = 3(3) = 9$

51. Replace x with 1 and z with 5.
$$\frac{z}{5x} = \frac{5}{5(1)} = \frac{5}{5} = 1$$

53. Replace x with 1.
$$3x - 2 = 3(1) - 2 = 3 - 2 = 1$$

55. Replace x with 1 and y with 3.
$$|2x + 3y| = |2(1) + 3(3)| = |2 + 9| = |11| = 11$$

57. Replace x with 1, y with 3, and z with 5.
$$xy + z = 1(3) + 5 = 3 + 5 = 8$$

59. Replace y with 3.
$$5y^2 = 5(3^2) = 5(9) = 45$$

61. Replace z with 3.
$$5z = 5(3) = 15$$

63. Replace x with 2, y with 6, and z with 3.
$$\frac{z}{xy} = \frac{3}{2(6)} = \frac{3}{12} = \frac{1}{4}$$

65. Replace x with 2 and y with 6.
$$\frac{y}{x} + \frac{y}{x} = \frac{6}{2} + \frac{6}{2} = 3 + 3 = 6$$

67.
$$3x - 6 = 9$$
$$3(5) - 6 \stackrel{?}{=} 9$$
$$15 - 6 \stackrel{?}{=} 9$$
$$9 = 9 \quad \text{True}$$
Since the result is true, 5 is a solution of the given equation.

69.
$$2x + 6 = 5x - 1$$
$$2(0) + 6 \stackrel{?}{=} 5(0) - 1$$
$$0 + 6 \stackrel{?}{=} 0 - 1$$
$$6 = -1 \quad \text{False}$$
Since the result is false, 0 is not a solution of the given equation.

71.
$$2x - 5 = 5$$
$$2(8) - 5 \stackrel{?}{=} 5$$
$$16 - 5 \stackrel{?}{=} 5$$
$$11 = 5 \quad \text{False}$$
Since the result is false, 8 is not a solution of the given equation.

73.
$$x + 6 = x + 6$$
$$2 + 6 \stackrel{?}{=} 2 + 6$$
$$8 = 8 \quad \text{True}$$
Since the result is true, 2 is a solution of the given equation.

75.
$$x = 5x + 15$$
$$0 \stackrel{?}{=} 5(0) + 15$$
$$0 \stackrel{?}{=} 0 + 15$$
$$0 = 15 \quad \text{False}$$
Since the result is false, 0 is not a solution of the given equation.

77.
$$\frac{1}{3}x = 9$$
$$\frac{1}{3}(27) \stackrel{?}{=} 9$$
$$9 = 9 \quad \text{True}$$
Since the result is true, 27 is a solution of the given equation.

79. Fifteen more than a number is written as $x + 15$.

81. Five subtracted from a number is written as $x - 5$.

83. The ratio of a number and 4 is written as $\frac{x}{4}$.

85. Three times a number, increased by 22 is written as $3x + 22$.

87. One increased by two equals the quotient of nine and three is written as $1 + 2 = 9 \div 3$.

89. Three is not equal to four divided by two is written as $3 \neq 4 \div 2$.

91. The sum of 5 and a number is 20 is written as $5 + x = 20$.

93. The product 7.6 and a number is 17 is written as $7.6x = 17$.

95. Thirteen minus three times a number is 13 is written as $13 - 3x = 13$.

97. No, the parentheses are not necessary.

99. a. $(6 + 2) \cdot (5 + 3) = 8 \cdot 8 = 64$

    b. $(6 + 2) \cdot 5 + 3 = 8 \cdot 5 + 3 = 40 + 3 = 43$

    c. $6 + 2 \cdot 5 + 3 = 6 + 10 + 3 = 19$

    d. $6 + 2 \cdot (5 + 3) = 6 + 2 \cdot 8 = 6 + 16 = 22$

101.

Length, l	Width, w	Perimeter of Rectangle: $2l + 2w$	Area of Rectangle: lw
4 in.	3 in.	$2l + 2w$ $= 2(4 \text{ in.}) + 2(3 \text{ in.})$ $= 8 \text{ in.} + 6 \text{ in.}$ $= 14 \text{ in.}$	lw $= (4 \text{ in.})(3 \text{ in.})$ $= 12 \text{ sq in.}$
6 in.	1 in.	$2l + 2w$ $= 2(6 \text{ in.}) + 2(1 \text{ in.})$ $= 12 \text{ in.} + 2 \text{ in.}$ $= 14 \text{ in.}$	lw $= (6 \text{ in.})(1 \text{ in.})$ $= 6 \text{ sq in.}$
5 in.	2 in.	$2l + 2w$ $= 2(5 \text{ in.}) + 2(2 \text{ in.})$ $= 10 \text{ in.} + 4 \text{ in.}$ $= 14 \text{ in.}$	lw $= (5 \text{ in.})(2 \text{ in.})$ $= 10 \text{ sq in.}$

103. $(20 - 4) \cdot 4 \div 2 = 16 \cdot 4 \div 2 = 64 \div 2 = 32$

105. answers may vary

107. a. $5x + 6$ is an expression since it does not contain the equal symbol, "=."

    b. $2a = 7$ is an equation since it contains the equal symbol.

    c. $3a + 2 = 9$ is an equation since it contains the equal symbol.

    d. $4x + 3y - 8z$ is an expression since it does not contain the equal symbol.

    e. $5^2 - 2(6 - 2)$ is an expression since it does not contain the equal symbol.

109. answers may vary

Section 1.4 Practice Problems

1.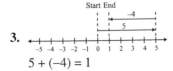

$-2 + (-4) = -6$

2.

$-5 + 8 = 3$

3.

$5 + (-4) = 1$

4. $-8 + (-5) = -13$

5. $(-14) + 6 = -8$

6. $(-17) + (-10) = -27$

7. $(-4) + 12 = 8$

8. $1.5 + (-3.2) = -1.7$

9. $-\dfrac{6}{12} + \left(-\dfrac{3}{12}\right) = -\dfrac{9}{12} = -\dfrac{3 \cdot 3}{3 \cdot 4} = -\dfrac{3}{4}$

10. $12.1 + (-3.6) = 8.5$

11. $-\dfrac{4}{5} + \dfrac{2}{3} = -\dfrac{12}{15} + \dfrac{10}{15} = -\dfrac{2}{15}$

12. a. $16 + (-9) + (-9) = 7 + (-9) = -2$

    b. $[3 + (-13)] + [-4 + (-7\} = [-10] + [-11]$
    $= -21$

13. The opposite of -35 is 35.

14. The opposite of 12 is -12.

15. The opposite of $-\dfrac{3}{11}$ is $\dfrac{3}{11}$.

16. The opposite of 1.9 is -1.9.

17. a. $-(-22) = 22$

    b. $-\left(-\dfrac{2}{7}\right) = \dfrac{2}{7}$

    c. $-(-x) = x$

    d. $-|-14| = -14$

    e. $-|2.3| = -2.3$

18. $-2 + (-1) + 3 + 3 = 3$
    The overall gain is \$3.

Exercise Set 1.4

1. $6 + (-3) = 3$

3. $-6 + (-8) = -14$

5. $8 + (-7) = 1$

7. $-14 + 2 = -12$

9. $-2 + (-3) = -5$

11. $-9 + (-3) = -12$

13. $-7 + 3 = -4$

15. $10 + (-3) = 7$

17. $5 + (-7) = -2$

19. $-16 + 16 = 0$

21. $27 + (-46) = -19$

23. $-18 + 49 = 31$

25. $-33 + (-14) = -47$

27. $6.3 + (-8.4) = -2.1$

29. $117 + (-79) = 38$

31. $-9.6 + (-3.5) = -13.1$

33. $-\dfrac{3}{8}+\dfrac{5}{8}=\dfrac{2}{8}=\dfrac{2\cdot1}{2\cdot4}=\dfrac{1}{4}$

35. $-\dfrac{7}{16}+\dfrac{1}{4}=-\dfrac{7}{16}+\dfrac{4}{16}=-\dfrac{3}{16}$

37. $-\dfrac{7}{10}+\left(-\dfrac{3}{5}\right)=-\dfrac{7}{10}+\left(-\dfrac{6}{10}\right)=-\dfrac{13}{10}$

39. $|-8|+(-16)=8+(-16)=-8$

41. $-15+9+(-2)=-6+(-2)=-8$

43. $-21+(-16)+(-22)=-37+(-22)=-59$

45. $-23+16+(-2)=-7+(-2)=-9$

47. $|5+(-10)|=|-5|=5$

49. $6+(-4)+9=2+9=11$

51. $[-17+(-4)]+[-12+15]=[-21]+[3]=-18$

53. $\begin{aligned}|9+(-12)|+|-16|&=|-3|+|-16|\\&=3+16\\&=19\end{aligned}$

55. $\begin{aligned}-13+[5+(-3)+4]&=-13+[2+4]\\&=-13+[6]\\&=-7\end{aligned}$

57. $-38+12=-26$
The sum of -38 and 12 is -26.

59. The additive inverse of 6 is -6.

61. The additive inverse of -2 is 2.

63. The additive inverse of 0 is 0.

65. Since $|-6|=6$, the additive inverse of $|-6|$ is -6.

67. $-|-2|=-2$

69. $-(-7)=7$

71. $-(-7.9)=7.9$

73. $-(-5z)=5z$

75. $\left|-\dfrac{2}{3}\right|=\dfrac{2}{3}$

77. $-35+142=107$
The highest recorded temperature in Massachusetts was 107°F.

79. $-411+316=-95$
Your elevation is -95 meters.

81. $-2.50+(-0.86)=-3.36$
The combined change is -3.36 points.

83. $\begin{aligned}-5+(-4)+(-3)+4&=-9+(-3)+4\\&=-12+4\\&=-8\end{aligned}$
His overall score was -8.

85. $\begin{aligned}-23+(-581)+93+155&=-604+93+155\\&=-511+155\\&=-356\end{aligned}$
The total net income was $-\$356$ million.

87. answers will vary

89. The highest bar corresponds to July, so the highest low temperature was in July.

91. The shortest bar above 0 corresponds to October.

93. $\dfrac{-9.1+14.4+8.8}{3}=\dfrac{5.3+8.8}{3}=\dfrac{14.1}{3}=4.7$
The average for the months of April, May, and October is 4.7°F.

95. If p is a positive number, then $-p$ is a negative number.

97. If p is a positive number, then $p+p$ is a positive number.

99. answers may vary

101. answers may vary

Section 1.5 Practice Problems

1. a. $-20 - 6 = -20 + (-6) = -26$

b. $3 - (-5) = 3 + 5 = 8$

c. $7 - 17 = 7 + (-17) = -10$

d. $-4 - (-9) = -4 + 9 = 5$

2. $9.6 - (-5.7) = 15.3$

3. $-\dfrac{4}{9} - \dfrac{2}{9} = -\dfrac{4}{9} + \left(-\dfrac{2}{9}\right) = -\dfrac{6}{9} = -\dfrac{2}{3}$

4. $-\dfrac{1}{4} - \left(-\dfrac{2}{5}\right) = -\dfrac{1}{4} + \dfrac{2}{5} = -\dfrac{5}{20} + \dfrac{8}{20} = \dfrac{3}{20}$

5. a. $-11 - 7 = -11 + (-7) = -18$

b. $35 - (-25) = 35 + 25 = 60$

6. a. $-20 - 5 + 12 - (-3) = -20 + (-5) + 12 + 3$
$= -10$

b. $5.2 - (-4.4) + (-8.8) = 5.2 + 4.4 + (-8.8)$
$= 0.8$

7. a. $-9 + [(-4 - 1) - 10]$
$= -9 + [(-4 + (-1)) - 10]$
$= -9 + [(-5) - 10]$
$= -9 + [-5 + (-10)]$
$= -9 + [-15]$
$= -24$

b. $5^2 - 20 + [-11 - (-3)]$
$= 5^2 - 20 + [-11 + 3]$
$= 5^2 - 20 + [-8]$
$= 25 - 20 + (-8)$
$= 25 + (-20) + (-8)$
$= 5 + (-8)$
$= -3$

8. a. Replace x with 1 and y with -4.
$$\frac{x - y}{14 + x} = \frac{1 - (-4)}{14 + 1} = \frac{1 + 4}{15} = \frac{5}{15} = \frac{1}{3}$$

b. Replace x with 1 and y with -4.
$$x^2 - y = (1)^2 - (-4)$$
$$= 1 - (-4)$$
$$= 1 + 4$$
$$= 5$$

9. $-1 + x = 1$
$-1 + (-2) \overset{?}{=} 1$
$-3 = 1$ False
-2 is not a solution.

10. $-23 - 14 = -23 + (-14) = -37$
The overall change in temperature was $-37°$.

11. a. These angles are supplementary, so their sum is $180°$. This means that $m\angle x$ is $180° - 78°$.
$m\angle x = 180° - 78° = 102°$

b. These angles are complementary, so their sum is $90°$. This means that $m\angle y$ is $90° - 81°$.
$m\angle y = 90° - 81° = 9°$

Exercise Set 1.5

1. $-6 - 4 = -6 + (-4) = -10$

3. $4 - 9 = 4 + (-9) = -5$

5. $16 - (-3) = 16 + 3 = 19$

7. $7 - (-4) = 7 + 4 = 11$

9. $-26 - (-18) = -26 + 18 = -8$

11. $-6 - 5 = -6 + (-5) = -11$

13. $16 - (-21) = 16 + 21 = 37$

15. $-6 - (-11) = -6 + 11 = 5$

17. $-44 - 27 = -44 + (-27) = -71$

19. $-21 - (-21) = -21 + 21 = 0$

21. $-\dfrac{3}{11} - \left(-\dfrac{5}{11}\right) = -\dfrac{3}{11} + \dfrac{5}{11} = \dfrac{2}{11}$

23. $9.7 - 16.1 = 9.7 + (-16.1) = -6.4$

25. $-2.6 - (-6.7) = -2.6 + 6.7 = 4.1$

27. $\dfrac{1}{2} - \dfrac{2}{3} = \dfrac{1}{2} + \left(-\dfrac{2}{3}\right) = \dfrac{3}{6} + \left(-\dfrac{4}{6}\right) = -\dfrac{1}{6}$

29. $-\dfrac{1}{6} - \dfrac{3}{4} = -\dfrac{1}{6} + \left(-\dfrac{3}{4}\right)$

$\qquad = -\dfrac{2}{12} + \left(-\dfrac{9}{12}\right)$

$\qquad = -\dfrac{11}{12}$

31. $8.3 - (-0.62) = 8.3 + 0.62 = 8.92$

33. $0 - 8.92 = 0 + (-8.92) = -8.92$

35. $8 - (-5) = 8 + 5 = 13$
-5 subtracted from 8 is 13.

37. $-6 - (-1) = -6 + 1 = -5$
The difference between -6 and -1 is -5.

39. $7 - 8 = 7 + (-8) = -1$
8 subtracted from 7 is -1.

41. $-8 - 15 = -8 + (-15) = -23$
-8 decreased by 15 is -23.

43. $-10 - (-8) + (-4) - 20$
$\quad = -10 + 8 + (-4) + (-20)$
$\quad = -2 + (-4) + (-20)$
$\quad = -6 + (-20)$
$\quad = -26$

45. $5 - 9 + (-4) - 8 - 8$
$\quad = 5 + (-9) + (-4) + (-8) + (-8)$
$\quad = -4 + (-4) + (-8) + (-8)$
$\quad = -8 + (-8) + (-8)$
$\quad = -16 + (-8)$
$\quad = -24$

47. $-6 - (2 - 11) = -6 - (-9) = -6 + 9 = 3$

49. $3^3 - 8 \cdot 9 = 27 - 72 = 27 + (-72) = -45$

51. $2 - 3(8 - 6) = 2 - 3[8 + (-6)]$
$\qquad\qquad\qquad = 2 - 3(2)$
$\qquad\qquad\qquad = 2 - 6$
$\qquad\qquad\qquad = 2 + (-6)$
$\qquad\qquad\qquad = -4$

53. $(3 - 6) + 4^2 = [3 + (-6)] + 16 = [-3] + 16 = 13$

55. $-2 + [(8 - 11) - (-2 - 9)]$
$\quad = -2 + [(8 + (-11)) - (-2 + (-9))]$
$\quad = -2 + [(-3) - (-11)]$
$\quad = -2 + [-3 + 11]$
$\quad = -2 + 8$
$\quad = 6$

57. $|-3| + 2^2 + [-4 - (-6)] = |-3| + 4 + [-4 + 6]$
$\qquad\qquad\qquad\qquad\qquad = 3 + 4 + 2$
$\qquad\qquad\qquad\qquad\qquad = 7 + 2$
$\qquad\qquad\qquad\qquad\qquad = 9$

59. Replace x with -5 and y with 4.
$x - y = -5 - 4 = -5 + (-4) = -9$

61. Replace x with -5 and y with 4.
$\dfrac{9 - x}{y + 6} = \dfrac{9 - (-5)}{4 + 6} = \dfrac{9 + 5}{10} = \dfrac{14}{10} = \dfrac{2 \cdot 7}{2 \cdot 5} = \dfrac{7}{5}$

63. Replace x with -5, y with 4, and t with 10.
$|x| + 2t - 8y = |-5| + 2(10) - 8(4)$
$\qquad\qquad\quad = 5 + 20 - 32$
$\qquad\qquad\quad = 25 - 32$
$\qquad\qquad\quad = 25 + (-32)$
$\qquad\qquad\quad = -7$

65. Replace x with -5 and y with 4.
$$y^2 - x = 4^2 - (-5) = 16 + 5 = 21$$

67. Replace x with -5 and t with 10.
$$\frac{|x - (-10)|}{2t} = \frac{|-5 - (-10)|}{2(10)}$$
$$= \frac{|-5 + 10|}{2(10)}$$
$$= \frac{|5|}{2(10)}$$
$$= \frac{5}{20}$$
$$= \frac{1 \cdot 5}{4 \cdot 5}$$
$$= \frac{1}{4}$$

69.
$$x - 9 = 5$$
$$-4 - 9 \stackrel{?}{=} 5$$
$$-4 + (-9) \stackrel{?}{=} 5$$
$$-13 = 5 \quad \text{False}$$
Since the result is false, -4 is not a solution of the given equation.

71.
$$-x + 6 = -x - 1$$
$$-(-2) + 6 \stackrel{?}{=} -(-2) - 1$$
$$2 + 6 \stackrel{?}{=} 2 - 1$$
$$8 \stackrel{?}{=} 2 + (-1)$$
$$8 = 1 \quad \text{False}$$
Since the result is false, -2 is not a solution of the given equation.

73.
$$-x - 13 = -15$$
$$-2 - 13 \stackrel{?}{=} -15$$
$$-2 + (-13) \stackrel{?}{=} -15$$
$$-15 = -15 \quad \text{True}$$
Since the result is true, 2 is a solution of the given equation.

75. $44 - (-56) = 44 + 56 = 100$
The total drop in temperature was $100°$.

77. $2 - 5 - 20 = 2 + (-5) + (-20) = -3 + (-20) = -23$
The total change in yardage is a loss of 23 yards.

79. $-475 - 94 = -475 + (-94) = -569$
He was born in -569 or 569 B.C.

81. Complementary angles sum to 90°.
$$90° - 60° = x$$
$$90° + (-60°) = x$$
$$30° = x$$

83. $-250 + 120 - 178 = -250 + 120 + (-178) = -130 + (-178) = -308$
The overall vertical change is -308 feet.

85. $19,340 - (-512) = 19,340 + 512 = 19,852$
Mt. Kilimanjaro is 19,852 feet higher than Lake Assal.

87. $y = 180° - 50°$
$y = 180° + (-50°)$
$y = 130°$

89.

Month	Monthly Increase or Decrease
February	$-23.7 - (-19.3) = -23.7 + 19.3 = -4.4°$
March	$-21.1 - (-23.7) = -21.1 + 23.7 = 2.6°$
April	$-9.1 - (-21.1) = -9.1 + 21.1 = 12°$
May	$14.4 - (-9.1) = 14.4 + 9.1 = 23.5°$
June	$29.7 - 14.4 = 29.7 + (-14.4) = 15.3°$
July	$33.6 - 29.7 + 33.6 + (-29.7) = 3.9°$
August	$33.3 - 33.6 = 33.3 + (-33.6) = -0.3°$
September	$27.0 - 33.3 = 27.0 + (-33.3) = -6.3°$
October	$8.8 - 27.0 = 8.8 + (-27.0) = -18.2°$
November	$-6.9 - 8.8 = -6.9 + (-8.8) = -15.7°$
December	$-17.2 - (-6.9) = -17.2 + 6.9 = -10.3°$

91. The largest negative number corresponds to October.

93. answers may vary

95. true; answers may vary

97. true; answers may vary

99. Since 4.362 is less than 7.0086, the answer is negative.
$4.362 - 7.0086 = -2.6466$

101. $a - b$ is sometimes positive and sometimes negative. For example $-9 - (-8) = -1$ and $-7 - (-11) = 4$.

Integrated Review

1. The opposite of a positive number is a
   <u>negative</u> number.

2. The sum of two negative numbers is a
   <u>negative</u> number.

3. The absolute value of a negative number is a
   <u>positive</u> number.

4. The absolute value of zero is <u>0</u>.

5. The reciprocal of a positive number is a
   <u>positive</u> number.

6. The sum of a number and its opposite is <u>0</u>.

7. The absolute value of a positive number is a
   <u>positive</u> number.

8. The reciprocal of a negative number is a
   <u>negative</u> number.

9. The opposite of $\frac{1}{7}$ is $-\frac{1}{7}$.

   The absolute value of $\frac{1}{7}$ is $\frac{1}{7}$.

10. The opposite of $-\frac{12}{5}$ is $\frac{12}{5}$.

    The absolute value of $-\frac{12}{5}$ is $\frac{12}{5}$.

11. The number whose opposite is -3 is 3.
    The absolute value of 3 is 3.

12. The number whose opposite is $\frac{9}{11}$ is $-\frac{9}{11}$.

    The absolute value of $-\frac{9}{11}$ is $\frac{9}{11}$.

13. $-19 + (-23) = -42$

14. $7 - (-3) = 7 + 3 = 10$

15. $-15 + 17 = 2$

16. $-8 - 10 = -8 + (-10) = -18$

17. $18 + (-25) = -7$

18. $-2 + (-37) = -39$

19. $-14 - (-12) = -14 + 12 = -2$

20. $5 - 14 = 5 + (-14) = -9$

21. $4.5 - 7.9 = 4.5 + (-7.9) = -3.4$

22. $-8.6 - 1.2 = -8.6 + (-1.2) = -9.8$

23. $-\dfrac{3}{4} - \dfrac{1}{7} = -\dfrac{3}{4} + \left(-\dfrac{1}{7}\right)$

    $= -\dfrac{21}{28} + \left(-\dfrac{4}{28}\right)$

    $= -\dfrac{25}{28}$

24. $\dfrac{2}{3} - \dfrac{7}{8} = \dfrac{2}{3} + \left(-\dfrac{7}{8}\right) = \dfrac{16}{24} + \left(-\dfrac{21}{24}\right) = -\dfrac{5}{24}$

25. $-9 - (-7) + 4 - 6 = -9 + 7 + 4 + (-6)$
    $\qquad = -2 + 4 + (-6)$
    $\qquad = 2 + (-6)$
    $\qquad = -4$

26. $11 - 20 + (-3) - 12$
    $= 11 + (-20) + (-3) + (-12)$
    $= -9 + (-3) + (-12)$
    $= -12 + (-12)$
    $= -24$

27. $24 - 6(14 - 11) = 24 - 6[14 + (-11)]$
    $\qquad = 24 - 6(3)$
    $\qquad = 24 + (-6)(3)$
    $\qquad = 24 + (-18)$
    $\qquad = 6$

28. $30 - 5(10 - 8) = 30 - 5[10 + (-8)]$
    $\qquad = 30 - 5(2)$
    $\qquad = 30 + (-5)(2)$
    $\qquad = 30 + (-10)$
    $\qquad = 20$

29. $(7-17)+4^2 = [7+(-17)]+4^2$
$$= -10+4^2$$
$$= -10+16$$
$$= 6$$

30. $9^2 + (10-30) = 9^2 + [10+(-30)]$
$$= 9^2 + (-20)$$
$$= 81 + (-20)$$
$$= 61$$

31. $|-9| + 3^2 + (-4-20)$
$$= |-9| + 3^2 + [-4+(-20)]$$
$$= 9 + 9 + (-24)$$
$$= 18 + (-24)$$
$$= -6$$

32. $|-4-5| + 5^2 + (-50)$
$$= |-4+(-5)| + 5^2 + (-50)$$
$$= |-9| + 5^2 + (-50)$$
$$= 9 + 25 + (-50)$$
$$= 34 + (-50)$$
$$= -16$$

33. $-7 + [(1-2)+(-2-9)]$
$$= -7 + [(1+(-2))+(-2+(-9))]$$
$$= -7 + [(-1)+(-11)]$$
$$= -7 + (-12)$$
$$= -19$$

34. $-6 + [(-3+7)+(4-15)]$
$$= -6 + [(-3+7)+(4+(-15))]$$
$$= -6 + [4+(-11)]$$
$$= -6 + (-7)$$
$$= -13$$

35. $1-5 = 1 + (-5) = -4$

36. $-3-(-2) = -3+2 = -1$

37. $\dfrac{1}{4} - \left(-\dfrac{2}{5}\right) = \dfrac{1}{4} + \dfrac{2}{5} = \dfrac{5}{20} + \dfrac{8}{20} = \dfrac{13}{20}$

38. $-\dfrac{5}{8} - \dfrac{1}{10} = -\dfrac{5}{8} + \left(-\dfrac{1}{10}\right)$
$$= -\dfrac{25}{40} + \left(-\dfrac{4}{40}\right)$$
$$= -\dfrac{29}{40}$$

39. $2(19-17)^3 - 3(-7+9)^2$
$$= 2[19+(-17)]^3 + (-3)(-7+9)^2$$
$$= 2(2)^3 + (-3)(2)^2$$
$$= 2(8) + (-3)(4)$$
$$= 16 + (-12)$$
$$= 4$$

40. $3(10-9)^2 + 6(20-19)^3$
$$= 3[10+(-9)]^2 + 6[20+(-19)]^3$$
$$= 3(1)^2 + 6(1)^3$$
$$= 3(1) + 6(1)$$
$$= 3 + 6$$
$$= 9$$

41. Replace x with -2 and y with -1.
$$x - y = -2 - (-1) = -2 + 1 = -1$$

42. Replace x with -2 and y with -1.
$$x + y = -2 + (-1) = -3$$

43. Replace y with -1 and z with 9.
$$y + z = -1 + 9 = 8$$

44. Replace y with -1 and z with 9.
$$z - y = 9 - (-1) = 9 + 1 = 10$$

45. Replace x with -2, y with -1, and z with 9.
$$\dfrac{|5z-x|}{y-x} = \dfrac{|5(9)-(-2)|}{-1-(-2)}$$
$$= \dfrac{|45+2|}{-1+2}$$
$$= \dfrac{|47|}{1}$$
$$= \dfrac{47}{1}$$
$$= 47$$

46. Replace x with -2, y with -1, and z with 9.

$$\frac{|-x-y+z|}{2z} = \frac{|-(-2)-(-1)+9|}{2(9)}$$

$$= \frac{|2+1+9|}{18}$$

$$= \frac{|12|}{18}$$

$$= \frac{12}{18}$$

$$= \frac{2\cdot 6}{3\cdot 6}$$

$$= \frac{2}{3}$$

Section 1.6 Practice Problems

1. $-8(3) = -24$

2. $5(-30) = -150$

3. $-4(-12) = 48$

4. $-\dfrac{5}{6}\cdot\dfrac{1}{4} = -\dfrac{5\cdot 1}{6\cdot 4} = -\dfrac{5}{24}$

5. $6(-2.3) = -13.8$

6. $-15(-2) = 30$

7. a. $5(0)(-3) = 0(-3) = 0$

   b. $(-1)(-6)(-7) = (6)(-7) = -42$

   c. $(-2)(4)(-8) = (-8)(-8) = 64$

8. a. $(-2)^4 = (-2)(-2)(-2)(-2) = 16$

   b. $-2^4 = -(2\cdot 2\cdot 2\cdot 2) = -16$

   c. $(-1)^5 = (-1)(-1)(-1)(-1)(-1) = -1$

   d. $-1^5 = -(1\cdot 1\cdot 1\cdot 1\cdot 1) = -1$

e. $\left(-\dfrac{7}{9}\right)^2 = \left(-\dfrac{7}{9}\right)\left(-\dfrac{7}{9}\right) = \dfrac{49}{81}$

9. a. The reciprocal of 13 is $\dfrac{1}{13}$ since

$$13\cdot\dfrac{1}{13} = 1.$$

   b. The reciprocal of $\dfrac{7}{15}$ is $\dfrac{15}{7}$ since

$$\dfrac{7}{15}\cdot\dfrac{15}{7} = 1.$$

   c. The reciprocal of -5 is $-\dfrac{1}{5}$ since

$$-5\cdot-\dfrac{1}{5} = 1.$$

   d. The reciprocal of $-\dfrac{8}{11}$ is $-\dfrac{11}{8}$ since

$$-\dfrac{8}{11}\cdot-\dfrac{11}{8} = 1.$$

   e. The reciprocal of 7.9 is $\dfrac{1}{7.9}$ since

$$7.9\cdot\dfrac{1}{7.9} = 1.$$

10. a. $-12\div 4 = -12\cdot\dfrac{1}{4} = -3$

   b. $\dfrac{-20}{-10} = -20\cdot-\dfrac{1}{10} = 2$

   c. $\dfrac{36}{-4} = 36\cdot-\dfrac{1}{4} = -9$

11. a. $\dfrac{-25}{5} = -5$

   b. $\dfrac{-48}{-6} = 8$

c. $\dfrac{50}{-2} = -25$

d. $\dfrac{-72}{0.2} = -360$

12. $-\dfrac{5}{9} \div \dfrac{2}{3} = -\dfrac{5}{9} \cdot \dfrac{3}{2} = -\dfrac{15}{18} = -\dfrac{3 \cdot 5}{3 \cdot 6} = -\dfrac{5}{6}$

13. $-\dfrac{2}{7} \div \left(-\dfrac{1}{5}\right) = -\dfrac{2}{7} \cdot \left(-\dfrac{5}{1}\right) = \dfrac{10}{7}$

14. a. $\dfrac{-7}{0}$ is undefined.

b. $\dfrac{0}{-2} = 0$

15. a. $\dfrac{0(-5)}{3} = \dfrac{0}{3} = 0$

b. $-3(-9) - 4(-4) = 27 - (-16)$
$= 27 + 16$
$= 43$

c. $(-3)^2 + 2\big[(5-15) - |-4-1|\big]$
$= (-3)^2 + 2\big[(-10) - |-5|\big]$
$= (-3)^2 + 2[-10 - 5]$
$= (-3)^2 + 2(-15)$
$= 9 + (-30)$
$= -21$

d. $\dfrac{-7(-4) + 2}{-10 - (-5)} = \dfrac{28 + 2}{-10 + 5} = \dfrac{30}{-5} = -6$

e. $\dfrac{5(-2)^3 + 52}{-4 + 1} = \dfrac{5(-8) + 52}{-3}$
$= \dfrac{-40 + 52}{-3}$
$= \dfrac{12}{-3}$
$= -4$

16. a. Replace x with -1 and y with -5.
$\dfrac{3y}{45x} = \dfrac{3(-5)}{45(-1)} = \dfrac{-15}{-45} = \dfrac{15 \cdot 1}{15 \cdot 3} = \dfrac{1}{3}$

b. Replace x with -1 and y with -5.
$x^2 - y^3 = (-1)^2 - (-5)^3$
$= 1 - (-125)$
$= 1 + 125$
$= 126$

c. Replace x with -1 and y with -5.
$\dfrac{x+y}{3x} = \dfrac{-1 + (-5)}{3(-1)} = \dfrac{-6}{-3} = 2$

17. $\dfrac{x}{4} - 3 = x + 3$
$\dfrac{-8}{4} - 3 \stackrel{?}{=} -8 + 3$
$-2 - 3 \stackrel{?}{=} -5$
$-5 = -5$ True
-8 is a solution.

Calculator Explorations

1. $-38(26 - 27) = 38$

2. $-59(-8) + 1726 = 2198$

3. $134 + 25(68 - 91) = -441$

4. $45(32) - 8(218) = -304$

5. $\dfrac{-50(294)}{175 - 205} = 490$

6. $\dfrac{-444 - 444.8}{-181 - (-181)}$ is undefined.

7. $9^5 - 4550 = 54,499$

8. $5^8 - 6259 = 384,366$

9. $(-125)^2 = 15,625$

10. $-125^2 = -15,625$

Mental Math

1. The product of two negative numbers is a <u>positive</u> number.

2. The quotient of two negative numbers is a <u>positive</u> number.

3. The quotient of a positive number and a negative number is a <u>negative</u> number.

4. The product of a positive number and a negative number is a <u>negative</u> number.

5. The reciprocal of a positive number is a <u>positive</u> number.

6. The opposite of a positive number is a <u>negative</u> number.

Exercise Set 1.6

1. $-6(4) = -24$

3. $2(-1) = -2$

5. $-5(-10) = 50$

7. $-3 \cdot 15 = -45$

9. $-\dfrac{1}{2}\left(-\dfrac{3}{5}\right) = \dfrac{3}{10}$

11. $5(-1.4) = -7$

13. $(-1)(-3)(-5) = 3(-5) = -15$

15. $(2)(-1)(-3)(0) = -2(-3)(0) = 6(0) = 0$

17. $(-4)^2 = (-4)(-4) = 16$

19. $-4^2 = -4 \cdot 4 = -16$

21. $\left(-\dfrac{3}{4}\right)^2 = \left(-\dfrac{3}{4}\right)\left(-\dfrac{3}{4}\right) = \dfrac{9}{16}$

23. $-0.7^2 = -0.7 \cdot 0.7 = -0.49$

25. The reciprocal of $\dfrac{2}{3}$ is $\dfrac{3}{2}$ since $\dfrac{2}{3} \cdot \dfrac{3}{2} = 1$.

27. The reciprocal of -14 is $-\dfrac{1}{14}$ since $-14 \cdot \left(-\dfrac{1}{14}\right) = 1$.

29. The reciprocal of $-\dfrac{3}{11}$ is $-\dfrac{11}{3}$ since $-\dfrac{3}{11} \cdot \left(-\dfrac{11}{3}\right) = 1$.

31. The reciprocal of 0.2 is $\dfrac{1}{0.2}$ since $0.2\left(\dfrac{1}{0.2}\right) = 1$.

33. $\dfrac{18}{-2} = -9$

35. $-48 \div 12 = -48 \cdot \dfrac{1}{12} = -4$

37. $\dfrac{0}{-4} = 0$

39. $\dfrac{5}{0}$ is undefined.

41. $\dfrac{6}{7} \div \left(-\dfrac{1}{3}\right) = \dfrac{6}{7} \cdot \left(-\dfrac{3}{1}\right) = -\dfrac{18}{7}$

43. $-3.2 \div -0.02 = \dfrac{-3.2}{-0.02} = 160$

45. $(-8)(-8) = 64$

47. $\dfrac{2}{3}\left(-\dfrac{4}{9}\right) = -\dfrac{8}{27}$

49. $\dfrac{-12}{-4} = 3$

51. $\dfrac{30}{-2} = -15$

53. $(-5)^3 = (-5)(-5)(-5) = 25(-5) = -125$

55. $(-0.2)^3 = (-0.2)(-0.2)(-0.2)$
$= 0.04(-0.2)$
$= -0.008$

57. $\left(-\dfrac{3}{4}\right)\left(-\dfrac{8}{9}\right) = \dfrac{24}{36} = \dfrac{2 \cdot 12}{3 \cdot 12} = \dfrac{2}{3}$

59. $-\dfrac{5}{9} \div \left(-\dfrac{3}{4}\right) = -\dfrac{5}{9} \cdot \left(-\dfrac{4}{3}\right) = \dfrac{20}{27}$

61. $-2.1(-0.4) = 0.84$

63. $\dfrac{-48}{1.2} = -40$

65. $(-3)^4 = (-3)(-3)(-3)(-3)$
$= 9(-3)(-3)$
$= -27(-3)$
$= 81$

67. $-1^7 = -1 \cdot 1 \cdot 1 \cdot 1 \cdot 1 \cdot 1 \cdot 1 = -1$

69. $-11 \cdot 11 = -121$

71. $-\dfrac{4}{9} \div \dfrac{4}{9} = -\dfrac{4}{9} \cdot \dfrac{9}{4} = -\dfrac{36}{36} = -1$

73. $-9 - 10 = -9 + (-10) = -19$

75. $-9(-10) = 90$

77. $7(-12) = -84$

79. $7 + (-12) = -5$

81. $\dfrac{-9(-3)}{-6} = \dfrac{27}{-6} = -\dfrac{9 \cdot 3}{2 \cdot 3} = -\dfrac{9}{2}$

83. $-3(2 - 8) = -3[2 + (-8)] = -3(-6) = 18$

85. $-7(-2) - 3(-1) = 14 - (-3) = 14 + 3 = 17$

87. $2^2 - 3[(2 - 8) - (-6 - 8)]$
$= 2^2 - 3[(-6) - (-14)]$
$= 2^2 - 3[-6 + 14]$
$= 4 - 3(8)$
$= 4 - 24$
$= 4 + (-24)$
$= -20$

89. $\dfrac{-6^2 + 4}{-2} = \dfrac{-36 + 4}{-2} = \dfrac{-32}{-2} = 16$

91. $\dfrac{-3 - 5^2}{2(-7)} = \dfrac{-3 - 25}{-14}$
$= \dfrac{-3 + (-25)}{-14}$
$= \dfrac{-28}{-14}$
$= \dfrac{-14 \cdot 2}{-14 \cdot 1}$
$= 2$

93. $\dfrac{22 + (3)(-2)^2}{-5 - 2} = \dfrac{22 + 3(4)}{-5 + (-2)}$
$= \dfrac{22 + 12}{-7}$
$= \dfrac{34}{-7}$
$= -\dfrac{34}{7}$

95. $\dfrac{(-4)^2 - 16}{4 - 12} = \dfrac{16 - 16}{4 - 12} = \dfrac{16 + (-16)}{4 + (-12)} = \dfrac{0}{-8} = 0$

97. $\dfrac{6 - 2(-3)}{4 - 3(-2)} = \dfrac{6 - (-6)}{4 - (-6)} = \dfrac{6 + 6}{4 + 6} = \dfrac{12}{10} = \dfrac{2 \cdot 6}{2 \cdot 5} = \dfrac{6}{5}$

99.
$$\frac{|5-9|+|10-15|}{|2(-3)|} = \frac{|5+(-9)|+|10+(-15)|}{|-6|}$$
$$= \frac{|-4|+|-5|}{|-6|}$$
$$= \frac{4+5}{6}$$
$$= \frac{9}{6}$$
$$= \frac{3\cdot 3}{3\cdot 2}$$
$$= \frac{3}{2}$$

101.
$$\frac{-7(-1)+(-3)4}{(-2)(5)+(-6)(-8)} = \frac{7+(-12)}{-10+48} = \frac{-5}{38} = -\frac{5}{38}$$

103. Replace x with -5 and y with -3.
$$\frac{2x-5}{y-2} = \frac{2(-5)-5}{-3-2}$$
$$= \frac{-10-5}{-3-2}$$
$$= \frac{-10+(-5)}{-3+(-2)}$$
$$= \frac{-15}{-5}$$
$$= 3$$

105. Replace x with -5 and y with -3.
$$\frac{6-y}{x-4} = \frac{6-(-3)}{-5-4} = \frac{6+3}{-5+(-4)} = \frac{9}{-9} = -1$$

107. Replace x with -5 and y with -3.
$$\frac{4-2x}{y+3} = \frac{4-2(-5)}{-3+3}$$
$$= \frac{4-(-10)}{0}$$
$$= \frac{4+10}{0}$$
$$= \frac{14}{0} \text{ is undefined.}$$

109. Replace x with -5 and y with -3.
$$\frac{x^2+y}{3y} = \frac{(-5)^2+(-3)}{3(-3)}$$
$$= \frac{25+(-3)}{-9}$$
$$= \frac{22}{-9}$$
$$= -\frac{22}{9}$$

111.
$$-3x-5 = -20$$
$$-3(5)-5 \stackrel{?}{=} -20$$
$$-15-5 \stackrel{?}{=} -20$$
$$-15+(-5) \stackrel{?}{=} -20$$
$$-20 = -20 \quad \text{True}$$
Since the result is true, 5 is a solution of the given equation.

113.
$$\frac{x}{5}+2 = -1$$
$$\frac{15}{5}+2 \stackrel{?}{=} -1$$
$$3+2 \stackrel{?}{=} -1$$
$$5 = -1 \quad \text{False}$$
Since the result is false, 15 is not a solution of the given equation.

115.
$$\frac{x-3}{7} = -2$$
$$\frac{-11-3}{7} \stackrel{?}{=} -2$$
$$\frac{-11+(-3)}{7} \stackrel{?}{=} -2$$
$$\frac{-14}{7} \stackrel{?}{=} -2$$
$$-2 = -2 \quad \text{True}$$
Since the result is true, -11 is a solution of the given equation.

117. True since the product of an odd number of negative numbers is negative.

119. False since the product of an even number of negative integers is positive.

121. $2(-81) = -162$
The surface temperature of Jupiter is $-162°F$.

123. answers may vary

125. Since $\dfrac{1}{1} = 1$ and $\dfrac{1}{-1} = -1,$ both 1 and -1 are their own reciprocals.

127. $\dfrac{0}{5} - 7 = 0 - 7 = 0 + (-7) = -7$

129. $-8(-5) + (-1) = 40 + (-1) = 39$

The Bigger Picture

1. $-0.2(25) = -5$

2. $86 - 100 = 86 + (-100) = -14$

3. $-\dfrac{1}{7} + \left(-\dfrac{3}{5}\right) = -\dfrac{5}{35} + \left(-\dfrac{21}{35}\right) = -\dfrac{26}{35}$

4. $\dfrac{-40}{-5} = 8$

5. $(-7)^2 = (-7)(-7) = 49$

6. $-7^2 = -(7 \cdot 7) = -49$

7. $\dfrac{|-42|}{-|-2|} = \dfrac{42}{-2} = -21$

8. $\dfrac{8.6}{0}$ is undefined.

9. $\dfrac{0}{8.6} = 0$

10. $-25 - (-13) = -25 + 13 = -12$

11. $-8.3 - 8.3 = -8.3 + (-8.3) = -16.6$

12. $-\dfrac{8}{9}\left(-\dfrac{3}{16}\right) = \dfrac{24}{144} = \dfrac{24 \cdot 1}{24 \cdot 6} = \dfrac{1}{6}$

13. $2 + 3(8 - 11)^3 = 2 + 3(-3)^3$
$= 2 + 3(-27)$
$= 2 + (-81)$
$= -79$

14. $-2\dfrac{1}{2} \div \left(-3\dfrac{1}{4}\right) = -\dfrac{5}{2} \div \left(-\dfrac{13}{4}\right)$
$= -\dfrac{5}{2} \cdot \left(-\dfrac{4}{13}\right)$
$= \dfrac{20}{26}$
$= \dfrac{10}{13}$

15. $20 \div 2 \cdot 5 = 10 \cdot 5 = 50$

16. $-2[(1 - 5) - (7 - 17)] = -2[(-4) - (-10)]$
$= -2[-4 + 10]$
$= -2[6]$
$= -12$

Section 1.7 Practice Problems

1. a. $7 \cdot y = y \cdot 7$

   b. $4 + x = x + 4$

2. a. $5 \cdot (-3 \cdot 6) = (5 \cdot -3) \cdot 6$

   b. $(-2 + 7) + 3 = -2 + (7 + 3)$

   c. $(q + r) + 17 = q + (r + 17)$

   d. $(ab) \cdot 21 = a \cdot (b \cdot 21)$

3. Since the order of two numbers was changed but their grouping was not, the statement is true by the commutative property of multiplication.

4. Since the grouping of the numbers was changed and their order was not, the statement is true by the associative property of addition.

5. $(-3+x)+17 = -3+(x+17)$
$\qquad = -3+(17+x)$
$\qquad = (-3+17)+x$
$\qquad = 14+x$

6. $4(5x) = (4 \cdot 5) \cdot x = 20x$

7. $5(x+y) = 5(x)+5(y) = 5x+5y$

8. $-3(2+7x) = -3(2)+(-3)(7x) = -6-21x$

9. $4(x+6y-2z) = 4(x)+4(6y)-4(2z)$
$\qquad = 4x+24y-8z$

10. $-1(3-a) = (-1)(3)-(-1)(a) = -3+a$

11. $-(8+a-b) = -1(8+a-b)$
$\qquad = (-1)(8)+(-1)(a)-(-1)(b)$
$\qquad = -8-a+b$

12. $\frac{1}{2}(2x+4)+9 = \frac{1}{2}(2x)+\frac{1}{2}(4)+9$
$\qquad = 1x+2+9$
$\qquad = x+11$

13. $9 \cdot 3 + 9 \cdot y = 9(3+y)$

14. $4x+4y = 4(x+y)$

15. $7(a+b) = 7 \cdot a + 7 \cdot b$ illustrates the distributive property.

16. $12+y = y+12$ illustrates the commutative property of addition.

17. $-4 \cdot (6 \cdot x) = (-4 \cdot 6) \cdot x$ illustrates the associative property of multiplication.

18. $6+(z+2) = 6+(2+z)$ illustrates the commutative property of addition.

19. $3\left(\frac{1}{3}\right) = 1$ illustrates the multiplicative inverse property.

20. $(x+0)+23 = x+23$ illustrates the identity element for addition.

21. $(7 \cdot y) \cdot 10 = y \cdot (7 \cdot 10)$ illustrates the commutative and associative properties of multiplication.

Exercise Set 1.7

1. $x+16 = 16+x$ by the commutative property of addition.

3. $-4 \cdot y = y \cdot (-4)$ by the commutative property of multiplication.

5. $xy = yx$ by the commutative property of multiplication.

7. $2x+13 = 13+2x$ by the commutative property of addition.

9. $(xy) \cdot z = x \cdot (yz)$ by the associative property of multiplication.

11. $2+(a+b) = (2+a)+b$ by the associative property of addition.

13. $4 \cdot (ab) = 4a \cdot (b)$ by the associative property of multiplication.

15. $(a+b)+c = a+(b+c)$ by the associative property of addition.

17. $8+(9+b) = (8+9)+b = 17+b$

19. $4(6y) = (4 \cdot 6)y = 24y$

21. $\frac{1}{5}(5y) = \left(\frac{1}{5} \cdot 5\right)y = 1y = y$

23. $(13+a)+13 = (a+13)+13$
$\qquad = a+(13+13)$
$\qquad = a+26$

25. $-9(8x) = (-9 \cdot 8)x = -72x$

27. $\frac{3}{4}\left(\frac{4}{3}s\right) = \left(\frac{3}{4} \cdot \frac{4}{3}\right)s = 1s = s$

29. $-\frac{1}{2}(5x) = \left(-\frac{1}{2} \cdot 5\right)x = -\frac{5}{2}x$

31. $4(x + y) = 4(x) + 4(y) = 4x + 4y$

33. $9(x - 6) = 9(x) - 9(6) = 9x - 54$

35. $2(3x + 5) = 2(3x) + 2(5) = 6x + 10$

37. $7(4x - 3) = 7(4x) - 7(3) = 28x - 21$

39. $3(6 + x) = 3(6) + 3(x) = 18 + 3x$

41. $-2(y - z) = -2(y) - (-2)z = -2y + 2z$

43. $-\dfrac{1}{3}(3y + 5) = -\dfrac{1}{3}(3y) - \dfrac{1}{3}(5) = -y - \dfrac{5}{3}$

45. $5(x + 4m + 2) = 5(x) + 5(4m) + 5(2)$
$$= 5x + 20m + 10$$

47. $-4(1 - 2m + n) + 4$
$$= -4(1) - 4(-2m) - 4(n) + 4$$
$$= -4 + 8m - 4n + 4$$
$$= (-4 + 4) + 8m - 4n$$
$$= 0 + 8m - 4n$$
$$= 8m - 4n$$

49. $-(5x + 2) = -1(5x + 2)$
$$= -1(5x) + (-1)(2)$$
$$= -5x - 2$$

51. $-(r - 3 - 7p) + 3 = -1(r - 3 - 7p) + 3$
$$= -1(r) - 1(-3) - 1(-7p) + 3$$
$$= -r + 3 + 7p + 3$$
$$= -r + (3 + 3) + 7p$$
$$= -r + 6 + 7p$$

53. $\dfrac{1}{2}(6x + 7) + \dfrac{1}{2} = \dfrac{1}{2}(6x) + \dfrac{1}{2}(7) + \dfrac{1}{2}$
$$= \left(\dfrac{1}{2} \cdot 6\right)x + \dfrac{7}{2} + \dfrac{1}{2}$$
$$= 3x + \dfrac{8}{2}$$
$$= 3x + 4$$

55. $-\dfrac{1}{3}(3x - 9y) = -\dfrac{1}{3}(3x) - \dfrac{1}{3}(-9y) = -x + 3y$

57. $3(2r + 5) - 7 = 3(2r) + 3(5) - 7$
$$= 6r + 15 - 7$$
$$= 6r + 8$$

59. $-9(4x + 8) + 2 = -9(4x) - 9(8) + 2$
$$= -36x - 72 + 2$$
$$= -36x - 70$$

61. $-0.4(4x + 5) - 0.5$
$$= -0.4(4x) + (-0.4)(5) - 0.5$$
$$= -1.6x - 2 - 0.5$$
$$= -1.6x - 2.5$$

63. $4 \cdot 1 + 4 \cdot y = 4(1 + y)$

65. $11x + 11y = 11 \cdot x + 11 \cdot y = 11(x + y)$

67. $(-1) \cdot 5 + (-1) \cdot x = -1(5 + x) = -(5 + x)$

69. $30a + 30b = 30 \cdot a + 30 \cdot b = 30(a + b)$

71. $3 \cdot 5 = 5 \cdot 3$ illustrates the commutative property of multiplication.

73. $2 + (x + 5) = (2 + x) + 5$ illustrates the associative property of addition.

75. $(x + 9) + 3 = (9 + x) + 3$ illustrates the commutative property of addition.

77. $(4 \cdot y) \cdot 9 = 4 \cdot (y \cdot 9)$ illustrates the associative property of multiplication.

79. $0 + 6 = 6$ illustrates the identity property of addition.

81. $-4(y + 7) = -4 \cdot y + (-4) \cdot 7$ illustrates the distributive property.

83. $6 \cdot \dfrac{1}{6} = 1$ illustrates the multiplicative inverse property.

85. $-6 \cdot 1 = -6$ illustrates the identity element for multiplication.

87. The opposite of 8 is −8.

The reciprocal of 8 is $\dfrac{1}{8}$.

89. The opposite of x is $-x$.

The reciprocal of x is $\dfrac{1}{x}$.

91. The expression is the reciprocal of $\dfrac{1}{2x}$ or $2x$.

The opposite of $2x$ is $-2x$.

93. False; the opposite of $-\dfrac{a}{2}$ is $\dfrac{a}{2}$. $-\dfrac{2}{a}$ is the reciprocal of $-\dfrac{a}{2}$.

95. "Taking a test" and "studying for the test" are not commutative, since the order in which they are performed affects the outcome.

97. "Putting on your left shoe" and "putting on your right shoe" are commutative, since the order in which they are performed does not affect the outcome.

99. "Mowing the lawn" and "trimming the hedges" are commutative, since the order in which they are performed does not affect the outcome.

101. "Feeding the dog" and "feeding the cat" are commutative, since the order in which they are performed does not affect the outcome.

103. a. The property illustrated is the commutative property of addition since the order in which they are added changed.

b. The property illustrated is the commutative property of addition since the order in which they are added changed.

c. The property illustrated is the associative property of addition since the grouping of addition changed.

105. answers may vary

107. answers may vary

Section 1.8 Practice Problems

1. a. The numerical coefficient of $-4x$ is -4.

b. The numerical coefficient of $15y^3$ is 15.

c. The numerical coefficient of x is 1, since $x = 1x$.

d. The numerical coefficient of $-y$ is -1, since $-y = -1y$.

e. The numerical coefficient of $\dfrac{z}{4}$ is $\dfrac{1}{4}$, since $\dfrac{z}{4}$ is $\dfrac{1}{4} \cdot z$.

2. a. $7x^2$ and $-6x^3$ are unlike terms, since the exponents on x are not the same.

b. $3x^2y^2$, $-x^2y^2$, and $4x^2y^2$ are like terms, since each variable and its exponent match.

c. $-5ab$ and $3ba$ are like terms, since $ab = ba$ by the commutative property.

d. $2x^3$ and $4y^3$ are unlike terms, since the variables are not the same.

e. $-7m^4$ and $7m^4$ are like terms, since the variable and its exponent match.

3. a. $9y - 4y = (9 - 4)y = 5y$

b. $11x^2 + x^2 = (11 + 1)x^2 = 12x^2$

c. $\begin{aligned} 5y - 3x + 4x &= 5y + (-3+4)x \\ &= 5y + 1x \\ &= 5y + x \end{aligned}$

d. $\begin{aligned} 14m^2 - m^2 + 3m^2 &= (14-1+3)m^2 \\ &= 16m^2 \end{aligned}$

4. $\begin{aligned} 7y + 2y + 6 + 10 &= (7+2)y + (6+10) \\ &= 9y + 16 \end{aligned}$

5. $\begin{aligned} -2x + 4 + x - 11 &= -2x + x + 4 - 11 \\ &= (-2+1)x + (4-11) \\ &= -x - 7 \end{aligned}$

6. The terms $3z$ and $-3z^2$ cannot be combined because they are unlike terms.

7. $\begin{aligned} 8.9y + 4.2y - 3 &= (8.9+4.2)y - 3 \\ &= 13.1y - 3 \end{aligned}$

8. $3(11y + 6) = 3(11y) + 3(6) = 33y + 18$

9. $\begin{aligned} &-4(x + 0.2y - 3) \\ &= -4(x) + (-4)(0.2y) - (-4)(3) \\ &= -4x - 0.8y + 12 \end{aligned}$

10. $\begin{aligned} &-(3x + 2y + z - 1) \\ &= -1(3x + 2y + z - 1) \\ &= -1(3x) + (-1)(2y) + (-1)(z) - (-1)(1) \\ &= -3x - 2y - z + 1 \end{aligned}$

11. $4(4x - 6) + 20 = 16x - 24 + 20 = 16x - 4$

12. $\begin{aligned} 5 - (3x + 9) + 6x &= 5 - 3x - 9 + 6x \\ &= -3x + 6x + 5 - 9 \\ &= 3x - 4 \end{aligned}$

13. $\begin{aligned} -3(7x + 1) - (4x - 2) &= -21x - 3 - 4x + 2 \\ &= -25x - 1 \end{aligned}$

14. $8 + 11(2y - 9) = 8 + 22y - 99 = -91 + 22y$

15. $\begin{aligned} (4x - 3) - (9x - 10) &= 4x - 3 - 9x + 10 \\ &= -5x + 7 \end{aligned}$

16. Three times a number subtracted from 10 is written as $10 - 3x$. This expression cannot be simplified.

17. The sum of a number and 2, divided by 5, is written as $(x + 2) \div 5$ or $\dfrac{x+2}{5}$. This expression cannot be simplified.

18. Three times a number, added to the sum of a number and 6, is written as $(x + 6) + 3x$. $(x + 6) + 3x = x + 6 + 3x = 4x + 6$

19. Seven times the difference of a number and 4 is written as $7(x - 4)$. $7(x - 4) = 7x - 28$

Mental Math

1. The numerical coefficient of $-7y$ is -7.

2. The numerical coefficient of $3x$ is 3.

3. The numerical coefficient of x is 1, since $x = 1x$.

4. The numerical coefficient of $-y$ is -1, since $-y = -1y$.

5. The numerical coefficient of $17x^2 y$ is 17.

6. The numerical coefficient of $1.2xyz$ is 1.2.

7. $5y$ and $-y$ are like terms, since the variable and its exponent match.

8. $-2x^2 y$ and $6xy$ are unlike terms, since the exponents on x are not the same.

9. $2z$ and $3z^2$ are unlike terms, since the exponents on z are not the same.

10. ab^2 and $-7ab^2$ are like terms, since each variable and its exponent match.

11. $8wz$ and $\dfrac{1}{7}zw$ are like terms, since $wz = zw$

by the commutative property.

12. $7.4p^3q^2$ and $6.2p^3q^2r$ are unlike terms, since the exponents on r are not the same.

Exercise Set 1.8

1. $7y + 8y = (7 + 8)y = 15y$

3. $8w - w + 6w = 8w - 1w + 6w$
$= (8 - 1 + 6)w$
$= 13w$

5. $3b - 5 - 10b - 4 = 3b - 10b - 5 - 4$
$= (3 - 10)b + (-5 - 4)$
$= -7b - 9$

7. $m - 4m + 2m - 6 = (1 - 4 + 2)m - 6$
$= -1m - 6$
$= -m - 6$

9. $5g - 3 - 5 - 5g = 5g - 5g - 3 - 5$
$= (5 - 5)g + (-3 - 5)$
$= 0g - 8$
$= -8$

11. $6.2x - 4 + x - 1.2 = 6.2x + 1x - 4 - 1.2$
$= (6.2 + 1)x + (-4 - 1.2)$
$= 7.2x - 5.2$

13. $2k - k - 6 = 2k - 1k - 6$
$= (2 - 1)k - 6$
$= 1k - 6$
$= k - 6$

15. $-9x + 4x + 18 - 10x = -9x + 4x - 10x + 18$
$= (-9 + 4 - 10)x + 18$
$= -15x + 18$

17. $6x - 5x + x - 3 + 2x = 6x - 5x + x + 2x - 3$
$= (6 - 5 + 1 + 2)x - 3$
$= 4x - 3$

19. $7x^2 + 8x^2 - 10x^2 = (7 + 8 - 10)x^2 = 5x^2$

21. $3.4m - 4 - 3.4m - 7 = 3.4m - 3.4m - 4 - 7$
$= (3.4 - 3.4)m + (-4 - 7)$
$= 0m - 11$
$= -11$

23. $6x + 0.5 - 4.3x - 0.4x + 3$
$= 6x - 4.3x - 0.4x + 0.5 + 3$
$= (6 - 4.3 - 0.4)x + (0.5 + 3)$
$= 1.3x + 3.5$

25. $5(y + 4) = 5(y) + 5(4) = 5y + 20$

27. $-2(x + 2) = -2(x) - 2(2) = -2x - 4$

29. $-5(2x - 3y + 6)$
$= -5(2x) - (-5)(3y) + (-5)(6)$
$= -10x + 15y - 30$

31. $-(3x - 2y + 1) = -1(3x - 2y + 1)$
$= -1(3x) - 1(-2y) - 1(1)$
$= -3x + 2y - 1$

33. $7(d - 3) + 10 = 7d - 21 + 10 = 7d - 11$

35. $-4(3y - 4) + 12y = -12y + 16 + 12y$
$= 0y + 16$
$= 16$

37. $3(2x - 5) - 5(x - 4) = 6x - 15 - 5x + 20$
$= 6x - 5x - 15 + 20$
$= x + 5$

39. $-2(3x - 4) + 7x - 6 = -6x + 8 + 7x - 6$
$= -6x + 7x + 8 - 6$
$= 1x + 2$
$= x + 2$

41. $5k - (3k - 10) = 5k - 3k + 10 = 2k + 10$

43. $3x + 4 - (6x - 1) = 3x + 4 - 6x + 1$
$= 3x - 6x + 4 + 1$
$= -3x + 5$

45. $5(x + 2) - (3x - 4) = 5x + 10 - 3x + 4$
$= 5x - 3x + 10 + 4$
$= 2x + 14$

47. $\frac{1}{3}(7y-1)+\frac{1}{6}(4y+7)=\frac{7}{3}y-\frac{1}{3}+\frac{4}{6}y+\frac{7}{6}$

$\qquad\qquad\qquad = \frac{7}{3}y+\frac{2}{3}y-\frac{2}{6}+\frac{7}{6}$

$\qquad\qquad\qquad = \frac{9}{3}y+\frac{5}{6}$

$\qquad\qquad\qquad = 3y+\frac{5}{6}$

49. $2+4(6x-6)=2+24x-24$

$\qquad\qquad\quad = 2-24+24x$

$\qquad\qquad\quad = -22+24x$

51. $0.5(m+2)+0.4m=0.5m+1+0.4m$

$\qquad\qquad\qquad = 0.5m+0.4m+1$

$\qquad\qquad\qquad = 0.9m+1$

53. $10-3(2x+3y)=10-6x-9y$

55. $6(3x-6)-2(x+1)-17x$

$= 18x-36-2x-2-17x$

$= 18x-2x-17x-36-2$

$= -1x-38$

$= -x-38$

57. $\frac{1}{2}(12x-4)-(x+5)=6x-2-x-5$

$\qquad\qquad\qquad = 6x-x-2-5$

$\qquad\qquad\qquad = 5x-7$

59. $(4x-10)+(6x+7)=4x-10+6x+7$

$\qquad\qquad\qquad = 4x+6x-10+7$

$\qquad\qquad\qquad = 10x-3$

61. $(3x-8)-(7x+1)=3x-8-7x-1$

$\qquad\qquad\qquad = 3x-7x-8-1$

$\qquad\qquad\qquad = -4x-9$

63. $(m-9)-(5m-6)=m-9-5m+6$

$\qquad\qquad\qquad = m-5m-9+6$

$\qquad\qquad\qquad = -4m-3$

65. Twice a number, decreased by four is written as $2x-4$.

67. Three-fourths of a number, increased by twelve is written as $\frac{3}{4}x+12$.

69. The sum of 5 times a number and -2, added to 7 times the number is written as $5x+(-2)+7x=5x+7x-2=12x-2$.

71. Eight times the sum of a number and six is written as $8(x+6)=8x+48$.

73. Double a number minus the sum of the number and ten is written as $2x-(x+10)=2x-x-10=x-10$.

75. Replace x with -1 and y with 3.
$y-x^2=3-(-1)^2=3-1=2$

77. Replace a with 2 and b with -5.
$a-b^2=2-(-5)^2=2-25=-23$

79. Replace y with -5 and z with 0.
$yz-y^2=-5(0)-(-5)^2=0-25=-25$

81. Since 1 cone balances 1 cube and 1 cylinder balances 2 cubes, 1 cone and 1 cylinder balances $1+2=3$ cubes. The scale shown is balanced.

83. Since 1 cylinder balances 2 cubes, the left side is equivalent to 5 cubes. Since 1 cone balances 1 cube, the right side is equivalent to 5 cubes. The scale is balanced.

85. answers may vary

87. $2[(4x-1)+5x]=2(4x-1+5x)$

$\qquad\qquad\qquad = 2(9x-1)$

$\qquad\qquad\qquad = 18x-2$

The perimeter is $(18x-2)$ feet.

89. The length of the first board in inches is $12(x+2)$.
$12(x+2)+(3x-1)=12x+24+3x-1$

$\qquad\qquad\qquad\qquad = 12x+3x+24-1$

$\qquad\qquad\qquad\qquad = 15x+23$

The total length is $(15x+23)$ inches.

91. answers may vary

Chapter 1 Vocabulary Check

1. The symbols $\neq$, $<$, and $>$ are called inequality symbols.

2. A mathematical statement that two expressions are equal is called and equation.

3. The absolute value of a number is the distance between that number and 0 on the number line.

4. A symbol used to represent a number is called a variable.

5. Two numbers that are the same distance from 0 but lie on opposite sides of 0 are called opposites.

6. The number in a fraction above the fraction bar is called the numerator.

7. A solution of an equation is a value for the variable that makes the equation a true statement.

8. Two numbers whose product is 1 are called reciprocals.

9. In 2^3, the 2 is called the base and the 3 is called the exponent.

10. The numerical coefficient of a term is its numerical factor.

11. The number in a fraction below the fraction bar is called the denominator.

12. Parentheses and brackets are examples of grouping symbols.

13. A term is a number or the product of a number and variables raised to powers.

14. Terms with the same variables raised to the same powers are called like terms.

15. If terms are not like terms, then they are unlike terms.

Chapter 1 Review

1. Since 8 is to the left of 10 on the number line, $8 < 10$.

2. Since 7 is to the right of 2 on the number line, $7 > 2$.

3. Since -4 is to the right of -5 on the number line, $-4 > -5$.

4. Since $\dfrac{12}{2} = 6$ is to the right of -8 on the number line, $\dfrac{12}{2} > -8$.

5. Since $|-7| = 7$ is to the left of $|-8| = 8$ on the number line, $|-7| < |-8|$.

6. Since $|-9| = 9$ is to the right of -9 on the number line, $|-9| > -9$.

7. $-|-1| = -1$

8. Since $|-14| = 14$ and $-(-14) = 14$, $|-14| = -(-14)$.

9. Since 1.2 is to the right of 1.02 on the number line, $1.2 > 1.02$.

10. Since $-\dfrac{3}{2} = -\dfrac{6}{4}$ and $-\dfrac{6}{4}$ is to the left of $-\dfrac{3}{4}$ on the number line, $-\dfrac{3}{2} < -\dfrac{3}{4}$.

11. Four is greater than or equal to negative three is written as $4 \geq -3$.

12. Six is not equal to five is written as $6 \neq 5$.

13. 0.03 is less than 0.3 is written as $0.03 < 0.3$.

14. Since 400 is to the right of 155 on the number line, $400 > 155$.

15. a. The natural numbers are 1, 3.

   b. The whole numbers are 0, 1, 3.

   c. The integers are −6, 0, 1, 3.

   d. The rational numbers are −6, 0, 1, $1\frac{1}{2}$, 3, 9.62.

   e. The irrational number is π.

   f. The real numbers are all numbers in the set.

16. a. The natural numbers are 2, 5.

   b. The whole numbers are 2, 5.

   c. The integers are −3, 2, 5.

   d. The rational numbers are −3, −1.6, 2, 5, $\frac{11}{2}$, 15.1.

   e. The irrational numbers are $\sqrt{5}$, 2π.

   f. The real numbers are all numbers in the set.

17. Since $-4 < -2$, the most negative number is −4 which corresponds to Friday. Thus, Friday showed the greatest loss.

18. The greatest positive number is +5 which corresponds to Wednesday. Thus, Wednesday showed the greatest gain.

19. $6 \cdot 3^2 + 2 \cdot 8 = 6 \cdot 9 + 2 \cdot 8 = 54 + 16 = 70$
The answer is c.

20. $68 - 5 \cdot 2^3 = 68 - 5 \cdot 8$
$$= 68 - 40$$
$$= 68 + (-40)$$
$$= 28$$
The answer is b.

21. $3(1 + 2 \cdot 5) + 4 = 3(1 + 10) + 4$
$$= 3(11) + 4$$
$$= 33 + 4$$
$$= 37$$

22. $8 + 3(2 \cdot 6 - 1) = 8 + 3(12 - 1)$
$$= 8 + 3[12 + (-1)]$$
$$= 8 + 3(11)$$
$$= 8 + 33$$
$$= 41$$

23. $\dfrac{4 + |6 - 2| + 8^2}{4 + 6 \cdot 4} = \dfrac{4 + |6 + (-2)| + 8^2}{4 + 6 \cdot 4}$
$$= \frac{4 + |4| + 8^2}{4 + 6 \cdot 4}$$
$$= \frac{4 + 4 + 64}{4 + 24}$$
$$= \frac{72}{28}$$
$$= \frac{18 \cdot 4}{7 \cdot 4}$$
$$= \frac{18}{7}$$

24. $5[3(2 + 5) - 5] = 5[3(7) - 5]$
$$= 5[21 - 5]$$
$$= 5[21 + (-5)]$$
$$= 5(16)$$
$$= 80$$

25. The difference of twenty and twelve is equal to the product of two and four is written as $20 - 12 = 2 \cdot 4$.

26. The quotient of nine and two is greater than negative five is written as $\dfrac{9}{2} > -5$.

27. Replace x with 6 and y with 2.
$2x + 3y = 2(6) + 3(2) = 12 + 6 = 18$

28. Replace x with 6, y with 2, and z with 8.
$$x(y+2z) = 6[2+2(8)]$$
$$= 6(2+16)$$
$$= 6(18)$$
$$= 108$$

29. Replace x with 6, y with 2, and z with 8.
$$\frac{x}{y}+\frac{z}{2y} = \frac{6}{2}+\frac{8}{2(2)} = \frac{6}{2}+\frac{8}{4} = 3+2 = 5$$

30. Replace x with 6 and y with 2.
$$x^2 - 3y^2 = 6^2 - 3(2)^2$$
$$= 36 - 3(4)$$
$$= 36 - 12$$
$$= 36 + (-12)$$
$$= 24$$

31. Replace a with 37 and b with 80.
$$180 - a - b = 180 - 37 - 80$$
$$= 180 + (-37) + (-80)$$
$$= 143 + (-80)$$
$$= 63$$
The measure of the unknown angle is 63°.

32.
$$7x - 3 = 18$$
$$7(3) - 3 \overset{?}{=} 18$$
$$21 - 3 \overset{?}{=} 18$$
$$21 + (-3) \overset{?}{=} 18$$
$$18 = 18$$
Since the results is true, 3 is a solution of the given equation.

33.
$$3x^2 + 4 = x - 1$$
$$3(1)^2 + 4 \overset{?}{=} 1 - 1$$
$$3(1) + 4 \overset{?}{=} 0$$
$$3 + 4 \overset{?}{=} 0$$
$$7 = 0$$
Since the result is false, 1 is not a solution of the given equation.

34. The opposite of −9 is 9.

35. The opposite of $\frac{2}{3}$ is $-\frac{2}{3}$.

36. The opposite of $|-2| = 2$ is −2.

37. The opposite of $-|-7| = -7$ is 7.

38. $-15 + 4 = -11$

39. $-6 + (-11) = -17$

40. $\frac{1}{16} + \left(-\frac{1}{4}\right) = \frac{1}{16} + \left(-\frac{4}{16}\right) = -\frac{3}{16}$

41. $-8 + |-3| = -8 + 3 = -5$

42. $-4.6 + (-9.3) = -13.9$

43. $-2.8 + 6.7 = 3.9$

44. $6 - 20 = 6 + (-20) = -14$

45. $-3.1 - 8.4 = -3.1 + (-8.4) = -11.5$

46. $-6 - (-11) = -6 + 11 = 5$

47. $4 - 15 = 4 + (-15) = -11$

48. $-21 - 16 + 3(8 - 2)$
$$= -21 + (-16) + 3[8 + (-2)]$$
$$= -21 + (-16) + 3(6)$$
$$= -21 + (-16) + 18$$
$$= -37 + 18$$
$$= -19$$

49. $\dfrac{11 - (-9) + 6(8 - 2)}{2 + 3 \cdot 4} = \dfrac{11 + 9 + 6[8 + (-2)]}{2 + 3 \cdot 4}$
$$= \frac{11 + 9 + 6(6)}{2 + 3 \cdot 4}$$
$$= \frac{11 + 9 + 36}{2 + 12}$$
$$= \frac{56}{14}$$
$$= 4$$

50. Replace x with 3, y with -6, and z with -9.

$$\begin{aligned}
2x^2 - y + z &= 2(3)^2 - (-6) + (-9) \\
&= 2(9) + 6 + (-9) \\
&= 18 + 6 + (-9) \\
&= 24 + (-9) \\
&= 15
\end{aligned}$$

The answer is a.

51. Replace x with 3 and y with -6.

$$\begin{aligned}
\frac{|y - 4x|}{2x} &= \frac{|-6 - 4(3)|}{2(3)} \\
&= \frac{|-6 - 12|}{6} \\
&= \frac{|-6 + (-12)|}{6} \\
&= \frac{|-18|}{6} \\
&= \frac{18}{6} \\
&= 3
\end{aligned}$$

The answer is a.

52.
$$\begin{aligned}
&50 + 1 + (-2) + 5 + 1 + (-4) \\
&= 51 + (-2) + 5 + 1 + (-4) \\
&= 49 + 5 + 1 + (-4) \\
&= 54 + 1 + (-4) \\
&= 55 + (-4) \\
&= 51
\end{aligned}$$

The price at the end of the week is \$51.

53. The reciprocal of -6 is $-\dfrac{1}{6}$.

54. The reciprocal of $\dfrac{3}{5}$ is $\dfrac{5}{3}$.

55. $6(-8) = -48$

56. $(-2)(-14) = 28$

57. $\dfrac{-18}{-6} = 3$

58. $\dfrac{42}{-3} = -14$

59. $-3(-6)(-2) = 18(-2) = -36$

60. $(-4)(-3)(0)(-6) = 12(0)(-6) = 0(-6) = 0$

61. $\dfrac{4(-3) + (-8)}{2 + (-2)} = \dfrac{-12 + (-8)}{2 + (-2)} = \dfrac{-20}{0}$ is undefined.

62.
$$\begin{aligned}
\frac{3(-2)^2 - 5}{-14} &= \frac{3(4) - 5}{-14} \\
&= \frac{12 - 5}{-14} \\
&= \frac{12 + (-5)}{-14} \\
&= \frac{7}{-14} \\
&= -\frac{1}{2}
\end{aligned}$$

63. $-6 + 5 = 5 + (-6)$ illustrates the commutative property of addition.

64. $6 \cdot 1 = 6$ illustrates the multiplicative identity property.

65. $3(8 - 5) = 3 \cdot 8 + 3 \cdot (-5)$ illustrates the distributive property.

66. $4 + (-4) = 0$ illustrates the additive inverse property.

67. $2 + (3 + 9) = (2 + 3) + 9$ illustrates the associative property of addition.

68. $2 \cdot 8 = 8 \cdot 2$ illustrates the commutative property of multiplication.

69. $6(8 + 5) = 6 \cdot 8 + 6 \cdot 5$ illustrates the distributive property.

70. $(3 \cdot 8) \cdot 4 = 3 \cdot (8 \cdot 4)$ illustrates the associative property of multiplication.

71. $4 \cdot \frac{1}{4} = 1$ illustrates the multiplicative inverse property.

72. $8 + 0 = 8$ illustrates the additive identity property.

73. $4(8 + 3) = 4(3 + 8)$ illustrates the commutative property of addition.

74. $5(2 + 1) = 5 \cdot 2 + 5 \cdot 1$ illustrates the distributive property.

75. $5x - x + 2x = 5x - 1x + 2x\,x$
$\qquad = (5 - 1 + 2)x$
$\qquad = 6x$

76. $0.2z - 4.6z - 7.4z = (0.2 - 4.6 - 7.4)z$
$\qquad\qquad\qquad\quad = -11.8z$

77. $\frac{1}{2}x + 3 + \frac{7}{2}x - 5 = \frac{1}{2}x + \frac{7}{2}x + 3 - 5$
$\qquad\qquad = \left(\frac{1}{2} + \frac{7}{2}\right)x + 3 + (-5)$
$\qquad\qquad = \frac{8}{2}x + (-2)$
$\qquad\qquad = 4x - 2$

78. $\frac{4}{5}y + 1 + \frac{6}{5}y + 2 = \frac{4}{5}y + \frac{6}{5}y + 1 + 2$
$\qquad\qquad = \left(\frac{4}{5} + \frac{6}{5}\right)y + 3$
$\qquad\qquad = \frac{10}{5}y + 3$
$\qquad\qquad = 2y + 3$

79. $2(n - 4) + n - 10 = 2n - 8 + n - 10$
$\qquad\qquad\qquad = 2n + n - 8 - 10$
$\qquad\qquad\qquad = 3n - 18$

80. $3(w + 2) - (12 - w) = 3w + 6 - 12 + w$
$\qquad\qquad\qquad = 3w + w + 6 - 12$
$\qquad\qquad\qquad = 4w - 6$

81. $(x + 5) - (7x - 2) = x + 5 - 7x + 2$
$\qquad\qquad\qquad = x - 7x + 5 + 2$
$\qquad\qquad\qquad = -6x + 7$

82. $(y - 0.7) - (1.4y - 3) = y - 0.7 - 1.4y + 3$
$\qquad\qquad\qquad = y - 1.4y - 0.7 + 3$
$\qquad\qquad\qquad = -0.4y + 2.3$

83. Three times a number decreased by 7 is written as $3x - 7$.

84. Twice the sum of a number and 2.8, added to 3 times the number is written as
$2(x + 2.8) + 3x = 2x + 5.6 + 3x$
$\qquad\qquad\quad = 2x + 3x + 5.6$
$\qquad\qquad\quad = 5x + 5.6.$

85. $-|-11| = -11$
$|11.4| = 11.4$
So, $-|-11| < |11.4|$.

86. Since $-1\frac{1}{2}$ is to the right of $-2\frac{1}{2}$ on the number line, $-1\frac{1}{2} > -2\frac{1}{2}$.

87. $-7.2 + (-8.1) = -15.3$

88. $14 - 20 = 14 + (-20) = -6$

89. $4(-20) = -80$

90. $\frac{-20}{4} = -5$

91. $-\frac{4}{5}\left(\frac{5}{16}\right) = -\frac{20}{80} = -\frac{1 \cdot 20}{4 \cdot 20} = -\frac{1}{4}$

92. $-0.5(-0.3) = 0.15$

93. $8 \div 2 \cdot 4 = 4 \cdot 4 = 16$

94. $(-2)^4 = (-2)(-2)(-2)(-2)$
$= 4(-2)(-2)$
$= -8(-2)$
$= 16$

95. $\dfrac{-3 - 2(-9)}{-15 - 3(-4)} = \dfrac{-3 + 18}{-15 + 12} = \dfrac{15}{-3} = -5$

96. $5 + 2[(7-5)^2 + (1-3)]$
$= 5 + 2[(7 + (-5))^2 + (1 + (-3))]$
$= 5 + 2[(2)^2 + (-2)]$
$= 5 + 2[4 + (-2)]$
$= 5 + 2(2)$
$= 5 + 4$
$= 9$

97. $-\dfrac{5}{8} \div \dfrac{3}{4} = -\dfrac{5}{8} \cdot \dfrac{4}{3} = -\dfrac{20}{24} = -\dfrac{5 \cdot 4}{6 \cdot 4} = -\dfrac{5}{6}$

98. $\dfrac{-15 + (-4)^2 + |-9|}{10 - 2 \cdot 5} = \dfrac{-15 + 16 + 9}{10 - 10} = \dfrac{10}{0}$ is undefined.

99. $7(3x - 3) - 5(x + 4) = 21x - 21 - 5x - 20$
$= 21x - 5x - 21 - 20$
$= 16x - 41$

100. $8 + 2(9x - 10) = 8 + 18x - 20$
$= 18x + 8 - 20$
$= 18x - 12$

Chapter 1 Test

1. The absolute value of negative seven is greater than five is written as $|-7| > 5$.

2. The sum of nine and five is greater than or equal to four is written as $(9 + 5) \geq 4$.

3. $-13 + 8 = -5$

4. $-13 - (-2) = -13 + 2 = -11$

5. $6 \cdot 3 - 8 \cdot 4 = 18 - 32 = 18 + (-32) = -14$

6. $(13)(-3) = -39$

7. $(-6)(-2) = 12$

8. $\dfrac{|-16|}{-8} = \dfrac{16}{-8} = -2$

9. $\dfrac{-8}{0}$ is undefined.

10. $\dfrac{|-6| + 2}{5 - 6} = \dfrac{6 + 2}{5 - 6} = \dfrac{8}{-1} = -8$

11. $\dfrac{1}{2} - \dfrac{5}{6} = \dfrac{3}{6} - \dfrac{5}{6} = -\dfrac{2}{6} = -\dfrac{1}{3}$

12. $-1\dfrac{1}{8} + 5\dfrac{3}{4} = -1\dfrac{1}{8} + 5\dfrac{6}{8} = 4\dfrac{5}{8}$

13. $-\dfrac{3}{5} + \dfrac{15}{8} = -\dfrac{24}{40} + \dfrac{75}{40} = \dfrac{51}{40}$ or $1\dfrac{11}{40}$

14. $3(-4)^2 - 80 = 3(16) - 80 = 48 - 80 = -32$

15. $6[5 + 2(3 - 8) - 3] = 6[5 + 2(3 + (-8)) - 3]$
$= 6[5 + 2(-5) - 3]$
$= 6[5 + (-10) + (-3)]$
$= 6[-5 + (-3)]$
$= 6[-8]$
$= -48$

16. $\dfrac{-12 + 3 \cdot 8}{4} = \dfrac{-12 + 24}{4} = \dfrac{12}{4} = 3$

17. $\dfrac{(-2)(0)(-3)}{-6} = \dfrac{0(-3)}{-6} = \dfrac{0}{-6} = 0$

18. Since -3 is to the right of -7 on the number line, $-3 > -7$.

19. Since 4 is to the right of -8 on the number line,
$4 > -8$.

20. Since $|-3| = 3$ is to the right of 2 on the number line, $|-3| > 2$.

21. $|-2| = 2$
$-1 - (-3) = -1 + 3 = 2$
Since $2 = 2$, $|-2| = -1 - (-3)$.

22. **a.** The natural numbers are 1, 7.

    b. The whole numbers are 0, 1, 7.

    c. The integers are $-5, -1, 0, 1, 7$.

    d. The rational numbers are $-5, -1, \dfrac{1}{4}, 0,$ 1, 7, 11.6.

    e. The irrational numbers are $\sqrt{7}, 3\pi$.

    f. The real numbers are $-5, -1, \dfrac{1}{4}, 0, 1,$ 7, 11.6, $\sqrt{7}, 3\pi$.

23. Replace x with 6 and y with -2.
$x^2 + y^2 = 6^2 + (-2)^2 = 36 + 4 = 40$

24. Replace x with 6, y with -2, and z with -3.
$x + yz = 6 + (-2)(-3) = 6 + 6 = 12$

25. Replace x with 6 and y with -2.
$2 + 3x - y = 2 + 3(6) - (-2)$
$ = 2 + 18 + 2$
$ = 20 + 2$
$ = 22$

26. Replace x with 6, y with -2, and z with -3.
$\dfrac{y + z - 1}{x} = \dfrac{-2 + (-3) - 1}{6}$
$\phantom{\dfrac{y + z - 1}{x}} = \dfrac{-5 - 1}{6}$
$\phantom{\dfrac{y + z - 1}{x}} = \dfrac{-5 + (-1)}{6}$
$\phantom{\dfrac{y + z - 1}{x}} = \dfrac{-6}{6}$
$\phantom{\dfrac{y + z - 1}{x}} = -1$

27. $8 + (9 + 3) = (8 + 9) + 3$ illustrates the associative property of addition.

28. $6 \cdot 8 = 8 \cdot 6$ illustrates the commutative property of multiplication.

29. $-6(2 + 4) = -6 \cdot 2 + (-6) \cdot 4$ illustrates the distributive property.

30. $\dfrac{1}{6}(6) = 1$ illustrates the multiplicative inverse property.

31. The opposite of -9 is 9.

32. The reciprocal of $-\dfrac{1}{3}$ of -3.

33. Losses of yardage occurred on the second and third downs. -10 indicates a loss of 10 yards while -2 indicates a loss of 2 yards, so the greatest loss of yardage occurred on the second down.

34. $5 + (-10) + (-2) + 29 = -5 + (-2) + 29$
$ = -7 + 29$
$ = 22$
Since the team was 22 yards from the goal, a touchdown was scored.

35. $-14 + 31 = 17$
The temperature was 17° at noon.

36. $-1.5(280) = -420$
She lost $420.

37. $2y - 6 - y - 4 = 2y - y - 6 - 4$
$ = 1y - 10$
$ = y - 10$

38. $2.7x + 6.1 + 3.2x - 4.9$
$= 2.7x + 3.2x + 6.1 - 4.9$
$= 5.9x + 1.2$

39. $4(x - 2) - 3(2x - 6) = 4x - 8 - 6x + 18$
$ = 4x - 6x - 8 + 18$
$ = -2x + 10$

40. $-5(y + 1) + 2(3 - 5y) = -5y - 5 + 6 - 10y$
$ = -5y - 10y - 5 + 6$
$ = -15y + 1$

Chapter 2

Section 2.1 Practice Problems

1.
$$x - 5 = 8$$
$$x - 5 + 5 = 8 + 5$$
$$x = 13$$
Check: $x - 5 = 8$
$$13 - 5 \stackrel{?}{=} 8$$
$$8 = 8 \quad \text{True}$$
The solution is 13.

2.
$$y + 1.7 = 0.3$$
$$y + 1.7 - 1.7 = 0.3 - 1.7$$
$$y = -1.4$$
Check: $y + 1.7 = 0.3$
$$-1.4 + 1.7 \stackrel{?}{=} 0.3$$
$$0.3 = 0.3 \quad \text{True}$$
The solution is -1.4.

3.
$$\frac{7}{8} = y - \frac{1}{3}$$
$$\frac{7}{8} + \frac{1}{3} = y - \frac{1}{3} + \frac{1}{3}$$
$$\frac{7}{8} \cdot \frac{3}{3} + \frac{1}{3} \cdot \frac{8}{8} = y$$
$$\frac{21}{24} + \frac{8}{24} = y$$
$$\frac{29}{24} = y$$
Check: $\dfrac{7}{8} = y - \dfrac{1}{3}$
$$\frac{7}{8} \stackrel{?}{=} \frac{29}{24} - \frac{1}{3}$$
$$\frac{7}{8} \stackrel{?}{=} \frac{29}{24} - \frac{8}{24}$$
$$\frac{7}{8} \stackrel{?}{=} \frac{21}{24}$$
$$\frac{7}{8} = \frac{7}{8} \quad \text{True}$$
The solution is $\dfrac{29}{24}$.

4.
$$3x + 10 = 4x$$
$$3x + 10 - 3x = 4x - 3x$$
$$10 = x$$
Check: $3x + 10 = 4x$
$$3(10) + 10 \stackrel{?}{=} 4(10)$$
$$30 + 10 \stackrel{?}{=} 40$$
$$40 = 40 \quad \text{True}$$
The solution is 10.

5.
$$10w + 3 - 4w + 4 = -2w + 3 + 7w$$
$$6w + 7 = 5w + 3$$
$$-5w + 6w + 7 = -5w + 5w + 3$$
$$w + 7 = 3$$
$$w + 7 - 7 = 3 - 7$$
$$w = -4$$
Check:
$$10w + 3 - 4w + 4 = -2w + 3 + 7w$$
$$10(-4) + 3 - 4(-4) + 4 \stackrel{?}{=} -2(-4) + 3 + 7(-4)$$
$$-40 + 3 + 16 + 4 \stackrel{?}{=} 8 + 3 - 28$$
$$-17 = -17 \quad \text{True}$$
The solution is -4.

6.
$$3(2w - 5) - (5w + 1) = -3$$
$$3(2w) - 3(5) - 1(5w) - 1(1) = -3$$
$$6w - 15 - 5w - 1 = -3$$
$$w - 16 = -3$$
$$w - 16 + 16 = -3 + 16$$
$$w = 13$$
Check: $3(2w - 5) - (5w + 1) = -3$
$$3(2 \cdot 13 - 5) - (5 \cdot 13 + 1) \stackrel{?}{=} -3$$
$$3(26 - 5) - (65 + 1) \stackrel{?}{=} -3$$
$$3(21) - 66 \stackrel{?}{=} -3$$
$$63 - 66 \stackrel{?}{=} -3$$
$$-3 = -3 \quad \text{True}$$
The solution is 13.

7. $12 - y = 9$
$12 - y - 12 = 9 - 12$
$-y = -3$
$y = 3$
Check: $12 - y = 9$
$12 - 3 \stackrel{?}{=} 9$
$9 = 9$ True
The solution is 3.

8. a.   If the sum of two numbers is 11 and one number is 4, find the other number by subtracting 4 from 11. The other number is 11 − 4, or 7.

   b.   If the sum of two numbers is 11 and one number is *x*, find the other number by subtracting *x* from 11. The other number is 11 − *x*.

   c.   If the sum of two numbers is 56 and one number is *a*, find the other number by subtracting *a* from 56. The other number is
56 − *a*.

9.   Lucille received 49,489 more votes than Wayne, who received *n* votes. So, she received (*n* + 49,489) votes.

Mental Math

1. $x + 4 = 6$
$x = 2$

2. $x + 7 = 17$
$x = 10$

3. $n + 18 = 30$
$n = 12$

4. $z + 22 = 40$
$z = 18$

5. $b - 11 = 6$
$b = 17$

6. $d - 16 = 5$
$d = 21$

Exercise Set 2.1

1. $x + 7 = 10$
$x + 7 - 7 = 10 - 7$
$x = 3$
Check: $x + 7 = 10$
$3 + 7 \stackrel{?}{=} 10$
$10 = 10$ True
The solution is 3.

3. $x - 2 = -4$
$x - 2 + 2 = -4 + 2$
$x = -2$
Check: $x - 2 = -4$
$-2 - 2 \stackrel{?}{=} -4$
$-4 = -4$ True
The solution is −2.

5. $-11 = 3 + x$
$-11 - 3 = 3 + x - 3$
$-14 = x$
Check: $-11 = 3 + x$
$-11 \stackrel{?}{=} 3 + (-14)$
$-11 = -11$ True
The solution is −14.

7. $r - 8.6 = -8.1$
$r - 8.6 + 8.6 = -8.1 + 8.6$
$r = 0.5$
Check: $r - 8.6 = -8.1$
$0.5 - 8.6 \stackrel{?}{=} -8.1$
$-8.1 = -8.1$ True
The solution is 0.5.

9. $x - \dfrac{2}{5} = -\dfrac{3}{20}$
$x - \dfrac{2}{5} + \dfrac{2}{5} = -\dfrac{3}{20} + \dfrac{2}{5}$
$x = -\dfrac{3}{20} + \dfrac{8}{20}$
$x = \dfrac{5}{20}$
$x = \dfrac{1}{4}$

Check: $x - \dfrac{2}{5} = -\dfrac{3}{20}$

$$\dfrac{1}{4} - \dfrac{2}{5} \overset{?}{=} -\dfrac{3}{20}$$

$$\dfrac{5}{20} - \dfrac{8}{20} \overset{?}{=} -\dfrac{3}{20}$$

$$-\dfrac{3}{20} = -\dfrac{3}{20} \quad \text{True}$$

The solution is $\dfrac{1}{4}$.

11. $\dfrac{1}{3} + f = \dfrac{3}{4}$

$$-\dfrac{1}{3} + \dfrac{1}{3} + f = -\dfrac{1}{3} + \dfrac{3}{4}$$

$$f = -\dfrac{4}{12} + \dfrac{9}{12}$$

$$f = \dfrac{5}{12}$$

Check: $\dfrac{1}{3} + f = \dfrac{3}{4}$

$$\dfrac{1}{3} + \dfrac{5}{12} \overset{?}{=} \dfrac{3}{4}$$

$$\dfrac{4}{12} + \dfrac{5}{12} \overset{?}{=} \dfrac{3}{4}$$

$$\dfrac{9}{12} \overset{?}{=} \dfrac{3}{4}$$

$$\dfrac{3}{4} = \dfrac{3}{4} \quad \text{True}$$

The solution is $\dfrac{5}{12}$.

13. $7x + 2x = 8x - 3$

$$9x = 8x - 3$$

$$9x - 8x = 8x - 3 - 8x$$

$$x = -3$$

Check: $7x + 2x = 8x - 3$

$$7(-3) + 2(-3) \overset{?}{=} 8(-3) - 3$$

$$-21 - 6 \overset{?}{=} -24 - 3$$

$$-27 = -27 \quad \text{True}$$

The solution is -3.

15. $\dfrac{5}{6}x + \dfrac{1}{6}x = -9$

$$\dfrac{6}{6}x = -9$$

$$x = -9$$

Check: $\dfrac{5}{6}x + \dfrac{1}{6}x = -9$

$$\dfrac{5}{6}(-9) + \dfrac{1}{6}(-9) \overset{?}{=} -9$$

$$-\dfrac{45}{6} - \dfrac{9}{6} \overset{?}{=} -9$$

$$-\dfrac{54}{6} \overset{?}{=} -9$$

$$-9 = -9 \quad \text{True}$$

The solution is -9.

17. $2y + 10 = 5y - 4y$

$$2y + 10 = y$$

$$2y + 10 - 2y = y - 2y$$

$$10 = -y$$

$$-10 = y$$

Check: $2y + 10 = 5y - 4y$

$$2(-10) + 10 \overset{?}{=} 5(-10) - 4(-10)$$

$$-20 + 10 \overset{?}{=} -50 + 40$$

$$-10 = -10 \quad \text{True}$$

The solution is -10.

19. $-5(n - 2) = 8 - 4n$

$$-5n + 10 = 8 - 4n$$

$$5n - 5n + 10 = 5n + 8 - 4n$$

$$10 = n + 8$$

$$10 - 8 = n + 8 - 8$$

$$2 = n$$

Check: $-5(n - 2) = 8 - 4n$

$$-5(2 - 2) \overset{?}{=} 8 - 4(2)$$

$$-5(0) \overset{?}{=} 8 - 8$$

$$0 = 0 \quad \text{True}$$

The solution is 2.

21.
$$\frac{3}{7}x + 2 = -\frac{4}{7}x - 5$$
$$\frac{3}{7}x + 2 + \frac{4}{7}x = -\frac{4}{7}x - 5 + \frac{4}{7}x$$
$$x + 2 = -5$$
$$x + 2 - 2 = -5 - 2$$
$$x = -7$$

Check:
$$\frac{3}{7}x + 2 = -\frac{4}{7}x - 5$$
$$\frac{3}{7}(-7) + 2 \stackrel{?}{=} -\frac{4}{7}(-7) - 5$$
$$-3 + 2 \stackrel{?}{=} 4 - 5$$
$$-1 = -1 \quad \text{True}$$

The solution is -7.

23.
$$5x - 6 = 6x - 5$$
$$-5x + 5x - 6 = -5x + 6x - 5$$
$$-6 = x - 5$$
$$-6 + 5 = x - 5 + 5$$
$$-1 = x$$

Check:
$$5x - 6 = 6x - 5$$
$$5(-1) - 6 \stackrel{?}{=} 6(-1) - 5$$
$$-5 - 6 \stackrel{?}{=} -6 - 5$$
$$-11 = -11 \quad \text{True}$$

The solution is -1.

25.
$$8y + 2 - 6y = 3 + y - 10$$
$$2y + 2 = y - 7$$
$$2y + 2 - y = y - 7 - y$$
$$y + 2 = -7$$
$$y + 2 - 2 = -7 - 2$$
$$y = -9$$

Check:
$$8y + 2 - 6y = 3 + y - 10$$
$$8(-9) + 2 - 6(-9) \stackrel{?}{=} 3 + (-9) - 10$$
$$-72 + 2 + 54 \stackrel{?}{=} 3 - 9 - 10$$
$$-16 = -16 \quad \text{True}$$

The solution is -9.

27.
$$-3(x - 4) = -4x$$
$$-3x + 12 = -4x$$
$$3x - 3x + 12 = 3x - 4x$$
$$12 = -x$$
$$-12 = x$$

Check:
$$-3(x - 4) = -4x$$
$$-3(-12 - 4) \stackrel{?}{=} -4(-12)$$
$$-3(-16) \stackrel{?}{=} 48$$
$$48 = 48 \quad \text{True}$$

The solution is -12.

29.
$$\frac{3}{8}x - \frac{1}{6} = -\frac{5}{8}x - \frac{2}{3}$$
$$\frac{3}{8}x - \frac{1}{6} + \frac{5}{8}x = -\frac{5}{8}x - \frac{2}{3} + \frac{5}{8}x$$
$$x - \frac{1}{6} = -\frac{2}{3}$$
$$x - \frac{1}{6} + \frac{1}{6} = -\frac{2}{3} + \frac{1}{6}$$
$$x = -\frac{4}{6} + \frac{1}{6}$$
$$x = -\frac{3}{6}$$
$$x = -\frac{1}{2}$$

Check:
$$\frac{3}{8}x - \frac{1}{6} = -\frac{5}{8}x - \frac{2}{3}$$
$$\frac{3}{8}\left(-\frac{1}{2}\right) - \frac{1}{6} \stackrel{?}{=} -\frac{5}{8}\left(-\frac{1}{2}\right) - \frac{2}{3}$$
$$-\frac{3}{16} - \frac{1}{6} \stackrel{?}{=} \frac{5}{16} - \frac{2}{3}$$
$$-\frac{9}{48} - \frac{8}{48} \stackrel{?}{=} \frac{15}{48} - \frac{32}{48}$$
$$-\frac{17}{48} = -\frac{17}{48} \quad \text{True}$$

The solution is $-\frac{1}{2}$.

31.
$$2(x-4) = x+3$$
$$2x-8 = x+3$$
$$-x+2x-8 = -x+x+3$$
$$x-8 = 3$$
$$x-8+8 = 3+8$$
$$x = 11$$
Check: $2(x-4) = x+3$
$$2(11-4) \overset{?}{=} 11+3$$
$$2(7) \overset{?}{=} 14$$
$$14 = 14 \quad \text{True}$$
The solution is 7.

33.
$$3(n-5)-(6-2n) = 4n$$
$$3n-15-6+2n = 4n$$
$$5n-21 = 4n$$
$$5n-21-5n = 4n-5n$$
$$-21 = -n$$
$$21 = n$$
Check: $3(n-5)-(6-2n) = 4n$
$$3(21-5)-(6-2\cdot21) \overset{?}{=} 4(21)$$
$$3(21-5)-(6-42) \overset{?}{=} 84$$
$$3(16)-(-36) \overset{?}{=} 84$$
$$48+36 \overset{?}{=} 84$$
$$84 = 84 \quad \text{True}$$
The solution is 21.

35.
$$-2(x+6)+3(2x-5) = 3(x-4)+10$$
$$-2x-12+6x-15 = 3x-12+10$$
$$4x-27 = 3x-2$$
$$-3x+4x-27 = -3x+3x-2$$
$$x-27 = -2$$
$$x-27+27 = -2+27$$
$$x = 25$$
Check:
$$-2(x+6)+3(2x-5) = 3(x-4)+10$$
$$-2(25+6)+3(2\cdot25-5) \overset{?}{=} 3(25-4)+10$$
$$-2(31)+3(50-5) \overset{?}{=} 3(21)+10$$
$$-62+3(45) \overset{?}{=} 63+10$$
$$-62+135 \overset{?}{=} 73$$
$$73 = 73 \quad \text{True}$$
The solution is 25.

37.
$$13x-3 = 14x$$
$$13x-3-13x = 14x-13x$$
$$-3 = x$$

39.
$$5b-0.7 = 6b$$
$$5b-0.7-5b = 6b-5b$$
$$-0.7 = b$$

41.
$$3x-6 = 2x+5$$
$$3x-6+6 = 2x+5+6$$
$$3x = 2x+11$$
$$3x-2x = 2x+11-2x$$
$$x = 11$$

43.
$$13x-9+2x-5 = 12x-1+2x$$
$$15x-14 = 14x-1$$
$$15x-14-14x = 14x-1-14x$$
$$x-14 = -1$$
$$x-14+14 = -1+14$$
$$x = 13$$

45.
$$7(6+w) = 6(2+w)$$
$$42+7w = 12+6w$$
$$42+7w-6w = 12+6w-6w$$
$$42+w = 12$$
$$42+w-42 = 12-42$$
$$w = -30$$

47.
$$n+4 = 3.6$$
$$n+4-4 = 3.6-4$$
$$n = -0.4$$

49.
$$10-(2x-4) = 7-3x$$
$$10-2x+4 = 7-3x$$
$$14-2x = 7-3x$$
$$14-2x+3x = 7-3x+3x$$
$$14+x = 7$$
$$14+x-14 = 7-14$$
$$x = -7$$

51.
$$\frac{1}{3} = x+\frac{2}{3}$$
$$\frac{1}{3}-\frac{2}{3} = x+\frac{2}{3}-\frac{2}{3}$$
$$-\frac{1}{3} = x$$

53. $-6.5 - 4x - 1.6 - 3x = -6x + 9.8$
$$-8.1 - 7x = -6x + 9.8$$
$$-8.1 - 7x + 7x = -6x + 9.8 + 7x$$
$$-8.1 = x + 9.8$$
$$-8.1 - 9.8 = x + 9.8 - 9.8$$
$$-17.9 = x$$

55. If the sum of the lengths of the two pieces is 10 feet and one piece is x feet, then the other piece has a length of $(10 - x)$ feet.

57. If the sum of the measures of two angles is 180° and one angle measures $x°$, then the other angle measures $(180 - x)°$.

59. If the number of undergraduate students is n, and the number of graduate students is 28,000 fewer than n, then the number of graduate students is $n - 28{,}000$.

61. If the area of the Gobi Desert is x square miles and the area of the Sahara Desert is 7 times the area of the Gobi Desert, then the area of the Sahara Desert is $7x$ square miles.

63. The multiplicative inverse of $\dfrac{5}{8}$ is $\dfrac{8}{5}$, since
$$\frac{5}{8} \cdot \frac{8}{5} = 1.$$

65. The multiplicative inverse of 2 is $\dfrac{1}{2}$, since
$$2 \cdot \frac{1}{2} = 1.$$

67. The multiplicative inverse of $-\dfrac{1}{9}$ is -9,

since $-\dfrac{1}{9} \cdot (-9) = 1.$

69. $\dfrac{3x}{3} = \dfrac{3 \cdot x}{3 \cdot 1} = \dfrac{x}{1} = x$

71. $-5\left(-\dfrac{1}{5}y\right) = \left[-5 \cdot \left(-\dfrac{1}{5}\right)\right]y = 1y = y$

73. $\dfrac{3}{5}\left(\dfrac{5}{3}x\right) = \left(\dfrac{3}{5} \cdot \dfrac{5}{3}\right)x = 1x = x$

75. answers may vary

77. $x - 4 = -9$
$$x - 4 + 4 = -9 + 4$$
$$x = -5$$

79. answers may vary

81. $180 - x - (2x + 7) = 180 - x - 2x - 7$
$$= 173 - 3x$$
The measure of the third angle is $(173 - 3x)°$.

83. answers may vary

85. $36.766 + x = -108.712$
$$36.766 + x - 36.766 = -108.712 - 36.766$$
$$x = -145.478$$

Section 2.2 Practice Problems

1. $\dfrac{3}{7}x = 9$
$$\frac{7}{3} \cdot \left(\frac{3}{7}x\right) = \frac{7}{3} \cdot 9$$
$$\left(\frac{7}{3} \cdot \frac{3}{7}\right)x = \frac{7}{3} \cdot 9$$
$$1x = 21$$
$$x = 21$$

Check: $\dfrac{3}{7}x = 9$
$$\frac{3}{7}(21) \stackrel{?}{=} 9$$
$$9 = 9 \quad \text{True}$$
The solution is 21.

2. $7x = 42$

$\dfrac{7x}{7} = \dfrac{42}{7}$

$1 \cdot x = 6$

$x = 6$

Check: $7x = 42$

$7 \cdot 6 \overset{?}{=} 42$

$42 = 42$ True

The solution is 6.

3. $-4x = 52$

$\dfrac{-4x}{-4} = \dfrac{52}{-4}$

$1x = -13$

$x = -13$

Check: $-4x = 52$

$-4(-13) \overset{?}{=} 52$

$52 = 52$ True

The solution is -13.

4. $\dfrac{y}{5} = 13$

$\dfrac{1}{5}y = 13$

$5 \cdot \dfrac{1}{5}y = 5 \cdot 13$

$1y = 65$

$y = 65$

Check: $\dfrac{y}{5} = 13$

$\dfrac{65}{5} \overset{?}{=} 13$

$13 = 13$ True

The solution is 65.

5. $2.6x = 13.52$

$\dfrac{2.6x}{2.6} = \dfrac{13.52}{2.6}$

$x = 5.2$

Check: $2.6x = 13.52$

$2.6(5.2) \overset{?}{=} 13.52$

$13.52 = 13.52$ True

The solution is 5.2.

6. $-\dfrac{5}{6}y = -\dfrac{3}{5}$

$-\dfrac{6}{5} \cdot -\dfrac{5}{6}y = -\dfrac{6}{5} \cdot -\dfrac{3}{5}$

$y = \dfrac{18}{25}$

Check: $-\dfrac{5}{6}y = -\dfrac{3}{5}$

$-\dfrac{5}{6}\left(\dfrac{18}{25}\right) \overset{?}{=} -\dfrac{3}{5}$

$-\dfrac{3}{5} = -\dfrac{3}{5}$ True

The solution is $\dfrac{18}{25}$.

7. $-x + 7 = -12$

$-x + 7 - 7 = -12 - 7$

$-x = -19$

$\dfrac{-x}{-1} = \dfrac{-19}{-1}$

$1x = 19$

$x = 19$

Check: $-x + 7 = -12$

$-19 + 7 \overset{?}{=} -12$

$-12 = -12$ True

The solution is 19.

8. $-7x + 2x + 3 - 20 = -2$

$-5x - 17 = -2$

$-5x - 17 + 17 = -2 + 17$

$-5x = 15$

$\dfrac{-5x}{-5} = \dfrac{15}{-5}$

$x = -3$

Check: $-7x + 2x + 3 - 20 = -2$

$-7(-3) + 2(-3) + 3 - 20 \overset{?}{=} -2$

$21 - 6 + 3 - 20 \overset{?}{=} -2$

$-2 = -2$ True

The solution is -3.

9. $10x - 4 = 7x + 14$

$10x - 4 - 7x = 7x + 14 - 7x$

$3x - 4 = 14$

$3x - 4 + 4 = 14 + 4$

$3x = 18$

$\dfrac{3x}{3} = \dfrac{18}{3}$

$x = 6$

Check: $10x - 4 = 7x + 14$

$10(6) - 4 \stackrel{?}{=} 7(6) + 14$

$60 - 4 \stackrel{?}{=} 42 + 14$

$56 = 56$ True

The solution is 6.

10. $4(3x - 2) = -1 + 4$

$4(3x) - 4(2) = -1 + 4$

$12x - 8 = 3$

$12x - 8 + 8 = 3 + 8$

$12x = 11$

$\dfrac{12x}{12} = \dfrac{11}{12}$

$x = \dfrac{11}{12}$

Check: $4(3x - 2) = -1 + 4$

$4\left(3 \cdot \dfrac{11}{12} - 2\right) \stackrel{?}{=} -1 + 4$

$4\left(\dfrac{11}{4} - 2\right) \stackrel{?}{=} -1 + 4$

$11 - 8 \stackrel{?}{=} 3$

$3 = 3$ True

The solution is $\dfrac{11}{12}$.

11. a. Let x be the first integer.
Then $x + 1$ is the second integer.
Their sum is
$x + (x + 1) = x + x + 1 = 2x + 1$.

b. Let x be the first even integer.
Then $x + 2$ is the second consecutive
even integer.
Their sum is
$x + (x + 2) = x + x + 2 = 2x + 2$.

Mental Math

1. $3a = 27$
$a = 9$

2. $9c = 54$
$c = 6$

3. $5b = 10$
$b = 2$

4. $7t = 14$
$t = 2$

5. $6x = -30$
$x = -5$

6. $8r = -64$
$r = -8$

Exercise Set 2.2

1. $-5x = -20$
$\dfrac{-5x}{-5} = \dfrac{-20}{-5}$
$x = 4$
Check: $-5x = -20$
$-5(4) \stackrel{?}{=} -20$
$-20 = -20$ True
The solution is 4.

3. $3x = 0$
$\dfrac{3x}{3} = \dfrac{0}{3}$
$x = 0$
Check: $3x = 0$
$3 \cdot 0 \stackrel{?}{=} 0$
$0 = 0$ True
The solution is 0.

5. $-x = -12$
$\dfrac{-x}{-1} = \dfrac{-12}{-1}$
$x = 12$
Check: $-x = -12$
$-12 = -12$ True
The solution is 12.

7. $\dfrac{2}{3}x = -8$

$\dfrac{3}{2} \cdot \dfrac{2}{3}x = \dfrac{3}{2} \cdot (-8)$

$\quad\quad x = -12$

Check: $\quad \dfrac{2}{3}x = -8$

$\quad\quad\quad \dfrac{2}{3}(-12) \overset{?}{=} -8$

$\quad\quad\quad\quad\quad -8 = -8 \quad$ True

The solution is -12.

9. $\dfrac{1}{6}d = \dfrac{1}{2}$

$6 \cdot \dfrac{1}{6}d = 6 \cdot \dfrac{1}{2}$

$\quad\quad d = 3$

Check: $\dfrac{1}{6}d = \dfrac{1}{2}$

$\quad\quad \dfrac{1}{6}(3) \overset{?}{=} \dfrac{1}{2}$

$\quad\quad\quad \dfrac{1}{2} = \dfrac{1}{2} \quad$ True

The solution is 3.

11. $\dfrac{a}{2} = 1$

$2 \cdot \dfrac{a}{2} = 2 \cdot 1$

$\quad\quad a = 2$

Check: $\dfrac{a}{2} = 1$

$\quad\quad \dfrac{2}{2} \overset{?}{=} 1$

$\quad\quad 1 = 1 \quad$ True

The solution is 2.

13. $\dfrac{k}{-7} = 0$

$-7\left(\dfrac{k}{-7}\right) = -7(0)$

$\quad\quad k = 0$

Check: $\dfrac{k}{-7} = 0$

$\quad\quad \dfrac{0}{-7} \overset{?}{=} 0$

$\quad\quad 0 = 0 \quad$ True

The solution is 0.

15. $1.7x = 10.71$

$\dfrac{1.7x}{1.7} = \dfrac{10.71}{1.7}$

$\quad\quad x = 6.3$

Check: $\quad 1.7x = 10.71$

$\quad\quad\quad 1.7(6.3) \overset{?}{=} 10.71$

$\quad\quad\quad\quad 10.71 = 10.71 \quad$ True

The solution is 6.3.

17. $2x - 4 = 16$

$2x - 4 + 4 = 16 + 4$

$\quad\quad 2x = 20$

$\quad\quad \dfrac{2x}{2} = \dfrac{20}{2}$

$\quad\quad x = 10$

Check: $\quad 2x - 4 = 16$

$\quad\quad\quad 2(10) - 4 \overset{?}{=} 16$

$\quad\quad\quad 20 - 4 \overset{?}{=} 16$

$\quad\quad\quad\quad 16 = 16 \quad$ True

The solution is 10.

19. $-x + 2 = 22$

$-x + 2 - 2 = 22 - 2$

$\quad\quad -x = 20$

$\quad\quad \dfrac{-x}{-1} = \dfrac{20}{-1}$

$\quad\quad x = -20$

Check: $\quad\quad -x + 2 = 22$

$\quad\quad\quad -(-20) + 2 \overset{?}{=} 22$

$\quad\quad\quad 20 + 2 \overset{?}{=} 22$

$\quad\quad\quad\quad 22 = 22 \quad$ True

The solution is -20.

21.
$$6a + 3 = 3$$
$$6a + 3 - 3 = 3 - 3$$
$$6a = 0$$
$$\frac{6a}{6} = \frac{0}{6}$$
$$a = 0$$
Check: $6a + 3 = 3$
$$6(0) + 3 \overset{?}{=} 3$$
$$0 + 3 \overset{?}{=} 3$$
$$3 = 3 \quad \text{True}$$
The solution is 0.

23.
$$\frac{x}{3} - 2 = -5$$
$$\frac{x}{3} - 2 + 2 = -5 + 2$$
$$\frac{x}{3} = -3$$
$$3 \cdot \frac{x}{3} = 3 \cdot (-3)$$
$$x = -9$$
Check: $\frac{x}{3} - 2 = -5$
$$\frac{-9}{3} - 2 \overset{?}{=} -5$$
$$-3 - 2 \overset{?}{=} -5$$
$$-5 = -5 \quad \text{True}$$
The solution is −9.

25.
$$6z - 8 - z + 3 = 0$$
$$5z - 5 = 0$$
$$5z - 5 + 5 = 0 + 5$$
$$5z = 5$$
$$\frac{5z}{5} = \frac{5}{5}$$
$$z = 1$$
Check: $6z - 8 - z + 3 = 0$
$$6(1) - 8 - 1 + 3 \overset{?}{=} 0$$
$$6 - 8 - 1 + 3 \overset{?}{=} 0$$
$$0 = 0 \quad \text{True}$$
The solution is 1.

27.
$$1 = 0.4x - 0.6x - 5$$
$$1 = -0.2x - 5$$
$$1 + 5 = -0.2x - 5 + 5$$
$$6 = -0.2x$$
$$\frac{6}{-0.2} = \frac{-0.2x}{-0.2}$$
$$-30 = x$$
Check: $1 = 0.4x - 0.6x - 5$
$$1 \overset{?}{=} 0.4(-30) - 0.6(-30) - 5$$
$$1 \overset{?}{=} -12 + 18 - 5$$
$$1 = 1 \quad \text{True}$$
The solution is −30.

29.
$$\frac{2}{3}y - 11 = -9$$
$$\frac{2}{3}y - 11 + 11 = -9 + 11$$
$$\frac{2}{3}y = 2$$
$$\frac{3}{2}\left(\frac{2}{3}y\right) = \frac{3}{2}(2)$$
$$y = 3$$
Check: $\frac{2}{3}y - 11 = -9$
$$\frac{2}{3}(3) - 11 \overset{?}{=} -9$$
$$2 - 11 \overset{?}{=} -9$$
$$-9 = -9 \quad \text{True}$$
The solution is 3.

31.
$$\frac{3}{4}t - \frac{1}{2} = \frac{1}{3}$$
$$\frac{3}{4}t - \frac{1}{2} + \frac{1}{2} = \frac{1}{3} + \frac{1}{2}$$
$$\frac{3}{4}t = \frac{2}{6} + \frac{3}{6}$$
$$\frac{3}{4}t = \frac{5}{6}$$
$$\frac{4}{3} \cdot \frac{3}{4}t = \frac{4}{3} \cdot \frac{5}{6}$$
$$t = \frac{10}{9}$$

Check: $\dfrac{3}{4}t - \dfrac{1}{2} = \dfrac{1}{3}$

$\dfrac{3}{4} \cdot \dfrac{10}{9} - \dfrac{1}{2} \overset{?}{=} \dfrac{1}{3}$

$\dfrac{5}{6} - \dfrac{3}{6} \overset{?}{=} \dfrac{1}{3}$

$\dfrac{2}{6} \overset{?}{=} \dfrac{1}{3}$

$\dfrac{1}{3} = \dfrac{1}{3}$ True

The solution is $\dfrac{10}{9}$.

33. $8x + 20 = 6x + 18$

$8x + 20 - 6x = 6x + 18 - 6x$

$2x + 20 = 18$

$2x + 20 - 20 = 18 - 20$

$2x = -2$

$\dfrac{2x}{2} = \dfrac{-2}{2}$

$x = -1$

35. $3(2x + 5) = -18 + 9$

$6x + 15 = -18 + 9$

$6x + 15 = -9$

$6x + 15 - 15 = -9 - 15$

$6x = -24$

$\dfrac{6x}{6} = \dfrac{-24}{6}$

$x = -4$

37. $2x - 5 = 20x + 4$

$2x - 5 - 20x = 20x + 4 - 20x$

$-18x - 5 = 4$

$-18x - 5 + 5 = 4 + 5$

$-18x = 9$

$\dfrac{-18x}{-18} = \dfrac{9}{-18}$

$x = -\dfrac{1}{2}$

39. $2 + 14 = -4(3x - 4)$

$2 + 14 = -12x + 16$

$16 = -12x + 16$

$16 - 16 = -12x + 16 - 16$

$0 = -12x$

$\dfrac{0}{-12} = \dfrac{-12x}{-12}$

$0 = x$

41. $-6y - 3 = -5y - 7$

$-6y - 3 + 6y = -5y - 7 + 6y$

$-3 = y - 7$

$-3 + 7 = y - 7 + 7$

$4 = y$

43. $\dfrac{1}{2}(2x - 1) = -\dfrac{1}{7} - \dfrac{3}{7}$

$x - \dfrac{1}{2} = -\dfrac{4}{7}$

$x - \dfrac{1}{2} + \dfrac{1}{2} = -\dfrac{4}{7} + \dfrac{1}{2}$

$x = -\dfrac{8}{14} + \dfrac{7}{14}$

$x = -\dfrac{1}{14}$

45. $-10z - 0.5 = -20z + 1.6$

$-10z - 0.5 + 20z = -20z + 1.6 + 20z$

$10z - 0.5 = 1.6$

$10z - 0.5 + 0.5 = 1.6 + 0.5$

$10z = 2.1$

$\dfrac{10z}{10} = \dfrac{2.1}{10}$

$z = 0.21$

47. $-4x + 20 = 4x - 20$

$-4x - 4x + 20 = -4x + 4x - 20$

$-8x + 20 = -20$

$-8x + 20 - 20 = -20 - 20$

$-8x = -40$

$\dfrac{-8x}{-8} = \dfrac{-40}{-8}$

$x = 5$

49. $42 = 7x$

$\dfrac{42}{7} = \dfrac{7x}{7}$

$6 = x$

51. $4.4 = -0.8x$

$\dfrac{4.4}{-0.8} = \dfrac{-0.8x}{-0.8}$

$-5.5 = x$

53. $6x + 10 = -20$

$6x + 10 - 10 = -20 - 10$

$6x = -30$

$\dfrac{6x}{6} = \dfrac{-30}{6}$

$x = -5$

55. $5 - 0.3k = 5$

$-5 + 5 - 0.3k = -5 + 5$

$-0.3k = 0$

$\dfrac{-0.3k}{-0.3} = \dfrac{0}{-0.3}$

$k = 0$

57. $13x - 5 = 11x - 11$

$13x - 5 + 5 = 11x - 11 + 5$

$13x = 11x - 6$

$13x - 11x = 11x - 11x - 6$

$2x = -6$

$\dfrac{2x}{2} = \dfrac{-6}{2}$

$x = -3$

59. $9(3x + 1) = 4x - 5x$

$27x + 9 = -x$

$-27x + 27x + 9 = -27x - x$

$9 = -28x$

$\dfrac{9}{-28} = \dfrac{-28x}{-28}$

$-\dfrac{9}{28} = x$

61. $-\dfrac{3}{7}p = -2$

$-\dfrac{7}{3}\left(-\dfrac{3}{7}p\right) = -\dfrac{7}{3}(-2)$

$p = \dfrac{14}{3}$

63. $-\dfrac{4}{3}x = 12$

$-\dfrac{3}{4} \cdot \left(-\dfrac{4}{3}x\right) = -\dfrac{3}{4} \cdot 12$

$x = -9$

65. $-2x - \dfrac{1}{2} = \dfrac{7}{2}$

$-2x - \dfrac{1}{2} + \dfrac{1}{2} = \dfrac{7}{2} + \dfrac{1}{2}$

$-2x = \dfrac{8}{2}$

$-2x = 4$

$\dfrac{-2x}{-2} = \dfrac{4}{-2}$

$x = -2$

67. $10 = 2x - 1$

$10 + 1 = 2x - 1 + 1$

$11 = 2x$

$\dfrac{11}{2} = \dfrac{2x}{2}$

$\dfrac{11}{2} = x$

69. $10 - 3x - 6 - 9x = 7$

$4 - 12x = 7$

$4 - 12x - 4 = 7 - 4$

$-12x = 3$

$\dfrac{-12x}{-12} = \dfrac{3}{-12}$

$x = -\dfrac{1}{4}$

71.
$$z - 5z = 7z - 9 - z$$
$$-4z = 6z - 9$$
$$-4z - 6z = 6z - 9 - 6z$$
$$-10z = -9$$
$$\frac{-10z}{-10} = \frac{-9}{-10}$$
$$z = \frac{9}{10}$$

73.
$$-x - \frac{4}{5} = x + \frac{1}{2} + \frac{2}{5}$$
$$-x - \frac{4}{5} = x + \frac{5}{10} + \frac{4}{10}$$
$$-x - \frac{4}{5} = x + \frac{9}{10}$$
$$-x - \frac{4}{5} + x = x + \frac{9}{10} + x$$
$$-\frac{4}{5} = 2x + \frac{9}{10}$$
$$-\frac{4}{5} - \frac{9}{10} = 2x$$
$$-\frac{8}{10} - \frac{9}{10} = 2x$$
$$-\frac{17}{10} = 2x$$
$$\frac{1}{2}\left(-\frac{17}{10}\right) = \frac{1}{2}(2x)$$
$$-\frac{17}{20} = x$$

75.
$$-15 + 37 = -2(x + 5)$$
$$22 = -2x - 10$$
$$22 + 10 = -2x - 10 + 10$$
$$32 = -2x$$
$$\frac{32}{-2} = \frac{-2x}{-2}$$
$$-16 = x$$

77. If x represents the first of two consecutive odd integers, then $x + 2$ represents the second. Thus, the sum is represented by $x + x + 2 = 2x + 2$.

79. If x represents the first integer, then $x + 1$, $x + 2$, and $x + 3$ represent the second, third, and fourth integers, respectively. The sum of the first and third integers is represented by $x + x + 2 = 2x + 2$.

81. If x represents the number on the first door, then the next four door numbers are represented by $x + 2$, $x + 4$, $x + 6$, and $x + 8$. The sum of the numbers is $x + x + 2 + x + 4 + x + 6 + x + 8 = 5x + 20$.

83.
$$5x + 2(x - 6) = 5x + 2 \cdot x + 2 \cdot (-6)$$
$$= 5x + 2x - 12$$
$$= 7x - 12$$

85.
$$6(2z + 4) + 20 = 6 \cdot 2z + 6 \cdot 4 + 20$$
$$= 12z + 24 + 20$$
$$= 12z + 44$$

87.
$$-(x - 1) + x = -x + 1 + x$$
$$= -x + x + 1$$
$$= 0 + 1$$
$$= 1$$

89. If the solution is -8, then replacing x by -8 results in a true statement.
$$6x = 6(-8) = -48$$
The missing number is -48.

91. answers may vary

93. answers may vary

95.
$$0.07x - 5.06 = -4.92$$
$$0.07x - 5.06 + 5.06 = -4.92 + 5.06$$
$$0.07x = 0.14$$
$$\frac{0.07x}{0.07} = \frac{0.14}{0.07}$$
$$x = 2$$

Section 2.3 Practice Problems

1. $5(3x-1)+2=12x+6$

$15x-5+2=12x+6$

$15x-3=12x+6$

$15x-3-12x=12x+6-12x$

$3x-3=6$

$3x-3+3=6+3$

$3x=9$

$\dfrac{3x}{3}=\dfrac{9}{3}$

$x=3$

Check: $5(3x-1)+2=12x+6$

$5[3(3)-1]+2\overset{?}{=}12(3)+6$

$5(9-1)+2\overset{?}{=}36+6$

$5(8)+2\overset{?}{=}42$

$40+2\overset{?}{=}42$

$42=42$ True

The solution is 3.

2. $9(5-x)=-3x$

$45-9x=-3x$

$45-9x+9x=-3x+9x$

$45=6x$

$\dfrac{45}{6}=\dfrac{6x}{6}$

$\dfrac{15}{2}=x$

Check: $9(5-x)=-3x$

$9\left(5-\dfrac{15}{2}\right)\overset{?}{=}-3\left(\dfrac{15}{2}\right)$

$9\left(\dfrac{10}{2}-\dfrac{15}{2}\right)\overset{?}{=}-\dfrac{45}{2}$

$9\left(-\dfrac{5}{2}\right)\overset{?}{=}-\dfrac{45}{2}$

$-\dfrac{45}{2}=-\dfrac{45}{2}$ True

The solution is $\dfrac{15}{2}$.

3. $\dfrac{5}{2}x-1=\dfrac{3}{2}x-4$

$2\left(\dfrac{5}{2}x-1\right)=2\left(\dfrac{3}{2}x-4\right)$

$5x-2=3x-8$

$5x-2-3x=3x-8-3x$

$2x-2=-8$

$2x-2+2=-8+2$

$2x=-6$

$\dfrac{2x}{2}=\dfrac{-6}{2}$

$x=-3$

Check: $\dfrac{5}{2}x-1=\dfrac{3}{2}x-4$

$\dfrac{5}{2}(-3)-1\overset{?}{=}\dfrac{3}{2}(-3)-4$

$-\dfrac{15}{2}-1\overset{?}{=}-\dfrac{9}{2}-4$

$-\dfrac{15}{2}-\dfrac{2}{2}\overset{?}{=}-\dfrac{9}{2}-\dfrac{8}{2}$

$-\dfrac{17}{2}=-\dfrac{17}{2}$ True

The solution is -3.

4. $\dfrac{3(x-2)}{5}=3x+6$

$5\cdot\dfrac{3(x-2)}{5}=5(3x+6)$

$3(x-2)=5(3x+6)$

$3x-6=15x+30$

$3x-6-3x=15x+30-3x$

$-6=12x+30$

$-6-30=12x+30-30$

$-36=12x$

$\dfrac{-36}{12}=\dfrac{12x}{12}$

$-3=x$

Check: $\dfrac{3(x-2)}{5} = 3x + 6$

$\dfrac{3(-3-2)}{5} \stackrel{?}{=} 3(-3) + 6$

$\dfrac{3(-5)}{5} \stackrel{?}{=} -9 + 6$

$\dfrac{-15}{5} \stackrel{?}{=} -3$

$-3 = -3$

The solution is -3.

5. $0.06x - 0.10(x-2) = -0.02(8)$

$100[0.06x - 0.10(x-2)] = 100[-0.02(8)]$

$6x - 10(x-2) = -2(8)$

$6x - 10x + 20 = -16$

$-4x + 20 = -16$

$-4x + 20 - 20 = -16 - 20$

$-4x = -36$

$\dfrac{-4x}{-4} = \dfrac{-36}{-4}$

$x = 9$

To check, replace x with 9 in the original equation. The solution is 9.

6. $5(2-x) + 8x = 3(x-6)$

$10 - 5x + 8x = 3x - 18$

$10 + 3x = 3x - 18$

$10 + 3x - 3x = 3x - 18 - 3x$

$10 = -18$

Since the statement $10 = -18$ is false, the equation has no solution.

7. $-6(2x+1) - 14 = -10(x+2) - 2x$

$-12x - 6 - 14 = -10x - 20 - 2x$

$-12x - 20 = -12x - 20$

$12x - 12x - 20 = 12x - 12x - 20$

$-20 = -20$

Since $-20 = -20$ is a true statement, every real number is a solution.

Calculator Explorations

1. $2x = 48 + 6x$

$\boxed{2}\boxed{\times}\boxed{-12}\boxed{=}$ Display: $\boxed{-24}$

$\boxed{48}\boxed{+}\boxed{6}\boxed{\times}\boxed{-12}\boxed{=}$ Display: $\boxed{-24}$

$x = -12$ is a solution.

2. $-3x - 7 = 3x - 1$

$\boxed{-3}\boxed{\times}\boxed{-1}\boxed{-}\boxed{7}\boxed{=}$ Display: $\boxed{-4}$

$\boxed{3}\boxed{\times}\boxed{-1}\boxed{-}\boxed{1}\boxed{=}$ Display: $\boxed{-4}$

Since the left side equals the right side, $x = -1$ is a solution.

3. $5x - 2.6 = 2(x + 0.8)$

$\boxed{5}\boxed{\times}\boxed{4.4}\boxed{-}\boxed{2.6}\boxed{=}$ Display: $\boxed{19.4}$

$\boxed{2}\boxed{(}\boxed{4.4}\boxed{+}\boxed{0.8}\boxed{)}\boxed{=}$ Display: $\boxed{10.4}$

Since the left side does not equal the right side, $x = 4.4$ is not a solution.

4. $-1.6x - 3.9 = -6.9x - 25.6$

$\boxed{-1.6}\boxed{\times}\boxed{5}\boxed{-}\boxed{3.9}\boxed{=}$ Display: $\boxed{-11.9}$

$\boxed{-6.9}\boxed{\times}\boxed{5}\boxed{-}\boxed{25.6}\boxed{=}$ Display: $\boxed{-60.1}$

Since the left side does not equal the right side, $x = 5$ is not a solution.

5. $\dfrac{564x}{4} = 200x - 11(649)$

$\boxed{(}\boxed{564}\boxed{\times}\boxed{121}\boxed{)}\boxed{\div}\boxed{4}\boxed{=}$ Display: $\boxed{17061}$

$\boxed{200}\boxed{\times}\boxed{121}\boxed{-}\boxed{11}\boxed{\times}\boxed{649}\boxed{=}$ Display: $\boxed{17061}$

Since the left side equals the right side, $x = 121$ is a solution.

6. $20(x - 39) = 5x - 432$

$\boxed{20}\boxed{(}\boxed{23.2}\boxed{-}\boxed{39}\boxed{)}\boxed{=}$ Display: $\boxed{-316}$

$\boxed{5}\boxed{\times}\boxed{23.2}\boxed{-}\boxed{432}\boxed{=}$ Display: $\boxed{-316}$

Since the left side equals the right side, $x = 23.2$ is a solution.

Exercise Set 2.3

1.
$$-4y + 10 = -2(3y + 1)$$
$$-4y + 10 = -6y - 2$$
$$-4y + 10 - 10 = -6y - 2 - 10$$
$$-4y = -6y - 12$$
$$-4y + 6y = -6y - 12 + 6y$$
$$2y = -12$$
$$\frac{2y}{2} = \frac{-12}{2}$$
$$y = -6$$

3.
$$15x - 8 = 10 + 9x$$
$$15x - 8 - 9x = 10 + 9x - 9x$$
$$6x - 8 = 10$$
$$6x - 8 + 8 = 10 + 8$$
$$6x = 18$$
$$\frac{6x}{6} = \frac{18}{6}$$
$$x = 3$$

5.
$$-2(3x - 4) = 2x$$
$$-6x + 8 = 2x$$
$$-6x + 8 + 6x = 2x + 6x$$
$$8 = 8x$$
$$\frac{8}{8} = \frac{8x}{8}$$
$$1 = x$$

7.
$$5(2x - 1) - 2(3x) = 1$$
$$10x - 5 - 6x = 1$$
$$-5 + 4x = 1$$
$$5 - 5 + 4x = 5 + 1$$
$$4x = 6$$
$$\frac{4x}{4} = \frac{6}{4}$$
$$x = \frac{3}{2}$$

9.
$$-6(x - 3) - 26 = -8$$
$$-6x + 18 - 26 = -8$$
$$-6x - 8 = -8$$
$$-6x - 8 + 8 = -8 + 8$$
$$-6x = 0$$
$$\frac{-6x}{-6} = \frac{0}{-6}$$
$$x = 0$$

11.
$$8 - 2(a + 1) = 9 + a$$
$$8 - 2a - 2 = 9 + a$$
$$-2a + 6 = 9 + a$$
$$-2a + 6 - a = 9 + a - a$$
$$-3a + 6 = 9$$
$$-3a + 6 - 6 = 9 - 6$$
$$-3a = 3$$
$$\frac{-3a}{-3} = \frac{3}{-3}$$
$$a = -1$$

13.
$$4x + 3 = -3 + 2x + 14$$
$$4x + 3 = 11 + 2x$$
$$4x + 3 - 2x = 11 + 2x - 2x$$
$$2x + 3 = 11$$
$$2x + 3 - 3 = 11 - 3$$
$$2x = 8$$
$$\frac{2x}{2} = \frac{8}{2}$$
$$x = 4$$

15.
$$-2y - 10 = 5y + 18$$
$$-2y - 10 + 10 = 5y + 18 + 10$$
$$-2y = 5y + 28$$
$$-2y - 5y = 5y + 28 - 5y$$
$$-7y = 28$$
$$\frac{-7y}{-7} = \frac{28}{-7}$$
$$y = -4$$

17.
$$\frac{2}{3}x + \frac{4}{3} = -\frac{2}{3}$$
$$3\left(\frac{2}{3}x + \frac{4}{3}\right) = 3\left(-\frac{2}{3}\right)$$
$$2x + 4 = -2$$
$$2x + 4 - 4 = -2 - 4$$
$$2x = -6$$
$$\frac{2x}{2} = \frac{-6}{2}$$
$$x = -3$$

19.
$$\frac{3}{4}x - \frac{1}{2} = 1$$
$$4\left(\frac{3}{4}x - \frac{1}{2}\right) = 4(1)$$
$$3x - 2 = 4$$
$$3x - 2 + 2 = 4 + 2$$
$$3x = 6$$
$$\frac{3x}{3} = \frac{6}{3}$$
$$x = 2$$

21.
$$0.50x + 0.15(70) = 35.5$$
$$50x + 15(70) = 3550$$
$$50x + 1050 = 3550$$
$$50x + 1050 - 1050 = 3550 - 1050$$
$$50x = 2500$$
$$\frac{50x}{50} = \frac{2500}{50}$$
$$x = 50$$

23.
$$\frac{2(x+1)}{4} = 3x - 2$$
$$4\left[\frac{2(x+1)}{4}\right] = 4(3x - 2)$$
$$2(x+1) = 4(3x - 2)$$
$$2x + 2 = 12x - 8$$
$$2x + 2 + 8 = 12x - 8 + 8$$
$$2x + 10 = 12x$$
$$2x + 10 - 2x = 12x - 2x$$
$$10 = 10x$$
$$\frac{10}{10} = \frac{10x}{10}$$
$$1 = x$$

25.
$$x + \frac{7}{6} = 2x - \frac{7}{6}$$
$$6\left(x + \frac{7}{6}\right) = 6\left(2x - \frac{7}{6}\right)$$
$$6x + 7 = 12x - 7$$
$$6x + 7 + 7 = 12x - 7 + 7$$
$$6x + 14 = 12x$$
$$6x + 14 - 6x = 12x - 6x$$
$$14 = 6x$$
$$\frac{14}{6} = \frac{6x}{6}$$
$$\frac{7}{3} = x$$

27.
$$0.12(y - 6) + 0.06y = 0.08y - 0.7$$
$$12(y - 6) + 6y = 8y - 70$$
$$12y - 72 + 6y = 8y - 70$$
$$18y - 72 = 8y - 70$$
$$18y - 72 - 8y = 8y - 70 - 8y$$
$$10y - 72 = -70$$
$$10y - 72 + 72 = -70 + 72$$
$$10y = 2$$
$$\frac{10y}{10} = \frac{2}{10}$$
$$y = 0.2$$

29.
$$4(3x + 2) = 12x + 8$$
$$12x + 8 = 12x + 8$$
Since both sides of the equation are identical, the equation is an identity and every real number is a solution.

31.
$$\frac{x}{4} + 1 = \frac{x}{4}$$
$$\frac{x}{4} + 1 - \frac{x}{4} = \frac{x}{4} - \frac{x}{4}$$
$$1 = 0$$
Since the statement $1 = 0$ is false, the equation has no solution.

33.
$$3x - 7 = 3(x + 1)$$
$$3x - 7 = 3x + 3$$
$$3x - 7 - 3x = 3x + 3 - 3x$$
$$-7 = 3$$
Since the statement $-7 = 3$ is false, the equation has no solution.

35.
$$-2(6x - 5) + 4 = -12x + 14$$
$$-12x + 10 + 4 = -12x + 14$$
$$-12x + 14 = -12x + 14$$
Since both sides of the equation are identical, the equation is an identity and every real number is a solution.

37.
$$\frac{6(3 - z)}{5} = -z$$
$$5 \cdot \frac{6(3 - z)}{5} = 5(-z)$$
$$6(3 - z) = -5z$$
$$18 - 6z = -5z$$
$$18 - 6z + 6z = -5z + 6z$$
$$18 = z$$

39.
$$-3(2t - 5) + 2t = 5t - 4$$
$$-6t + 15 + 2t = 5t - 4$$
$$-4t + 15 = 5t - 4$$
$$-4t + 15 + 4t = 5t - 4 + 4t$$
$$15 = 9t - 4$$
$$15 + 4 = 9t - 4 + 4$$
$$19 = 9t$$
$$\frac{19}{9} = \frac{9t}{9}$$
$$\frac{19}{9} = t$$

41.
$$5y + 2(y - 6) = 4(y + 1) - 2$$
$$5y + 2y - 12 = 4y + 4 - 2$$
$$7y - 12 = 4y + 2$$
$$7y - 12 + 12 = 4y + 2 + 12$$
$$7y = 4y + 14$$
$$7y - 4y = 4y + 14 - 4y$$
$$3y = 14$$
$$\frac{3y}{3} = \frac{14}{3}$$
$$y = \frac{14}{3}$$

43.
$$\frac{3(x - 5)}{2} = \frac{2(x + 5)}{3}$$
$$6\left[\frac{3(x - 5)}{2}\right] = 6\left[\frac{2(x + 5)}{3}\right]$$
$$9(x - 5) = 4(x + 5)$$
$$9x - 45 = 4x + 20$$
$$9x - 45 + 45 = 4x + 20 + 45$$
$$9x = 4x + 65$$
$$9x - 4x = 4x + 65 - 4x$$
$$5x = 65$$
$$\frac{5x}{5} = \frac{65}{5}$$
$$x = 13$$

45.
$$0.7x - 2.3 = 0.5$$
$$7x - 23 = 5$$
$$7x - 23 + 23 = 5 + 23$$
$$7x = 28$$
$$\frac{7x}{7} = \frac{28}{7}$$
$$x = 4$$

47.
$$5x - 5 = 2(x + 1) + 3x - 7$$
$$5x - 5 = 2x + 2 + 3x - 7$$
$$5x - 5 = 5x - 5$$
Since both sides of the equation are identical, the equation is an identity and every real number is a solution.

49.
$$4(2n+1) = 3(6n+3)+1$$
$$8n+4 = 18n+9+1$$
$$8n+4 = 18n+10$$
$$8n+4-10 = 18n+10-10$$
$$8n-6 = 18n$$
$$8n-6-8n = 18n-8n$$
$$-6 = 10n$$
$$\frac{-6}{10} = \frac{10n}{10}$$
$$-\frac{3}{5} = n$$

51.
$$x+\frac{5}{4} = \frac{3}{4}x$$
$$4\left(x+\frac{5}{4}\right) = 4\left(\frac{3}{4}x\right)$$
$$4x+5 = 3x$$
$$4x+5-4x = 3x-4x$$
$$5 = -x$$
$$\frac{5}{-1} = \frac{-x}{-1}$$
$$-5 = x$$

53.
$$\frac{x}{2}-1 = \frac{x}{5}+2$$
$$10\left(\frac{x}{2}-1\right) = 10\left(\frac{x}{5}+2\right)$$
$$5x-10 = 2x+20$$
$$5x-10+10 = 2x+20+10$$
$$5x = 2x+30$$
$$5x-2x = 2x+30-2x$$
$$3x = 30$$
$$\frac{3x}{3} = \frac{30}{3}$$
$$x = 10$$

55.
$$2(x+3)-5 = 5x-3(1+x)$$
$$2x+6-5 = 5x-3-3x$$
$$2x+1 = 2x-3$$
$$2x+1-2x = 2x-3-2x$$
$$1 = -3$$
Since the statement $1 = -3$ is false, the equation has no solution.

57.
$$0.06-0.01(x+1) = -0.02(2-x)$$
$$6-1(x+1) = -2(2-x)$$
$$6-x-1 = -4+2x$$
$$5-x = -4+2x$$
$$5-x+x = -4+2x+x$$
$$5 = -4+3x$$
$$5+4 = -4+3x+4$$
$$9 = 3x$$
$$\frac{9}{3} = \frac{3x}{3}$$
$$3 = x$$

59.
$$\frac{9}{2}+\frac{5}{2}y = 2y-4$$
$$2\left(\frac{9}{2}+\frac{5}{2}y\right) = 2(2y-4)$$
$$9+5y = 4y-8$$
$$9+5y-4y = 4y-8-4y$$
$$9+y = -8$$
$$9+y-9 = -8-9$$
$$y = -17$$

61. The perimeter is the sum of the lengths of the sides.
$$x+(2x-3)+(3x-5) = x+2x-3+3x-5$$
$$= 6x-8$$
The perimeter is $(6x-8)$ meters.

63. A number subtracted from -8 is $-8-x$.

65. The sum of -3 and twice a number is $-3 + 2x$.

67. The product of 9 and the sum of a number and 20 is $9(x + 20)$.

69. a. Since both sides of the equation are identical, the equation is an identity and every real number is a solution.

    b. answers may vary

    c. answers may vary

71. $5x + 1 = 5x + 1$
Since both sides of the equation are identical, the equation is an identity and every real number is a solution. The choice is a.

73. $2x - 6x - 10 = -4x + 3 - 10$
$$-4x - 10 = -4x - 7$$
$$-4x - 10 + 4x = -4x - 7 + 4x$$
$$-10 = -7$$
Since the statement $-10 = -7$ is false, the equation has no solution. The choice is b.

75. $9x - 20 = 8x - 20$
$$9x - 20 - 8x = 8x - 20 - 8x$$
$$x - 20 = -20$$
$$x - 20 + 20 = -20 + 20$$
$$x = 0$$
The choice is c.

77. answers may vary

79. a. The perimeter is the sum of the lengths of the sides.
$$x + x + x + 2x + 2x = 28$$

  b. $x + x + x + 2x + 2x = 28$
$$7x = 28$$
$$\frac{7x}{7} = \frac{28}{7}$$
$$x = 4$$

  c. The sides of length x are 4 cm and the sides of length $2x$ are $2(4) = 8$ cm.

81. answers may vary

83. $1000(7x - 10) = 50(412 + 100x)$
$$7000x - 10,000 = 20,600 + 5000x$$
$$7000x - 10,000 - 5000x = 20,600 + 5000x - 5000x$$
$$2000x - 10,000 = 20,600$$
$$2000x - 10,000 + 10,000 = 20,600 + 10,000$$
$$2000x = 30,600$$
$$\frac{2000x}{2000} = \frac{30,600}{2000}$$
$$x = 15.3$$

85.
$$0.035x + 5.112 = 0.010x + 5.107$$
$$35x + 5112 = 10x + 5107$$
$$35x + 5112 - 10x = 10x + 5107 - 10x$$
$$25x + 5112 = 5107$$
$$25x + 5112 - 5112 = 5107 - 5112$$
$$25x = -5$$
$$\frac{25x}{25} = \frac{-5}{25}$$
$$x = -\frac{1}{5}$$
$$x = -0.2$$

Integrated Review

1.
$$x - 10 = -4$$
$$x - 10 + 10 = -4 + 10$$
$$x = 6$$

2.
$$y + 14 = -3$$
$$y + 14 - 14 = -3 - 14$$
$$y = -17$$

3.
$$9y = 108$$
$$\frac{9y}{9} = \frac{108}{9}$$
$$y = 12$$

4.
$$-3x = 78$$
$$\frac{-3x}{-3} = \frac{78}{-3}$$
$$x = -26$$

5.
$$-6x + 7 = 25$$
$$-6x + 7 - 7 = 25 - 7$$
$$-6x = 18$$
$$\frac{-6x}{-6} = \frac{18}{-6}$$
$$x = -3$$

6.
$$5y - 42 = -47$$
$$5y - 42 + 42 = -47 + 42$$
$$5y = -5$$
$$\frac{5y}{5} = \frac{-5}{5}$$
$$y = -1$$

7.
$$\frac{2}{3}x = 9$$
$$\frac{3}{2} \cdot \frac{2}{3}x = \frac{3}{2} \cdot 9$$
$$x = \frac{27}{2}$$

8.
$$\frac{4}{5}z = 10$$
$$\frac{5}{4} \cdot \frac{4}{5}z = \frac{5}{4} \cdot 10$$
$$z = \frac{50}{4}$$
$$z = \frac{25}{2}$$

9.
$$\frac{r}{-4} = -2$$
$$-4 \cdot \frac{r}{-4} = -4 \cdot (-2)$$
$$r = 8$$

10.
$$\frac{y}{-8} = 8$$
$$-8 \cdot \frac{y}{-8} = -8 \cdot 8$$
$$y = -64$$

11.
$$6 - 2x + 8 = 10$$
$$-2x + 14 = 10$$
$$-2x + 14 - 14 = 10 - 14$$
$$-2x = -4$$
$$\frac{-2x}{-2} = \frac{-4}{-2}$$
$$x = 2$$

12.
$$-5 - 6y + 6 = 19$$
$$-6y + 1 = 19$$
$$-6y + 1 - 1 = 19 - 1$$
$$-6y = 18$$
$$\frac{-6y}{-6} = \frac{18}{-6}$$
$$y = -3$$

13.
$$2x - 7 = 6x - 27$$
$$2x - 7 + 7 = 6x - 27 + 7$$
$$2x = 6x - 20$$
$$2x - 6x = 6x - 20 - 6x$$
$$-4x = -20$$
$$\frac{-4x}{-4} = \frac{-20}{-4}$$
$$x = 5$$

14.
$$3 + 8y = 3y - 2$$
$$3 + 8y - 3y = 3y - 2 - 3y$$
$$3 + 5y = -2$$
$$-3 + 3 + 5y = -3 - 2$$
$$5y = -5$$
$$\frac{5y}{5} = \frac{-5}{5}$$
$$y = -1$$

15.
$$9(3x - 1) = -4 + 49$$
$$27x - 9 = 45$$
$$27x - 9 + 9 = 45 + 9$$
$$27x = 54$$
$$\frac{27x}{27} = \frac{54}{27}$$
$$x = 2$$

16.
$$12(2x + 1) = -6 + 66$$
$$24x + 12 = 60$$
$$24x + 12 - 12 = 60 - 12$$
$$24x = 48$$
$$\frac{24x}{24} = \frac{48}{24}$$
$$x = 2$$

17.
$$-3a + 6 + 5a = 7a - 8a$$
$$6 + 2a = -a$$
$$6 + 2a - 2a = -a - 2a$$
$$6 = -3a$$
$$\frac{6}{-3} = \frac{-3a}{-3}$$
$$-2 = a$$

18.
$$4b - 8 - b = 10b - 3b$$
$$3b - 8 = 7b$$
$$-3b + 3b - 8 = -3b + 7b$$
$$-8 = 4b$$
$$\frac{-8}{4} = \frac{4b}{4}$$
$$-2 = b$$

19.
$$-\frac{2}{3}x = \frac{5}{9}$$
$$-\frac{3}{2} \cdot \left(-\frac{2}{3}x\right) = -\frac{3}{2} \cdot \frac{5}{9}$$
$$x = -\frac{15}{18}$$
$$x = -\frac{5}{6}$$

20.
$$-\frac{3}{8}y = -\frac{1}{16}$$
$$-\frac{8}{3} \cdot \left(-\frac{3}{8}y\right) = -\frac{8}{3} \cdot \left(-\frac{1}{16}\right)$$
$$y = \frac{1}{6}$$

21.
$$10 = -6n + 16$$
$$10 - 16 = -6n + 16 - 16$$
$$-6 = -6n$$
$$\frac{-6}{-6} = \frac{-6n}{-6}$$
$$1 = n$$

22.
$$-5 = -2m + 7$$
$$-5 - 7 = -2m + 7 - 7$$
$$-12 = -2m$$
$$\frac{-12}{-2} = \frac{-2m}{-2}$$
$$6 = m$$

23. $3(5c-1)-2=13c+3$
$15c-3-2=13c+3$
$15c-5=13c+3$
$15c-5+5=13c+3+5$
$15c=13c+8$
$15c-13c=13c+8-13c$
$2c=8$
$\dfrac{2c}{2}=\dfrac{8}{2}$
$c=4$

24. $4(3t+4)-20=3+5t$
$12t+16-20=3+5t$
$12t-4=3+5t$
$12t-4-5t=3+5t-5t$
$7t-4=3$
$7t-4+4=3+4$
$7t=7$
$\dfrac{7t}{7}=\dfrac{7}{7}$
$t=1$

25. $\dfrac{2(z+3)}{3}=5-z$
$3\left[\dfrac{2(z+3)}{3}\right]=3(5-z)$
$2(z+3)=3(5-z)$
$2z+6=15-3z$
$2z+6+3z=15-3z+3z$
$6+5z=15$
$6+5z-6=15-6$
$5z=9$
$\dfrac{5z}{5}=\dfrac{9}{5}$
$z=\dfrac{9}{5}$

26. $\dfrac{3(w+2)}{4}=2w+3$
$4\left[\dfrac{3(w+2)}{4}\right]=4(2w+3)$
$3(w+2)=4(2w+3)$
$3w+6=8w+12$
$3w+6-6=8w+12-6$
$3w=8w+6$
$3w-8w=8w+6-8w$
$-5w=6$
$\dfrac{-5w}{-5}=\dfrac{6}{-5}$
$w=-\dfrac{6}{5}$

27. $-2(2x-5)=-3x+7-x+3$
$-4x+10=-4x+10$
Since both sides of the equation are identical, the equation is an identity and every real number is a solution.

28. $-4(5x-2)=-12x+4-8x+4$
$-20x+8=-20x+8$
Since both sides of the equation are identical, the equation is an identity and every real number is a solution.

29. $0.02(6t-3)=0.04(t-2)+0.02$
$2(6t-3)=4(t-2)+2$
$12t-6=4t-8+2$
$12t-6=4t-6$
$12t-6-4t=4t-6-4t$
$8t-6=-6$
$8t-6+6=-6+6$
$8t=0$
$\dfrac{8t}{8}=\dfrac{0}{8}$
$t=0$

30.
$$0.03(m+7) = 0.02(5-m) + 0.03$$
$$3(m+7) = 2(5-m) + 3$$
$$3m + 21 = 10 - 2m + 3$$
$$3m + 21 = 13 - 2m$$
$$3m + 21 + 2m = 13 - 2m + 2m$$
$$5m + 21 = 13$$
$$5m + 21 - 21 = 13 - 21$$
$$5m = -8$$
$$\frac{5m}{5} = \frac{-8}{5}$$
$$m = -1.6$$

31.
$$-3y = \frac{4(y-1)}{5}$$
$$5(-3y) = 5\left[\frac{4(y-1)}{5}\right]$$
$$-15y = 4(y-1)$$
$$-15y = 4y - 4$$
$$-15y - 4y = 4y - 4 - 4y$$
$$-19y = -4$$
$$\frac{-19y}{-19} = \frac{-4}{-19}$$
$$y = \frac{4}{19}$$

32.
$$-4x = \frac{5(1-x)}{6}$$
$$6(-4x) = 6 \cdot \frac{5(1-x)}{6}$$
$$-24x = 5(1-x)$$
$$-24x = 5 - 5x$$
$$-24x + 5x = 5 - 5x + 5x$$
$$-19x = 5$$
$$\frac{-19x}{-19} = \frac{5}{-19}$$
$$x = -\frac{5}{19}$$

33.
$$\frac{5}{3}x - \frac{7}{3} = x$$
$$3\left(\frac{5}{3}x - \frac{7}{3}\right) = 3x$$
$$5x - 7 = 3x$$
$$-5x + 5x - 7 = -5x + 3x$$
$$-7 = -2x$$
$$\frac{-7}{-2} = \frac{-2x}{-2}$$
$$\frac{7}{2} = x$$

34.
$$\frac{7}{5}n + \frac{3}{5} = -n$$
$$5\left(\frac{7}{5}n + \frac{3}{5}\right) = 5(-n)$$
$$7n + 3 = -5n$$
$$-7n + 7n + 3 = -7n - 5n$$
$$3 = -12n$$
$$\frac{3}{-12} = \frac{-12n}{-12}$$
$$-\frac{1}{4} = n$$

35.
$$\frac{1}{10}(3x-7) = \frac{3}{10}x + 5$$
$$\frac{3}{10}x - \frac{7}{10} = \frac{3}{10}x + 5$$
$$-\frac{3}{10}x + \frac{3}{10}x - \frac{7}{10} = -\frac{3}{10}x + \frac{3}{10}x + 5$$
$$-\frac{7}{10} = 5$$

Since the statement $-\dfrac{7}{10} = 5$ is false, the equation has no solution.

36. $\frac{1}{7}(2x-5) = \frac{2}{7}x+1$

$$7 \cdot \frac{1}{7}(2x-5) = 7\left(\frac{2}{7}x+1\right)$$
$$2x-5 = 2x+7$$
$$2x-5-2x = 2x+7-2x$$
$$-5 = 7$$

Since the statement $-5 = 7$ is false, the equation has no solution.

37. $5+2(3x-6) = -4(6x-7)$

$$5+6x-12 = -24x+28$$
$$6x-7 = -24x+28$$
$$24x+6x-7 = 24x-24x+28$$
$$30x-7 = 28$$
$$30x-7+7 = 28+7$$
$$30x = 35$$
$$\frac{30x}{30} = \frac{35}{30}$$
$$x = \frac{7}{6}$$

38. $3+5(2x-4) = -7(5x+2)$

$$3+10x-20 = -35x-14$$
$$10x-17 = -35x-14$$
$$10x-17+35x = -35x-14+35x$$
$$45x-17 = -14$$
$$45x-17+17 = -14+17$$
$$45x = 3$$
$$\frac{45x}{45} = \frac{3}{45}$$
$$x = \frac{1}{15}$$

Section 2.4 Practice Problems

1. Let x represent the number.

$$3(x-5) = 2x-3$$
$$3x-15 = 2x-3$$
$$3x-15-2x = 2x-3-2x$$
$$x-15 = -3$$
$$x-15+15 = -3+15$$
$$x = 12$$

The number is 12.

2. If x is the first even integer, then $x + 2$ and $x + 4$ are the next two even integers.

$$x+x+2+x+4 = 144$$
$$3x+6 = 144$$
$$3x+6-6 = 144-6$$
$$3x = 138$$
$$\frac{3x}{3} = \frac{138}{3}$$
$$x = 46$$

If $x = 46$, then $x + 2 = 48$ and $x + 4 = 50$. The integers are 46, 48, 50.

3. Let x represent the length of the shorter piece. Then $5x$ represents the length of the longer piece. Their sum is 18 feet.

$$x+5x = 18$$
$$6x = 18$$
$$\frac{6x}{6} = \frac{18}{6}$$
$$x = 3$$

The shorter piece is 3 feet and the longer piece is $5(3) = 15$ feet.

4. Let x represent the number of votes for Texas. Then $x + 21$ represents the number of votes for California. Their sum is 89.

$$x+x+21 = 89$$
$$2x+21 = 89$$
$$2x+21-21 = 89-21$$
$$2x = 68$$
$$\frac{2x}{2} = \frac{68}{2}$$
$$x = 34$$

Texas has 34 electoral votes and California had $34 + 21 = 55$ electoral votes.

5. Let x represent the number of miles driven. The cost for x miles is $0.15x$. The daily cost is \$28.

$$0.15x+28 = 52$$
$$0.15x+28-28 = 52-28$$
$$0.15x = 24$$
$$\frac{0.15x}{0.15} = \frac{24}{0.15}$$
$$x = 160$$

You drove 160 miles.

6. Let x represent the measure of the smallest angle. Then $2x$ represents the measure of the second angle and $3x$ represents the measure of the third angle. The sum of the measures of the angles of a triangle equals 180.
$$x + 2x + 3x = 180$$
$$6x = 180$$
$$\frac{6x}{6} = \frac{180}{6}$$
$$x = 30$$
If $x = 30$, then $2x = 2(30) = 60$ and $3x = 3(30) = 90$.
The smallest is $30°$, second is $60°$, and third is $90°$.

Exercise Set 2.4

1.
$$2(x - 8) = 3(x + 3)$$
$$2x - 16 = 3x + 9$$
$$2x - 16 - 2x = 3x + 9 - 2x$$
$$-16 = x + 9$$
$$-16 - 9 = x + 9 - 9$$
$$-25 = x$$
The number is -25.

3.
$$2x(3) = 5x - \frac{3}{4}$$
$$6x = 5x - \frac{3}{4}$$
$$6x - 5x = 5x - \frac{3}{4} - 5x$$
$$x = -\frac{3}{4}$$

The number is $-\frac{3}{4}$.

5. If x is the first integer, the next consecutive integer is $x + 1$.
$$x + x + 1 = 469$$
$$2x + 1 = 469$$
$$2x + 1 - 1 = 469 - 1$$
$$2x = 468$$
$$\frac{2x}{2} = \frac{468}{2}$$
$$x = 234$$
The page numbers are 234 and $234 + 1 = 235$.

7. If x is the first integer, the next two consecutive integers are $x + 1$ and $x + 2$.
$$x + x + 1 + x + 2 = 99$$
$$3x + 3 = 99$$
$$3x + 3 - 3 = 99 - 3$$
$$3x = 96$$
$$\frac{3x}{3} = \frac{96}{3}$$
$$x = 32$$
The code for Belgium is 32, France is $32 + 1 = 33$, and Spain is $32 + 2 = 34$.

9. The sum of the three lengths is 25 inches.
$$x + 2x + 1 + 5x = 25$$
$$1 + 8x = 25$$
$$1 + 8x - 1 = 25 - 1$$
$$8x = 24$$
$$\frac{8x}{8} = \frac{24}{8}$$
$$x = 3$$
$2x = 2(3) = 6$
$1 + 5x = 1 + 5(3) = 1 + 15 = 16$
The lengths are 3 inches, 6 inches, and 16 inches.

11. Let x be the length of the first piece. Then the second piece is $2x$ and the third piece is $5x$. The sum of the lengths is 40 inches.
$$x + 2x + 5x = 40$$
$$8x = 40$$
$$\frac{8x}{8} = \frac{40}{8}$$
$$x = 5$$
$2x = 2(5) = 10$
$5x = 5(5) = 25$
The 1st piece is 5 inches, 2nd piece is 10 inches, and 3rd piece is 25 inches.

13. Let x represent the salary of the Governor of Florida. Then $x + 50,425$ represents the salary of the Governor of California.

$$x + x + 50,425 = 299,575$$
$$2x + 50,425 = 299,575$$
$$2x + 50,425 - 50,425 = 299,575 - 50,425$$
$$2x = 249,150$$
$$\frac{2x}{2} = \frac{249,150}{2}$$
$$x = 124,575$$

$x + 50,425 = 124,575 + 50,425 = 175,000$

The salary of the Governor of Florida is $124,575, while that of the Governor of California is $175,000.

15. Let x be the number of miles. Then the cost for x miles is $0.29x$. Each day costs $24.95.

$$0.29x + 2(24.95) = 100$$
$$0.29x + 49.9 = 100$$
$$0.29x + 49.9 - 49.9 = 100 - 49.9$$
$$0.29x = 50.1$$
$$\frac{0.29x}{0.29} = \frac{50.1}{0.29}$$
$$x \approx 172.8$$

You can drive 172 whole miles on a $100 budget.

17. Let x be the number of miles. Then the total fare is $3 + 0.8x + 4.5$.

$$3 + 0.8x + 4.5 = 27.5$$
$$30 + 8x + 45 = 275$$
$$8x + 75 = 275$$
$$8x + 75 - 75 = 275 - 75$$
$$8x = 200$$
$$\frac{8x}{8} = \frac{200}{8}$$
$$x = 25$$

You can travel 25 miles from the airport by taxi for $27.50.

19. Let x be the measure of each of the two equal angles. Then $2x + 30$ is the measure of the third angle. Their sum is $180°$.

$$x + x + 2x + 30 = 180$$
$$4x + 30 = 180$$
$$4x + 30 - 30 = 180 - 30$$
$$4x = 150$$
$$\frac{4x}{4} = \frac{150}{4}$$
$$x = 37.5$$

$2x + 30 = 2(37.5) + 30 = 75 + 30 = 105$

The 1st angle measures $37.5°$, the 2nd angle measures $37.5°$, and the 3rd angle measures $105°$.

21. Angles A and D both measure $x°$, while angles C and B both measure $(2x)°$. The sum of the angle measures is $360°$.

$$x + 2x + x + 2x = 360$$
$$6x = 360$$
$$\frac{6x}{6} = \frac{360}{6}$$
$$x = 60$$

$2x = 2(60) = 120$

Angles A and D measure $60°$; angles B and C measure $120°$.

23. Let x be the length of the shorter piece. Then $2x + 2$ is the length of the longer piece. The measures sum to 17 feet.

$$x + 2x + 2 = 17$$
$$3x + 2 = 17$$
$$3x + 2 - 2 = 17 - 2$$
$$3x = 5$$
$$\frac{3x}{3} = \frac{15}{3}$$
$$x = 5$$

$2x + 2 = 2(5) + 2 = 10 + 2 = 12$

The pieces measure 5 feet and 12 feet.

25. Let x represent the number of prescriptions written in 1997, in millions. Then the number written in 2001 was $(x + 5.5)$ million.

$$x + x + 5.5 = 35.7$$
$$2x + 5.5 = 35.7$$
$$2x + 5.5 - 5.5 = 35.7 - 5.5$$
$$2x = 30.2$$
$$\frac{2x}{2} = \frac{30.2}{2}$$
$$x = 15.1$$

$x + 5.5 = 15.1 + 5.5 = 20.6$

There were 15.1 million prescriptions for ADHD drugs written in 1997, and 20.6 million prescriptions written in 2001.

27. Let x be the measure of the smaller angle. Then the larger angle measures $3x$. Their sum is $180°$.

$$x + 3x = 180$$
$$4x = 180$$
$$\frac{4x}{4} = \frac{180}{4}$$
$$x = 45$$

$3x = 3(45) = 135$

The angles measure $45°$ and $135°$.

29. Let x be the first even integer. Then the next two consecutive even integers are $x + 2$ and $x + 4$. The sum of the measures of the angles of a triangle is $180°$.

$$x + x + 2 + x + 4 = 180$$
$$3x + 6 = 180$$
$$3x + 6 - 6 = 180 - 6$$
$$3x = 174$$
$$\frac{3x}{3} = \frac{174}{3}$$
$$x = 58$$

$x + 2 = 58 + 2 = 60$
$x + 4 = 58 + 4 = 62$

The angles measure $58°$, $60°$, and $62°$.

31.
$$\frac{1}{5} + 2x = 3x - \frac{4}{5}$$
$$\frac{1}{5} + 2x - 2x = 3x - \frac{4}{5} - 2x$$
$$\frac{1}{5} = x - \frac{4}{5}$$
$$\frac{1}{5} + \frac{4}{5} = x - \frac{4}{5} + \frac{4}{5}$$
$$\frac{5}{5} = x$$
$$1 = x$$

The number is 1.

33. Let x be the number of miles. Then the charge for driving x miles in one day is $39 + 0.2x$.

$$39 + 0.2x = 95$$
$$390 + 2x = 950$$
$$390 + 2x - 390 = 950 - 390$$
$$2x = 560$$
$$\frac{2x}{2} = \frac{560}{2}$$
$$x = 280$$

You drove 280 miles.

35. Let x be the number of points earned by Johnson. Then $x + 90$ is the number of points earned by Kenseth. Together they earned 9954 points.

$$x + x + 90 = 9954$$
$$2x + 90 = 9954$$
$$2x + 90 - 90 = 9954 - 90$$
$$2x = 9864$$
$$\frac{2x}{2} = \frac{9864}{2}$$
$$x = 4932$$

$x + 90 = 4932 + 90 = 5022$

Johnson earned 4932 points and Kenseth earned 5022 points.

37. Let x represent the number of counties in Montana. Then $x + 2$ represents the number of counties in California.

$$x + x + 2 = 114$$
$$2x + 2 = 114$$
$$2x + 2 - 2 = 114 - 2$$
$$2x = 112$$
$$\frac{2x}{2} = \frac{112}{2}$$
$$x = 56$$
$$x + 2 = 56 + 2 = 58$$

Montana has 56 counties and California has 58 counties.

39. Let x represent the number of moons for Neptune. Then $x + 13$ represents the number of moons for Uranus and $2x + 2$ represents the number of moons for Saturn. The total is 47.

$$x + x + 13 + 2x + 2 = 47$$
$$4x + 15 = 47$$
$$4x + 15 - 15 = 47 - 15$$
$$4x = 32$$
$$\frac{4x}{4} = \frac{32}{4}$$
$$x = 8$$
$$x + 13 = 8 + 13 = 21$$
$$2x + 2 = 2(8) + 2 = 16 + 2 = 18$$

Neptune has 8 moons, Uranus has 21 moons, and Saturn has 18 moons.

41.
$$3(x + 5) = 2x - 1$$
$$3x + 15 = 2x - 1$$
$$3x + 15 - 2x = 2x - 1 - 2x$$
$$x + 15 = -1$$
$$x + 15 - 15 = -1 - 15$$
$$x = -16$$

The number is -16.

43. Let x represent the area of the Gobi Desert, in square miles. Then $7x$ represents the area of the Sahara Desert.

$$x + 7x = 4,000,000$$
$$8x = 4,000,000$$
$$\frac{8x}{8} = \frac{4,000,000}{8}$$
$$x = 500,000$$

$$7x = 7(500,000) = 3,500,000$$

The Gobi Desert's area is 500,000 square miles and the Sahara Desert's area is 3,500,000 square miles.

45. Let x represent the number of gold medals won by Korea. Then Italy won $x + 1$ gold medals and France won $x + 2$ gold medals.

$$x + x + 1 + x + 2 = 30$$
$$3x + 3 = 30$$
$$3x + 3 - 3 = 30 - 3$$
$$3x = 27$$
$$\frac{3x}{3} = \frac{27}{3}$$
$$x = 9$$
$$x + 1 = 9 + 1 = 10$$
$$x + 2 = 9 + 2 = 11$$

Korea won 9 gold medals, Italy won 10, and France won 11.

47. Let x be the number of votes for Bill Randall. Then $x + 13,288$ is the number of votes for Corrine Brown.

$$x + x + 13,288 = 119,436$$
$$2x + 13,288 = 119,436$$
$$2x + 13,288 - 13,288 = 119,436 - 13,288$$
$$2x = 106,148$$
$$\frac{2x}{2} = \frac{106,148}{2}$$
$$x = 53,074$$

$$x + 13,288 = 53,074 + 13,288 = 66,362$$

Randall received 53,074 votes and Brown received 66,362 votes.

49. The tallest bar represents the amount spent by Illinois, so Illinois spends the most on tourism.

51. Let x be the amount spent by Florida. Then $x + 2.2$ is the amount spent by Texas.

$$x + x + 2.2 = 56.6$$
$$2x + 2.2 = 56.6$$
$$2x + 2.2 - 2.2 = 56.6 - 2.2$$
$$2x = 54.4$$
$$\frac{2x}{2} = \frac{54.4}{2}$$
$$x = 27.2$$

$x + 2.2 = 27.2 + 2.2 = 29.4$
Florida spent \$27.2 million and Texas spent \$29.4 million.

53. answers may vary

55. Replace W by 7 and L by 10.
$2W + 2L = 2(7) + 2(10) = 14 + 20 = 34$

57. Replace r by 15.
$\pi r^2 = \pi(15)^2 = \pi(225) = 225\pi$

59. Let x represent the width. Then $1.6x$ represents the length. The perimeter is $2 \cdot \text{length} + 2 \cdot \text{width}$.
$$2(1.6x) + 2x = 78$$
$$3.2x + 2x = 78$$
$$5.2x = 78$$
$$\frac{5.2x}{5.2} = \frac{78}{5.2}$$
$$x = 15$$
$1.6x = 1.6(15) = 24$
The dimensions of the garden are 15 feet by 24 feet.

61. One blink every 5 seconds is $\dfrac{1 \text{ blink}}{5 \text{ sec}}$.
There are $60 \cdot 60 = 3600$ seconds in one hour.
$$\frac{1 \text{ blink}}{5 \text{ sec}} \cdot 3600 \text{ sec} = 720 \text{ blinks}$$
The average eye blinks 720 times each hour.
$16 \cdot 720 = 11{,}520$
The average eye blinks 11,520 times while awake for a 16-hour day.
$11{,}520 \cdot 365 = 4{,}204{,}800$
The average eye blinks 4,204,800 times in one year.

63. answers may vary

65. answers may vary

Section 2.5 Practice Problems

1. Use $d = rt$ when $d = 1180$ and $r = 50$.
$$d = rt$$
$$1180 = 50t$$
$$\frac{1180}{50} = \frac{50t}{50}$$
$$23.6 = t$$
They will spend 23.6 hours driving.

2. Use $A = lw$ when $w = 18$.
$$A = lw$$
$$450 = l \cdot 18$$
$$\frac{450}{18} = \frac{18l}{18}$$
$$25 = l$$
The length of the deck is 25 feet.

3. Use $F = \dfrac{9}{5}C + 32$ with $C = 5$.
$$F = \frac{9}{5}C + 32$$
$$F = \frac{9}{5} \cdot 5 + 32$$
$$F = 9 + 32$$
$$F = 41$$
Thus, 5°C is equivalent to 41°F.

4. Let x be the width. Then $4x + 1$ is the length. The perimeter is 52 meters.
$$P = 2l + 2w$$
$$52 = 2(4x + 1) + 2x$$
$$52 = 8x + 2 + 2x$$
$$52 = 10x + 2$$
$$52 - 2 = 10x + 2 - 2$$
$$50 = 10x$$
$$\frac{50}{10} = \frac{10x}{10}$$
$$5 = x$$
$4x + 1 = 4(5) + 1 = 20 + 1 = 21$
The width is 5 meters and the length is 21 meters.

5. $C = 2\pi r$

$$\frac{C}{2\pi} = \frac{2\pi r}{2\pi}$$

$$\frac{C}{2\pi} = r \text{ or } r = \frac{C}{2\pi}$$

6. $P = 2l + 2w$

$P - 2w = 2l + 2w - 2w$

$P - 2w = 2l$

$$\frac{P - 2w}{2} = \frac{2l}{2}$$

$$\frac{P - 2w}{2} = l \text{ or } l = \frac{P - 2w}{2}$$

7. $P = 2a + b - c$

$P + c = 2a + b - c + c$

$P + c = 2a + b$

$P + c - b = 2a + b - b$

$P + c - b = 2a$

$$\frac{P + c - b}{2} = a \text{ or } a = \frac{P - b - c}{2}$$

8. $A = \dfrac{a + b}{2}$

$2A = 2 \cdot \dfrac{a + b}{2}$

$2A = a + b$

$2A - a = a + b - a$

$2A - a = b \text{ or } b = 2A - a$

Exercise Set 2.5

1. Use $A = bh$ when $A = 45$ and $b = 15$.

$A = bh$

$45 = 15 \cdot h$

$$\frac{45}{15} = \frac{15h}{15}$$

$3 = h$

3. Use $S = 4lw + 2wh$ when $S = 102$, $l = 7$, and $w = 3$.

$S = 4lw + 2wh$

$102 = 4 \cdot 7 \cdot 3 + 2 \cdot 3 \cdot h$

$102 = 84 + 6h$

$102 - 84 = 84 + 6h - 84$

$18 = 6h$

$$\frac{18}{6} = \frac{6h}{6}$$

$3 = h$

5. Use $A = \dfrac{1}{2} h(B + b)$ when $A = 180$, $B = 11$, and $b = 7$.

$A = \dfrac{1}{2} h(B + b)$

$180 = \dfrac{1}{2} h(11 + 7)$

$180 = \dfrac{1}{2} h(18)$

$180 = 9h$

$$\frac{180}{9} = \frac{9h}{9}$$

$20 = h$

7. Use $P = a + b + c$ when $P = 30$, $a = 8$, and $b = 10$.

$P = a + b + c$

$30 = 8 + 10 + c$

$30 = 18 + c$

$30 - 18 = 18 + c - 18$

$12 = c$

9. Use $C = 2\pi r$ when $C = 15.7$ and 3.14 is used as an approximation for π.

$C = 2\pi r$

$15.7 = 2(3.14)r$

$15.7 = 6.28r$

$$\frac{15.7}{6.28} = \frac{6.28r}{6.28}$$

$2.5 = r$

11. $f = 5gh$

$$\frac{f}{5g} = \frac{5gh}{5g}$$

$$\frac{f}{5g} = h$$

13. $V = lwh$

$$\frac{V}{lh} = \frac{lwh}{lh}$$

$$\frac{V}{lh} = w$$

15. $3x + y = 7$

$$3x + y - 3x = 7 - 3x$$

$$y = 7 - 3x$$

17. $A = P + PRT$

$$A - P = P + PRT - P$$

$$A - P = PRT$$

$$\frac{A - P}{PT} = \frac{PRT}{PT}$$

$$\frac{A - P}{PT} = R$$

19. $V = \frac{1}{3}Ah$

$$3V = 3 \cdot \frac{1}{3}Ah$$

$$3V = Ah$$

$$\frac{3V}{h} = \frac{Ah}{h}$$

$$\frac{3V}{h} = A$$

21. $P = a + b + c$

$$P - b - c = a + b + c - b - c$$

$$P - b - c = a$$

23. $S = 2\pi rh + 2\pi r^2$

$$S - 2\pi r^2 = 2\pi rh + 2\pi r^2 - 2\pi r^2$$

$$S - 2\pi r^2 = 2\pi rh$$

$$\frac{S - 2\pi r^2}{2\pi r} = \frac{2\pi rh}{2\pi r}$$

$$\frac{S - 2\pi r^2}{2\pi r} = h$$

25. a. Area $= l \cdot w = (11.5)(9) = 103.5$

Perimeter $= 2l + 2w$

$$= 2(11.5) + 2(9)$$

$$= 23 + 18$$

$$= 41$$

The area is 103.5 square feet and the perimeter is 41 feet.

b. The baseboard goes around the edges of the room, so it involves the perimeter. The carpet covers the floor of the room, so it involves area.

27. a. Area $= \frac{1}{2}h(B + b)$

$$= \frac{1}{2} \cdot 12(56 + 24)$$

$$= 6(80)$$

$$= 480$$

Perimeter $= 24 + 20 + 56 + 20 = 120$

The area is 480 square inches and the perimeter is 120 inches.

b. The frame goes around the edges of the picture, so it involves perimeter. The glass covers the picture, so it involves area.

29. Use $A = lw$ when $A = 3990$ and $w = 57$.

$$A = lw$$

$$3990 = l(57)$$

$$\frac{3990}{57} = \frac{57l}{57}$$

$$70 = l$$

The length (height) of the billboard was 70 feet.

31. Use $F = \dfrac{9}{5}C + 32$ when $F = 14$.

$$F = \frac{9}{5}C + 32$$

$$14 = \frac{9}{5}C + 32$$

$$14 - 32 = \frac{9}{5}C + 32 - 32$$

$$-18 = \frac{9}{5}C$$

$$\frac{5}{9} \cdot (-18) = \frac{5}{9} \cdot \frac{9}{5}C$$

$$-10 = C$$

Thus, $14°F$ is equivalent to $-10°C$.

33. Use $d = rt$ when $d = 25,000$ and $r = 4000$.

$$d = rt$$

$$25,000 = 4000t$$

$$\frac{25,000}{4000} = \frac{4000t}{4000}$$

$$6.25 = t$$

It will take the X-30 6.25 hours to travel around the Earth.

35. Let x be the length. Then $\dfrac{2}{3}x$ is the width.

Use $P = 2 \cdot \text{length} + 2 \cdot \text{width}$ when $P = 260$.

$$P = 2 \cdot \text{length} + 2 \cdot \text{width}$$

$$260 = 2x + 2 \cdot \frac{2}{3}x$$

$$260 = 2x + \frac{4}{3}x$$

$$260 = \frac{6}{3}x + \frac{4}{3}x$$

$$260 = \frac{10}{3}x$$

$$\frac{3}{10} \cdot 260 = \frac{3}{10} \cdot \frac{10}{3}x$$

$$78 = x$$

The length is 78 feet and the width is

$$\frac{2}{3} \cdot 78 = 52 \text{ feet.}$$

37. Let x represent the length of the shortest side. Then the second side has length $2x$ and the third side has length $30 + x$. The perimeter is the sum of the lengths of the sides.

$$x + 2x + 30 + x = 102$$

$$4x + 30 = 102$$

$$4x + 30 - 30 = 102 - 30$$

$$4x = 72$$

$$\frac{4x}{4} = \frac{72}{4}$$

$$x = 18$$

$2x = 2(18) = 36$

$30 + x = 30 + 18 = 48$

The flower bed has sides of length 18 feet, 36 feet, and 48 feet.

39. Use $d = rt$ when $r = 55$ and $t = 2\dfrac{1}{2}$.

$$d = rt$$

$$d = 55 \cdot 2\frac{1}{2}$$

$$d = 55 \cdot 2.5$$

$$d = 137.5$$

The distance between Bar Harbor and Yarmouth is 137.5 miles.

41. To find the amount of water in the tank, use $V = lwh$ with $l = 8$, $w = 3$, and $h = 6$.

$V = lwh = 8 \cdot 3 \cdot 6 = 144$

The tank holds 144 cubic feet of water. Let x represent the number of piranhas the tank could hold. Then $1.5x = 144$.

$$1.5x = 144$$

$$\frac{1.5x}{1.5} = \frac{144}{1.5}$$

$$x = 96$$

The tank could hold 96 piranhas.

43. Use $A = \dfrac{1}{2}h(B + b)$ to find the area of the lawn.

$$A = \frac{1}{2}h(B + b)$$

$$A = \frac{1}{2}(60)(130 + 70) = 30(200) = 6000$$

Let x be the number of bags of fertilizer.

$$4000x = 6000$$

$$\frac{4000x}{4000} = \frac{6000}{4000}$$

$$x = 1.5$$

Since $\frac{1}{2}$ bag cannot be purchased, 2 bags

must be purchased to cover the lawn.

45. Use $A = \pi r^2$ to find the area of a pizza.

For the 16-inch pizza, $r = \frac{16}{2} = 8.$

$$A = \pi r^2 = \pi(8)^2 = 64\pi$$

For a 10-inch pizza, $r = \frac{10}{2} = 5.$

$$A = \pi r^2 = \pi(5)^2 = 25\pi$$

Two 10-inch pizzas have an area of
$2 \cdot 25\pi = 50\pi$ square inches. Since
$50\pi < 64\pi$, you get more pizza by buying
the 16-inch pizza.

47. Use $d = rt$ when $r = 552$ and $d = 42.8$.

$$d = rt$$

$$42.8 = 552t$$

$$\frac{42.8}{552} = \frac{552t}{552}$$

$$0.0775 = t$$

It would last 0.0775 hour or
$0.0775(60) \approx 465$ minutes.

49. Let s represent the length of one side of the
square. Then the perimeter of the square is
$4s$. A side of the triangle is $s + 5$ and the
triangle's perimeter is $3(s + 5)$.

$$3(s + 5) = 4s + 7$$

$$3s + 15 = 4s + 7$$

$$3s + 15 - 3s = 4s + 7 - 3s$$

$$15 = s + 7$$

$$15 - 7 = s + 7 - 7$$

$$8 = s$$

$$s + 5 = 8 + 5 = 13$$

Each side of the triangle has length
13 inches.

51. Use $d = rt$ when $d = 135$ an $r = 60$.

$$d = rt$$

$$135 = 60t$$

$$\frac{135}{60} = \frac{60t}{60}$$

$$2.25 = t$$

It will take 2.25 hours.

53. Use $A = lw$ when $A = 1,813,500$ and
$w = 150$.

$$A = lw$$

$$1,813,500 = l(150)$$

$$\frac{1,813,500}{150} = \frac{150l}{150}$$

$$12,090 = l$$

The length of the runway is 12,090 feet
(more than 2 miles!).

55. Use $F = \frac{9}{5}C + 32$ when $F = 122$.

$$122 = \frac{9}{5}C + 32$$

$$122 - 32 = \frac{9}{5}C + 32 - 32$$

$$90 = \frac{9}{5}C$$

$$\frac{5}{9} \cdot 90 = \frac{5}{9} \cdot \frac{9}{5}C$$

$$50 = C$$

Thus, 122°F is equivalent to 50°C.

57. Use $V = lwh$ when $l = 199$, $w = 78.5$, and
$h = 33$.
$V = lwh = 199(78.5)(33) = 515,509.5$
The smallest possible shipping crate has a
volume of 515,509.5 cubic inches.

59. Use $V = \frac{4}{3}\pi r^3$ when $r = \frac{9.5}{2} = 4.75$ and

$\pi = 3.14$.

$$V = \frac{4}{3}\pi r^3 = \frac{4}{3}(3.14)(4.75)^3 \approx 449$$

The volume of the sphere is 449 cubic
inches.

61. Use $F = \frac{9}{5}C + 32$ when $C = 167$.

$$F = \frac{9}{5}C + 32$$
$$= \frac{9}{5}(167) + 32$$
$$= 300.6 + 32$$
$$= 332.6$$
$$\approx 333$$

The average temperature on the planet Mercury is 333°F.

63. $32\% = 0.32$

65. $200\% = 2.00$ or 2

67. $0.17 = 0.17(100\%) = 17\%$

69. $7.2 = 7.2(100\%) = 720\%$

71.
$$N = R + \frac{V}{G}$$
$$N - R = R + \frac{V}{G} - R$$
$$N - R = \frac{V}{G}$$
$$G(N - R) = G \cdot \frac{V}{G}$$
$$G(N - R) = V$$

73. Use $V = lwh$. If the length is doubled, the new length is $2l$. If the width and height are doubled, the new width and height are $2w$ and $2h$, respectively.
$V = (2l)(2w)(2h) = 2 \cdot 2 \cdot 2lwh = 8lwh$
The volume of the box is multiplied by 8.

75. Let x be the temperature. Use $F = \frac{9}{5}C + 32$ when $F = C = x$.

$$F = \frac{9}{5}C + 32$$
$$x = \frac{9}{5}x + 32$$
$$x - \frac{9}{5}x = \frac{9}{5}x + 32 - \frac{9}{5}x$$
$$\frac{5}{5}x - \frac{9}{5}x = 32$$
$$-\frac{4}{5}x = 32$$
$$-\frac{5}{4} \cdot \left(-\frac{4}{5}x\right) = -\frac{5}{4} \cdot 32$$
$$x = -40$$

They are the same when the temperature is $-40°$.

77.

79. $\dfrac{20 \text{ miles}}{1 \text{ hour}} \cdot \dfrac{5280 \text{ feet}}{1 \text{ mile}} \cdot \dfrac{1 \text{ hour}}{60 \text{ minutes}} \cdot \dfrac{1 \text{ minute}}{60 \text{ seconds}}$

$= \dfrac{20 \cdot 5280 \text{ feet}}{60 \cdot 60 \text{ seconds}}$

≈ 29.3 feet/second

Use $d = rt$ when $d = 1300$ and $r = 29.3$.
$$d = rt$$
$$1300 = 29.3t$$
$$\frac{1300}{29.3} = \frac{29.3t}{29.3}$$
$$44.3 \approx t$$

It took 44.3 seconds to travel that distance.

81. Use $I = PRT$ when $I = 1{,}056{,}000$, $R = 0.055$, and $T = 6$.

$$I = PRT$$
$$1{,}056{,}000 = P(0.055)(6)$$
$$1{,}056{,}000 = 0.33P$$
$$\frac{1{,}056{,}000}{0.33} = \frac{0.33P}{0.33}$$
$$3{,}200{,}000 = P$$

83. Use $V = \frac{4}{3}\pi r^3$ when $r = 3$.

$$V = \frac{4}{3}\pi \cdot 3^3$$
$$V \approx 113.1$$

Section 2.6 Practice Problems

1. Let x be the unknown percent.

$$22 = x \cdot 40$$
$$22 = 40x$$
$$\frac{22}{40} = \frac{40x}{40}$$
$$0.55 = x$$
$$55\% = x$$

The number 22 is 55% of 40.

2. Let x be the unknown number.

$$150 = 40\% \cdot x$$
$$150 = 0.4x$$
$$\frac{150}{0.4} = \frac{0.4x}{0.4}$$
$$375 = x$$

The number 150 is 40% of 375.

3. a. From the graph, we see 66% are for solely pleasure.

b. From the graph, 66% are for pleasure and 4% are for combined business/pleasure.
The sum is $66\% + 4\% = 70\%$.

c. Find 66% of 250.
$0.66(250) = 165$
We expect 165 people to be traveling solely for pleasure.

4. discount = percent · original price

$$= 40\% \cdot \$400$$
$$= 0.40 \cdot \$400$$
$$= \$160$$

new price = original price − discount

$$= \$400 − \$160$$
$$= \$240$$

The discount in price is $160 and the new price is $240.

5. increase = new − old = $200 − 120 = 80$

Let x be the percent increase.

$$80 = x \cdot 120$$
$$\frac{80}{120} = \frac{120x}{120}$$
$$0.667 \approx x$$
$$66.7\% \approx x$$

The percent increase is 66.7%.

6. Let x be the original price.

$$x − 0.20x = 46$$
$$0.8x = 46$$
$$\frac{0.8x}{0.8} = \frac{46}{0.8}$$
$$x = 57.5$$

The original price is $57.50.

7. Let x represent the liters of 20% solution.

	Number of Liters	Dye Strength	Amount
20% solution	x	20%	$0.2x$
50% solution	$6 − x$	50%	$0.5(6 − x)$
40% solution	6	40%	$0.4(6)$

$$0.2x + 0.5(6 − x) = 0.4(6)$$
$$0.2x + 3 − 0.5x = 2.4$$
$$−0.3x + 3 = 2.4$$
$$−0.3x + 3 − 3 = 2.4 − 3$$
$$−0.3x = −0.6$$
$$\frac{−0.3x}{−0.3} = \frac{−0.6}{−0.3}$$
$$x = 2$$

$6 − x = 6 − 2 = 4$

If 2 liters of 20% solution are mixed with 4 liters of 50% solution, the result is 6 liters of 40% solution.

Mental Math

1. no; $25\% + 25\% + 40\% \neq 100\%$

2. no; $30\% + 30\% + 30\% \neq 100\%$

3. yes; $25\% + 25\% + 25\% + 25\% = 100\%$

4. yes; $40\% + 50\% + 10\% = 100\%$

Exercise Set 2.6

1. Let x be the unknown number.
$$x = 16\% \cdot 70$$
$$x = 0.16 \cdot 70$$
$$x = 11.2$$
11.2 is 16% of 70.

3. Let x be the unknown percent.
$$28.6 = x \cdot 52$$
$$\frac{28.6}{52} = \frac{52x}{52}$$
$$0.55 = x$$
$$55\% = x$$
The number 28.6 is 55% of 52.

5. Let x be the unknown number.
$$45 = 25\% \cdot x$$
$$45 = 0.25 \cdot x$$
$$\frac{45}{0.25} = \frac{0.25x}{0.25}$$
$$180 = x$$
45 is 25% of 180.

7. From the graph, 4% of adults spend more than 121 minutes on the phone each day.

9. 37% of adults talk 16–60 minutes on the phone each day.
$37\% \cdot 27,000 = 0.37 \cdot 27,000 = 9990$
You would expect 9990 of the adults in Florence to talk 16–60 minutes each day.

11. discount = percent · original price
$$= 8\% \cdot \$18,500$$
$$= 0.08 \cdot \$18,500$$
$$= \$1480$$
new price = original price − discount
$$= \$18,500 - \$1480$$
$$= \$17,020$$
The discount is $1480 and the new price is $17,020.

13. $15\% \cdot 40.50 = 0.15 \cdot 40.5 = 6.075$
The tip is $6.08.
$40.5 + 6.08 = 46.58$
The total cost of the meal is $46.58.

15. percent increase $= \dfrac{\text{amount of increase}}{\text{original amount}}$
$$= \frac{380,000 - 220,000}{220,000}$$
$$= \frac{160,000}{220,000}$$
$$\approx 0.73$$
The number of complaints increased by 73%.

17. percent decrease $= \dfrac{\text{amount of decrease}}{\text{original amount}}$
$$= \frac{40 - 28}{40}$$
$$= \frac{12}{40}$$
$$= 0.3$$
The area decreased by 30%.

19. Let x represent the original price.
$$x - 25\% \cdot x = 78$$
$$x - 0.25x = 78$$
$$0.75x = 78$$
$$\frac{0.75x}{0.75} = \frac{78}{0.75}$$
$$x = 104$$
The original price of the shoes was $104.

21. Let x represent last year's salary.
$$x + 4\% \cdot x = 44,200$$
$$x + 0.04x = 44,200$$
$$1.04x = 44,200$$
$$\frac{1.04x}{1.04} = \frac{44,200}{1.04}$$
$$x = 42,500$$

Last year's salary was $42,500.

23. Let x represent the number of gallons of pure acid.

	Number of Gallons $\cdot$	Acid Strength $=$	Amount of Acid
Pure Acid	x	100%	$1x$
40% Acid Solution	2	40%	$0.4x$
70% Acid Solution Needed	$x + 2$	70%	$0.7(x + 2)$

The amount of acid being combined must be the same as that in the mixture.
$$x + 0.4x = 0.7(x + 2)$$
$$1.4x = 0.7x + 1.4$$
$$1.4x - 0.7x = 0.7x + 1.4 - 0.7x$$
$$0.7x = 1.4$$
$$\frac{0.7x}{0.7} = \frac{1.4}{0.7}$$
$$x = 2$$

Thus, 2 gallons of pure acid should be used.

25. Let x represent the number of pounds of coffee worth $7 a pound.

	Number of pounds $\cdot$	Cost per pound $=$	Value
$7/lb coffee	x	7	$7x$
$4/lb coffee	14	4	$4 \cdot 14 = 56$
$5/lb coffee wanted	$x + 14$	5	$5(x + 14)$

The value of the coffee being combined must be the same as the value of the mixture.

$$7x + 56 = 5(x + 14)$$
$$7x + 56 = 5x + 70$$
$$7x + 56 - 5x = 5x + 70 - 5x$$
$$2x + 56 = 70$$
$$2x + 56 - 56 = 70 - 56$$
$$2x = 14$$
$$\frac{2x}{2} = \frac{14}{2}$$
$$x = 7$$

7 pounds of the $4 a pound coffee should be used.

27. $23\% \cdot 20 = 0.23 \cdot 20 = 4.6$

29. Let x represent the unknown number.
$$40 = 80\% \cdot x$$
$$40 = 0.80 \cdot x$$
$$\frac{40}{0.8} = \frac{0.8x}{0.8}$$
$$50 = x$$
40 is 80% of 50.

31. Let x represent the unknown percent.
$$144 = x \cdot 480$$
$$\frac{144}{480} = \frac{480x}{480}$$
$$0.3 = x$$
$$30\% = x$$
144 is 30% of 480.

33. From the graph, it appears that 71% of the population of Fairbanks, Alaska shops by catalog.

35. $65\% \cdot 270{,}951 = 0.65 \cdot 270{,}951 \approx 176{,}118$
We predict 176,118 catalog shoppers live in Anchorage.

37.

	Ford Motor Company Model Year 2004 Vehicle Sales Worldwide	
	Thousands of vehicles	Percent of Total (Rounded to nearest percent)
North America	3277	$\dfrac{3277}{5462} \approx 0.59996 \approx 60\%$
Europe	1474	$\dfrac{1474}{5462} \approx 0.26986 \approx 27\%$
Asia-Pacific	328	$\dfrac{328}{5462} \approx 0.06005 \approx 6\%$
Rest of the world	383	$\dfrac{383}{5462} \approx 0.07012 \approx 7\%$
Total	5462	

39. percent increase $= \dfrac{\text{amount of increase}}{\text{original amount}}$

$$= \frac{70-40}{40}$$

$$= \frac{30}{40}$$

$$= 0.75$$

The price increased by 75%.

41. Let x represent the amount Charles paid for the car.

$$x + 20\% \cdot x = 4680$$
$$x + 0.20x = 4680$$
$$1.2x = 4680$$
$$\frac{1.2x}{1.2} = \frac{4680}{1.2}$$
$$x = 3900$$

Charles paid $3900 for the car.

43. percent increase $= \dfrac{\text{amount of increase}}{\text{original amount}}$

$$= \frac{144-36}{36}$$

$$= \frac{108}{36}$$

$$= 3$$

The area increased by 300%.

45. Markup $= 5\% \cdot 2.20 = 0.05 \cdot 2.2 = 0.11$
New price $= 2.20 + 0.11 = 2.31$
The markup is $0.11 and the new price is $2.31.

47. Let x be the ounces of alloy that is 20% copper.

	ounces	concentration	amount
20% copper	x	20%	$0.2x$
50% copper	200	50%	$0.5(200)$
30% copper	$200 + x$	30%	$0.3(200 + x)$

The amount of copper being combined must be the same as that in the mixture.
$$0.2x + 0.5(200) = 0.3(200 + x)$$
$$0.2x + 100 = 60 + 0.3x$$
$$0.2x + 100 - 0.2x = 60 + 0.3x - 0.2x$$
$$100 = 60 + 0.1x$$
$$100 - 60 = 60 + 0.1x - 60$$
$$40 = 0.1x$$
$$\frac{40}{0.1} = \frac{0.1x}{0.1}$$
$$400 = x$$
Thus 400 ounces should be used.

49. percent decrease $= \dfrac{\text{amount of decrease}}{\text{original amount}}$
$$= \frac{151 - 73}{151}$$
$$= \frac{78}{151}$$
$$\approx 0.517$$
The number of decisions by the Supreme Court decreased by 51.7%.

51. Let x be the prior number of employees.
$$x - 0.35x = 78$$
$$0.65x = 78$$
$$\frac{0.65x}{0.65} = \frac{78}{0.65}$$
$$x = 120$$
There were 120 employees prior to the layoffs.

53. decrease $= 25\% \cdot 256 = 0.25 \cdot 256 = 64$
$256 - 64 = 192$
The price of the coat decreased by $64. The sale price was $192.

55. increase $= 48\% \cdot 577 = 0.48 \cdot 577 = 276.96$
$577 + 276.96 = 853.96$
The Naga Jolokia pepper measures 854 thousand Scoville units.

57. $42\% \cdot 860 = 0.42 \cdot 860 = 361.2$
You would expect 361 students to rank flexible hours as their top priority.

59. Let x be the ounces of self-tanning lotion.

	ounces	cost ($)	value
self-tanning	x	3	$3x$
everyday	800	0.30	$0.3(800)$
experimental	$800 + x$	1.20	$1.2(800 + x)$

The value of those being combined must be the same as the value of the mixture.
$$3x + 0.3(800) = 1.2(800 + x)$$
$$3x + 240 = 960 + 1.2x$$
$$3x + 240 - 1.2x = 960 + 1.2x - 1.2x$$
$$1.8x + 240 = 960$$
$$1.8x + 240 - 240 = 960 - 240$$
$$1.8x = 720$$
$$\frac{1.8x}{1.8} = \frac{720}{1.8}$$
$$x = 400$$
Therefore, 400 ounces of the self-tanning lotion should be used.

61. $-5 > -7$ since -5 is to the right of -7 on a number line.

63. $|-5| = 5$
$-(-5) = 5$
$|-5| = -(-5)$ since $5 = 5$.

65. $(-3)^2 = (-3)(-3) = 9$
$-3^2 = -(3 \cdot 3) = -9$
Since $9 > -9$, $(-3)^2 > -3^2$.

67. no; answers may vary

69. 230 mg is what percent of 2400 mg?
Let x represent the unknown percent.
$$x \cdot 2400 = 230$$
$$\frac{2400x}{2400} = \frac{230}{2400}$$
$$x = 0.0958\overline{3}$$
This food contains 9.6% of the daily value of sodium in one serving.

71. 35 is what percent of 130? Let x be the unknown percent.
$$35 = x \cdot 130$$
$$\frac{35}{130} = \frac{130x}{130}$$
$$0.269 \approx x$$
The percent calories from fat is 26.9%. Yes, this food satisfies the recommendation since $26.9\% \le 30\%$.

73. 12 g $\cdot$ 4 calories/gram = 48 calories
48 of the 280 calories come from protein.
$$\frac{48}{280} \approx 0.171$$
17.1% of the calories in this food come from protein.

Section 2.7 Practice Problems

1. $x \ge -2$
$-5\,-4\,-3\,-2\,-1\ \ 0\ \ 1\ \ 2\ \ 3\ \ 4\ \ 5$

2. $5 > x$ or $x < 5$
$-5\,-4\,-3\,-2\,-1\ \ 0\ \ 1\ \ 2\ \ 3\ \ 4\ \ 5$

3. $-3 \le x < 1$
$-5\,-4\,-3\,-2\,-1\ \ 0\ \ 1\ \ 2\ \ 3\ \ 4\ \ 5$

4. $x - 6 \ge -11$
$$x - 6 + 6 \ge -11 + 6$$
$$x \ge -5$$
$-5\,-4\,-3\,-2\,-1\ \ 0\ \ 1\ \ 2\ \ 3\ \ 4\ \ 5$

5. $-3x \le 12$
$$\frac{-3x}{-3} \ge \frac{12}{-3}$$
$$x \ge -4$$
$-5\,-4\,-3\,-2\,-1\ \ 0\ \ 1\ \ 2\ \ 3\ \ 4\ \ 5$

6. $5x > -20$
$$\frac{5x}{5} > \frac{-20}{5}$$
$$x > -4$$
$-5\,-4\,-3\,-2\,-1\ \ 0\ \ 1\ \ 2\ \ 3\ \ 4\ \ 5$

7. $-3x + 11 \le -13$
$$-3x + 11 - 11 \le -13 - 11$$
$$-3x \le -24$$
$$\frac{-3x}{-3} \ge \frac{-24}{-3}$$
$$x \ge 8$$
$\{x \mid x \ge 8\}$
$-10\,-8\,-6\,-4\,-2\ \ 0\ \ 2\ \ 4\ \ 6\ \ 8\ \ 10$

8. $2x - 3 > 4(x - 1)$
$$2x - 3 > 4x - 4$$
$$2x - 3 - 4x > 4x - 4 - 4x$$
$$-2x - 3 > -4$$
$$-2x - 3 + 3 > -4 + 3$$
$$-2x > -1$$
$$\frac{-2x}{-2} < \frac{-1}{-6}$$
$$x < \frac{1}{2}$$
$\left\{ x \mid x < \dfrac{1}{2} \right\}$
$-5\,-4\,-3\,-2\,-1\ \ 0\ \ 1\ \ 2\ \ 3\ \ 4\ \ 5$

9.
$$3(x+5)-1 \geq 5(x-1)+7$$
$$3x+15-1 \geq 5x-5+7$$
$$3x+14 \geq 5x+2$$
$$3x+14-5x \geq 5x+2-5x$$
$$-2x+14 \geq 2$$
$$-2x+14-14 \geq 2-14$$
$$-2x \geq -12$$
$$\frac{-2x}{-2} \leq \frac{-12}{-2}$$
$$x \leq 6$$
$$\{x | x \leq 6\}$$

10. Let x be the unknown number.
$$35-2x > 15$$
$$35-2x-35 > 15-35$$
$$-2x > -20$$
$$\frac{-2x}{-2} < \frac{-20}{-2}$$
$$x < 10$$

11. Let x represent the minimum sales.
$$600+0.04x \geq 3000$$
$$0.04x \geq 2400$$
$$x \geq 60,000$$
Alex must have minimum sales of \$60,000.

Mental Math

1. $5x > 10$
$x > 2$

2. $4x < 20$
$x < 5$

3. $2x \geq 16$
$x \geq 8$

4. $9x \leq 63$
$x \leq 7$

5. $x \geq -3$
-5 is not a solution.

6. $x < 6$
$|-6| = 6$ is not a solution.

7. $x < 4.01$
4.1 is not a solution.

8. $x \geq -3$
-4 is not a solution.

Exercise Set 2.7

1.

3.

5.

7.

9.

11.

13.
$$x-2 \geq -7$$
$$x-2+2 \geq -7+2$$
$$x \geq -5$$
$$\{x | x \geq -5\}$$

15.
$$-9+y < 0$$
$$9-9+y < 9+0$$
$$y < 9$$
$$\{y | y < 9\}$$

17.
$$3x-5 > 2x-8$$
$$3x-5-2x > 2x-8-2x$$
$$x-5 > -8$$
$$x-5+5 > -8+5$$
$$x > -3$$
$$\{x | x > -3\}$$

19.
$$4x - 1 \leq 5x - 2x$$
$$4x - 1 \leq 3x$$
$$4x - 1 - 4x \leq 3x - 4x$$
$$-1 \leq -x$$
$$\frac{-1}{-1} \geq \frac{-x}{-1}$$
$$1 \geq x \text{ or } x \leq 1$$
$$\{x | x \leq 1\}$$

21. $2x < -6$
$$\frac{2x}{2} < \frac{-6}{2}$$
$$x < -3$$
$$\{x | x < -3\}$$

23. $-8x \leq 16$
$$\frac{-8x}{-8} \geq \frac{16}{-8}$$
$$x \geq -2$$
$$\{x | x \geq -2\}$$

25.
$$-x > 0$$
$$(-1)(-x) < (-1)(0)$$
$$x < 0$$
$$\{x | x < 0\}$$

27. $\frac{3}{4}y \geq -2$
$$\frac{4}{3} \cdot \frac{3}{4}y \geq \frac{4}{3} \cdot (-2)$$
$$y \geq -\frac{8}{3}$$
$$\left\{ y \middle| y \geq -\frac{8}{3} \right\}$$

29. $-0.6y < -1.8$
$$\frac{-0.6y}{-0.6} > \frac{-1.8}{-0.6}$$
$$y > 3$$
$$\{y | y > 3\}$$

31.
$$-8 < x + 7$$
$$-8 - 7 < x + 7 - 7$$
$$-15 < x$$
$$\{x | x > -15\}$$

33. $7(x + 1) - 6x \geq -4$
$$7x + 7 - 6x \geq -4$$
$$x + 7 \geq -4$$
$$x + 7 - 7 \geq -4 - 7$$
$$x \geq -11$$
$$\{x | x \geq -11\}$$

35. $4x > 1$
$$\frac{4x}{4} > \frac{1}{4}$$
$$x > \frac{1}{4}$$
$$\left\{ x \middle| x > \frac{1}{4} \right\}$$

37. $-\frac{2}{3}y \leq 8$
$$-\frac{3}{2}\left(-\frac{2}{3}y\right) \geq -\frac{3}{2}(8)$$
$$y \geq -12$$
$$\{y | y \geq -12\}$$

39. $4(2z + 1) < 4$
$$8z + 4 < 4$$
$$8z + 4 - 4 < 4 - 4$$
$$8z < 0$$
$$\frac{8z}{8} < \frac{0}{8}$$
$$z < 0$$
$$\{z | z < 0\}$$

41.
$$3x - 7 < 6x + 2$$
$$3x - 7 - 3x < 6x + 2 - 3x$$
$$-7 < 3x + 2$$
$$-7 - 2 < 3x + 2 - 2$$
$$-9 < 3x$$
$$\frac{-9}{3} < \frac{3x}{3}$$
$$-3 < x$$
$$\{x \mid x > -3\}$$

43.
$$5x - 7x \le x + 2$$
$$-2x \le x + 2$$
$$-2x - x \le x + 2 - x$$
$$-3x \le 2$$
$$\frac{-3x}{-3} \ge \frac{2}{-3}$$
$$x \ge -\frac{2}{3}$$
$$\left\{ x \mid x \ge -\frac{2}{3} \right\}$$

45.
$$-6x + 2 \ge 2(5 - x)$$
$$-6x + 2 \ge 10 - 2x$$
$$-6x + 2 + 6x \ge 10 - 2x + 6x$$
$$2 \ge 10 + 4x$$
$$2 - 10 \ge 10 + 4x - 10$$
$$-8 \ge 4x$$
$$\frac{-8}{4} \ge \frac{4x}{4}$$
$$-2 \ge x$$
$$\{x \mid x \le -2\}$$

47.
$$3(x - 5) < 2(2x - 1)$$
$$3x - 15 < 4x - 2$$
$$3x - 15 - 3x < 4x - 2 - 3x$$
$$-15 < x - 2$$
$$-15 + 2 < x - 2 + 2$$
$$-13 < x$$
$$\{x \mid x > -13\}$$

49.
$$4(3x - 1) \le 5(2x - 4)$$
$$12x - 4 \le 10x - 20$$
$$12x - 4 - 10x \le 10x - 20 - 10x$$
$$2x - 4 \le -20$$
$$2x - 4 + 4 \le -20 + 4$$
$$2x \le -16$$
$$\frac{2x}{2} \le \frac{-16}{2}$$
$$x \le -8$$
$$\{x \mid x \le -8\}$$

51.
$$3(x + 2) - 6 > -2(x - 3) + 14$$
$$3x + 6 - 6 > -2x + 6 + 14$$
$$3x > -2x + 20$$
$$3x + 2x > -2x + 20 + 2x$$
$$5x > 20$$
$$\frac{5x}{5} > \frac{20}{5}$$
$$x > 4$$
$$\{x \mid x > 4\}$$

53.
$$-5(1 - x) + x \le -(6 - 2x) + 6$$
$$-5 + 5x + x \le -6 + 2x + 6$$
$$-5 + 6x \le 2x$$
$$-5 + 6x - 6x \le 2x - 6x$$
$$-5 \le -4x$$
$$\frac{-5}{-4} \ge \frac{-4x}{-4}$$
$$\frac{5}{4} \ge x$$
$$\left\{ x \mid x \le \frac{5}{4} \right\}$$

55.
$$\frac{1}{4}(x+4) < \frac{1}{5}(2x+3)$$
$$20 \cdot \frac{1}{4}(x+4) < 20 \cdot \frac{1}{5}(2x+3)$$
$$5(x+4) < 4(2x+3)$$
$$5x+20 < 8x+12$$
$$5x+20-5x < 8x+12-5x$$
$$20 < 3x+12$$
$$20-12 < 3x+12-12$$
$$8 < 3x$$
$$\frac{8}{3} < \frac{3x}{3}$$
$$\frac{8}{3} < x$$
$$\left\{ x \middle| x > \frac{8}{3} \right\}$$

57.
$$-5x+4 \le -4(x-1)$$
$$-5x+4 \le -4x+4$$
$$-5x+4+4x \le -4x+4+4x$$
$$-x+4 \le 4$$
$$-x+4-4 \le 4-4$$
$$-x \le 0$$
$$-1(-x) \ge -1(0)$$
$$x \ge 0$$
$$\{x|x \ge 0\}$$

59. Let x be the number.
$$2x+6 > -14$$
$$2x+6-6 > -14-6$$
$$2x > -20$$
$$\frac{2x}{2} > \frac{-20}{2}$$
$$x > -10$$
All numbers greater than -10 make this statement true.

61. Use $P = 2l + 2w$ when $w = 15$ and $P \le 100$.
$$2l + 2(15) \le 100$$
$$2l + 30 \le 100$$
$$2l + 30 - 30 \le 100 - 30$$
$$2l \le 70$$
$$\frac{2l}{2} \le \frac{70}{2}$$
$$l \le 35$$
The maximum length of the rectangle is 35 cm.

63. Let x be the score in his third game.
$$\frac{146+201+x}{3} \ge 180$$
$$\frac{347+x}{3} \ge 180$$
$$3 \cdot \frac{347+x}{3} \ge 3 \cdot 180$$
$$347+x \ge 540$$
$$347+x-347 \ge 540-347$$
$$x \ge 193$$
He must bowl at least 193 on the third game.

65. Let x represent the number of people. Then the cost is $50 + 34x$.
$$50 + 34x \le 3000$$
$$50 + 34x - 50 \le 3000 - 50$$
$$34x \le 2950$$
$$\frac{34x}{34} \le \frac{2950}{34}$$
$$x \le \frac{2950}{34} \approx 86.76$$
They can invite at most 86 people.

67. Let x represent the number of minutes.
$$5.8x \ge 200$$
$$\frac{5.8x}{5.8} \ge \frac{200}{5.8}$$
$$x \ge \frac{200}{5.8} \approx 35$$
The person must walk at least 35 minutes.

69. $3^4 = 3 \cdot 3 \cdot 3 \cdot 3 = 81$

71. $1^8 = 1 \cdot 1 \cdot 1 \cdot 1 \cdot 1 \cdot 1 \cdot 1 \cdot 1 = 1$

73. $\left(\frac{7}{8}\right)^2 = \left(\frac{7}{8}\right)\left(\frac{7}{8}\right) = \frac{49}{64}$

75. There were about 120 Krispy Kreme locations in 1998.

77. The greatest increase occurred between 2003 and 2004.

79. During 2001 the number of Krispy Kreme locations rose above 200.

81. Since $3 < 5$, $3(-4) > 5(-4)$.

83. If $m \le n$, then $-2m \ge -2n$.

85. Reverse the direction of the inequality symbol when multiplying or dividing by a negative number.

87. Let x be the score on his final exam. Since the final counts as two tests, his final course average is $\dfrac{75 + 83 + 85 + 2x}{5}$.

$$\frac{75 + 83 + 85 + 2x}{5} \ge 80$$
$$\frac{243 + 2x}{5} \ge 80$$
$$5\left(\frac{243 + 2x}{5}\right) \ge 5(80)$$
$$243 + 2x \ge 400$$
$$243 + 2x - 243 \ge 400 - 243$$
$$2x \ge 157$$
$$\frac{2x}{2} \ge \frac{157}{2}$$
$$x \ge 78.5$$

His final exam score must be at least 78.5 for him to get a B.

The Bigger Picture

1.
$$-5x = 15$$
$$\frac{-5x}{-5} = \frac{15}{-5}$$
$$x = -3$$

2.
$$-5x > 15$$
$$\frac{-5x}{-5} < \frac{15}{-5}$$
$$x < -3$$
$$\{x | x < -3\}$$

3.
$$9y - 14 = -12$$
$$9y - 14 + 14 = -12 + 14$$
$$9y = 2$$
$$\frac{9y}{9} = \frac{2}{9}$$
$$y = \frac{2}{9}$$

4.
$$9x - 3 = 5x - 4$$
$$9x - 3 + 3 = 5x - 4 + 3$$
$$9x = 5x - 1$$
$$9x - 5x = 5x - 1 - 5x$$
$$4x = -1$$
$$\frac{4x}{4} = \frac{-1}{4}$$
$$x = -\frac{1}{4}$$

5.
$$4(x - 2) \le 5x + 7$$
$$4x - 8 \le 5x + 7$$
$$4x - 8 - 5x \le 5x + 7 - 5x$$
$$-x - 8 \le 7$$
$$-x - 8 + 8 \le 7 + 8$$
$$-x \le 15$$
$$\frac{-x}{-1} \ge \frac{15}{-1}$$
$$x \ge -15$$
$$\{x | x \ge -15\}$$

6.
$$5(4x - 1) = 2(10x - 1)$$
$$20x - 5 = 20x - 2$$
$$20x - 5 - 20x = 20x - 2 - 20x$$
$$-5 = -2$$

Since the statement $-5 = -2$ is false, there is no solution.

7.
$$-5.4 = 0.6x - 9.6$$
$$-5.4 + 9.6 = 0.6x - 9.6 + 9.6$$
$$4.2 = 0.6x$$
$$\frac{4.6}{0.6} = \frac{0.6x}{0.6}$$
$$7 = x$$

8. $\dfrac{1}{3}(x-4) < \dfrac{1}{4}(x+7)$

$12 \cdot \dfrac{1}{3}(x-4) < 12 \cdot \dfrac{1}{4}(x+7)$

$4(x-4) < 3(x+7)$

$4x-16 < 3x+21$

$4x-16-3x < 3x+21-3x$

$x-16 < 21$

$x-16+16 < 21+16$

$x < 37$

$\{x|x<37\}$

9. $3y-5(y-4) = -2(y-10)$

$3y-5y+20 = -2y+20$

$-2y+20 = -2y+20$

$-2y+20+2y = -2y+20+2y$

$20 = 20$

Since the statement $20 = 20$ is an identity, all real numbers satisfy the equation.

10. $\dfrac{7(x-1)}{3} = \dfrac{2(x+1)}{5}$

$15\left[\dfrac{7(x-1)}{3}\right] = 15\left[\dfrac{2(x+1)}{5}\right]$

$35(x-1) = 6(x+1)$

$35x-35 = 6x+6$

$35x-35-6x = 6x+6-6x$

$29x-35 = 6$

$29x-35+35 = 6+35$

$29x = 41$

$\dfrac{29x}{29} = \dfrac{41}{29}$

$x = \dfrac{41}{29}$

Chapter 2 Vocabulary Check

1. A <u>linear equation in one variable</u> can be written in the form $ax + b = c$.

2. Equations that have the same solution are called <u>equivalent equations</u>.

3. An equation that describes a known relationship among quantities is called a <u>formula</u>.

4. A <u>linear inequality in one variable</u> can be written in the form $ax + b < c$, (or $>$, $\le$, $\ge$).

5. The solution(s) to the equation $x + 5 = x + 5$ is/are <u>all real numbers</u>.

6. The solution(s) to the equation $x + 5 = x + 4$ is/are <u>no solution</u>.

7. If both sides of an inequality are multiplied or divided by the same positive number, the direction of the inequality symbol is <u>the same</u>.

8. If both sides of an inequality are multiplied by the same negative number, the direction of the inequality symbol is <u>reversed</u>.

Chapter 2 Review

1. $8x+4 = 9x$

$8x+4-8x = 9x-8x$

$4 = x$

2. $5y-3 = 6y$

$5y-3-5y = 6y-5y$

$-3 = y$

3. $\dfrac{2}{7}x + \dfrac{5}{7}x = 6$

$\dfrac{7}{7}x = 6$

$1x = 6$

$x = 6$

4. $3x-5 = 4x+1$

$3x-5-3x = 4x+1-3x$

$-5 = x+1$

$-5-1 = x+1-1$

$-6 = x$

5. $2x-6 = x-6$

$2x-6-x = x-6-x$

$x-6 = -6$

$x-6+6 = -6+6$

$x = 0$

6.
$$4(x+3) = 3(1+x)$$
$$4x + 12 = 3 + 3x$$
$$4x + 12 - 3x = 3 + 3x - 3x$$
$$12 + x = 3$$
$$-12 + 12 + x = -12 + 3$$
$$x = -9$$

7.
$$6(3 + n) = 5(n - 1)$$
$$18 + 6n = 5n - 5$$
$$18 + 6n - 5n = 5n - 5 - 5n$$
$$18 + n = -5$$
$$-18 + 18 + n = -18 - 5$$
$$n = -23$$

8.
$$5(2 + x) - 3(3x + 2) = -5(x - 6) + 2$$
$$10 + 5x - 9x - 6 = -5x + 30 + 2$$
$$-4x + 4 = -5x + 32$$
$$5x - 4x + 4 = 5x - 5x + 32$$
$$x + 4 = 32$$
$$x + 4 - 4 = 32 - 4$$
$$x = 28$$

9. If the sum is 10 and one number is x, then the other number is $10 - x$. The choice is b.

10. Since Mandy is 5 inches taller than Melissa, and x represents Mandy's height, then $x - 5$ represents Melissa's height. The choice is a.

11. Complementary angles sum to 90°. The complement of angle x is $90 - x$. The choice is b.

12. Supplementary angles sum to 180°. The supplement to $(x + 5)°$ is
$180 - (x + 5) = 180 - x - 5 = 175 - x$.
The choice is c.

13.
$$\frac{3}{4}x = -9$$
$$\frac{4}{3} \cdot \frac{3}{4}x = \frac{4}{3} \cdot (-9)$$
$$x = -12$$

14.
$$\frac{x}{6} = \frac{2}{3}$$
$$6 \cdot \frac{x}{6} = 6 \cdot \frac{2}{3}$$
$$x = 4$$

15.
$$-5x = 0$$
$$\frac{-5x}{-5} = \frac{0}{-5}$$
$$x = 0$$

16.
$$-y = 7$$
$$\frac{-y}{-1} = \frac{7}{-1}$$
$$y = -7$$

17.
$$0.2x = 0.15$$
$$20x = 15$$
$$\frac{20x}{20} = \frac{15}{20}$$
$$x = 0.75$$

18.
$$\frac{-x}{3} = 1$$
$$-3\left(\frac{-x}{3}\right) = -3(1)$$
$$x = -3$$

19.
$$-3x + 1 = 19$$
$$-3x + 1 - 1 = 19 - 1$$
$$-3x = 18$$
$$\frac{-3x}{-3} = \frac{18}{-3}$$
$$x = -6$$

20.
$$5x + 25 = 20$$
$$5x + 25 - 25 = 20 - 25$$
$$5x = -5$$
$$\frac{5x}{5} = \frac{-5}{5}$$
$$x = -1$$

21. $7(x-1)+9=5x$

$7x-7+9=5x$

$7x+2=5x$

$-7x+7x+2=-7x+5x$

$2=-2x$

$\dfrac{2}{-2}=\dfrac{-2x}{-2}$

$-1=x$

22. $7x-6=5x-3$

$7x-6-5x=5x-3-5x$

$2x-6=-3$

$2x-6+6=-3+6$

$2x=3$

$\dfrac{2x}{2}=\dfrac{3}{2}$

$x=\dfrac{3}{2}$ or $1\dfrac{1}{2}$

23. $-5x+\dfrac{3}{7}=\dfrac{10}{7}$

$7\left(-5x+\dfrac{3}{7}\right)=7\cdot\dfrac{10}{7}$

$-35x+3=10$

$-35x+3-3=10-3$

$-35x=7$

$\dfrac{-35x}{-35}=\dfrac{7}{-35}$

$x=-\dfrac{1}{5}$

24. $5x+x=9+4x-1+6$

$6x=4x+14$

$6x-4x=4x+14-4x$

$2x=14$

$\dfrac{2x}{2}=\dfrac{14}{2}$

$x=7$

25. Let x be the first integer. Then $x + 1$ and $x + 2$ are the next two consecutive integers. Their sum is $x + x + 1 + x + 2 = 3x + 3$.

26. Let x be the first even integer. Then $x + 2$, $x + 4$, and $x + 6$ are the 2nd, 3rd, and 4th

consecutive even integers. The sum of the first and fourth is $x + x + 6 = 2x + 6$.

27. $\dfrac{5}{3}x+4=\dfrac{2}{3}x$

$3\left(\dfrac{5}{3}x+4\right)=3\left(\dfrac{2}{3}x\right)$

$5x+12=2x$

$5x+12-5x=2x-5x$

$12=-3x$

$\dfrac{12}{-3}=\dfrac{-3x}{-3}$

$-4=x$

28. $\dfrac{7}{8}x+1=\dfrac{5}{8}x$

$8\left(\dfrac{7}{8}x+1\right)=8\left(\dfrac{5}{8}x\right)$

$7x+8=5x$

$7x+8-7x=5x-7x$

$8=-2x$

$\dfrac{8}{-2}=\dfrac{-2x}{-2}$

$-4=x$

29. $-(5x+1)=-7x+3$

$-5x-1=-7x+3$

$-5x-1+7x=-7x+3+7x$

$2x-1=3$

$2x-1+1=3+1$

$2x=4$

$\dfrac{2x}{2}=\dfrac{4}{2}$

$x=2$

30. $-4(2x+1)=-5x+5$

$-8x-4=-5x+5$

$-8x-4+8x=-5x+5+8x$

$-4=3x+5$

$-4-5=3x+5-5$

$-9=3x$

$\dfrac{-9}{3}=\dfrac{3x}{3}$

$-3=x$

31.
$$-6(2x-5) = -3(9+4x)$$
$$-12x+30 = -27-12x$$
$$12x-12x+30 = 12x-27-12x$$
$$30 = -27$$

Since the statement $30 = -27$ is false, the equation has no solution.

32.
$$3(8y-1) = 6(5+4y)$$
$$24y-3 = 30+24y$$
$$24y-3-24y = 30+24y-24y$$
$$-3 = 30$$

Since the statement $-3 = 30$ is false, the equation has no solution.

33.
$$\frac{3(2-z)}{5} = z$$
$$5\left[\frac{3(2-z)}{5}\right] = 5\cdot z$$
$$3(2-z) = 5z$$
$$6-3z = 5z$$
$$6-3z+3z = 5z+3z$$
$$6 = 8z$$
$$\frac{6}{8} = \frac{8z}{8}$$
$$\frac{3}{4} = z$$

34.
$$\frac{4(n+2)}{5} = -n$$
$$5\left[\frac{4(n+2)}{5}\right] = 5(-n)$$
$$4(n+2) = -5n$$
$$4n+8 = -5n$$
$$4n+8-4n = -5n-4n$$
$$8 = -9n$$
$$\frac{8}{-9} = \frac{-9n}{-9}$$
$$-\frac{8}{9} = n$$

35.
$$0.5(2n-3) - 0.1 = 0.4(6+2n)$$
$$5(2n-3) - 1 = 4(6+2n)$$
$$10n-15-1 = 24+8n$$
$$10n-16 = 24+8n$$
$$10n-16-8n = 24+8n-8n$$
$$2n-16 = 24$$
$$2n-16+16 = 24+16$$
$$2n = 40$$
$$\frac{2n}{2} = \frac{40}{2}$$
$$n = 20$$

36.
$$-9-5a = 3(6a-1)$$
$$-9-5a = 18a-3$$
$$9-5a+5a = 18a-3+5a$$
$$-9 = 23a-3$$
$$-9+3 = 23a-3+3$$
$$-6 = 23a$$
$$\frac{-6}{23} = \frac{23a}{23}$$
$$-\frac{6}{23} = a$$

37.
$$\frac{5(c+1)}{6} = 2c-3$$
$$6\left[\frac{5(c+1)}{6}\right] = 6(2c-3)$$
$$5(c+1) = 6(2c-3)$$
$$5c+5 = 12c-18$$
$$5c+5-5c = 12c-18-5c$$
$$5 = 7c-18$$
$$5+18 = 7c-18+18$$
$$23 = 7c$$
$$\frac{23}{7} = \frac{7c}{7}$$
$$\frac{23}{7} = c$$

38.
$$\frac{2(8-a)}{3} = 4-4a$$
$$3\left[\frac{2(8-a)}{3}\right] = 3(4-4a)$$
$$2(8-a) = 3(4-4a)$$
$$16-2a = 12-12a$$
$$16-2a+12a = 12-12a+12a$$
$$16+10a = 12$$
$$16+10a-16 = 12-16$$
$$10a = -4$$
$$\frac{10a}{10} = \frac{-4}{10}$$
$$a = -\frac{2}{5}$$

39.
$$200(70x-3560) = -179(150x-19,300)$$
$$14,000x-712,000 = -26,850x+3,454,700$$
$$14,000x-712,000+26,850x = -26,850x+3,454,700+26,850x$$
$$40,850x-712,000 = 3,454,700$$
$$40,850x-712,000+712,000 = 3,454,700+712,000$$
$$40,850x = 4,166,700$$
$$\frac{40,850x}{40,850} = \frac{4,166,700}{40,850}$$
$$x = 102$$

40.
$$1.72y-0.04y = 0.42$$
$$172y-4y = 42$$
$$168y = 42$$
$$\frac{168y}{168} = \frac{42}{168}$$
$$y = 0.25$$

41. Let x be the length of the side of the square base. Then the height is $10x + 50.5$. The sum is 7327.
$$x+10x+50.5 = 7327$$
$$11x+50.5 = 7327$$
$$11x+50.5-50.5 = 7327-50.5$$
$$11x = 7276.5$$
$$\frac{11x}{11} = \frac{7276.5}{11}$$
$$x = 661.5$$
$$10x+50.5 = 10(661.5)+50.5$$
$$= 6615+50.5$$
$$= 6665.5$$
The height is 6665.5 inches.

42. Let x be the length of the short piece. Then $2x$ is the length of the long piece. The lengths sum to 12.

$$x + 2x = 12$$
$$3x = 12$$
$$\frac{3x}{3} = \frac{12}{3}$$
$$x = 4$$
$$2x = 2(4) = 8$$

The short piece is 4 feet and the long piece is 8 feet.

43. Let x be the number of Keebler plants. Then $2x - 1$ is the number of Kellogg plants. The total number of plants is 53.

$$x + 2x - 1 = 53$$
$$3x - 1 = 53$$
$$3x - 1 + 1 = 53 + 1$$
$$3x = 54$$
$$\frac{3x}{3} = \frac{54}{3}$$
$$x = 18$$
$$2x - 1 = 2(18) - 1 = 36 - 1 = 35$$

Keebler has 18 plants and Kellogg has 35 plants.

44. Let x be the first integer. Then $x + 1$ and $x + 2$ are the next two consecutive integers. Their sum is -114.

$$x + x + 1 + x + 2 = -114$$
$$3x + 3 = -114$$
$$3x + 3 - 3 = -114 - 3$$
$$3x = -117$$
$$\frac{3x}{3} = \frac{-117}{3}$$
$$x = -39$$
$$x + 1 = -39 + 1 = -38$$
$$x + 2 = -39 + 2 = -37$$

The integers are -39, -38, and -37.

45.
$$\frac{x}{3} = x - 2$$
$$3 \cdot \frac{x}{3} = 3(x - 2)$$
$$x = 3x - 6$$
$$x - 3x = 3x - 6 - 3x$$
$$-2x = -6$$
$$\frac{-2x}{-2} = \frac{-6}{-2}$$
$$x = 3$$

The number is 3.

46.
$$2(x + 6) = -x$$
$$2x + 12 = -x$$
$$-2x + 2x + 12 = -2x - x$$
$$12 = -3x$$
$$\frac{12}{-3} = \frac{-3x}{-3}$$
$$-4 = x$$

The number is -4.

47. Use $P = 2l + 2w$ when $P = 46$ and $l = 14$.
$$P = 2l + 2w$$
$$46 = 2(14) + 2w$$
$$46 = 28 + 2w$$
$$46 - 28 = 28 + 2w - 28$$
$$18 = 2w$$
$$\frac{18}{2} = \frac{2w}{2}$$
$$9 = w$$

48. Use $V = lwh$ when $V = 192$, $l = 8$, and $w = 6$.
$$V = lwh$$
$$192 = 8 \cdot 6 \cdot h$$
$$192 = 48h$$
$$\frac{192}{48} = \frac{48h}{48}$$
$$4 = h$$

49.
$$y = mx + b$$
$$y - b = mx + b - b$$
$$y - b = mx$$
$$\frac{y-b}{x} = \frac{mx}{x}$$
$$\frac{y-b}{x} = m$$

50.
$$r = vst - 5$$
$$r + 5 = vst - 5 + 5$$
$$r + 5 = vst$$
$$\frac{r+5}{vt} = \frac{vst}{vt}$$
$$\frac{r+5}{vt} = s$$

51.
$$2y - 5x = 7$$
$$-2y + 2y - 5x = -2y + 7$$
$$-5x = -2y + 7$$
$$\frac{-5x}{-5} = \frac{-2y+7}{-5}$$
$$x = \frac{2y-7}{5}$$

52.
$$3x - 6y = -2$$
$$-3x + 3x - 6y = -3x - 2$$
$$-6y = -3x - 2$$
$$\frac{-6y}{-6} = \frac{-3x-2}{-6}$$
$$y = \frac{3x+2}{6}$$

53.
$$C = \pi D$$
$$\frac{C}{D} = \frac{\pi D}{D}$$
$$\frac{C}{D} = \pi$$

54.
$$C = 2\pi r$$
$$\frac{C}{2r} = \frac{2\pi r}{2r}$$
$$\frac{C}{2r} = \pi$$

55. Use $V = lwh$ when $V = 900$, $l = 20$ and $h = 3$.
$$V = lwh$$
$$900 = 20 \cdot w \cdot 3$$
$$900 = 60w$$
$$\frac{900}{60} = \frac{60w}{60}$$
$$15 = w$$
The width is 15 meters.

56. Let x be the width. Then the length is $x + 6$.
Use $P = 2 \cdot \text{length} + 2 \cdot \text{width}$ when $P = 60$.
$$P = 2 \cdot \text{length} + 2 \cdot \text{width}$$
$$60 = 2(x + 6) + 2x$$
$$60 = 2x + 12 + 2x$$
$$60 = 4x + 12$$
$$60 - 12 = 4x + 12 - 12$$
$$48 = 4x$$
$$\frac{48}{4} = \frac{4x}{4}$$
$$12 = x$$
$$x + 6 = 12 + 6 = 18$$
The dimensions of the billboard are 12 feet by 18 feet.

57. Use $d = rt$ when $d = 10K$ or $10,000$ m and $r = 125$.
$$d = rt$$
$$10,000 = 125t$$
$$\frac{10,000}{125} = \frac{125t}{125}$$
$$80 = t$$
The time is 80 minutes or $\frac{80}{60} = 1\frac{1}{3}$ hours or 1 hour and 20 minutes.

58. Use $F = \dfrac{9}{5}C + 32$ when $F = 104$.

$$F = \frac{9}{5}C + 32$$
$$104 = \frac{9}{5}C + 32$$
$$104 - 32 = \frac{9}{5}C + 32 - 32$$
$$72 = \frac{9}{5}C$$
$$\frac{5}{9} \cdot 72 = \frac{5}{9} \cdot \frac{9}{5}C$$
$$40 = C$$

Thus, 104°F is equivalent to 40°C.

59. Let x be the unknown percent.

$$9 = x \cdot 45$$
$$\frac{9}{45} = \frac{45x}{45}$$
$$0.2 = x$$
$$20\% = x$$

9 is 20% of 45.

60. Let x be the unknown percent.

$$59.5 = x \cdot 85$$
$$\frac{59.5}{85} = \frac{85x}{85}$$
$$0.7 = x$$
$$70\% = x$$

59.5 is 70% of 85.

61. Let x be the unknown number.

$$137.5 = 125\% \cdot x$$
$$137.5 = 1.25x$$
$$\frac{137.5}{1.25} = \frac{1.25x}{1.25}$$
$$110 = x$$

137.5 is 125% of 110.

62. Let x be the unknown number.

$$768 = 60\% \cdot x$$
$$768 = 0.6x$$
$$\frac{768}{0.6} = \frac{0.6x}{0.6}$$
$$1280 = x$$

768 is 60% of 1280.

63. increase $= 11\% \cdot 1900 = 0.11 \cdot 1900 = 209$
new price $= 1900 + 209 = 2109$
The mark-up is \$209 and the new price is \$2109.

64. Find 66.9% of 76,000.
$66.9\% \cdot 76{,}000 = 0.669 \cdot 76{,}000 = 50{,}844$
You would expect 50,844 to use the Internet.

65. Let x be the number of gallons of 40% solution. Then $30 - x$ is the number of gallons of 10% solution.

	gallons	concentration	amount
40% solution	x	40%	$0.4x$
10% solution	$30 - x$	10%	$0.1(30 - x)$
20% solution	30	20%	$0.2(30)$

The amount of acid in the combined solutions must be the same as in the mixture.
$$0.4x + 0.1(30 - x) = 0.2(30)$$
$$0.4x + 3 - 0.1x = 6$$
$$3 + 0.3x = 6$$
$$3 + 0.3x - 3 = 6 - 3$$
$$0.3x = 3$$
$$\frac{0.3x}{0.3} = \frac{3}{0.3}$$
$$x = 10$$
$30 - x = 30 - 10 = 20$
Mix 10 gallons of 40% solution with 20 gallons of 10% solution.

66. percent increase $= \dfrac{\text{amount of increase}}{\text{original amount}}$
$$= \frac{21.0 - 20.7}{20.7}$$
$$= \frac{0.3}{20.7}$$
$$\approx 0.0145$$
The percent increase is 1.45%.

67. From the graph, 18% of motorists who use a cell phone while driving have almost hit another car.

68. The tallest bar represents the most common effect. Therefore, swerving is the most common effect of cell phone use on driving.

69. 21% of drivers cut off someone. Find 21% of 4600.
$21\% \cdot 4600 = 0.21 \cdot 4600 = 966$
You expect 966 customers to cut someone off while driving and talking on their cell phones.

70. $46\% + 41\% + 21\% + 18\% = 126\%$
No, the percents do not sum to 100%.
Answers may vary.

71.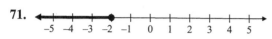

72.

73. $x - 5 \leq -4$
$x - 5 + 5 \leq -4 + 5$
$x \leq 1$
$\{x | x \leq 1\}$

74. $x + 7 > 2$
$x + 7 - 7 > 2 - 7$
$x > -5$
$\{x | x > -5\}$

75. $-2x \geq -20$
$\dfrac{-2x}{-2} \leq \dfrac{-20}{-2}$
$x \leq 10$
$\{x | x \leq 10\}$

76. $-3x > 12$
$\dfrac{-3x}{-3} < \dfrac{12}{-3}$
$x < -4$
$\{x | x < -4\}$

77. $5x - 7 > 8x + 5$
$5x - 7 - 8x > 8x + 5 - 8x$
$-3x - 7 > 5$
$-3x - 7 + 7 > 5 + 7$
$-3x > 12$
$\dfrac{-3x}{-3} < \dfrac{12}{-3}$
$x < -4$
$\{x | x < -4\}$

78. $x + 4 \geq 6x - 16$
$x + 4 - 6x \geq 6x - 16 - 6x$
$-5x + 4 \geq -16$
$-5x + 4 - 4 \geq -16 - 4$
$-5x \geq -20$
$\dfrac{-5x}{-5} \leq \dfrac{-20}{-5}$
$x \leq 4$
$\{x | x \leq 4\}$

79. $\dfrac{2}{3}y > 6$
$\dfrac{3}{2} \cdot \dfrac{2}{3}y > \dfrac{3}{2} \cdot 6$
$y > 9$
$\{y | y > 9\}$

80. $-0.5y \leq 7.5$
$\dfrac{-0.5y}{-0.5} \geq \dfrac{7.5}{-0.5}$
$y \geq -15$
$\{y | y \geq -15\}$

81. $-2(x - 5) > 2(3x - 2)$
$-2x + 10 > 6x - 4$
$-2x + 10 - 6x > 6x - 4 - 6x$
$-8x + 10 > -4$
$-8x + 10 - 10 > -4 - 10$
$-8x > -14$
$\dfrac{-8x}{-8} < \dfrac{-14}{-8}$
$x < \dfrac{7}{4}$
$\left\{ x \middle| x < \dfrac{7}{4} \right\}$

82.
$$4(2x-5) \le 5x-1$$
$$8x-20 \le 5x-1$$
$$8x-20-5x \le 5x-1-5x$$
$$3x-20 \le -1$$
$$3x-20+20 \le -1+20$$
$$3x \le 19$$
$$\frac{3x}{3} \le \frac{19}{3}$$
$$x \le \frac{19}{3}$$
$$\left\{ x \middle| x \le \frac{19}{3} \right\}$$

83. Let x be the sales. Her weekly earnings are $175 + 0.05x$.
$$175 + 0.05x \ge 300$$
$$175 + 0.05x - 175 \ge 300 - 175$$
$$0.05x \ge 125$$
$$\frac{0.05x}{0.05} \ge \frac{125}{0.05}$$
$$x \ge 2500$$
She must have weekly sales of at least $2500.

84. Let x be his score on the fourth round.
$$\frac{76+82+79+x}{4} < 80$$
$$\frac{237+x}{4} < 80$$
$$4 \cdot \frac{237+x}{4} < 4 \cdot 80$$
$$237 + x < 320$$
$$237 + x - 237 < 320 - 237$$
$$x < 83$$
His score must be less than 83.

85.
$$6x + 2x - 1 = 5x + 11$$
$$8x - 1 = 5x + 11$$
$$8x - 1 - 5x = 5x + 11 - 5x$$
$$3x - 1 = 11$$
$$3x - 1 + 1 = 11 + 1$$
$$3x = 12$$
$$\frac{3x}{3} = \frac{12}{3}$$
$$x = 4$$

86.
$$2(3y - 4) = 6 + 7y$$
$$6y - 8 = 6 + 7y$$
$$6y - 8 - 6y = 6 + 7y - 6y$$
$$-8 = 6 + y$$
$$-8 - 6 = 6 + y - 6$$
$$-14 = y$$

87.
$$4(3 - a) - (6a + 9) = -12a$$
$$12 - 4a - 6a - 9 = -12a$$
$$3 - 10a = -12a$$
$$3 - 10a + 10a = -12a + 10a$$
$$3 = -2a$$
$$\frac{3}{-2} = \frac{-2a}{-2}$$
$$-\frac{3}{2} = a$$

88.
$$\frac{x}{3} - 2 = 5$$
$$\frac{x}{3} - 2 + 2 = 5 + 2$$
$$\frac{x}{3} = 7$$
$$3 \cdot \frac{x}{3} = 3 \cdot 7$$
$$x = 21$$

89.
$$2(y + 5) = 2y + 10$$
$$2y + 10 = 2y + 10$$
Since both sides of the equation are identical, the equation is an identity and every real number is a solution.

90.
$$7x - 3x + 2 = 2(2x - 1)$$
$$4x + 2 = 4x - 2$$
$$4x + 2 - 4x = 4x - 2 - 4x$$
$$2 = -2$$
Since the statement $2 = -2$ is false, there is no solution.

91. Let x be the number.
$$6 + 2x = x - 7$$
$$6 + 2x - x = x - 7 - x$$
$$6 + x = -7$$
$$6 + x - 6 = -7 - 6$$
$$x = -13$$
The number is -13.

92. Let x be the length of the shorter piece.
Then $4x + 3$ is the length of the longer piece.
The lengths sum to 23.
$$x + 4x + 3 = 23$$
$$5x + 3 = 23$$
$$5x + 3 - 3 = 23 - 3$$
$$5x = 20$$
$$\frac{5x}{5} = \frac{20}{5}$$
$$x = 4$$
$4x + 3 = 4(4) + 3 = 16 + 3 = 19$
The shorter piece is 4 inches and the longer piece is 19 inches.

93. $V = \frac{1}{3}Ah$
$$3V = 3 \cdot \frac{1}{3}Ah$$
$$3V = Ah$$
$$\frac{3V}{A} = \frac{Ah}{A}$$
$$\frac{3V}{A} = h$$

94. Let x be the number.
$$x = 26\% \cdot 85$$
$$x = 0.26 \cdot 85$$
$$x = 22.1$$
22.1 is 26% of 85.

95. Let x be the unknown number.
$$72 = 45\% \cdot x$$
$$72 = 0.45x$$
$$\frac{72}{0.45} = \frac{0.45x}{0.45}$$
$$160 = x$$
72 is 45% of 160.

96. percent increase $= \dfrac{\text{amount of increase}}{\text{original amount}}$
$$= \frac{282 - 235}{235}$$
$$= \frac{47}{235}$$
$$= 0.2$$
The percent increase is 20%.

97.
$$4x - 7 > 3x + 2$$
$$4x - 7 - 3x > 3x + 2 - 3x$$
$$x - 7 > 2$$
$$x - 7 + 7 > 2 + 7$$
$$x > 9$$
$\{x | x > 9\}$

98. $-5x < 20$
$$\frac{-5x}{5} > \frac{20}{-5}$$
$$x > -4$$
$\{x | x > -4\}$

99. $-3(1 + 2x) + x \geq -(3 - x)$
$$-3 - 6x + x \geq -3 + x$$
$$-3 - 5x \geq -3 + x$$
$$-3 - 5x - x \geq -3 + x - x$$
$$-3 - 6x \geq -3$$
$$-3 - 6x + 3 \geq -3 + 3$$
$$-6x \geq 0$$
$$\frac{-6x}{-6} \leq \frac{0}{-6}$$
$$x \leq 0$$
$\{x | x \leq 0\}$

Chapter 2 Test

1.
$$-\frac{4}{5}x = 4$$
$$-\frac{5}{4}\left(-\frac{4}{5}x\right) = -\frac{5}{4}(4)$$
$$x = -5$$

2.
$$4(n-5) = -(4-2n)$$
$$4n-20 = -4+2n$$
$$4n-20-2n = -4+2n-2n$$
$$2n-20 = -4$$
$$2n-20+20 = -4+20$$
$$2n = 16$$
$$\frac{2n}{2} = \frac{16}{2}$$
$$n = 8$$

3.
$$5y-7+y = -(y+3y)$$
$$6y-7 = -y-3y$$
$$6y-7 = -4y$$
$$6y-7-6y = -4y-6y$$
$$-7 = -10y$$
$$\frac{-7}{-10} = \frac{-10y}{-10}$$
$$\frac{7}{10} = y$$

4.
$$4z+1-z = 1+z$$
$$3z+1 = 1+z$$
$$3z+1-z = 1+z-z$$
$$2z+1 = 1$$
$$2z+1-1 = 1-1$$
$$2z = 0$$
$$\frac{2z}{2} = \frac{0}{2}$$
$$z = 0$$

5.
$$\frac{2(x+6)}{3} = x-5$$
$$3\left(\frac{2(x+6)}{3}\right) = 3(x-5)$$
$$2(x+6) = 3(x-5)$$
$$2x+12 = 3x-15$$
$$2x+12-2x = 3x-15-2x$$
$$12 = x-15$$
$$12+15 = x-15+15$$
$$27 = x$$

6.
$$\frac{4(y-1)}{5} = 2y+3$$
$$5\left[\frac{4(y-1)}{5}\right] = 5(2y+3)$$
$$4(y-1) = 5(2y+3)$$
$$4y-4 = 10y+15$$
$$4y-4-10y = 10y+15-10y$$
$$-6y-4 = 15$$
$$-6y-4+4 = 15+4$$
$$-6y = 19$$
$$\frac{-6y}{-6} = \frac{19}{-6}$$
$$y = -\frac{19}{6}$$

7.
$$\frac{1}{2}-x+\frac{3}{2} = x-4$$
$$-x+\frac{4}{2} = x-4$$
$$-x+2 = x-4$$
$$-x+2+x = x-4+x$$
$$2 = 2x-4$$
$$2+4 = 2x-4+4$$
$$6 = 2x$$
$$\frac{6}{2} = \frac{2x}{2}$$
$$3 = x$$

8. $\dfrac{1}{3}(y+3) = 4y$

$3 \cdot \dfrac{1}{3}(y+3) = 3 \cdot 4y$

$y + 3 = 12y$

$y + 3 - y = 12y - y$

$3 = 11y$

$\dfrac{3}{11} = \dfrac{11y}{11}$

$\dfrac{3}{11} = y$

9. $-0.3(x-4) + x = 0.5(3-x)$

$-0.3(x-4) + 1.0x = 0.5(3-x)$

$-3(x-4) + 10x = 5(3-x)$

$-3x + 12 + 10x = 15 - 5x$

$7x + 12 = 15 - 5x$

$7x + 12 + 5x = 15 - 5x + 5x$

$12x + 12 = 15$

$12x + 12 - 12 = 15 - 12$

$12x = 3$

$\dfrac{12x}{12} = \dfrac{3}{12}$

$x = \dfrac{1}{4} = 0.25$

10. $-4(a+1) - 3a = -7(2a-3)$

$-4a - 4 - 3a = -14a + 21$

$-4 - 7a = -14a + 21$

$-4 - 7a + 14a = -14a + 21 + 14a$

$-4 + 7a = 21$

$-4 + 7a + 4 = 21 + 4$

$7a = 25$

$\dfrac{7a}{7} = \dfrac{25}{7}$

$a = \dfrac{25}{7}$

11. $-2(x-3) = x + 5 - 3x$

$-2x + 6 = -2x + 5$

$2x - 2x + 6 = 2x - 2x + 5$

$6 = 5$

Since the statement $6 = 5$ is false, there is no solution.

12. Let x be the number.

$x + \dfrac{2}{3}x = 35$

$\dfrac{3}{3}x + \dfrac{2}{3}x = 35$

$\dfrac{5}{3}x = 35$

$\dfrac{3}{5} \cdot \dfrac{5}{3}x = \dfrac{3}{5} \cdot 35$

$x = 21$

The number is 21.

13. $A = lw = (35)(20) = 700$

The area of the deck is 700 square feet. To paint two coats of water seal means covering $2 \cdot 700 = 1400$ square feet.

$$1400 \text{ sq ft} \cdot \dfrac{1 \text{ gal}}{200 \text{ sq ft}} = 7 \text{ gal}$$

7 gallons of water seal are needed.

14. Use $y = mx + b$ when $y = -14$, $m = -2$, and $b = -2$.

$y = mx + b$

$-14 = -2x + (-2)$

$-14 + 2 = -2x + (-2) + 2$

$-12 = -2x$

$\dfrac{-12}{-2} = \dfrac{-2x}{-2}$

$6 = x$

15. $V = \pi r^2 h$

$\dfrac{V}{\pi r^2} = \dfrac{\pi r^2 h}{\pi r^2}$

$\dfrac{V}{\pi r^2} = h$

16. $3x - 4y = 10$

$3x - 4y - 3x = 10 - 3x$

$-4y = 10 - 3x$

$\dfrac{-4y}{-4} = \dfrac{10 - 3x}{-4}$

$y = \dfrac{3x - 10}{4}$

17.
$$3x - 5 > 7x + 3$$
$$3x - 5 - 3x > 7x + 3 - 3x$$
$$-5 > 4x + 3$$
$$-5 - 3 > 4x + 3 - 3$$
$$-8 > 4x$$
$$\frac{-8}{4} > \frac{4x}{4}$$
$$-2 > x$$
$$\{x \mid x < -2\}$$

18.
$$x + 6 > 4x - 6$$
$$x + 6 - 4x > 4x - 6 - 4x$$
$$-3x + 6 > -6$$
$$-3x + 6 - 6 > -6 - 6$$
$$-3x > -12$$
$$\frac{-3x}{-3} < \frac{-12}{-3}$$
$$x < 4$$
$$\{x \mid x < 4\}$$

19.
$$-0.3x \geq 2.4$$
$$\frac{-0.3x}{-0.3} \leq \frac{2.4}{-0.3}$$
$$x \leq -8$$
$$\{x \mid x \leq -8\}$$

20.
$$-5(x - 1) + 6 \leq -3(x + 4) + 1$$
$$-5x + 5 + 6 \leq -3x - 12 + 1$$
$$-5x + 11 \leq -3x - 11$$
$$-5x + 11 + 3x \leq -3x - 11 + 3x$$
$$-2x + 11 \leq -11$$
$$-2x + 11 - 11 \leq -11 - 11$$
$$-2x \leq -22$$
$$\frac{-2x}{-2} \geq \frac{-22}{-2}$$
$$x \geq 11$$
$$\{x \mid x \geq 11\}$$

21.
$$\frac{2(5x + 1)}{3} > 2$$
$$3 \cdot \frac{2(5x + 1)}{3} > 3(2)$$
$$2(5x + 1) > 6$$
$$10x + 2 > 6$$
$$10x + 2 - 2 > 6 - 2$$
$$10x > 4$$
$$\frac{10x}{10} > \frac{4}{10}$$
$$x > \frac{2}{5}$$
$$\left\{x \mid x > \frac{2}{5}\right\}$$

22. From the graph, 29% of tornadoes occurring in the U.S. are classified as strong, that is F2 or F3.

23. From the graph, 69% are classified as weak. Find 69% of 800.
$$69\% \cdot 800 = 0.69 \cdot 800 = 552$$
You would expect 552 of the 800 to be classified as weak.

24. Let x be the unknown percent.
$$72 = x \cdot 180$$
$$\frac{72}{180} = \frac{180x}{180}$$
$$0.4 = x$$
72 is 40% of 180.

25. Let x represent the number of public libraries in Indiana. Then there are $x + 650$ public libraries in New York.
$$x + x + 650 = 1504$$
$$2x + 650 = 1504$$
$$2x + 650 - 650 = 1504 - 650$$
$$2x = 854$$
$$\frac{2x}{2} = \frac{854}{2}$$
$$x = 427$$
$$x + 650 = 427 + 650 = 1077$$
Indiana has 427 public libraries and New York has 1077.

Cumulative Review Chapters 1–2

1. Since $8 = 8$, the statement $8 \geq 8$ is true.

2. Since -4 is to the right of -6 on the number line, the statement $-4 < -6$ is false.

3. Since $8 = 8$, the statement $8 \leq 8$ is true.

4. Since 3 is to the right of -3 on the number line, the statement $3 > -3$ is true.

5. Since neither $23 < 0$ nor $23 = 0$ is true, the statement $23 \leq 0$ is false.

6. Since $-8 = -8$, the statement $-8 \geq -8$ is true.

7. Since $0 < 23$ is true, the statement $0 \leq 23$ is true.

8. Since $-8 = -8$, the statement $-8 \leq -8$ is true.

9. **a.** $|0| < 2$ since $|0| = 0$ and $0 < 2$.

   b. $|-5| = 5$

   c. $|-3| > |-2|$ since $3 > 2$.

   d. $|-9| < |-9.7|$ since $9 < -9.7$.

   e. $\left|-7\dfrac{1}{6}\right| > |7|$ since $7\dfrac{1}{6} > 7$.

10. **a.** $|5| = 5$ since 5 is 5 units from 0 on the number line.

   b. $|-8| = 8$ since -8 is 8 units from 0 on the number line.

   c. $\left|-\dfrac{2}{3}\right| = \dfrac{2}{3}$ since $-\dfrac{2}{3}$ is $\dfrac{2}{3}$ unit from 0 on the number line.

11. $\dfrac{3 + |4 - 3| + 2^2}{6 - 3} = \dfrac{3 + |1| + 2^2}{6 - 3}$

$= \dfrac{3 + 1 + 2^2}{3}$

$= \dfrac{3 + 1 + 4}{3}$

$= \dfrac{8}{3}$

12. $1 + 2(9 - 7)^3 + 4^2 = 1 + 2(2)^3 + 4^2$

$= 1 + 2(8) + 16$

$= 1 + 16 + 16$

$= 33$

13. $(-8) + (-11) = -19$

14. $-2 + (-8) = -10$

15. $(-2) + 10 = 8$

16. $-10 + 20 = 10$

17. $0.2 + (-0.5) = -0.3$

18. $1.2 + (-1.2) = 0$

19. **a.** $-3 + [(-2 - 5) - 2]$

$= -3 + [(-2 + (-5)) - 2]$

$= -3 + [(-7) - 2]$

$= -3 + [-7 + (-2)]$

$= -3 + [-9]$

$= -12$

   b. $2^3 - 10 + [-6 - (-5)] = 2^3 - 10 + [-6 + 5]$

$= 2^3 - 10 + [-1]$

$= 8 - 10 + (-1)$

$= 8 + (-10) + (-1)$

$= -2 + (-1)$

$= -3$

20. **a.** $-(-5) = 5$

   b. $-\left(-\dfrac{2}{3}\right) = \dfrac{2}{3}$

c. $-(-a) = a$

d. $-|-3| = -3$

21. a. $7(0)(-6) = 0(-6) = 0$

b. $(-2)(-3)(-4) = (6)(-4) = -24$

c. $(-1)(5)(-9) = (-5)(-9) = 45$

22. a. $-2.7 - 8.4 = -2.7 + (-8.4) = -11.1$

b. $-\dfrac{4}{5} - \left(-\dfrac{3}{5}\right) = -\dfrac{4}{5} + \dfrac{3}{5} = -\dfrac{1}{5}$

c. $\dfrac{1}{4} - \left(-\dfrac{1}{2}\right) = \dfrac{1}{4} + \dfrac{1}{2} = \dfrac{1}{4} + \dfrac{2}{4} = \dfrac{3}{4}$

23. a. $-18 \div 3 = -18 \cdot \dfrac{1}{3} = -6$

b. $\dfrac{-14}{-2} = -14 \cdot -\dfrac{1}{2} = 7$

c. $\dfrac{20}{-4} = 20 \cdot -\dfrac{1}{4} = -5$

24. a. $(4.5)(-0.08) = -0.36$

b. $-\dfrac{3}{4} \cdot -\dfrac{8}{17} = \dfrac{3 \cdot 8}{4 \cdot 17} = \dfrac{6}{17}$

25. $-5(-3 + 2z) = -5(-3) + (-5)(2z) = 15 - 10z$

26. $2x(x^2 - 3x + 4) = 2x(x^2) - 2x(3x) + 2x(4)$
$= 2x^3 - 6x^2 + 8x$

27. $\dfrac{1}{2}(6x + 14) + 10 = \dfrac{1}{2}(6x) + \dfrac{1}{2}(14) + 10$
$= 3x + 7 + 10$
$= 3x + 17$

28. $-(x + 4) + 3(x + 4)$
$= -1(x + 4) + 3(x + 4)$
$= -1 \cdot x + (-1)(4) + 3 \cdot x + 3 \cdot 4$
$= -x - 4 + 3x + 12$
$= -x + 3x - 4 + 12$
$= 2x + 8$

29. a. $2x$ and $3x^2$ are unlike terms, since the exponents on x are not the same.

b. $4x^2y$, xy^2, and $-2x^2y$ are like terms, since each variable and its exponent match.

c. $-2yz$ and $-3zy$ are like terms, since $zy = yz$ by the commutative property.

d. $-x^4$ and x^4 are like terms. The variable and its exponent match.

e. $-8a^5$ and $8a^5$ are like terms. The variable and its exponent match.

30. a. $\dfrac{-32}{8} = -4$

b. $\dfrac{-108}{-12} = 9$

c. $\dfrac{-5}{7} \div \left(\dfrac{-9}{2}\right) = \dfrac{-5}{7} \cdot \left(\dfrac{2}{-9}\right) = \dfrac{5 \cdot 2}{7 \cdot 9} = \dfrac{10}{63}$

31. $(2x - 3) - (4x - 2) = 2x - 3 - 4x + 2$
$= -2x - 1$

32. $(-5x + 1) - (10x + 3) = -5x + 1 - 10x - 3$
$= -15x - 2$

33. $\qquad x - 7 = 10$
$x - 7 + 7 = 10 + 7$
$\qquad x = 17$

34.
$$\frac{5}{6} + x = \frac{2}{3}$$
$$\frac{5}{6} + x - \frac{5}{6} = \frac{2}{3} - \frac{5}{6}$$
$$x = \frac{4}{6} - \frac{5}{6}$$
$$x = -\frac{1}{6}$$

35.
$$-z - 4 = 6$$
$$-z - 4 + 4 = 6 + 4$$
$$-z = 10$$
$$\frac{-z}{-1} = \frac{10}{-1}$$
$$z = -10$$

36. $-3x + 1 - (-4x - 6) = 10$
$$-3x + 1 + 4x + 6 = 10$$
$$x + 7 = 10$$
$$x + 7 - 7 = 10 - 7$$
$$x = 3$$

37.
$$\frac{2(a+3)}{3} = 6a + 2$$
$$3 \cdot \frac{2(a+3)}{3} = 3(6a + 2)$$
$$2(a+3) = 3(6a + 2)$$
$$2a + 6 = 18a + 6$$
$$2a + 6 - 18a = 18a + 6 - 18a$$
$$-16a + 6 = 6$$
$$-16a + 6 - 6 = 6 - 6$$
$$-16a = 0$$
$$\frac{-16a}{-16} = \frac{0}{-16}$$
$$a = 0$$

38.
$$\frac{x}{4} = 18$$
$$4 \cdot \frac{x}{4} = 4 \cdot 18$$
$$x = 72$$

39. Let x be the number of Democrats. Then $x + 15$ is the number of Republicans. The total number is 431.
$$x + x + 15 = 431$$
$$2x + 15 = 431$$
$$2x + 15 - 15 = 431 - 15$$
$$2x = 416$$
$$\frac{2x}{2} = \frac{416}{2}$$
$$x = 208$$
$x + 15 = 208 + 15 = 223$
There are 208 Democrats and 223 Republicans.

40.
$$6x + 5 = 4(x + 4) - 1$$
$$6x + 5 = 4x + 16 - 1$$
$$6x + 5 = 4x + 15$$
$$6x + 5 - 4x = 4x + 15 - 4x$$
$$2x + 5 = 15$$
$$2x + 5 - 5 = 15 - 5$$
$$2x = 10$$
$$\frac{2x}{2} = \frac{10}{2}$$
$$x = 5$$

41. Use $d = rt$ when $d = 31,680$ and $r = 400$.
$$d = rt$$
$$31,680 = 400t$$
$$\frac{31,680}{400} = \frac{400t}{400}$$
$$79.2 = t$$
It will take the ice 79.2 years to reach the lake.

42.
$$x + 4 = 3x - 8$$
$$x + 4 - 3x = 3x - 8 - 3x$$
$$-2x + 4 = -8$$
$$-2x + 4 - 4 = -8 - 4$$
$$-2x = -12$$
$$\frac{-2x}{-2} = \frac{-12}{-2}$$
$$x = 6$$
The number is 6.

43. Let x be the unknown percent.

$$63 = x \cdot 72$$
$$\frac{63}{72} = \frac{72x}{72}$$
$$0.875 = x$$
$$87.5\% = x$$

63 is 87.5% of 72.

44.
$$C = 2\pi r$$
$$\frac{C}{2\pi} = \frac{2\pi r}{2\pi}$$
$$\frac{C}{2\pi} = r \text{ or } r = \frac{C}{2\pi}$$

45.
$$5(2x + 3) = -1 + 7$$
$$5(2x) + 5(3) = -1 + 7$$
$$10x + 15 = 6$$
$$10x + 15 - 15 = 6 - 15$$
$$10x = -9$$
$$\frac{10x}{10} = \frac{-9}{10}$$
$$x = -\frac{9}{10}$$

46.
$$x - 3 > 2$$
$$x - 3 + 3 > 2 + 3$$
$$x > 5$$
$$\{x | x > 5\}$$

47. $-1 > x$ or $x < -1$

48.
$$3x - 4 \le 2x - 14$$
$$3x - 4 - 2x \le 2x - 14 - 2x$$
$$x - 4 \le -14$$
$$x - 4 + 4 \le -14 + 4$$
$$x \le -10$$
$$\{x | x \le -10\}$$

49.
$$2(x - 3) - 5 \le 3(x + 2) - 18$$
$$2x - 6 - 5 \le 3x + 6 - 18$$
$$2x - 11 \le 3x - 12$$
$$-x - 11 \le -12$$
$$-x \le -1$$
$$\frac{-x}{-1} \ge \frac{-1}{-1}$$
$$x \ge 1$$
$$\{x | x \ge 1\}$$

50.
$$-3x \ge 9$$
$$\frac{-3x}{-3} \le \frac{9}{-3}$$
$$x \le -3$$
$$\{x | x \le -3\}$$

Chapter 3

1. $3^4 = 3 \cdot 3 \cdot 3 \cdot 3 = 81$

2. $7^1 = 7$

3. $(-2)^3 = (-2)(-2)(-2) = -8$

4. $-2^3 = -(2 \cdot 2 \cdot 2) = -8$

5. $\left(\dfrac{2}{3}\right)^2 = \dfrac{2}{3} \cdot \dfrac{2}{3} = \dfrac{4}{9}$

6. $5 \cdot 6^2 = 5 \cdot 36 = 180$

7. a. When x is 4, $3x^2 = 3 \cdot 4^2$
$$= 3 \cdot (4 \cdot 4)$$
$$= 3 \cdot 16$$
$$= 48.$$

b. When x is -2, $\dfrac{x^4}{-8} = \dfrac{(-2)^4}{-8}$
$$= \dfrac{(-2)(-2)(-2)(-2)}{-8}$$
$$= \dfrac{16}{-8}$$
$$= -2.$$

8. $7^3 \cdot 7^2 = 7^{3+2} = 7^5$

9. $x^4 \cdot x^9 = x^{4+9} = x^{13}$

10. $r^5 \cdot r = r^{5+1} = r^6$

11. $s^6 \cdot s^2 \cdot s^3 = s^{6+2+3} = s^{11}$

12. $(-3)^9 \cdot (-3) = (-3)^{9+1} = (-3)^{10}$

13. $(6x^3)(-2x^9) = (6 \cdot x^3) \cdot (-2 \cdot x^9)$
$$= (6 \cdot -2) \cdot (x^3 \cdot x^9)$$
$$= -12x^{12}$$

14. $(m^5 n^{10})(mn^8) = (m^5 \cdot m) \cdot (n^{10} \cdot n^8)$
$$= m^6 \cdot n^{18}$$
$$= m^6 n^{18}$$

15. $(-x^9 y)(4x^2 y^{11}) = (-1 \cdot 4) \cdot (x^9 \cdot x^2) \cdot (y \cdot y^{11})$
$$= -4x^{11} y^{12}$$

16. $(9^4)^{10} = 9^{4 \cdot 10} = 9^{40}$

17. $(z^6)^3 = z^{6 \cdot 3} = z^{18}$

18. $(xy)^7 = x^7 \cdot y^7 = x^7 y^7$

19. $(3y)^4 = 3^4 \cdot y^4 = 81y^4$

20. $(-2p^4 q^2 r)^3 = (-2)^3 \cdot (p^4)^3 \cdot (q^2)^3 \cdot (r^1)^3$
$$= -8p^{12} q^6 r^3$$

21. $(-a^4 b)^7 = (-1a^4 b)^7$
$$= (-1)^7 \cdot (a^4)^7 \cdot (b^1)^7$$
$$= -1a^{28} b^7$$
$$= -a^{28} b^7$$

22. $\left(\dfrac{r}{s}\right)^6 = \dfrac{r^6}{s^6}$, $s \neq 0$

23. $\left(\dfrac{5x^6}{9y^3}\right)^2 = \dfrac{5^2 \cdot (x^6)^2}{9^2 \cdot (y^3)^2} = \dfrac{25x^{12}}{81y^6}$, $y \neq 0$

24. $\dfrac{y^7}{y^3} = y^{7-3} = y^4$

25. $\dfrac{5^9}{5^6} = 5^{9-6} = 5^3 = 125$

26. $\dfrac{(-2)^{14}}{(-2)^{10}} = (-2)^{14-10} = (-2)^4 = 16$

27. $\dfrac{7a^4b^{11}}{ab} = 7 \cdot \dfrac{a^4}{a^1} \cdot \dfrac{b^{11}}{b^1}$

$\qquad = 7 \cdot (a^{4-1}) \cdot (b^{11-1})$

$\qquad = 7a^3b^{10}$

28. $8^0 = 1$

29. $(2r^2s)^0 = 1$

30. $(-7)^0 = 1$

31. $-7^0 = -1 \cdot 7^0 = -1 \cdot 1 = -1$

32. $7y^0 = 7 \cdot y^0 = 7 \cdot 1 = 7$

33. a. $\dfrac{x^7}{x^4} = x^{7-4} = x^3$

    b. $(3y^4)^4 = 3^4 \cdot (y^4)^4 = 81y^{16}$

    c. $\left(\dfrac{x}{4}\right)^3 = \dfrac{x^3}{4^3} = \dfrac{x^3}{64}$

Mental Math

1. In 3^2, base is 3 and exponent is 2.

2. In 5^4, base is 5 and exponent is 4.

3. In $(-3)^6$, base is -3 and exponent is 6.

4. In -3^7, base is 3 and exponent is 7.

5. In -4^2, base is 4 and exponent is 2.

6. In $(-4)^3$, base is -4 and exponent is 3.

7. In $5 \cdot 3^4$, one base is 5 and exponent is 1 and the other base is 3 and exponent is 4.

8. In $9 \cdot 7^6$, one base is 9 and exponent is 1 and other base is 7 and exponent is 6.

9. In $5x^2$, one base is 5 and exponent is 1 and the other base is x and exponent is 2.

10. In $(5x)^2$, base is $5x$ and exponent is 2.

Exercise Set 3.1

1. $7^2 = 7 \cdot 7 = 49$

3. $(-5)^1 = -5$

5. $-2^4 = -(2 \cdot 2 \cdot 2 \cdot 2) = -16$

7. $(-2)^4 = (-2)(-2)(-2)(-2) = 16$

9. $\left(\dfrac{1}{3}\right)^3 = \dfrac{1}{3} \cdot \dfrac{1}{3} \cdot \dfrac{1}{3} = \dfrac{1}{27}$

11. $7 \cdot 2^4 = 7(2 \cdot 2 \cdot 2 \cdot 2) = 7 \cdot 16 = 112$

13. When x is -2, $x^2 = (-2)^2 = (-2)(-2) = 4$.

15. When x is 3,
$5x^3 = 5(3)^3 = 5(3)(3)(3) = 5(27) = 135$.

17. When $x = 3$ and $y = -5$,
$2xy^2 = 2(3)(-5)^2$
$\qquad = 2(3)(-5)(-5)$
$\qquad = 2(3)(25)$
$\qquad = 150$.

19. When z is -2,

$$\frac{5z^4}{7} = \frac{5(-2)^4}{7}$$
$$= \frac{5(-2)(-2)(-2)(-2)}{7}$$
$$= \frac{5(16)}{7}$$
$$= \frac{80}{7}.$$

21. $x^2 \cdot x^5 = x^{2+5} = x^7$

23. $(-3)^3 \cdot (-3)^9 = (-3)^{3+9} = (-3)^{12}$

25. $(5y^4)(3y) = (5 \cdot 3)(y^4 \cdot y) = 15y^{4+1} = 15y^5$

27. $(x^9 y)(x^{10} y^5) = (x^9 \cdot x^{10})(y \cdot y^5)$
$$= (x^{9+10})(y^{1+5})$$
$$= x^{19} y^6$$

29. $(-8mn^6)(9m^2 n^2) = (-8 \cdot 9)(m \cdot m^2)(n^6 \cdot n^2)$
$$= -72m^{1+2} n^{6+2}$$
$$= -72m^3 n^8$$

31. $(4z^{10})(-6z^7)(z^3) = (4 \cdot -6 \cdot 1)(z^{10} \cdot z^7 \cdot z^3)$
$$= -24z^{10+7+3}$$
$$= -24z^{20}$$

33. Area = (length)(width)
$$= (5x^3 \text{ feet})(4x^2 \text{ feet})$$
$$= (5 \cdot 4)(x^3 \cdot x^2) \text{ square feet}$$
$$= 20x^{3+2} \text{ square feet}$$
$$= 20x^5 \text{ square feet}$$

35. $(x^9)^4 = x^{9 \cdot 4} = x^{36}$

37. $(pq)^8 = p^8 \cdot q^8 = p^8 q^8$

39. $(2a^5)^3 = (2)^3(a^5)^3 = 2^3 a^{15} = 8a^{15}$

41. $(x^2 y^3)^5 = (x^2)^5(y^3)^5 = x^{10} y^{15}$

43. $(-7a^2 b^5 c)^2 = (-7)^2(a^2)^2(b^5)^2(c)^2$
$$= 49a^4 b^{10} c^2$$

45. $\left(\dfrac{r}{s}\right)^9 = \dfrac{r^9}{s^9}$

47. $\left(\dfrac{mp}{n}\right)^9 = \dfrac{m^9 p^9}{n^9}$

49. $\left(\dfrac{-2xz}{y^5}\right)^2 = \dfrac{(-2)^2(x)^2(z)^2}{(y^5)^2}$
$$= \dfrac{4x^2 z^2}{y^{5 \cdot 2}}$$
$$= \dfrac{4x^2 z^2}{y^{10}}$$

51. Area = (length)(length)
$$= (8z^5 \text{ decimeters})(8z^5 \text{ decimeters})$$
$$= (8 \cdot 8)(z^5 \cdot z^5) \text{ square decimeters}$$
$$= 64z^{5+5} \text{ square decimeters}$$
$$= 64z^{10} \text{ square decimeters}$$

53. Volume = (length)(width)(height)
$$= (3y^4 \text{ feet})(3y^4 \text{ feet})(3y^4 \text{ feet})$$
$$= (3)^3(y^4)^3 \text{ cubic feet}$$
$$= 27y^{4 \cdot 3} \text{ cubic feet}$$
$$= 27y^{12} \text{ cubic feet}$$

55. $\dfrac{x^3}{x} = x^{3-1} = x^2$

57. $\dfrac{(-4)^6}{(-4)^3} = (-4)^{6-3} = (-4)^3 = -64$

59. $\dfrac{p^7 q^{20}}{pq^{15}} = \dfrac{p^7}{p} \cdot \dfrac{q^{20}}{q^{15}} = p^{7-1} \cdot q^{20-15} = p^6 q^5$

61. $\dfrac{7x^2 y^6}{14x^2 y^3} = \dfrac{7}{14} \cdot \dfrac{x^2}{x^2} \cdot \dfrac{y^6}{y^3}$

$\qquad = \dfrac{1}{2} \cdot x^{2-2} \cdot y^{6-3}$

$\qquad = \dfrac{1}{2} x^0 y^3$

$\qquad = \dfrac{y^3}{2}$

63. $7^0 = 1$

65. $(2x)^0 = 1$

67. $-7x^0 = -7 \cdot x^0 = -7 \cdot 1 = -7$

69. $5^0 + y^0 = 1 + 1 = 2$

71. $-9^2 = -(9)^2 = -(9 \cdot 9) = -81$

73. $\left(\dfrac{1}{4}\right)^3 = \dfrac{1}{4} \cdot \dfrac{1}{4} \cdot \dfrac{1}{4} = \dfrac{1}{64}$

75. $b^4 b^2 = b^{4+2} = b^6$

77. $a^2 a^3 a^4 = a^{2+3+4} = a^9$

79. $(2x^3)(-8x^4) = (2 \cdot -8)(x^3 \cdot x^4)$

$\qquad = -16x^{3+4}$

$\qquad = -16x^7$

81. $(a^7 b^{12})(a^4 b^8) = a^7 a^4 \cdot b^{12} b^8$

$\qquad = a^{7+4} b^{12+8}$

$\qquad = a^{11} b^{20}$

83. $(-2mn^6)(-13m^8 n)$

$\qquad = (-2)(-13)(m \cdot m^8)(n^6 \cdot n)$

$\qquad = 26m^{1+8} n^{6+1}$

$\qquad = 26m^9 n^7$

85. $(z^4)^{10} = z^{4 \cdot 10} = z^{40}$

87. $(4ab)^3 = (4)^3 a^3 b^3 = 64a^3 b^3$

89. $(-6xyz^3)^2 = (-6)^2 x^2 y^2 (z^3)^2$

$\qquad = 36x^2 y^2 z^{3 \cdot 2}$

$\qquad = 36x^2 y^2 z^6$

91. $\dfrac{z^{12}}{z^4} = z^{12-4} = z^8$

93. $\dfrac{3x^5}{x^4} = 3 \cdot \dfrac{x^5}{x^4} = 3x^{5-4} = 3x^1 = 3x$

95. $(6b)^0 = 1$

97. $(9xy)^2 = 9^2 \cdot x^2 y^2 = 81x^2 y^2$

99. $2^3 + 2^5 = (2 \cdot 2 \cdot 2) + (2 \cdot 2 \cdot 2 \cdot 2 \cdot 2)$

$\qquad = 8 + 32$

$\qquad = 40$

101. $\left(\dfrac{3y^5}{6x^4}\right)^3 = \left(\dfrac{y^5}{2x^4}\right)^3$

$\qquad = \dfrac{(y^5)^3}{2^3 (x^4)^3}$

$\qquad = \dfrac{y^{5 \cdot 3}}{8x^{4 \cdot 3}}$

$\qquad = \dfrac{y^{15}}{8x^{12}}$

103. $\dfrac{2x^3 y^2 z}{xyz} = 2 \cdot \dfrac{x^3}{x} \cdot \dfrac{y^2}{y} \cdot \dfrac{z}{z}$

$\qquad = 2x^{3-1} y^{2-1} z^{1-1}$

$\qquad = 2x^2 y^1 z^0$

$\qquad = 2x^2 y$

105. $5 - 7 = 5 + (-7) = -2$

107. $3 - (-2) = 3 + 2 = 5$

109. $-11 - (-4) = -11 + 4 = -7$

111. The expression $(x^{14})^{23}$ can be simplified by multiplying the exponents; c.

113. The expression $x^{14} + x^{23}$ cannot be simplified by adding subtracting, multiplying, or dividing the exponents; e.

115. answers may vary

117. answers may vary

119. $V = x^3$
$= (7 \text{ meters})^3$
$= 7^3 \text{ cubic meters}$
$= 343 \text{ cubic meters}$

121. The volume of a cube measures the amount of material that the cube can hold, so to find the amount of water that a swimming pool can hold, the formula for volume should be used.

123. answers may vary

125. answers may vary

127. $x^{5a} x^{4a} = x^{5a+4a} = x^{9a}$

129. $(a^b)^5 = a^{b \cdot 5} = a^{5b}$

131. $\dfrac{x^{9a}}{x^{4a}} = x^{9a-4a} = x^{5a}$

Section 3.2 Practice Problems

1. $5^{-3} = \dfrac{1}{5^3} = \dfrac{1}{125}$

2. $7x^{-4} = 7^1 \cdot \dfrac{1}{x^4} = \dfrac{7^1}{x^4} \text{ or } \dfrac{7}{x^4}$

3. $5^{-1} + 3^{-1} = \dfrac{1}{5} + \dfrac{1}{3} = \dfrac{3}{15} + \dfrac{5}{15} = \dfrac{8}{15}$

4. $(-3)^{-4} = \dfrac{1}{(-3)^4} = \dfrac{1}{(-3)(-3)(-3)(-3)} = \dfrac{1}{81}$

5. $\left(\dfrac{6}{7}\right)^{-2} = \dfrac{6^{-2}}{7^{-2}}$
$= \dfrac{6^{-2}}{1} \cdot \dfrac{1}{7^{-2}}$
$= \dfrac{1}{6^2} \cdot \dfrac{7^2}{1}$
$= \dfrac{7^2}{6^2}$
$= \dfrac{49}{36}$

6. $\dfrac{x}{x^{-4}} = \dfrac{x^1}{x^{-4}} = x^{1-(-4)} = x^5$

7. $\dfrac{y^{-9}}{z^{-5}} = y^{-9} \cdot \dfrac{1}{z^{-5}} = \dfrac{1}{y^9} \cdot z^5 = \dfrac{z^5}{y^9}$

8. $\dfrac{y^{-4}}{y^6} = y^{-4-6} = y^{-10} = \dfrac{1}{y^{10}}$

9. $\dfrac{(x^5)^3 x}{x^4} = \dfrac{x^{15} \cdot x}{x^4}$
$= \dfrac{x^{15+1}}{x^4}$
$= \dfrac{x^{16}}{x^4}$
$= x^{16-4}$
$= x^{12}$

10. $\left(\dfrac{9x^3}{y}\right)^{-2} = \dfrac{9^{-2}(x^3)^{-2}}{y^{-2}}$

$\qquad = \dfrac{9^{-2}x^{-6}}{y^{-2}}$

$\qquad = \dfrac{y^2}{9^2 x^6}$

$\qquad = \dfrac{y^2}{81x^6}$

11. $(a^{-4}b^7)^{-5} = (a^{-4})^{-5}(b^7)^{-5}$

$\qquad = a^{20}b^{-35}$

$\qquad = \dfrac{a^{20}}{b^{35}}$

12. $\dfrac{(2x)^4}{x^8} = \dfrac{2^4 x^4}{x^8} = 2^4 x^{4-8} = 2^4 x^{-4} = \dfrac{16}{x^4}$

13. $\dfrac{y^{-10}}{(y^5)^4} = \dfrac{y^{-10}}{y^{20}} = y^{-10-20} = y^{-30} = \dfrac{1}{y^{30}}$

14. $(4a^2)^{-3} = 4^{-3}(a^2)^{-3}$

$\qquad = 4^{-3}a^{-6}$

$\qquad = \dfrac{1}{4^3 a^6}$

$\qquad = \dfrac{1}{64a^6}$

15. $-\dfrac{32x^{-3}y^{-6}}{8x^{-5}y^{-2}} = -\dfrac{32}{8} \cdot x^{-3-(-5)}y^{-6-(-2)}$

$\qquad = -4x^2 y^{-4}$

$\qquad = -\dfrac{4x^2}{y^4}$

16. $\dfrac{(3x^{-2}y)^{-2}}{(2x^7 y)^3} = \dfrac{3^{-2}(x^{-2})^{-2}y^{-2}}{2^3(x^7)^3 y^3}$

$\qquad = \dfrac{3^{-2}x^4 y^{-2}}{2^3 x^{21}y^3}$

$\qquad = \dfrac{3^{-2}}{2^3} \cdot x^{4-21}y^{-2-3}$

$\qquad = \dfrac{3^{-2}}{2^3}x^{-17}y^{-5}$

$\qquad = \dfrac{1}{2^3 3^2 x^{17}y^5}$

$\qquad = \dfrac{1}{72x^{17}y^5}$

17. a. $420,000 = 4.2 \times 10^5$

$\quad$ **b.** $0.00017 = 1.7 \times 10^{-4}$

$\quad$ **c.** $9,060,000,000 = 9.06 \times 10^9$

$\quad$ **d.** $0.000007 = 7.0 \times 10^{-6}$

18. a. $3.062 \times 10^{-4} = 0.0003062$

$\quad$ **b.** $5.21 \times 10^4 = 52,100$

$\quad$ **c.** $9.6 \times 10^{-5} = 0.000096$

$\quad$ **d.** $6.002 \times 10^6 = 6,002,000$

19. a. $(9 \times 10^7)(4 \times 10^{-9}) = 9 \cdot 4 \cdot 10^7 \cdot 10^{-9}$

$\qquad\qquad\qquad\qquad\quad = 36 \times 10^{-2}$

$\qquad\qquad\qquad\qquad\quad = 0.36$

$\quad$ **b.** $\dfrac{8 \times 10^4}{2 \times 10^{-3}} = \dfrac{8}{2} \times 10^{4-(-3)}$

$\qquad\qquad\qquad = 4 \times 10^7$

$\qquad\qquad\qquad = 40,000,000$

Calculator Explorations

1. 5.31×10^3 5.31 EE 3

2. -4.8×10^{14} -4.8 EE 14

3. 6.6×10^{-9} 6.6 EE -9

4. -9.9811×10^{-2} -9.9811 EE -2

5. $3,000,000 \times 5,000,000 = 1.5 \times 10^{13}$

6. $230,000 \times 1,000 = 2.3 \times 10^8$

7. $(3.26 \times 10^6)(2.5 \times 10^{13}) = 8.15 \times 10^{19}$

8. $(8.76 \times 10^{-4})(1.237 \times 10^9) = 1.083612 \times 10^6$

Mental Math

1. $5x^{-2} = 5 \cdot \dfrac{1}{x^2} = \dfrac{5}{x^2}$

2. $3x^{-3} = 3 \cdot \dfrac{1}{x^3} = \dfrac{3}{x^3}$

3. $\dfrac{1}{y^{-6}} = y^6$

4. $\dfrac{1}{x^{-3}} = x^3$

5. $\dfrac{4}{y^{-3}} = 4 \cdot \dfrac{1}{y^{-3}} = 4y^3$

6. $\dfrac{16}{y^{-7}} = 16 \cdot \dfrac{1}{y^{-7}} = 16y^7$

Exercise Set 3.2

1. $4^{-3} = \dfrac{1}{4^3} = \dfrac{1}{64}$

3. $7x^{-3} = 7 \cdot \dfrac{1}{x^3} = \dfrac{7}{x^3}$

5. $\left(-\dfrac{1}{4}\right)^{-3} = \dfrac{(-1)^{-3}}{4^{-3}}$

$= (-1)^{-3} \cdot \dfrac{1}{4^{-3}}$

$= \dfrac{1}{(-1)^3} \cdot 4^3$

$= \dfrac{1}{-1} \cdot 64$

$= -64$

7. $3^{-1} + 2^{-1} = \dfrac{1}{3} + \dfrac{1}{2} = \dfrac{2}{6} + \dfrac{3}{6} = \dfrac{5}{6}$

9. $\dfrac{1}{p^{-3}} = p^3$

11. $\dfrac{p^{-5}}{q^{-4}} = \dfrac{1}{p^5} \cdot \dfrac{q^4}{1} = \dfrac{q^4}{p^5}$

13. $\dfrac{x^{-2}}{x} = \dfrac{x^{-2}}{x^1} = x^{-2-1} = x^{-3} = \dfrac{1}{x^3}$

15. $\dfrac{z^{-4}}{z^{-7}} = z^{-4-(-7)} = z^{-4+7} = z^3$

17. $3^{-2} + 3^{-1} = \dfrac{1}{3^2} + \dfrac{1}{3} = \dfrac{1}{9} + \dfrac{1}{3} = \dfrac{4}{9}$

19. $(-3)^{-2} = \dfrac{1}{(-3)^2} = \dfrac{1}{9}$

21. $\dfrac{-1}{p^{-4}} = -1 \cdot \dfrac{1}{p^{-4}} = -1 \cdot p^4 = -p^4$

23. $-2^0 - 3^0 = -(2^0) - (3^0) = -1 - 1 = -2$

25. $\dfrac{x^2 x^5}{x^3} = \dfrac{x^{2+5}}{x^3} = \dfrac{x^7}{x^3} = x^{7-3} = x^4$

27. $\dfrac{p^2 p}{p^{-1}} = \dfrac{p^{2+1}}{p^{-1}} = \dfrac{p^3}{p^{-1}} = p^{3-(-1)} = p^{3+1} = p^4$

29. $\dfrac{(m^5)^4 m}{m^{10}} = \dfrac{m^{5 \cdot 4} m^1}{m^{10}}$

$= \dfrac{m^{20} m^1}{m^{10}}$

$= \dfrac{m^{20+1}}{m^{10}}$

$= \dfrac{m^{21}}{m^{10}}$

$= m^{21-10}$

$= m^{11}$

31. $\dfrac{r}{r^{-3} r^{-2}} = \dfrac{r}{r^{(-3)+(-2)}}$

$= \dfrac{r}{r^{-5}}$

$= r^{1-(-5)}$

$= r^{1+5}$

$= r^6$

33. $(x^5 y^3)^{-3} = (x^5)^{-3} (y^3)^{-3}$

$= x^{-15} y^{-9}$

$= \dfrac{1}{x^{15} y^9}$

35. $\dfrac{(x^2)^3}{x^{10}} = \dfrac{x^6}{x^{10}} = x^{6-10} = x^{-4} = \dfrac{1}{x^4}$

37. $\dfrac{(a^5)^2}{(a^3)^4} = \dfrac{a^{10}}{a^{12}} = a^{10-12} = a^{-2} = \dfrac{1}{a^2}$

39. $\dfrac{8k^4}{2k} = \dfrac{8}{2} \cdot \dfrac{k^4}{k} = 4 \cdot k^{4-1} = 4k^3$

41. $\dfrac{-6m^4}{-2m^3} = \dfrac{-6}{-2} \cdot \dfrac{m^4}{m^3} = 3 \cdot m^{4-3} = 3m^1 = 3m$

43. $\dfrac{-24a^6 b}{6ab^2} = \dfrac{-24}{6} \cdot \dfrac{a^6}{a} \cdot \dfrac{b}{b^2}$

$= -4a^{6-1} b^{1-2}$

$= -4a^5 b^{-1}$

$= -\dfrac{4a^5}{b}$

45. $\dfrac{6x^2 y^3}{-7x^2 y^5} = \dfrac{6}{-7} \cdot \dfrac{x^2}{x^2} \cdot \dfrac{y^3}{y^5}$

$= -\dfrac{6}{7} \cdot x^{2-2} y^{3-5}$

$= -\dfrac{6}{7} x^0 y^{-2}$

$= -\dfrac{6}{7} \cdot 1 \cdot \dfrac{1}{y^2}$

$= -\dfrac{6}{7y^2}$

47. $(3a^2 b^{-4})^3 = 3^3 a^{2 \cdot 3} b^{-4 \cdot 3}$

$= 27a^6 b^{-12}$

$= \dfrac{27a^6}{b^{12}}$

49. $(a^{-5} b^2)^{-6} = (a^{-5})^{-6} (b^2)^{-6}$

$= a^{30} b^{-12}$

$= \dfrac{a^{30}}{b^{12}}$

51. $\left(\dfrac{x^{-2} y^4}{x^3 y^7}\right)^2 = \dfrac{x^{-2 \cdot 2}}{x^{3 \cdot 2}} \cdot \dfrac{y^{4 \cdot 2}}{y^{7 \cdot 2}}$

$= \dfrac{x^{-4}}{x^6} \cdot \dfrac{y^8}{y^{14}}$

$= x^{-4-6} y^{8-14}$

$= x^{-10} y^{-6}$

$= \dfrac{1}{x^{10} y^6}$

53. $\dfrac{4^2 z^{-3}}{4^3 z^{-5}} = \dfrac{4^2}{4^3} \cdot \dfrac{z^{-3}}{z^{-5}}$

$\quad = 4^{2-3} z^{-3-(-5)}$

$\quad = 4^{-1} z^{-3+5}$

$\quad = \dfrac{1}{4} \cdot z^2$

$\quad = \dfrac{z^2}{4}$

55. $\dfrac{3^{-1} x^4}{3^3 x^{-7}} = \dfrac{3^{-1}}{3^3} \cdot \dfrac{x^4}{x^{-7}}$

$\quad = 3^{-1-3} x^{4-(-7)}$

$\quad = 3^{-4} x^{4+7}$

$\quad = 3^{-4} x^{11}$

$\quad = \dfrac{x^{11}}{3^4}$

$\quad = \dfrac{x^{11}}{81}$

57. $\dfrac{7ab^{-4}}{7^{-1} a^{-3} b^2} = \dfrac{7}{7^{-1}} \cdot \dfrac{a^1}{a^{-3}} \cdot \dfrac{b^{-4}}{b^2}$

$\quad = 7^{1-(-1)} a^{1-(-3)} b^{-4-2}$

$\quad = 7^{1+1} a^{1+3} b^{-6}$

$\quad = 7^2 a^4 b^{-6}$

$\quad = \dfrac{49 a^4}{b^6}$

59. $\dfrac{-12 m^5 n^{-7}}{4 m^{-2} n^{-3}} = \dfrac{-12}{4} \cdot \dfrac{m^5}{m^{-2}} \cdot \dfrac{n^{-7}}{n^{-3}}$

$\quad = -3 m^{5-(-2)} n^{-7-(-3)}$

$\quad = -3 m^{5+2} n^{-7+3}$

$\quad = -3 m^7 n^{-4}$

$\quad = -\dfrac{3 m^7}{n^4}$

61. $\left(\dfrac{a^{-5} b}{ab^3} \right)^{-4} = \dfrac{(a^{-5})^{-4} b^{-4}}{a^{-4} (b^3)^{-4}}$

$\quad = \dfrac{a^{20} b^{-4}}{a^{-4} b^{-12}}$

$\quad = a^{20-(-4)} b^{-4-(-12)}$

$\quad = a^{20+4} b^{-4+12}$

$\quad = a^{24} b^8$

63. $(5^2)(8)(2^0) = (25)(8)(1) = 200$

65. $\dfrac{(xy^3)^5}{(xy)^{-4}} = \dfrac{x^5 (y^3)^5}{x^{-4} y^{-4}}$

$\quad = \dfrac{x^5 y^{15}}{x^{-4} y^{-4}}$

$\quad = x^{5-(-4)} y^{15-(-4)}$

$\quad = x^{5+4} y^{15+4}$

$\quad = x^9 y^{19}$

67. $\dfrac{(-2xy^{-3})^{-3}}{(xy^{-1})^{-1}} = \dfrac{-2^{-3} x^{-3} (y^{-3})^{-3}}{x^{-1} (y^{-1})^{-1}}$

$\quad = \dfrac{-2^{-3} x^{-3} y^9}{x^{-1} y}$

$\quad = -2^{-3} x^{-3-(-1)} y^{9-1}$

$\quad = -2^{-3} x^{-3+1} y^8$

$\quad = -2^{-3} x^{-2} y^8$

$\quad = -\dfrac{y^8}{2^3 x^2}$

$\quad = -\dfrac{y^8}{8 x^2}$

69. $\dfrac{(a^4 b^{-7})^{-5}}{(5a^2 b^{-1})^{-2}} = \dfrac{(a^4)^{-5}(b^{-7})^{-5}}{5^{-2}(a^2)^{-2}(b^{-1})^{-2}}$

$\qquad = \dfrac{a^{-20}b^{35}}{5^{-2}a^{-4}b^2}$

$\qquad = 5^2 a^{-20-(-4)}b^{35-2}$

$\qquad = 25a^{-20+4}b^{33}$

$\qquad = 25a^{-16}b^{33}$

$\qquad = \dfrac{25b^{33}}{a^{16}}$

71. $V = s^3$

$\qquad = \left(\dfrac{3x^{-2}}{z} \text{ inches}\right)^3$

$\qquad = \left(\dfrac{3^3(x^{-2})^3}{z^3}\right) \text{ cubic inches}$

$\qquad = \dfrac{27x^{-6}}{z^3} \text{ cubic inches}$

$\qquad = \dfrac{27}{z^3 x^6} \text{ cubic inches}$

73. $78,000 = 7.8 \times 10^4$

75. $0.00000167 = 1.67 \times 10^{-6}$

77. $0.00635 = 6.35 \times 10^{-3}$

79. $1,160,000 = 1.16 \times 10^6$

81. $13,600 = 1.36 \times 10^4$

83. $8.673 \times 10^{-10} = 0.0000000008673$

85. $3.3 \times 10^{-2} = 0.033$

87. $2.032 \times 10^4 = 20,320$

89. $7.0 \times 10^8 = 700,000,000$

91. $940,000,000 = 9.4 \times 10^8$

93. $1.23 \times 10^{12} = 1,230,000,000,000$

95. Tallest Bar = Yahoo!; 115,000,000;
$\quad 1.15 \times 10^8$
$\quad$ Shortest Bar = eBay; 58,000,000; 5.8×10^7

97. $(1.2 \times 10^{-3})(3 \times 10^{-2}) = 1.2 \cdot 3 \cdot 10^{-3} \cdot 10^{-2}$
$\qquad\qquad\qquad\qquad\quad = 3.6 \times 10^{-5}$
$\qquad\qquad\qquad\qquad\quad = 0.000036$

99. $(4 \times 10^{-10})(7 \times 10^{-9})$
$\quad = (4 \cdot 7)(10^{-10} \cdot 10^{-9})$
$\quad = 28 \cdot 10^{-19}$
$\quad = 0.0000000000000000028$

101. $\dfrac{8 \times 10^{-1}}{16 \times 10^5} = \dfrac{8}{16} \times 10^{-1-5}$
$\qquad\qquad = 0.5 \times 10^{-6}$
$\qquad\qquad = 5.0 \times 10^{-7}$
$\qquad\qquad = 0.0000005$

103. $\dfrac{1.4 \times 10^{-2}}{7 \times 10^{-8}} = \dfrac{1.4}{7} \cdot \dfrac{10^{-2}}{10^{-8}}$
$\qquad\qquad = 0.2 \cdot 10^{-2-(-8)}$
$\qquad\qquad = 0.2 \cdot 10^{-2+8}$
$\qquad\qquad = 0.2 \cdot 10^6$
$\qquad\qquad = 200,000$

105. $7.5 \times 10^5 \cdot 3600 = 7.5 \times 10^5 \cdot 3.6 \times 10^3$
$\qquad\qquad\qquad = 7.5 \cdot 3.6 \cdot 10^5 \cdot 10^3$
$\qquad\qquad\qquad = 27 \times 10^8$
$\qquad\qquad\qquad = 2.7 \times 10^9$

107. $3x - 5x + 2 = -2x + 7$

109. $y - 10 + y = y + y - 10 = 2y - 10$

111. $7x + 2 - 8x - 6 = 7x - 8x + 2 - 6 = -x - 4$

113. $(2a^3)^3 a^4 + a^5 a^8 = 2^3 (a^3)^3 a^4 + a^5 a^8$
$$= 8a^9 a^4 + a^5 a^8$$
$$= 8a^{13} + a^{13}$$
$$= 9a^{13}$$

115. answers may vary

117. answers may vary

119. a. 9.7×10^{-2} or $1.3 \times 10^1 \Rightarrow 1.3 \times 10^1$

     b. 8.6×10^5 or $4.4 \times 10^7 \Rightarrow 4.4 \times 10^7$

     c. 6.1×10^{-2} or $5.6 \times 10^{-4} \Rightarrow 6.1 \times 10^{-2}$

121. a. $5^{-1} = \dfrac{1}{5}$
$$5^{-2} = \dfrac{1}{25}$$
Since $\dfrac{1}{5} > \dfrac{1}{25}$, the statement
"$5^{-1} < 5^{-2}$" is false.

     b. $\left(\dfrac{1}{5}\right)^{-1} = \dfrac{1}{5^{-1}} = 5$
$$\left(\dfrac{1}{5}\right)^{-2} = \dfrac{1}{5^{-2}} = 5^2 = 25$$
Since $5 < 25$, the statement
"$\left(\dfrac{1}{5}\right)^{-1} < \left(\dfrac{1}{5}\right)^{-2}$" is true.

     c. From part a, the statement "$a^{-1} < a^{-2}$ for all nonzero numbers" is false.

123. $(x^{-3s})^3 = x^{-9s} = \dfrac{1}{x^{9s}}$

125. $a^{4m+1} \cdot a^4 = a^{4m+1+4} = a^{4m+5}$

Section 3.3 Practice Problems

1. $-6x^6 + 4x^5 + 7x^3 - 9x^2 - 1$

Term	Coefficient
$7x^3$	7
$-9x^2$	-9
$-6x^6$	-6
$4x^5$	4
-1	-1

2. $-15x^3 + 2x^2 - 5$
The term $-15x^3$ has degree 3.
The term $2x^2$ has degree 2.
The term -5 has degree 0 since -5 is $-5x^0$.

3. a. The degree of the binomial $-6x + 14$ is 1.

     b. The degree of the polynomial $9x - 3x^6 + 5x^4 + 2$ is 6. The polynomial is neither a monomial, binomial, or trinomial.

     c. The degree of the trinomial $10x^2 - 6x - 6$ is 2.

4. a. When $x = -1$,
$-2x + 10 = -2(-1) + 10 = 2 + 10 = 12.$

     b. When $x = -1$,
$$6x^2 + 11x - 20 = 6(-1)^2 + 11(-1) - 20$$
$$= 6 - 11 - 20$$
$$= -25.$$

5. When $t = 2$ seconds,
$$-16t^2 + 592.1 = -16(2)^2 + 592.1$$
$$= -16(4) + 592.1$$
$$= -64 + 592.1$$
$$= 528.1 \text{ feet.}$$

When $t = 4$ seconds,
$$-16t^2 + 592.1 = -16(4)^2 + 592.1$$
$$= -16(16) + 592.1$$
$$= -256 + 592.1$$
$$= 336.1 \text{ feet.}$$

6. $-6y + 8y = (-6 + 8)y = 2y$

7. $14y^2 + 3 - 10y^2 - 9 = 14y^2 - 10y^2 + 3 - 9$
$$= 4y^2 - 6$$

8. $7x^3 + x^3 = 7x^3 + 1x^3 = 8x^3$

9. $23x^2 - 6x - x - 15 = 23x^2 - 7x - 15$

10. $\dfrac{2}{7}x^3 - \dfrac{1}{4}x + 2 - \dfrac{1}{2}x^3 + \dfrac{3}{8}x$
$$= \dfrac{2}{7}x^3 - \dfrac{1}{2}x^3 - \dfrac{1}{4}x + \dfrac{3}{8}x + 2$$
$$= \dfrac{4}{14}x^3 - \dfrac{7}{14}x^3 - \dfrac{2}{8}x + \dfrac{3}{8}x + 2$$
$$= -\dfrac{3}{14}x^3 + \dfrac{1}{8}x + 2$$

11. Area $= 5 \cdot x + x \cdot x + 4 \cdot 5 + x \cdot x + 8 \cdot x$
$$= 5x + x^2 + 20 + x^2 + 8x$$
$$= x^2 + x^2 + 5x + 8x + 20$$
$$= 2x^2 + 13x + 20$$

12.

Terms of Polynomial	Degree of Term
$-2x^3y^2$	3 + 2 or 5
4	0
$-8xy$	1 + 1 or 2
$3x^3y$	3 + 1 or 4
$5xy^2$	1 + 2 or 3

The degree of the polynomial is 5.

13. $11ab - 6a^2 - ba + 8b^2$
$$= (11 - 1)ab - 6a^2 + 8b^2$$
$$= 10ab - 6a^2 + 8b^2$$

14. $7x^2y^2 + 2y^2 - 4y^2x^2 + x^2 - y^2 + 5x^2$
$$= 7x^2y^2 - 4x^2y^2 + 2y^2 - y^2 + x^2 + 5x^2$$
$$= 3x^2y^2 + y^2 + 6x^2$$

15. a. $x^2 + 9 = x^2 + 0x^1 + 9$ or $x^2 + 0x + 9$

b. $9m^3 + m^2 - 5 = 9m^3 + m^2 + 0m^1 - 5$
$$= 9m^3 + m^2 + 0m - 5$$

c. $-3a^3 + a^4 = a^4 - 3a^3 + 0a^2 + 0a^1 + 0a^0$
$$= a^4 - 3a^3 + 0a^2 + 0a + 0a^0$$

Exercise Set 3.3

1.

Term	Coefficient
x^2	1
$-3x$	-3
5	5

3.

Term	Coefficient
$-5x^4$	-5
$3.2x^2$	3.2
x	1
-5	-5

5. $x + 2 = x^1 + 2$
This is a binomial of degree 1.

7. $9m^3 - 5m^2 + 4m - 8$
This is a polynomial of degree 3. None of these.

9. $12x^4 - x^6 - 12x^2 = -x^6 + 12x^4 - 12x^2$
This is a trinomial of degree 6.

11. $3z - 5z^4 = -5z^4 + 3z$
This is a binomial of degree 4.

13. a. $5x - 6 = 5(0) - 6 = 0 - 6 = -6$

   b. $5x - 6 = 5(-1) - 6 = -5 - 6 = -11$

15. a. $x^2 - 5x - 2 = (0)^2 - 5(0) - 2$
$$= 0 - 0 - 2$$
$$= -2$$

   b. $x^2 - 5x - 2 = (-1)^2 - 5(-1) - 2$
$$= 1 + 5 - 2$$
$$= 4$$

17. a. $-x^3 + 4x^2 - 15 = -(0)^3 + 4(0)^2 - 15$
$$= 0 + 0 - 15$$
$$= -15$$

   b. $-x^3 + 4x^2 - 15 = -(-1)^3 + 4(-1)^2 - 15$
$$= -(-1) + 4(1) - 15$$
$$= 1 + 4 - 15$$
$$= -10$$

19. $-16t^2 + 200t = -16(1)^2 + 200(1)$
$$= -16 + 200$$
$$= 184$$
After 1 second, the height of the rocket is 184 feet.

21. $-16t^2 + 200t = -16(7.6)^2 + 200(7.6)$
$$= -16(57.76) + 1520$$
$$= -924.16 + 1520$$
$$= 595.84$$
After 7.6 seconds, the height of the rocket is 595.84 feet.

23. $-24x^2 + 336x - 132$
$$= -24(7)^2 + 336(7) - 132$$
$$= -24(49) + 336(7) - 132$$
$$= -1176 + 2352 - 132$$
$$= 1044 \text{ thousand or } 1,044,000 \text{ visitors}$$

25. $9x - 20x = (9 - 20)x = -11x$

27. $14x^3 + 9x^3 = (14 + 9)x^3 = 23x^3$

29. $7x^2 + 3 + 9x^2 - 10 = 7x^2 + 9x^2 + 3 - 10$
$$= (7 + 9)x^2 + 3 - 10$$
$$= 16x^2 - 7$$

31. $15x^2 - 3x^2 - 13 = (15 - 3)x^2 - 13 = 12x^2 - 13$

33. $8s - 5s + 4s = (8 - 5 + 4)s = 7s$

35. $0.1y^2 - 1.2y^2 + 6.7 - 1.9$
$$= (0.1 - 1.2)y^2 + 6.7 - 1.9$$
$$= -1.1y^2 + 4.8$$

37. $\dfrac{2}{3}x^4 + 12x^3 + \dfrac{1}{6}x^4 - 19x^3 - 19$
$$= \dfrac{2}{3}x^4 + \dfrac{1}{6}x^4 + 12x^3 - 19x^3 - 19$$
$$= \left(\dfrac{4}{6} + \dfrac{1}{6}\right)x^4 + (12 - 19)x^3 - 19$$
$$= \dfrac{5}{6}x^4 - 7x^3 - 19$$

39. $\dfrac{3}{20}x^3 + \dfrac{1}{10} - \dfrac{3}{10}x - \dfrac{1}{5} - \dfrac{7}{20}x + 6x^2$
$$= \dfrac{3}{20}x^3 + 6x^2 - \dfrac{3}{10}x - \dfrac{7}{20}x + \dfrac{1}{10} - \dfrac{1}{5}$$
$$= \dfrac{3}{20}x^3 + 6x^2 + \left(-\dfrac{6}{20} - \dfrac{7}{20}\right)x + \dfrac{1}{10} - \dfrac{2}{10}$$
$$= \dfrac{3}{20}x^3 + 6x^2 - \dfrac{13}{20}x - \dfrac{1}{10}$$

41. $9ab = 9a^1b^1$ has degree $1 + 1 = 2$.

$-6a = -6a^1$ has degree 1.

$5b = 5b^1$ has degree 1.

$-3 = -3a^0b^0$ has degree 0.

$9ab - 6a + 5b - 3$ is a polynomial of degree 2.

43. $x^3y = x^3y^1$ has degree $3 + 1 = 4$.

$-6 = -6x^0y^0$ has degree 0.

$2x^2y^2$ has degree $2 + 2 = 4$.

$5y^3$ has degree 3.

$x^3y - 6 + 2x^2y^2 + 5y^3$ is a polynomial of degree 4.

45. $3ab - 4a + 6ab - 7a = 3ab + 6ab - 4a - 7a$
$$= (3+6)ab - (4+7)a$$
$$= 9ab - 11a$$

47. $4x^2 - 6xy + 3y^2 - xy$
$$= 4x^2 - 6xy - xy + 3y^2$$
$$= 4x^2 + (-6-1)xy + 3y^2$$
$$= 4x^2 - 7xy + 3y^2$$

49. $5x^2y + 6xy^2 - 5yx^2 + 4 - 9y^2x$
$$= 5x^2y - 5x^2y + 6xy^2 - 9xy^2 + 4$$
$$= (5-5)x^2y + (6-9)xy^2 + 4$$
$$= 0x^2y - 3xy^2 + 4$$
$$= -3xy^2 + 4$$

51. $14y^3 - 9 + 3a^2b^2 - 10 - 19b^2a^2$
$$= 14y^3 - 9 - 10 + 3a^2b^2 - 19a^2b^2$$
$$= 14y^3 + (-9-10) + (3-19)a^2b^2$$
$$= 14y^3 - 19 - 16a^2b^2$$

53. $7x^2 + 3 = 7x^2 + 0x + 3$

55. $x^3 - 64 = x^3 + 0x^2 + 0x - 64$

57. $5y^3 + 2y - 10 = 5y^3 + 0y^2 + 2y - 10$

59. $8y + 2y^4 = 2y^4 + 0y^3 + 0y^2 + 8y + 0$

61. $6x^5 + x^3 - 3x + 15$
$$= 6x^5 + 0x^4 + x^3 + 0x^2 - 3x + 15$$

63. $4x^2 + 7x + x^2 + 5x = 4x^2 + x^2 + 7x + 5x$
$$= (4+1)x^2 + (7+5)x$$
$$= 5x^2 + 12x$$

65. $5x + 3 + 4x + 3 + 2x + 6 + 3x + 7x$
$$= 5x + 4x + 2x + 3x + 7x + 3 + 3 + 6$$
$$= (5 + 4 + 2 + 3 + 7)x + 12$$
$$= 21x + 12$$

67. $4 + 5(2x + 3) = 4 + 10x + 15 = 10x + 19$

69. $2(x - 5) + 3(5 - x) = 2x - 10 + 15 - 3x$
$$= 2x - 3x - 10 + 15$$
$$= (2-3)x + 5$$
$$= -x + 5$$

71. answers may vary

73. answers may vary

75. $x^4 \cdot x^9 = x^{4+9} = x^{13}$

77. $a \cdot b^3 \cdot a^2 \cdot b^7 = a^1 \cdot a^2 \cdot b^3 \cdot b^7$
$$= a^{1+2}b^{3+7}$$
$$= a^3b^{10}$$

79. $(y^5)^4 + (y^2)^{10} = y^{20} + y^{20} = 2y^{20}$

81. answers may vary

83. answers may vary

85. $1.85x^2 - 3.76x + 9.25x^2 + 10.76 - 4.21x = 1.85x^2 + 9.25x^2 - 3.76x - 4.21x + 10.76$
$$= (1.85 + 9.25)x^2 - (3.76 + 4.21)x + 10.76$$
$$= 11.1x^2 - 7.97x + 10.76$$

Section 3.4 Practice Problems

1. $(3x^5 - 7x^3 + 2x - 1) + (3x^3 - 2x) = 3x^5 - 7x^3 + 2x - 1 + 3x^3 - 2x$
$$= 3x^5 + (-7x^3 + 3x^3) + (2x - 2x) - 1$$
$$= 3x^5 - 4x^3 - 1$$

2. $(5x^2 - 2x + 1) + (-6x^2 + x - 1) = 5x^2 - 2x + 1 - 6x^2 + x - 1$
$$= (5x^2 - 6x^2) + (-2x + x) + (1 - 1)$$
$$= -x^2 - x$$

3. $\quad 9y^2 - 6y + 5$
$\qquad\qquad 4y + 3$
$\overline{\quad 9y^2 - 2y + 8}$

4. $(9x + 5) - (4x - 3) = (9x + 5) + [-(4x - 3)]$
$$= (9x + 5) + (-4x + 3)$$
$$= 9x + 5 - 4x + 3$$
$$= 5x + 8$$

5. $(4x^3 - 10x^2 + 1) - (-4x^3 + x^2 - 11) = (4x^3 - 10x^2 + 1) + (4x^3 - x^2 + 11)$
$$= 4x^3 - 10x^2 + 1 + 4x^3 - x^2 + 11$$
$$= 4x^3 + 4x^3 - 10x^2 - x^2 + 1 + 11$$
$$= 8x^3 - 11x^2 + 12$$

6. $\quad 2y^2 - 2y + 7 \qquad\qquad 2y^2 - 2y + 7$
$\quad \underline{-(6y^2 - 3y + 2)} \quad \Rightarrow \quad \underline{-6y^2 + 3y - 2}$
$\qquad\qquad\qquad\qquad\qquad\quad -4y^2 + \ y + 5$

7. $[(4x - 3) + (12x - 5)] - (3x + 1) = 4x - 3 + 12x - 5 - 3x - 1$
$$= 4x + 12x - 3x - 3 - 5 - 1$$
$$= 13x - 9$$

8. $(2a^2 - ab + 6b^2) + (-3a^2 + ab - 7b^2) = 2a^2 - ab + 6b^2 - 3a^2 + ab - 7b^2$
$$= -a^2 - b^2$$

9. $(5x^2y^2 + 3 - 9x^2y + y^2) - (-x^2y^2 + 7 - 8xy^2 + 2y^2) = 5x^2y^2 + 3 - 9x^2y + y^2 + x^2y^2 - 7 + 8xy^2 - 2y^2$

$$= 6x^2y^2 - 4 - 9x^2y + 8xy^2 - y^2$$

Exercise Set 3.4

1. $(3x + 7) + (9x + 5) = 3x + 7 + 9x + 5$
$$= 3x + 9x + 7 + 5$$
$$= 12x + 12$$

3. $(-7x + 5) + (-3x^2 + 7x + 5) = -7x + 5 - 3x^2 + 7x + 5$
$$= -3x^2 - 7x + 7x + 5 + 5$$
$$= -3x^2 + 10$$

5. $(-5x^2 + 3) + (2x^2 + 1) = -5x^2 + 3 + 2x^2 + 1$
$$= -5x^2 + 2x^2 + 3 + 1$$
$$= -3x^2 + 4$$

7. $(-3y^2 - 4y) + (2y^2 + y - 1) = -3y^2 - 4y + 2y^2 + y - 1$
$$= -3y^2 + 2y^2 - 4y + y - 1$$
$$= -y^2 - 3y - 1$$

9. $(1.2x^3 - 3.4x + 7.9) + (6.7x^3 + 4.4x^2 - 10.9) = 1.2x^3 - 3.4x + 7.9 + 6.7x^3 + 4.4x^2 - 10.9$
$$= 1.2x^3 + 6.7x^3 + 4.4x^2 - 3.4x + 7.9 - 10.9$$
$$= 7.9x^3 + 4.4x^2 - 3.4x - 3$$

11. $\left(\dfrac{3}{4}m^2 - \dfrac{2}{5}m + \dfrac{1}{8}\right) + \left(-\dfrac{1}{4}m^2 - \dfrac{3}{10}m + \dfrac{11}{16}\right) = \dfrac{3}{4}m^2 - \dfrac{2}{5}m + \dfrac{1}{8} - \dfrac{1}{4}m^2 - \dfrac{3}{10}m + \dfrac{11}{16}$
$$= \dfrac{3}{4}m^2 - \dfrac{1}{4}m^2 - \dfrac{2}{5}m - \dfrac{3}{10}m + \dfrac{1}{8} + \dfrac{11}{16}$$
$$= \dfrac{3}{4}m^2 - \dfrac{1}{4}m^2 - \dfrac{4}{10}m - \dfrac{3}{10}m + \dfrac{2}{16} + \dfrac{11}{16}$$
$$= \dfrac{2}{4}m^2 - \dfrac{7}{10}m + \dfrac{13}{16}$$
$$= \dfrac{1}{2}m^2 - \dfrac{7}{10}m + \dfrac{13}{16}$$

13. $3t^2 + 4$
$\underline{5t^2 - 8}$
$8t^2 - 4$

15. $10a^3 - 8a^2 + 4a + 9$

$\underline{5a^3 + 9a^2 - 7a + 7}$

$15a^3 + a^2 - 3a + 16$

17. $(2x+5)-(3x-9) = (2x+5)+(-3x+9)$

$ = 2x+5-3x+9$

$ = 2x-3x+5+9$

$ = -x+14$

19. $(5x^2+4)-(-2y^2+4)$

$= (5x^2+4)+(2y^2-4)$

$= 5x^2+4+2y^2-4$

$= 5x^2+2y^2+4-4$

$= 5x^2+2y^2$

21. $3x-(5x-9) = 3x+(-5x+9)$

$ = 3x-5x+9$

$ = -2x+9$

23. $(2x^2+3x-9)-(-4x+7)$

$= (2x^2+3x-9)+(4x-7)$

$= 2x^2+3x-9+4x-7$

$= 2x^2+3x+4x-9-7$

$= 2x^2+7x-16$

25. $(5x+8)-(-2x^2-6x+8)$

$= (5x+8)+(2x^2+6x-8)$

$= 5x+8+2x^2+6x-8$

$= 2x^2+5x+6x+8-8$

$= 2x^2+11x$

27. $(0.7x^2+0.2x-0.8)-(0.9x^2+1.4)$

$= (0.7x^2+0.2x-0.8)+(-0.9x^2-1.4)$

$= 0.7x^2+0.2x-0.8-0.9x^2-1.4$

$= 0.7x^2-0.9x^2+0.2x-0.8-1.4$

$= -0.2x^2+0.2x-2.2$

29. $\left(\dfrac{1}{4}z^2 - \dfrac{1}{5}z\right) - \left(-\dfrac{3}{20}z^2 + \dfrac{1}{10}z - \dfrac{7}{20}\right)$

$= \left(\dfrac{1}{4}z^2 - \dfrac{1}{5}z\right) + \left(\dfrac{3}{20}z^2 - \dfrac{1}{10}z + \dfrac{7}{20}\right)$

$= \dfrac{1}{4}z^2 - \dfrac{1}{5}z + \dfrac{3}{20}z^2 - \dfrac{1}{10}z + \dfrac{7}{20}$

$= \dfrac{5}{20}z^2 + \dfrac{3}{20}z^2 - \dfrac{2}{10}z - \dfrac{1}{10}z + \dfrac{7}{20}$

$= \dfrac{8}{20}z^2 - \dfrac{3}{10}z + \dfrac{7}{20}$

$= \dfrac{2}{5}z^2 - \dfrac{3}{10}z + \dfrac{7}{20}$

31. $\begin{array}{r} 4z^2-8z+3 \\ -\ (6z^2+8z-3) \\ \hline \end{array} \quad \Rightarrow \quad \begin{array}{r} 4z^2-8z+3 \\ -6z^2-8z+3 \\ \hline -2z^2-16z+6 \end{array}$

33. $\begin{array}{r} 5u^5-4u^2+3u-7 \\ -(3u^5+6u^2-8u+2) \\ \end{array} \quad \begin{array}{r} 5u^5-4u^2\ +3u-7 \\ -3u^5-6u^2\ +8u-2 \\ \hline 2u^5-10u^2+11u-9 \end{array}$

35. $(3x+5)+(2x-14) = 3x+5+2x-14$

$ = 3x+2x+5-14$

$ = 5x-9$

37. $(9x-1)-(5x+2) = (9x-1)+(-5x-2)$

$ = 9x-1-5x-2$

$ = 4x-3$

39. $(14y+12)+(-3y-5) = 14y+12-3y-5$

$ = 11y+7$

41. $(x^2+2x+1)-(3x^2-6x+2)$

$= (x^2+2x+1)+(-3x^2+6x-2)$

$= x^2+2x+1-3x^2+6x-2$

$= -2x^2+8x-1$

43. $(3x^2 + 5x - 8) + (5x^2 + 9x + 12) - (8x^2 - 14) = (3x^2 + 5x - 8) + (5x^2 + 9x + 12) + (-8x^2 + 14)$
$$= 3x^2 + 5x - 8 + 5x^2 + 9x + 12 - 8x^2 + 14$$
$$= 14x + 18$$

45. $(-a^2 + 1) - (a^2 - 3) + (5a^2 - 6a + 7) = (-a^2 + 1) + (-a^2 + 3) + (5a^2 - 6a + 7)$
$$= -a^2 + 1 - a^2 + 3 + 5a^2 - 6a + 7$$
$$= 3a^2 - 6a + 11$$

47. $(7x - 3) - 4x = 7x - 3 - 4x = 3x - 3$

49. $(4x^2 - 6x + 1) + (3x^2 + 2x + 1) = 4x^2 - 6x + 1 + 3x^2 + 2x + 1$
$$= 7x^2 - 4x + 2$$

51. $(7x^2 + 3x + 9) - (5x + 7) = (7x^2 + 3x + 9) + (-5x - 7)$
$$= 7x^2 + 3x + 9 - 5x - 7$$
$$= 7x^2 - 2x + 2$$

53. $(8y^2 + 7) + (6y + 9) - (4y^2 - 6y - 3) = (8y^2 + 7) + (6y + 9) + (-4y^2 + 6y + 3)$
$$= 8y^2 + 7 + 6y + 9 - 4y^2 + 6y + 3$$
$$= 4y^2 + 12y + 19$$

55. $(x^2 - 9x + 2) + (2x^2 - 6x + 1) - (3x^2 - 4) = (x^2 - 9x + 2) + (2x^2 - 6x + 1) + (-3x^2 + 4)$
$$= x^2 - 9x + 2 + 2x^2 - 6x + 1 - 3x^2 + 4$$
$$= -15x + 7$$

57. $(9a + 6b - 5) + (-11a - 7b + 6) = 9a + 6b - 5 - 11a - 7b + 6$
$$= -2a - b + 1$$

59. $(4x^2 + y^2 + 3) - (x^2 + y^2 - 2) = 4x^2 + y^2 + 3 - x^2 - y^2 + 2$
$$= 3x^2 + 5$$

61. $(x^2 + 2xy - y^2) + (5x^2 - 4xy + 20y^2) = x^2 + 2xy - y^2 + 5x^2 - 4xy + 20y^2$
$$= 6x^2 - 2xy + 19y^2$$

63. $(11r^2s + 16rs - 3 - 2r^2s^2) - (3sr^2 + 5 - 9r^2s^2) = 11r^2s + 16rs - 3 - 2r^2s^2 - 3r^2s - 5 + 9r^2s^2$
$$= 8r^2s + 16rs - 8 + 7r^2s^2$$

65. $(2x^2 + 5) + (4x - 1) + (-x^2 + 3x) = 2x^2 + 5 + 4x - 1 - x^2 + 3x$
$$= x^2 + 7x + 4$$

The perimeter is $(x^2 + 7x + 4)$ feet.

67. $(2x-3)+\left(\dfrac{4}{5}x\right)+\left(\dfrac{7}{10}x-1\right)+(2x-2)+(x+4)+(3x+5)=2x-3+\dfrac{4}{5}x+\dfrac{7}{10}x-1+2x-2+x+4+3x+5$

$$=\dfrac{19}{2}x+3$$

The perimeter is $\left(\dfrac{19}{2}x+3\right)$ units.

69. $(4y^2+4y+1)-(y^2-10)=(4y^2+4y+1)+(-y^2+10)$

$$=4y^2+4y+1-y^2+10$$
$$=3y^2+4y+11$$

The remaining piece is $(3y^2+4y+11)$ meters long.

71. $[(1.2x^2-3x+9.1)-(7.8x^2-3.1+8)]+(1.2x-6)=(1.2x^2-3x+9.1)+(-7.8x^2+3.1-8)+(1.2x-6)$

$$=1.2x^2-3x+9.1-7.8x^2+3.1-8+1.2x-6$$
$$=-6.6x^2-1.8x-1.8$$

73. $3x(2x)=(3\cdot2)(x\cdot x)=6x^2$

75. $(12x^3)(-x^5)=(12\cdot-1)(x^3\cdot x^5)=-12x^8$

77. $10x^2(20xy^2)=10\cdot20\cdot(x^2\cdot x)(y^2)=200x^3y^2$

79. Since $3+4=7$, $3x^2+4x^2=7x^2$ is a true statement.

81. Since $2+4=6$ and $3-5=-2$, $2x^4+3x^3-5x^3+4x^4=6x^4-2x^3$ is a true statement.

83. $10y-6y^2-y=9y-6y^2$; b

85. $(5x-3)+(5x-3)=5x-3+5x-3=10x-6$; e

87. a. $z+3z=1z+3z=4z$

    b. $z\cdot3z=z^1\cdot3z^1=3z^{1+1}=3z^2$

    c. $-z-3z=-1z-3z=-4z$

    d. $(-z)(-3z)=(-z^1)(-3z^1)=3z^{1+1}=3z^2$; answers may vary

89. a. $m\cdot m\cdot m=m^1\cdot m^1\cdot m^1=m^{1+1+1}=m^3$

    b. $m+m+m=1m+1m+1m=(1+1+1)m=3m$

c. $(-m)(-m)(-m) = (-1 \cdot m^1)(-1 \cdot m^1)(-1 \cdot m^1)$
$$= (-1)(-1)(-1)(m \cdot m \cdot m)$$
$$= -1m^3$$
$$= -m^3$$

d. $-m - m - m = -1m - 1m - 1m$
$$= (-1 - 1 - 1)m$$
$$= -3m; \text{ answers may vary}$$

91. $(-0.35x^2 + 0.49x + 71.75) + (0.025x^2 + 9.65x + 11.83)$
$$= -0.35x^2 + 0.49x + 71.75 + 0.025x^2 + 9.65x + 11.83$$
$$= -0.325x^2 + 10.14x + 83.58$$

Section 3.5 Practice Problems

1. $10x \cdot 9x = (10 \cdot 9)(x \cdot x) = 90x^2$

2. $8x^3(-11x^7) = (8 \cdot -11)(x^3 \cdot x^7) = -88x^{10}$

3. $(-5x^4)(-x) = (-5 \cdot -1)(x^4 \cdot x) = 5x^5$

4. $4x(x^2 + 4x + 3) = 4x(x^2) + 4x(4x) + 4x(3)$
$$= 4x^3 + 16x^2 + 12x$$

5. $8x(7x^4 + 1) = 8x(7x^4) + 8x(1) = 56x^5 + 8x$

6. $-2x^3(3x^2 - x + 2) = -2x^3(3x^2) - 2x^3(-x) - 2x^3(2)$
$$= -6x^5 + 2x^4 - 4x^3$$

7. a. $(x + 5)(x + 10) = x(x + 10) + 5(x + 10)$
$$= x \cdot x + x \cdot 10 + 5 \cdot x + 5 \cdot 10$$
$$= x^2 + 10x + 5x + 50$$
$$= x^2 + 15x + 50$$

b. $(4x + 5)(3x - 4) = 4x(3x - 4) + 5(3x - 4)$
$$= 4x(3x) + 4x(-4) + 5(3x) + 5(-4)$$
$$= 12x^2 - 16x + 15x - 20$$
$$= 12x^2 - x - 20$$

8. $(3x - 2y)^2 = (3x - 2y)(3x - 2y)$
$$= 3x(3x) + 3x(-2y) + (-2y)(3x) + (-2y)(-2y)$$
$$= 9x^2 - 6xy - 6xy + 4y^2$$
$$= 9x^2 - 12xy + 4y^2$$

9. $(x + 3)(2x^2 - 5x + 4) = x(2x^2) + x(-5x) + x(4) + 3(2x^2) + 3(-5x) + 3(4)$
$$= 2x^3 - 5x^2 + 4x + 6x^2 - 15x + 12$$
$$= 2x^3 + x^2 - 11x + 12$$

10.
$$
\begin{array}{r}
y^2 - 4y + 5 \\
3y^2 + 1 \\
\hline
y^2 - 4y + 5 \\
3y^4 - 12y^3 + 15y^2 \\
\hline
3y^4 - 12y^3 + 16y^2 - 4y + 5
\end{array}
$$

11.
$$
\begin{array}{r}
4x^2 - x - 1 \\
3x^2 + 6x - 2 \\
\hline
-8x^2 + 2x + 2 \\
24x^3 - 6x^2 - 6x \\
12x^4 - 3x^3 - 3x^2 \\
\hline
12x^4 + 21x^3 - 17x^2 - 4x + 2
\end{array}
$$

Mental Math

1. $x^3 \cdot x^5 = x^{3+5} = x^8$

2. $x^2 \cdot x^6 = x^{2+6} = x^8$

3. $x^3 + x^5$ Cannot be simplified.

4. $x^2 + x^6$ Cannot be simplified.

5. $y^4 \cdot y = y^{4+1} = y^5$

6. $y^9 \cdot y = y^{9+1} = y^{10}$

7. $x^7 \cdot x^7 = x^{7+7} = x^{14}$

8. $x^{11} \cdot x^{11} = x^{11+11} = x^{22}$

9. $x^7 + x^7 = 2x^7$

10. $x^{11} + x^{11} = 2x^{11}$

Exercise Set 3.5

1. $8x^2 \cdot 3x = (8 \cdot 3)(x^2 \cdot x) = 24x^3$

3. $(-x^3)(-x) = (-1 \cdot -1)(x^3 \cdot x) = x^4$

5. $-4n^3 \cdot 7n^7 = (-4 \cdot 7)(n^3 \cdot n^7) = -28n^{10}$

7. $(-3.1x^3)(4x^9) = (-3.1 \cdot 4)(x^3 \cdot x^9)$
$\qquad = -12.4x^{12}$

9. $\left(-\dfrac{1}{3}y^2\right)\left(\dfrac{2}{5}y\right) = \left(-\dfrac{1}{3}\right)\left(\dfrac{2}{5}\right)(y^2 \cdot y)$
$\qquad = -\dfrac{2}{15}y^3$

11. $(2x)(-3x^2)(4x^5) = (2 \cdot -3 \cdot 4)(x \cdot x^2 \cdot x^5)$
$\qquad = -24x^8$

13. $3x(2x+5) = 3x(2x) + 3x(5) = 6x^2 + 15x$

15. $7x(x^2 + 2x - 1) = 7x(x^2) + 7x(2x) + 7x(-1)$
$\qquad = 7x^3 + 14x^2 - 7x$

17. $-2a(a+4) = -2a(a) + (-2a)(4) = -2a^2 - 8a$

19. $3x(2x^2 - 3x + 4)$
$\quad = 3x(2x^2) + 3x(-3x) + 3x(4)$
$\quad = 6x^3 - 9x^2 + 12x$

21. $3a^2(4a^3 + 15) = 3a^2(4a^3) + 3a^2(15)$
$\qquad = 12a^5 + 45a^2$

23. $-2a^2(3a^2 - 2a + 3)$
$\quad = -2a^2(3a^2) - 2a^2(-2a) - 2a^2(3)$
$\quad = -6a^4 + 4a^3 - 6a^2$

25. $3x^2y(2x^3 - x^2y^2 + 8y^3)$
$\quad = 3x^2y(2x^3) + 3x^2y(-x^2y^2) + 3x^2y(8y^3)$
$\quad = 6x^5y - 3x^4y^3 + 24x^2y^4$

27. $-y(4x^3 - 7x^2y + xy^2 + 3y^3)$
$\quad = -y(4x^3) - y(-7x^2y) - y(xy^2) - y(3y^3)$
$\quad = -4x^3y + 7x^2y^2 - xy^3 - 3y^4$

29. $\dfrac{1}{2}x^2(8x^2 - 6x + 1)$
$\quad = \dfrac{1}{2}x^2(8x^2) + \dfrac{1}{2}x^2(-6x) + \dfrac{1}{2}x^2(1)$
$\quad = 4x^4 - 3x^3 + \dfrac{1}{2}x^2$

31. $(x+4)(x+3) = x(x+3) + 4(x+3)$
$\qquad = x(x) + x(3) + 4(x) + 4(3)$
$\qquad = x^2 + 3x + 4x + 12$
$\qquad = x^2 + 7x + 12$

33. $(a+7)(a-2) = a(a-2) + 7(a-2)$
$\qquad = a(a) + a(-2) + 7(a) + 7(-2)$
$\qquad = a^2 - 2a + 7a - 14$
$\qquad = a^2 + 5a - 14$

35. $\left(x + \dfrac{2}{3}\right)\left(x - \dfrac{1}{3}\right)$
$\quad = x\left(x - \dfrac{1}{3}\right) + \dfrac{2}{3}\left(x - \dfrac{1}{3}\right)$
$\quad = x(x) + x\left(-\dfrac{1}{3}\right) + \dfrac{2}{3}(x) + \dfrac{2}{3}\left(-\dfrac{1}{3}\right)$
$\quad = x^2 - \dfrac{1}{3}x + \dfrac{2}{3}x - \dfrac{2}{9}$
$\quad = x^2 + \dfrac{1}{3}x - \dfrac{2}{9}$

37. $(3x^2 + 1)(4x^2 + 7)$
$\quad = 3x^2(4x^2 + 7) + 1(4x^2 + 7)$
$\quad = 3x^2(4x^2) + 3x^2(7) + 1(4x^2) + 1(7)$
$\quad = 12x^4 + 21x^2 + 4x^2 + 7$
$\quad = 12x^4 + 25x^2 + 7$

39. $(4x-3)(3x-5) = 4x(3x-5)+(-3)(3x-5)$
$$= 4x(3x)+4x(-5)+(-3)(3x)+(-3)(-5)$$
$$= 12x^2 -20x-9x+15$$
$$= 12x^2 -29x+15$$

41. $(1-3a)(1-4a) = 1(1-4a)+(-3a)(1-4a)$
$$= 1(1)+1(-4a)+(-3a)(1)+(-3a)(-4a)$$
$$= 1-4a-3a+12a^2$$
$$= 1-7a+12a^2$$

43. $(2y-4)^2 = (2y-4)(2y-4)$
$$= 2y(2y-4)+(-4)(2y-4)$$
$$= 2y(2y)+2y(-4)+(-4)(2y)+(-4)(-4)$$
$$= 4y^2 -8y-8y+16$$
$$= 4y^2 -16y+16$$

45. $(x-2)(x^2 -3x+7) = x(x^2 -3x+7)+(-2)(x^2 -3x+7)$
$$= x(x^2)+x(-3x)+x(7)+(-2)(x^2)+(-2)(-3x)+(-2)(7)$$
$$= x^3 -3x^2 +7x-2x^2 +6x-14$$
$$= x^3 -5x^2 +13x-14$$

47. $(x+5)(x^3 -3x+4) = x(x^3 -3x+4)+5(x^3 -3x+4)$
$$= x(x^3)+x(-3x)+x(4)+5(x^3)+5(-3x)+5(4)$$
$$= x^4 -3x^2 +4x+5x^3 -15x+20$$
$$= x^4 +5x^3 -3x^2 -11x+20$$

49. $(2a-3)(5a^2 -6a+4) = 2a(5a^2 -6a+4)+(-3)(5a^2 -6a+4)$
$$= 2a(5a^2)+2a(-6a)+2a(4)+(-3)(5a^2)+(-3)(-6a)+(-3)(4)$$
$$= 10a^3 -12a^2 +8a-15a^2 +18a-12$$
$$= 10a^3 -27a^2 +26a-12$$

51. $(7xy-y)^2 = (7xy-y)(7xy-y)$
$$= 7xy(7xy-y)+(-y)(7xy-y)$$
$$= 7xy(7xy)+7xy(-y)+(-y)(7xy)+(-y)(-y)$$
$$= 49x^2 y^2 -7xy^2 -7xy^2 +y^2$$
$$= 49x^2 y^2 -14xy^2 +y^2$$

53.
$$
\begin{array}{r}
2x - 11 \\
6x + \ 1 \\
\hline
2x - 11 \\
12x^2 - 66x \\
\hline
12x^2 - 64x - 11
\end{array}
$$

55.
$$
\begin{array}{r}
2x^2 + 4x - 1 \\
x + 3 \\
\hline
6x^2 + 12x - 3 \\
2x^3 + \ 4x^2 - \ x \\
\hline
2x^3 + 10x^2 + 11x - 3
\end{array}
$$

57.
$$
\begin{array}{r}
x^2 \ + 5x \ - 7 \\
2x^2 \ - 7x \ - 9 \\
\hline
-9x^2 - 45x + 63 \\
-7x^3 - 35x^2 + 49x \\
2x^4 + 10x^3 - 14x^2 \\
\hline
2x^4 + 3x^3 - 58x^2 \ + 4x + 63
\end{array}
$$

59. $-1.2y(-7y^6) = (-1.2 \cdot -7)(y \cdot y^6) = 8.4y^7$

61. $-3x(x^2 + 2x - 8) = -3x(x^2) + (-3x)(2x) + (-3x)(-8)$
$$= -3x^3 - 6x^2 + 24x$$

63. $(x + 19)(2x + 1) = x(2x + 1) + 19(2x + 1)$
$$= x(2x) + x(1) + 19(2x) + 19(1)$$
$$= 2x^2 + x + 38x + 19$$
$$= 2x^2 + 39x + 19$$

65. $\left(x + \dfrac{1}{7}\right)\left(x - \dfrac{3}{7}\right) = x\left(x - \dfrac{3}{7}\right) + \dfrac{1}{7}\left(x - \dfrac{3}{7}\right)$
$$= x(x) + x\left(-\dfrac{3}{7}\right) + \dfrac{1}{7}(x) + \dfrac{1}{7}\left(-\dfrac{3}{7}\right)$$
$$= x^2 - \dfrac{3}{7}x + \dfrac{1}{7}x - \dfrac{3}{49}$$
$$= x^2 - \dfrac{2}{7}x - \dfrac{3}{49}$$

67. $\begin{aligned}(3y+5)^2 &= (3y+5)(3y+5)\\ &= 3y(3y+5)+5(3y+5)\\ &= 3y(3y)+3y(5)+5(3y)+5(5)\\ &= 9y^2+15y+15y+25\\ &= 9y^2+30y+25\end{aligned}$

69. $\begin{aligned}(a+4)(a^2-6a+6) &= a(a^2-6a+6)+4(a^2-6a+6)\\ &= a(a^2)+a(-6a)+a(6)+4(a^2)+4(-6a)+4(6)\\ &= a^3-6a^2+6a+4a^2-24a+24\\ &= a^3-2a^2-18a+24\end{aligned}$

71. $\begin{aligned}(2x+5)(2x-5) &= 2x(2x-5)+5(2x-5)\\ &= 2x(2x)+2x(-5)+5(2x)+5(-5)\\ &= 4x^2-10x+10x-25\\ &= 4x^2-25\end{aligned}$

The area is $(4x^2-25)$ square yards.

73. $\begin{aligned}\text{Area} &= \frac{1}{2}(\text{base})(\text{height})\\ &= \frac{1}{2}(3x-2)(4x)\\ &= 2x(3x-2)\\ &= 2x(3x)+2x(-2)\\ &= 6x^2-4x\end{aligned}$

The area is $(6x^2-4x)$ square inches.

75. $(5x)^2 = 5^2x^2 = 25x^2$

77. $(-3y^3)^2 = (-3)^2(y^3)^2 = 9y^6$

79. a. $\begin{aligned}(3x+5)+(3x+7) &= 3x+5+3x+7\\ &= 6x+12\end{aligned}$

    b. $\begin{aligned}(3x+5)(3x+7) &= 3x(3x+7)+5(3x+7)\\ &= 3x(3x)+3x(7)+5(3x)+5(7)\\ &= 9x^2+21x+15x+35\\ &= 9x^2+36x+35\end{aligned}$

answers may vary

81. $(3x-1)+(10x-6) = 3x-1+10x-6 = 13x-7$

83. $(3x-1)(10x-6)$
$= 3x(10x-6)+(-1)(10x-6)$
$= 3x(10x)+3x(-6)+(-1)(10x)+(-1)(-6)$
$= 30x^2 -18x-10x+6$
$= 30x^2 -28x+6$

85. $(3x-1)-(10x-6) = (3x-1)+(-10x+6)$
$= 3x-1-10x+6$
$= -7x+5$

87. The areas of the smaller rectangles are:
$x \cdot x = x^2$
$x \cdot 3 = 3x$
The area of the figure is $x^2 + 3x.$

89. The areas of the smaller rectangles are:
$x \cdot x = x^2$
$x \cdot 3 = 3x$
$2 \cdot x = 2x$
$2 \cdot 3 = 6$
The area of the figure is
$x^2 + 3x + 2x + 6 = x^2 + 5x + 6.$

91. $5a + 6a = (5+6)a = 11a$

93. $(5x)^2 + (2y)^2 = 5^2 x^2 + 2^2 y^2 = 25x^2 + 4y^2$

95. a. $(a+b)(a-b)$
$= a(a-b)+b(a-b)$
$= a(a)+a(-b)+b(a)+b(-b)$
$= a^2 -ab+ab-b^2$
$= a^2 -b^2$

b. $(2x+3y)(2x-3y)$
$= 2x(2x-3y)+3y(2x-3y)$
$= 2x(2x)+2x(-3y)+3y(2x)+3y(-3y)$
$= 4x^2 -6xy+6xy-9y^2$
$= 4x^2 -9y^2$

c. $(4x+7)(4x-7)$
$= 4x(4x-7)+7(4x-7)$
$= 4x(4x)+4x(-7)+7(4x)+7(-7)$
$= 16x^2 -28x+28x-49$
$= 16x^2 -49$

d. answers may vary

Section 3.6 Practice Problems

1. $(x+7)(x-5)$
$= (x)(x)+(x)(-5)+(7)(x)+(7)(-5)$
$= x^2 -5x+7x-35$
$= x^2 +2x-35$

2. $(6x-1)(x-4)$
$= 6x(x)+6x(-4)+(-1)(x)+(-1)(-4)$
$= 6x^2 -24x-x+4$
$= 6x^2 -25x+4$

3. $(2y^2 +3)(y-4) = 2y^3 -8y^2 +3y-12$

4. $(2x+9)^2 = (2x+9)(2x+9)$
$= (2x)(2x)+(2x)(9)+9(2x)+9(9)$
$= 4x^2 +18x+18x+81$
$= 4x^2 +36x+81$

5. $(y+3)^2 = y^2 +2(y)(3)+3^2 = y^2 +6y+9$

6. $(r-s)^2 = r^2 -2(r)(s)+s^2 = r^2 -2rs+s^2$

7. $(6x+5)^2 = (6x)^2 +2(6x)(5)+5^2$
$= 36x^2 +60x+25$

8. $(x^2 -3y)^2 = (x^2)^2 -2(x^2)(3y)+(3y)^2$
$= x^4 -6x^2 y+9y^2$

9. $(x+9)(x-9) = x^2 -9^2 = x^2 -81$

10. $(5+4y)(5-4y) = 5^2 -(4y)^2 = 25-16y^2$

11. $\left(x-\dfrac{1}{3}\right)\left(x+\dfrac{1}{3}\right) = x^2 - \left(\dfrac{1}{3}\right)^2 = x^2 - \dfrac{1}{9}$

12. $(3a-b)(3a+b) = (3a)^2 - b^2 = 9a^2 - b^2$

13. $(2x^2-6y)(2x^2+6y) = (2x^2)^2 - (6y)^2$
$$= 4x^4 - 36y^2$$

14. $(7x-1)^2 = (7x)^2 - 2(7x)(1) + 1^2$
$$= 49x^2 - 14x + 1$$

15. $(5y+3)(2y-5)$
$= (5y)(2y) + (5y)(-5) + (3)(2y) + (3)(-5)$
$= 10y^2 - 25y + 6y - 15$
$= 10y^2 - 19y - 15$

16. $(2a-1)(2a+1) = (2a)^2 - (1)^2 = 4a^2 - 1$

17. $\left(5y-\dfrac{1}{9}\right)^2 = (5y)^2 - 2(5y)\left(\dfrac{1}{9}\right) + \left(\dfrac{1}{9}\right)^2$
$$= 25y^2 - \dfrac{10}{9}y + \dfrac{1}{81}$$

Exercise Set 3.6

1. $(x+3)(x+4) = x^2 + 4x + 3x + 12$
$$= x^2 + 7x + 12$$

3. $(x-5)(x+10) = x^2 + 10x - 5x - 50$
$$= x^2 + 5x - 50$$

5. $(5x-6)(x+2) = 5x^2 + 10x - 6x - 12$
$$= 5x^2 + 4x - 12$$

7. $(y-6)(4y-1) = 4y^2 - y - 24y + 6$
$$= 4y^2 - 25y + 6$$

9. $(2x+5)(3x-1) = 6x^2 - 2x + 15x - 5$
$$= 6x^2 + 13x - 5$$

11. $(y^2+7)(6y+4) = 6y^3 + 4y^2 + 42y + 28$

13. $\left(x-\dfrac{1}{3}\right)\left(x+\dfrac{2}{3}\right) = x^2 + \dfrac{2}{3}x - \dfrac{1}{3}x - \dfrac{2}{9}$
$$= x^2 + \dfrac{1}{3}x - \dfrac{2}{9}$$

15. $(0.4-3a)(0.2-5a)$
$= 0.08 - 2.0a - 0.6a + 15a^2$
$= 0.08 - 2.6a + 15a^2$

17. $(x+5y)(2x-y) = 2x^2 - xy + 10xy - 5y^2$
$$= 2x^2 + 9xy - 5y^2$$

19. $(x+2)^2 = x^2 + 2(2)(x) + 2^2 = x^2 + 4x + 4$

21. $(2x-1)^2 = (2x)^2 - 2(2x)(1) + (1)^2$
$$= 4x^2 - 4x + 1$$

23. $(3a-5)^2 = (3a)^2 - 2(3a)(5) + 5^2$
$$= 9a^2 - 30a + 25$$

25. $(x^2+0.5)^2 = (x^2)^2 + 2(x^2)(0.5) + (0.5)^2$
$$= x^4 + x^2 + 0.25$$

27. $\left(y-\dfrac{2}{7}\right)^2 = y^2 - 2(y)\left(\dfrac{2}{7}\right) + \left(\dfrac{2}{7}\right)^2$
$$= y^2 - \dfrac{4}{7}y + \dfrac{4}{49}$$

29. $(2a-3)^2 = (2a)^2 - 2(2a)(3) + (3)^2$
$$= 4a^2 - 12a + 9$$

31. $(5x+9)^2 = (5x)^2 + 2(5x)(9) + 9^2$
$$= 25x^2 + 90x + 81$$

33. $(3x-7y)^2 = (3x)^2 - 2(3x)(7y) + (7y)^2$
$$= 9x^2 - 42xy + 49y^2$$

35. $(4m+5n)^2 = (4m)^2 + 2(4m)(5n) + (5n)^2$
$= 16m^2 + 40mn + 25n^2$

37. $(5x^4 - 3)^2 = (5x^4)^2 - 2(5x^4)(3) + (3)^2$
$= 25x^8 - 30x^4 + 9$

39. $(a-7)(a+7) = a^2 - 7^2 = a^2 - 49$

41. $(x+6)(x-6) = (x)^2 - (6)^2 = x^2 - 36$

43. $(3x-1)(3x+1) = (3x)^2 - 1^2 = 9x^2 - 1$

45. $(x^2+5)(x^2-5) = (x^2)^2 - (5)^2 = x^4 - 25$

47. $(2y^2-1)(2y^2+1) = (2y^2)^2 - 1^2 = 4y^4 - 1$

49. $(4-7x)(4+7x) = (4)^2 - (7x)^2 = 16 - 49x^2$

51. $\left(3x-\dfrac{1}{2}\right)\left(3x+\dfrac{1}{2}\right) = (3x)^2 - \left(\dfrac{1}{2}\right)^2 = 9x^2 - \dfrac{1}{4}$

53. $(9x+y)(9x-y) = (9x)^2 - (y)^2 = 81x^2 - y^2$

55. $(2m+5n)(2m-5n) = (2m)^2 - (5n)^2$
$= 4m^2 - 25n^2$

57. $(a+5)(a+4) = a^2 + 4a + 5a + 20$
$= a^2 + 9a + 20$

59. $(a-7)^2 = a^2 - 2(a)(7) + 7^2 = a^2 - 14a + 49$

61. $(4a+1)(3a-1) = 12a^2 - 4a + 3a - 1$
$= 12a^2 - a - 1$

63. $(x+2)(x-2) = x^2 - 2^2 = x^2 - 4$

65. $(3a+1)^2 = (3a)^2 + 2(3a)(1) + (1)^2$
$= 9a^2 + 6a + 1$

67. $(x+y)(4x-y) = 4x^2 - xy + 4xy - y^2$
$= 4x^2 + 3xy - y^2$

69. $\left(a-\dfrac{1}{2}y\right)\left(a+\dfrac{1}{2}y\right) = (a)^2 - \left(\dfrac{1}{2}y\right)^2$
$= a^2 - \dfrac{1}{4}y^2$

71. $(3b+7)(2b-5) = 6b^2 - 15b + 14b - 35$
$= 6b^2 - b - 35$

73. $(x^2+10)(x^2-10) = (x^2)^2 - (10)^2$
$= x^4 - 100$

75. $(4x+5)(4x-5) = (4x)^2 - 5^2 = 16x^2 - 25$

77. $(5x-6y)^2 = (5x)^2 - 2(5x)(6y) + (6y)^2$
$= 25x^2 - 60xy + 36y^2$

79. $(2r-3s)(2r+3s) = (2r)^2 - (3s)^2$
$= 4r^2 - 9s^2$

81. $(2x+1)^2 = (2x)^2 + 2(2x)(1) + (1)^2$
$= 4x^2 + 4x + 1$

The area of the rug is $(4x^2 + 4x + 1)$ square feet.

83. $\dfrac{50b^{10}}{70b^5} = \dfrac{50}{70} \cdot \dfrac{b^{10}}{b^5} = \dfrac{5}{7}b^{10-5} = \dfrac{5}{7}b^5 = \dfrac{5b^5}{7}$

85. $\dfrac{8a^{17}b^5}{-4a^7b^{10}} = \dfrac{8}{-4} \cdot \dfrac{a^{17}}{a^7} \cdot \dfrac{b^5}{b^{10}}$
$= -2a^{17-7}b^{5-10}$
$= -2a^{10}b^{-5}$
$= -\dfrac{2a^{10}}{b^5}$

87. $\dfrac{2x^4y^{12}}{3x^4y^4} = \dfrac{2}{3} \cdot \dfrac{x^4}{x^4} \cdot \dfrac{y^{12}}{y^4}$

$= \dfrac{2}{3}x^{4-4}y^{12-4}$

$= \dfrac{2}{3}x^0y^8$

$= \dfrac{2y^8}{3}$

89. $(a-b)^2 = (a)^2 - 2(a)(b) + (b)^2$
$= a^2 - 2ab + b^2$
which is choice c.

91. $(a+b)^2 = (a)^2 + 2(a)(b) + (b)^2$ which is
$= a^2 + 2ab + b^2$
choice d.

93. $(x^2+7)(x^2+3) = x^4 + 3x^2 + 7x^2 + 21$
$= x^4 + 10x^2 + 21$

95. $(5x-3)^2 - (x+1)^2$
$= ((5x)^2 - 2(5x)(3) + 3^2) - (x^2 + 2x + 1^2)$
$= (25x^2 - 30x + 9) + (-x^2 - 2x - 1)$
$= 25x^2 - 30x + 9 - x^2 - 2x - 1$
$= 24x^2 - 32x + 8$
The area is $(24x^2 - 32x + 8)$ square meters.

97. answers may vary

Integrated Review

1. $(5x^2)(7x^3) = (5 \cdot 7)(x^2 \cdot x^3) = 35x^5$

3. $-4^2 = -(4^2) = -16$

5. $(x-5)(2x+1) = 2x^2 + 1x - 10x - 5$
$= 2x^2 - 9x - 5$

7. $(x-5) + (2x+1) = x - 5 + 2x + 1 = 3x - 4$

9. $\dfrac{7x^9y^{12}}{x^3y^{10}} = 7 \cdot x^{9-3} \cdot y^{12-10} = 7x^6y^2$

11. $(12m^7n^6)^2 = 12^2 \cdot m^{7 \cdot 2}n^{6 \cdot 2} = 144m^{14}n^{12}$

13. $(4y-3)(4y+3) = (4y)^2 - 3^2 = 16y^2 - 9$

15. $(x^{-7}y^5)^9 = x^{-7 \cdot 9}y^{5 \cdot 9} = x^{-63}y^{45} = \dfrac{y^{45}}{x^{63}}$

17. $(3^{-1}x^9)^3 = 3^{-1 \cdot 3}x^{9 \cdot 3} = 3^{-3}x^{27} = \dfrac{x^{27}}{3^3} = \dfrac{x^{27}}{27}$

19. $(7x^2 - 2x + 3) - (5x^2 + 9)$
$= (7x^2 - 2x + 3) + (-5x^2 - 9)$
$= 7x^2 - 2x + 3 - 5x^2 - 9$
$= 2x^2 - 2x - 6$

21. $0.7y^2 - 1.2 + 1.8y^2 - 6y + 1$
$= 2.5y^2 - 6y - 0.2$

23. $(3y^2 - 6y + 1) - (y^2 + 2)$
$= (3y^2 - 6y + 1) + (-y^2 - 2)$
$= 3y^2 - 6y + 1 - y^2 - 2$
$= 2y^2 - 6y - 1$

25. $(x+4)^2 = x^2 + 2(x)(4) + 4^2 = x^2 + 8x + 16$

27. $(x+4) + (x+4) = x + 4 + x + 4 = 2x + 8$

29. $7x^2 - 6xy + 4(y^2 - xy)$
$= 7x^2 - 6xy + 4y^2 - 4xy$
$= 7x^2 - 10xy + 4y^2$

31. $(x-3)(x^2+5x-1) = x(x^2+5x-1)+(-3)(x^2+5x-1)$
$$= x(x^2)+x(5x)+x(-1)+(-3)(x^2)+(-3)(5x)+(-3)(-1)$$
$$= x^3+5x^2-x-3x^2-15x+3$$
$$= x^3+2x^2-16x+3$$

33. $(2x-7)(3x+10) = 6x^2+20x-21x-70$
$$= 6x^2-x-70$$

35. $(2x-7)(x^2-6x+1) = 2x(x^2-6x+1)+(-7)(x^2-6x+1)$
$$= 2x(x^2)+2x(-6x)+2x(1)+(-7)(x^2)+(-7)(-6x)+(-7)(1)$$
$$= 2x^3-12x^2+2x-7x^2+42x-7$$
$$= 2x^3-19x^2+44x-7$$

37. $\left(2x+\dfrac{5}{9}\right)\left(2x-\dfrac{5}{9}\right) = (2x)^2-\left(\dfrac{5}{9}\right)^2 = 4x^2-\dfrac{25}{81}$

Section 3.7 Practice Problems

1. $\dfrac{25x^3+5x^2}{5x^2} = \dfrac{25x^3}{5x^2}+\dfrac{5x^2}{5x^2} = 5x+1$

2. $\dfrac{24x^7+12x^2-4x}{4x^2} = \dfrac{24x^7}{4x^2}+\dfrac{12x^2}{4x^2}-\dfrac{4x}{4x^2}$
$$= 6x^5+3-\dfrac{1}{x}$$

3. $\dfrac{12x^3y^3-18xy+6y}{3xy} = \dfrac{12x^3y^3}{3xy}-\dfrac{18xy}{3xy}+\dfrac{6y}{3xy}$
$$= 4x^2y^2-6+\dfrac{2}{x}$$

4.
$$\begin{array}{r} x+7 \\ x+5 \overline{\smash{\big)}\; x^2+12x+35} \\ \underline{x^2+5x} \\ 7x+35 \\ \underline{7x+35} \\ 0 \end{array}$$

Thus, $\dfrac{x^2+12x+35}{x+5} = x+7$.

5.

$$2x-1 \overline{\big)\ 8x^2+2x-7}$$

$$\begin{array}{r} 4x+3 \\ \underline{8x^2-4x} \\ 6x-7 \\ \underline{6x-3} \\ -4 \end{array}$$

Thus, $\dfrac{8x^2+2x-7}{2x-1} = 4x+3+\dfrac{-4}{2x-1}$.

6.

$$x-3 \overline{\big)\ -2x^2+0x+15}$$

$$\begin{array}{r} -2x-6 \\ \underline{-2x^2+6x} \\ -6x+15 \\ \underline{-6x+18} \\ -3 \end{array}$$

Thus, $\dfrac{-2x^2+15}{x-3} = -2x-6+\dfrac{-3}{x-3}$.

7. $\dfrac{5-x+9x^3}{3x+2} = \dfrac{9x^3+0x^2-x+5}{3x+2}$

$$3x+2 \overline{\big)\ 9x^3+0x^2-\ x+5}$$

$$\begin{array}{r} 3x^2-2x+1 \\ \underline{9x^3+6x^2} \\ -6x^2-\ x \\ \underline{-6x^2-4x} \\ 3x+5 \\ \underline{3x+2} \\ 3 \end{array}$$

Thus, $\dfrac{5-x+9x^3}{3x+2} = 3x^2-2x+1+\dfrac{3}{3x+2}$.

8.

$$x-1 \overline{\big)\ x^3+0x^2+0x-1}$$

$$\begin{array}{r} x^2+x+1 \\ \underline{x^3-\ x^2} \\ x^2+0x \\ \underline{x^2-\ x} \\ x-1 \\ \underline{x-1} \\ 0 \end{array}$$

Thus, $\dfrac{x^3-1}{x-1} = x^2+x+1$.

Mental Math

1. $\dfrac{a^6}{a^4} = a^{6-4} = a^2$

2. $\dfrac{y^2}{y} = y^{2-1} = y^1 = y$

3. $\dfrac{a^3}{a} = a^{3-1} = a^2$

4. $\dfrac{p^8}{p^3} = p^{8-3} = p^5$

5. $\dfrac{k^5}{k^2} = k^{5-2} = k^3$

6. $\dfrac{k^7}{k^5} = k^{7-5} = k^2$

Exercise Set 3.7

1. $\dfrac{12x^4+3x^2}{x} = \dfrac{12x^4}{x} + \dfrac{3x^2}{x} = 12x^3+3x$

3. $\dfrac{20x^3-30x^2+5x+5}{5}$

$= \dfrac{20x^3}{5} - \dfrac{30x^2}{5} + \dfrac{5x}{5} + \dfrac{5}{5}$

$= 4x^3-6x^2+x+1$

5. $\dfrac{15p^3 + 18p^2}{3p} = \dfrac{15p^3}{3p} + \dfrac{18p^2}{3p} = 5p^2 + 6p$

7. $\dfrac{-9x^4 + 18x^5}{6x^5} = \dfrac{-9x^4}{6x^5} + \dfrac{18x^5}{6x^5} = \dfrac{-3}{2x} + 3$

9. $\dfrac{-9x^5 + 3x^4 - 12}{3x^3} = \dfrac{-9x^5}{3x^3} + \dfrac{3x^4}{3x^3} - \dfrac{12}{3x^3}$

$\qquad = -3x^2 + x - \dfrac{4}{x^3}$

11. $\dfrac{4x^4 - 6x^3 + 7}{-4x^4} = \dfrac{4x^4}{-4x^4} - \dfrac{6x^3}{-4x^4} + \dfrac{7}{-4x^4}$

$\qquad = -1 + \dfrac{3}{2x} - \dfrac{7}{4x^4}$

13.
$$\begin{array}{r} x+1 \\ x+3 \overline{\smash{)}\ x^2 + 4x + 3} \\ \underline{x^2 + 3x} \\ x + 3 \\ \underline{x + 3} \\ 0 \end{array}$$

$\dfrac{x^2 + 4x + 3}{x + 3} = x + 1$

15.
$$\begin{array}{r} 2x+3 \\ x+5 \overline{\smash{)}\ 2x^2 + 13x + 15} \\ \underline{2x^2 + 10x} \\ 3x + 15 \\ \underline{3x + 15} \\ 0 \end{array}$$

$\dfrac{2x^2 + 13x + 15}{x + 5} = 2x + 3$

17.
$$\begin{array}{r} 2x+1 \\ x-4 \overline{\smash{)}\ 2x^2 - 7x + 3} \\ \underline{2x^2 - 8x} \\ x + 3 \\ \underline{x - 4} \\ 7 \end{array}$$

$\dfrac{2x^2 - 7x + 3}{x - 4} = 2x + 1 + \dfrac{7}{x - 4}$

19.
$$\begin{array}{r} 3a^2 - 3a + 1 \\ 3a+2 \overline{\smash{)}\ 9a^3 - 3a^2 - 3a + 4} \\ \underline{9a^3 + 6a^2} \\ -9a^2 - 3a \\ \underline{-9a^2 - 6a} \\ 3a + 4 \\ \underline{3a + 2} \\ 2 \end{array}$$

$\dfrac{9a^3 - 3a^2 - 3a + 4}{3a + 2} = 3a^2 - 3a + 1 + \dfrac{2}{3a + 2}$

21.
$$\begin{array}{r} 4x+3 \\ 2x+1 \overline{\smash{)}\ 8x^2 + 10x + 1} \\ \underline{8x^2 + 4x} \\ 6x + 1 \\ \underline{6x + 3} \\ -2 \end{array}$$

$\dfrac{8x^2 + 10x + 1}{2x + 1} = 4x + 3 - \dfrac{2}{2x + 1}$

23.
$$\begin{array}{r} 2x^2 + 6x - 5 \\ x-2 \overline{\smash{)}\ 2x^3 + 2x^2 - 17x + 8} \\ \underline{2x^3 - 4x^2} \\ 6x^2 - 17x \\ \underline{6x^2 - 12x} \\ -5x + 8 \\ \underline{-5x + 10} \\ -2 \end{array}$$

$\dfrac{2x^3 + 2x^2 - 17x + 8}{x - 2} = 2x^2 + 6x - 5 - \dfrac{2}{x - 2}$

25.
$$\begin{array}{r} x+6 \\ x-6 \overline{\smash{)}\ x^2 + 0x - 36} \\ \underline{x^2 - 6x} \\ 6x - 36 \\ \underline{6x - 36} \\ 0 \end{array}$$

$\dfrac{x^2 - 36}{x - 6} = x + 6$

27.

$$\require{enclose}\begin{array}{r}x^2+3x+9 \\ x-3\enclose{longdiv}{x^3+0x^2+0x-27} \\ \underline{x^3-3x^2} \\ 3x^2+0x \\ \underline{3x^2-9x} \\ 9x-27 \\ \underline{9x-27} \\ 0\end{array}$$

$$\frac{x^3-27}{x-3}=x^2+3x+9$$

29. $1-3x^2=-3x^2+0x+1$

$$\begin{array}{r}-3x+6 \\ x+2\enclose{longdiv}{-3x^2+0x\ +1} \\ \underline{-3x^2-6x} \\ 6x+\ 1 \\ \underline{6x+12} \\ -11\end{array}$$

$$\frac{1-3x^2}{x+2}=-3x+6-\frac{11}{x+2}$$

31.

$$\begin{array}{r}2b-1 \\ 2b-1\enclose{longdiv}{-4b^2-4b-5} \\ \underline{4b^2-2b} \\ -2b-5 \\ \underline{-2b+1} \\ -6\end{array}$$

$$\frac{-4b+4b^2-5}{2b-1}=2b-1-\frac{6}{2b-1}$$

33. $\dfrac{a^2b^2-ab^3}{ab}=\dfrac{a^2b^2}{ab}-\dfrac{ab^3}{ab}=ab-b^2$

35.

$$\begin{array}{r}4x+9 \\ 2x-3\enclose{longdiv}{8x^2+\ 6x-27} \\ \underline{8x^2-12x} \\ 18x-27 \\ \underline{18x-27} \\ 0\end{array}$$

$$\frac{8x^2+6x-27}{2x-3}=4x+9$$

37. $\dfrac{2x^2y+8x^2y^2-xy^2}{2xy}=\dfrac{2x^2y}{2xy}+\dfrac{8x^2y^2}{2xy}-\dfrac{xy^2}{2xy}$

$$=x+4xy-\frac{y}{2}$$

39.

$$\begin{array}{r}2b^2+b+2 \\ b+4\enclose{longdiv}{2b^3+9b^2+6b-4} \\ \underline{2b^3+8b^2} \\ b^2+6b \\ \underline{b^2+4b} \\ 2b-4 \\ \underline{2b+8} \\ -12\end{array}$$

$$\frac{2b^3+9b^2+6b-4}{b+4}=2b^2+b+2-\frac{12}{b+4}$$

41.

$$\begin{array}{r}y^2+5y+10 \\ y-2\enclose{longdiv}{y^3+3y^2+\ 0y+\ 4} \\ \underline{y^3-2y^2} \\ 5y^2+\ 0y \\ \underline{5y^2-10y} \\ 10y+\ 4 \\ \underline{10y-20} \\ 24\end{array}$$

$$\frac{y^3+3y^2+4}{y-2}=y^2+5y+10+\frac{24}{y-2}$$

43.

$$\begin{array}{r}-6x-12 \\ x-2\enclose{longdiv}{-6x^2+\ 0x+\ 5} \\ \underline{-6x^2+12x} \\ -12x+\ 5 \\ \underline{-12x+24} \\ -19\end{array}$$

$$\frac{5-6x^2}{x-2}=-6x-12-\frac{19}{x-2}$$

45.

$$
\begin{array}{r}
x^3 - x^2 + x \\
x^2 + x \overline{)x^5 + 0x^4 + 0x^3 + x^2} \\
\underline{x^5 + \ x^4} \\
-x^4 + 0x^3 \\
\underline{-x^4 - \ x^3} \\
x^3 + x^2 \\
\underline{x^3 + x^2} \\
0
\end{array}
$$

$$\frac{x^5 + x^2}{x^2 + x} = x^3 - x^2 + x$$

47. $\dfrac{12}{4} = 3$, so $12 = 4 \cdot 3$.

49. $\dfrac{20}{-5} = -4$, so $20 = -5 \cdot -4$.

51. $\dfrac{9x^2}{3x} = 3x$, so $9x^2 = 3x \cdot 3x$.

53. $\dfrac{36x^2}{4x} = 9x$, so $36x^2 = 4x \cdot 9x$.

55.
$$\frac{12x^3 + 4x - 16}{4} = \frac{12x^3}{4} + \frac{4x}{4} - \frac{16}{4}$$
$$= 3x^3 + x - 4$$

The length of each side of the square is $(3x^3 + x - 4)$ feet.

57.

$$
\begin{array}{r}
2x + 5 \\
5x + 3 \overline{)10x^2 + 31x + 15} \\
\underline{10x^2 + \ 6x} \\
25x + 15 \\
\underline{25x + 15} \\
0
\end{array}
$$

The height of the parallelogram is $(2x + 5)$ meters.

59. answers may vary

61. $\dfrac{a+7}{7} = \dfrac{a}{7} + \dfrac{7}{7} = \dfrac{a}{7} + 1$ which is choice c.

The Bigger Picture

1. $-5.7 + (-0.23) = -5.93$

2. $\dfrac{1}{2} - \dfrac{9}{10} = \dfrac{5}{10} - \dfrac{9}{10} = -\dfrac{4}{10} = -\dfrac{2}{5}$

3. $(-5x^2 y^3)(-x^7 y) = 5x^{2+7} y^{3+1} = 5x^9 y^4$

4. $2^{-3} a^{-7} a^3 = \dfrac{1}{8} a^{-7+3} = \dfrac{1}{8} a^{-4} = \dfrac{1}{8a^4}$

5. $\begin{aligned}(7y^3 - 6y + 2) - (y^3 + 2y^2 + 2) &= (7y^3 - 6y + 2) + (-y^3 - 2y^2 - 2) \\ &= 7y^3 - 6y + 2 - y^3 - 2y^2 - 2 \\ &= 6y^3 - 2y^2 - 6y \end{aligned}$

6. $\begin{aligned}(9y^2 - 3y) - (y^2 + 7) &= (9y^2 - 3y) + (-y^2 - 7) \\ &= 9y^2 - 3y - y^2 - 7 \\ &= 8y^2 - 3y - 7 \end{aligned}$

7. $\begin{aligned}(x - 3)(4x^2 - x + 7) &= x(4x^2 - x + 7) + (-3)(4x^2 - x + 7) \\ &= x(4x^2) + x(-x) + x(7) + (-3)(4x^2) + (-3)(-x) + (-3)(7) \\ &= 4x^3 - x^2 + 7x - 12x^2 + 3x - 21 \\ &= 4x^3 - 13x^2 + 10x - 21 \end{aligned}$

8. $\begin{aligned}(6m - 5)^2 &= (6m)^2 - 2(6m)(5) + (5)^2 \\ &= 36m^2 - 60m + 25 \end{aligned}$

9. $\dfrac{20n^2 - 5n + 10}{5n} = \dfrac{20n^2}{5n} - \dfrac{5n}{5n} + \dfrac{10}{5n} = 4n - 1 + \dfrac{2}{n}$

10.
$$\begin{array}{r} 2x - 6 \\ 3x-1 \overline{)6x^2 - 20x + 20} \\ \underline{6x^2 - 2x} \\ -18x + 20 \\ \underline{-18x + 6} \\ 14 \end{array}$$

$\dfrac{6x^2 - 20x + 20}{3x - 1} = 2x - 6 + \dfrac{14}{3x - 1}$

11. $-6x = 3.6$

$$x = \frac{3.6}{-6}$$

$$x = -0.6$$

12. $-6x < 3.6$

$$x > \frac{3.6}{-6}$$

$$x > -0.6$$

$$\{x | x > -0.6\}$$

13. 　$6x + 6 \geq \ \ 8x + 2$

$\underline{-6x - 2 \ \ -6x - 2}$

　　　　$4 \geq 2x$

　　　$2 \geq x$ or $x \leq 2$

　　$\{x | x \leq 2\}$

14. 　$7y + 3(y-1) = 4(y+1) - 3$

　　$7y + 3y - 3 = 4y + 4 - 3$

　　　$10y - 3 = 4y + 1$

　　　　$6y - 3 = 1$

　　　　　$6y = 4$

　　　　　$y = \dfrac{4}{6} = \dfrac{2}{3}$

Chapter 3 Vocabulary Check

1. A <u>term</u> is a number or the product of numbers and variables raised to powers.

2. The <u>FOIL</u> method may be used when multiplying two binomials.

3. A polynomial with exactly 3 terms is called a <u>trinomial</u>.

4. The <u>degree of polynomial</u> is the greatest degree of any term of the polynomial.

5. A polynomial with exactly 2 terms is called a <u>binomial</u>.

6. The <u>coefficient</u> of a term is its numerical factor.

7. The <u>degree of a term</u> is the sum of the exponents on the variables in the term.

8. A polynomial with exactly 1 term is called a <u>monomial</u>.

9. Monomials, binomials, and trinomials are all examples of <u>polynomials</u>.

Chapter 3 Review

1. In 3^2, the base is 3 and the exponent is 2.

2. In $(-5)^4$, the base is -5 and the exponent is 4.

3. In -5^4, the base is 5 and the exponent is 4.

4. In x^6, the base is x and the exponent is 6.

5. $8^3 = 8 \cdot 8 \cdot 8 = 512$

6. $(-6)^2 = (-6)(-6) = 36$

7. $-6^2 = -(6 \cdot 6) = -36$

8. $-4^3 - 4^0 = -64 - 1 = -65$

9. $(3b)^0 = 1$

10. $\dfrac{8b}{8b} = 1$

11. $y^2 \cdot y^7 = y^{2+7} = y^9$

12. $x^9 \cdot x^5 = x^{9+5} = x^{14}$

13. $(2x^5)(-3x^6) = -6x^{5+6} = -6x^{11}$

14. $(-5y^3)(4y^4) = -20y^{3+4} = -20y^7$

15. $(x^4)^2 = x^{4 \cdot 2} = x^8$

16. $(y^3)^5 = y^{3 \cdot 5} = y^{15}$

17. $(3y^6)^4 = 3^4 \cdot y^{6 \cdot 4} = 81y^{24}$

18. $(2x^3)^3 = 2^3 \cdot x^{3 \cdot 3} = 8x^9$

19. $\dfrac{x^9}{x^4} = x^{9-4} = x^5$

20. $\dfrac{z^{12}}{z^5} = z^{12-5} = z^7$

21. $\dfrac{a^5 b^4}{ab} = a^{5-1} b^{4-1} = a^4 b^3$

22. $\dfrac{x^4 y^6}{xy} = x^{4-1} y^{6-1} = x^3 y^5$

23. $\dfrac{12xy^6}{3x^4 y^{10}} = 4x^{1-4} y^{6-10} = 4x^{-3} y^{-4} = \dfrac{4}{x^3 y^4}$

24. $\dfrac{2x^7 y^8}{8xy^2} = \dfrac{1}{4} x^{7-1} y^{8-2} = \dfrac{x^6 y^6}{4}$

25. $5a^7 (2a^4)^3 = 5a^7 (2^3 \cdot a^{4 \cdot 3})$
$= 5a^7 (8a^{12})$
$= 40a^{7+12}$
$= 40a^{19}$

26. $(2x)^2 (9x) = (2^2 x^2)(9x)$
$= 4x^2 (9x)$
$= 36x^{2+1}$
$= 36x^3$

27. $(-5a)^0 + 7^0 + 8^0 = 1 + 1 + 1 = 3$

28. $8x^0 + 9^0 = 8 \cdot 1 + 1 = 8 + 1 = 9$

29. $\left(\dfrac{3x^4}{4y}\right)^3 = \dfrac{3^3 x^{4 \cdot 3}}{4^3 y^{1 \cdot 3}} = \dfrac{27x^{12}}{64y^3}$; b

30. $\left(\dfrac{5a^6}{b^3}\right)^2 = \dfrac{5^2 a^{6 \cdot 2}}{b^{3 \cdot 2}} = \dfrac{25a^{12}}{b^6}$; c

31. $7^{-2} = \dfrac{1}{7^2} = \dfrac{1}{49}$

32. $-7^{-2} = -\dfrac{1}{7^2} = -\dfrac{1}{49}$

33. $2x^{-4} = \dfrac{2}{x^4}$

34. $(2x)^{-4} = \dfrac{1}{(2x)^4} = \dfrac{1}{2^4 x^4} = \dfrac{1}{16x^4}$

35. $\left(\dfrac{1}{5}\right)^{-3} = (5)^3 = 125$

36. $\left(\dfrac{-2}{3}\right)^{-2} = \left(\dfrac{3}{-2}\right)^2 = \dfrac{9}{4}$

37. $2^0 + 2^{-4} = 1 + \dfrac{1}{2^4} = 1 + \dfrac{1}{16} = \dfrac{17}{16}$

38. $6^{-1} - 7^{-1} = \dfrac{1}{6} - \dfrac{1}{7} = \dfrac{7}{42} - \dfrac{6}{42} = \dfrac{1}{42}$

39. $\dfrac{x^5}{x^{-3}} = x^{5-(-3)} = x^8$

40. $\dfrac{z^4}{z^{-4}} = z^{4-(-4)} = z^8$

41. $\dfrac{r^{-3}}{r^{-4}} = r^{-3-(-4)} = r^1 = r$

42. $\dfrac{y^{-2}}{y^{-5}} = y^{-2-(-5)} = y^3$

43. $\left(\dfrac{bc^{-2}}{bc^{-3}}\right)^4 = (b^{1-1}c^{-2-(-3)})^4 = (b^0 c^1)^4 = c^4$

44. $\left(\dfrac{x^{-3}y^{-4}}{x^{-2}y^{-5}}\right)^{-3} = (x^{-3-(-2)}y^{-4-(-5)})^{-3}$

$\qquad = (x^{-1}y^1)^{-3}$

$\qquad = x^{-1\cdot -3}y^{1\cdot -3}$

$\qquad = x^3 y^{-3}$

$\qquad = \dfrac{x^3}{y^3}$

45. $\dfrac{x^{-4}y^{-6}}{x^2 y^7} = x^{-4-2}y^{-6-7} = x^{-6}y^{-13} = \dfrac{1}{x^6 y^{13}}$

46. $\dfrac{a^5 b^{-5}}{a^{-5}b^5} = a^{5-(-5)}b^{(-5)-5} = a^{10}b^{-10} = \dfrac{a^{10}}{b^{10}}$

47. $0.00027 = 2.7 \times 10^{-4}$

48. $0.8868 = 8.868 \times 10^{-1}$

49. $80,800,000 = 8.08 \times 10^7$

50. $868,000 = 8.68 \times 10^5$

51. $112,400,000 = 1.124 \times 10^8$

52. $150,000 = 1.5 \times 10^5$

53. $8.67 \times 10^5 = 867,000$

54. $3.86 \times 10^{-3} = 0.00386$

55. $8.6 \times 10^{-4} = 0.00086$

56. $8.936 \times 10^5 = 893,600$

57. $1.43128 \times 10^{15} = 1,431,280,000,000,000$

58. $1 \times 10^{-10} = 0.0000000001$

59. $(8 \times 10^4)(2 \times 10^{-7}) = 16 \times 10^{-3} = 0.016$

60. $\dfrac{8 \times 10^4}{2 \times 10^{-7}} = 4 \times 10^{11} = 400,000,000,000$

61. The degree of $(y^5 + 7x - 8x^4)$ is 5.

62. The degree of $(9y^2 + 30y + 25)$ is 2.

63. The degree of
$(-14x^2 y - 28x^2 y^3 - 42x^2 y^2)$ is
$2 + 3$ or 5.

64. The degree of $(6x^2 y^2 z^2 + 5x^2 y^3 - 12xyz)$ is
$2 + 2 + 2$ or 6.

65. $2x^2 + 20x$
$\qquad = 2(1)^2 + 20(1) = 2 + 20 = 22$
$\qquad = 2(3)^2 + 20(3) = 18 + 60 = 78$
$\qquad = 2(5.1)^2 + 20(5.1) = 52.02 + 102$
$\qquad\qquad = 154.02$
$\qquad = 2(10)^2 + 20(10) = 200 + 200 = 400$

x	1	3	5.1	10
$2x^2 + 20x$	22	78	154.02	400

66. $7a^2 - 4a^2 - a^2 = (7 - 4 - 1)a^2 = 2a^2$

67. $9y + y - 14y = (9 + 1 - 14)y = -4y$

68. $6a^2 + 4a + 9a^2 = 6a^2 + 9a^2 + 4a$
$\qquad\qquad = 15a^2 + 4a$

69. $21x^2 + 3x + x^2 + 6 = 21x^2 + x^2 + 3x + 6$
$\qquad\qquad = 22x^2 + 3x + 6$

70. $4a^2b - 3b^2 - 8q^2 - 10a^2b + 7q^2$
$= 4a^2b - 10a^2b - 3b^2 - 8q^2 + 7q^2$
$= -6a^2b - 3b^2 - q^2$

71. $2s^{14} + 3s^{13} + 12s^{12} - s^{10}$ cannot be combined.

72. $(3x^2 + 2x + 6) + (5x^2 + x)$
$= 3x^2 + 2x + 6 + 5x^2 + x$
$= 3x^2 + 5x^2 + 2x + x + 6$
$= 8x^2 + 3x + 6$

73. $(2x^5 + 3x^4 + 4x^3 + 5x^2) + (4x^2 + 7x + 6)$
$= 2x^5 + 3x^4 + 4x^3 + 5x^2 + 4x^2 + 7x + 6$
$= 2x^5 + 3x^4 + 4x^3 + 9x^2 + 7x + 6$

74. $(-5y^2 + 3) - (2y^2 + 4)$
$= (-5y^2 + 3) + (-2y^2 - 4)$
$= -5y^2 + 3 - 2y^2 - 4$
$= -7y^2 - 1$

75. $(2m^7 + 3x^4 + 7m^6) - (8m^7 + 4m^2 + 6x^4)$
$= (2m^7 + 3x^4 + 7m^6) + (-8m^7 - 4m^2 - 6x^4)$
$= 2m^7 + 3x^4 + 7m^6 - 8m^7 - 4m^2 - 6x^4$
$= -6m^7 - 3x^4 + 7m^6 - 4m^2$

76. $(3x^2 - 7xy + 7y^2) - (4x^2 - xy + 9y^2)$
$= (3x^2 - 7xy + 7y^2) + (-4x^2 + xy - 9y^2)$
$= 3x^2 - 7xy + 7y^2 - 4x^2 + xy - 9y^2$
$= -x^2 - 6xy - 2y^2$

77. $(-9x^2 + 6x + 2) + (4x^2 - x - 1)$
$= -9x^2 + 6x + 2 + 4x^2 - x - 1$
$= -5x^2 + 5x + 1$

78. $[(x^2 + 7x + 9) + (x^2 + 4)] - (4x^2 + 8x - 7)$
$= (x^2 + 7x + 9) + (x^2 + 4) + (-4x^2 - 8x + 7)$
$= x^2 + 7x + 9 + x^2 + 4 - 4x^2 - 8x + 7$
$= -2x^2 - x + 20$

79. $6(x + 5) = 6(x) + 6(5) = 6x + 30$

80. $9(x - 7) = 9(x) + 9(-7) = 9x - 63$

81. $4(2a + 7) = 4(2a) + 4(7) = 8a + 28$

82. $9(6a - 3) = 9(6a) + 9(-3) = 54a - 27$

83. $-7x(x^2 + 5) = (-7x)(x^2) + (-7x)(5)$
$= -7x^3 - 35x$

84. $-8y(4y^2 - 6) = (-8y)(4y^2) + (-8y)(-6)$
$= -32y^3 + 48y$

85. $-2(x^3 - 9x^2 + x)$
$= (-2)(x^3) + (-2)(-9x^2) + (-2)(x)$
$= -2x^3 + 18x^2 - 2x$

86. $-3a(a^2b + ab + b^2)$
$= (-3a)(a^2b) + (-3a)(ab) + (-3a)(b^2)$
$= -3a^3b - 3a^2b - 3ab^2$

87. $(3a^3 - 4a + 1)(-2a)$
$= (3a^3)(-2a) + (-4a)(-2a) + (1)(-2a)$
$= -6a^4 + 8a^2 - 2a$

88. $(6b^3 - 4b + 2)(7b)$
$= (6b^3)(7b) + (-4b)(7b) + (2)(7b)$
$= 42b^4 - 28b^2 + 14b$

89. $(2x + 2)(x - 7) = 2x^2 - 14x + 2x - 14$
$= 2x^2 - 12x - 14$

90. $(2x - 5)(3x + 2) = 6x^2 + 4x - 15x - 10$
$= 6x^2 - 11x - 10$

91. $(4a-1)(a+7) = 4a^2 + 28a - a - 7$
$$= 4a^2 + 27a - 7$$

92. $(6a-1)(7a+3) = 42a^2 + 18a - 7a - 3$
$$= 42a^2 + 11a - 3$$

93. $(x+7)(x^3+4x-5) = x(x^3+4x-5) + 7(x^3+4x-5)$
$$= x(x^3) + x(4x) + x(-5) + 7(x^3) + 7(4x) + 7(-5)$$
$$= x^4 + 4x^2 - 5x + 7x^3 + 28x - 35$$
$$= x^4 + 7x^3 + 4x^2 + 23x - 35$$

94. $(x+2)(x^5+x+1) = x(x^5+x+1) + 2(x^5+x+1)$
$$= x(x^5) + x(x) + x(1) + 2(x^5) + 2(x) + 2(1)$$
$$= x^6 + x^2 + x + 2x^5 + 2x + 2$$
$$= x^6 + 2x^5 + x^2 + 3x + 2$$

95. $(x^2+2x+4)(x^2+2x-4) = x^2(x^2+2x-4) + 2x(x^2+2x-4) + 4(x^2+2x-4)$
$$= x^4 + 2x^3 - 4x^2 + 2x^3 + 4x^2 - 8x + 4x^2 + 8x - 16$$
$$= x^4 + 4x^3 + 4x^2 - 16$$

96. $(x^3+4x+4)(x^3+4x-4) = x^3(x^3+4x-4) + 4x(x^3+4x-4) + 4(x^3+4x-4)$
$$= x^6 + 4x^4 - 4x^3 + 4x^4 + 16x^2 - 16x + 4x^3 + 16x - 16$$
$$= x^6 + 8x^4 + 16x^2 - 16$$

97. $(x+7)^3 = (x+7)(x+7)^2$
$$= (x+7)(x^2 + 2(x)(7) + 7^2)$$
$$= (x+7)(x^2 + 14x + 49)$$
$$= x(x^2 + 14x + 49) + 7(x^2 + 14x + 49)$$
$$= x^3 + 14x^2 + 49x + 7x^2 + 98x + 343$$
$$= x^3 + 21x^2 + 147x + 343$$

98. $(2x-5)^3 = (2x-5)(2x-5)^2$
$$= (2x-5)((2x)^2 - (2)(2x)(5) + 5^2)$$
$$= (2x-5)(4x^2 - 20x + 25)$$
$$= 2x(4x^2 - 20x + 25) + (-5)(4x^2 - 20x + 25)$$
$$= 8x^3 - 40x^2 + 50x - 20x^2 + 100x - 125$$
$$= 8x^3 - 60x^2 + 150x - 125$$

99. $(x+7)^2 = x^2 + 2(x)(7) + 7^2 = x^2 + 14x + 49$

100. $(x-5)^2 = x^2 - 2(x)(5) + 5^2 = x^2 - 10x + 25$

101. $(3x-7)^2 = (3x)^2 - 2(3x)(7) + 7^2$
$$= 9x^2 - 42x + 49$$

102. $(4x+2)^2 = (4x)^2 + 2(4x)(2) + 2^2$
$$= 16x^2 + 16x + 4$$

103. $(5x-9)^2 = (5x)^2 - 2(5x)(9) + 9^2$
$$= 25x^2 - 90x + 81$$

104. $(5x+1)(5x-1) = (5x)^2 - 1^2 = 25x^2 - 1$

105. $(7x+4)(7x-4) = (7x)^2 - 4^2 = 49x^2 - 16$

106. $(a+2b)(a-2b) = a^2 - (2b)^2 = a^2 - 4b^2$

107. $(2x-6)(2x+6) = (2x)^2 - 6^2 = 4x^2 - 36$

108. $(4a^2 - 2b)(4a^2 + 2b) = (4a^2)^2 - (2b)^2$
$$= 16a^4 - 4b^2$$

109. $(3x-1)^2 = (3x)^2 - 2(3x)(1) + 1^2$
$$= 9x^2 - 6x + 1$$
The area is $(9x^2 - 6x + 1)$ square meters.

110. $(5x+2)(x-1) = 5x^2 - 5x + 2x - 2$
$$= 5x^2 - 3x - 2$$
The area is $(5x^2 - 3x - 2)$ square miles.

111. $\dfrac{x^2 + 21x + 49}{7x^2} = \dfrac{x^2}{7x^2} + \dfrac{21x}{7x^2} + \dfrac{49}{7x^2}$
$$= \dfrac{1}{7} + \dfrac{3}{x} + \dfrac{7}{x^2}$$

112. $\dfrac{5a^3 b - 15ab^2 + 20ab}{-5ab}$
$$= \dfrac{5a^3 b}{-5ab} + \dfrac{-15ab^2}{-5ab} + \dfrac{20ab}{-5ab}$$
$$= -a^2 + 3b - 4$$

113.
$$
\begin{array}{r}
a+1 \\
a-2\overline{)a^2 - a + 4} \\
\underline{a^2 - 2a} \\
a + 4 \\
\underline{a - 2} \\
6
\end{array}
$$

$$\dfrac{a^2 - a + 4}{a - 2} = a + 1 + \dfrac{6}{a-2}$$

114.
$$
\begin{array}{r}
4x \\
x+5\overline{)4x^2 + 20x + 7} \\
\underline{4x^2 + 20x} \\
7
\end{array}
$$

$$\dfrac{4x^2 + 20x + 7}{x + 5} = 4x + \dfrac{7}{x+5}$$

115.
$$
\begin{array}{r}
a^2 + 3a + 8 \\
a-2\overline{)a^3 + a^2 + 2a + 6} \\
\underline{a^3 - 2a^2} \\
3a^2 + 2a \\
\underline{3a^2 - 6a} \\
8a + 6 \\
\underline{8a - 16} \\
22
\end{array}
$$

$$\dfrac{a^3 + a^2 + 2a + 6}{a - 2} = a^2 + 3a + 8 + \dfrac{22}{a-2}$$

116.

$$3b-2 \overline{\smash{\big)}\ 9b^3 - 18b^2 + 8b - 1} \quad \overset{\displaystyle 3b^2 - 4b}{}$$

$$\underline{9b^3 - 6b^2}$$
$$-12b^2 + 8b$$
$$\underline{-12b^2 + 8b}$$
$$-1$$

$$\frac{9b^3 - 18b^2 + 8b - 1}{3b - 2} = 3b^2 - 4b - \frac{1}{3b - 2}$$

117.

$$2x-1 \overline{\smash{\big)}\ 4x^4 - 4x^3 + x^2 + 4x - 3} \quad \overset{\displaystyle 2x^3 - x^2 + 0x + 2}{}$$

$$\underline{4x^4 - 2x^3}$$
$$-2x^3 + x^2$$
$$\underline{-2x^3 + x^2}$$
$$4x - 3$$
$$\underline{4x - 2}$$
$$-1$$

$$\frac{4x^4 - 4x^3 + x^2 + 4x - 3}{2x - 1}$$
$$= 2x^3 - x^2 + 2 - \frac{1}{2x - 1}$$

118.

$$x-6 \overline{\smash{\big)}\ -x^3 - 10x^2 - 21x + 18} \quad \overset{\displaystyle x^2 - 16x - 117}{}$$

$$\underline{-x^3 + 6x^2}$$
$$-16x^2 - 21x$$
$$\underline{-16x^2 + 96x}$$
$$-117x + 18$$
$$\underline{-117x + 702}$$
$$-684$$

$$\frac{-x^3 - 10x^2 - 21x + 18}{x - 6}$$
$$= -x^2 - 16x - 117 - \frac{684}{x - 6}$$

119.

$$\frac{15x^3 - 3x^2 + 60}{3x^2} = \frac{15x^3}{3x^2} - \frac{3x^2}{3x^2} + \frac{60}{3x^2}$$
$$= 5x - 1 + \frac{20}{x^2}$$

The width is $\left(5x - 1 + \frac{20}{x^2}\right)$ feet.

120.

$$\frac{21a^3 b^6 + 3a - 3}{3} = \frac{21a^3 b^6}{3} + \frac{3a}{3} - \frac{3}{3}$$
$$= 7a^3 b^6 + a - 1$$

The length of a side is $(7a^3 b^6 + a - 1)$ units.

121. $\left(-\dfrac{1}{2}\right)^3 = \left(-\dfrac{1}{2}\right)\left(-\dfrac{1}{2}\right)\left(-\dfrac{1}{2}\right) = -\dfrac{1}{8}$

122. $(4xy^2)(x^3 y^5) = 4x^{1+3} y^{2+5} = 4x^4 y^7$

123. $\dfrac{18x^9}{27x^3} = \dfrac{2}{3} x^{9-3} = \dfrac{2x^6}{3}$

124. $\left(\dfrac{3a^4}{b^2}\right)^3 = \dfrac{3^3 a^{4 \cdot 3}}{b^{2 \cdot 3}} = \dfrac{27a^{12}}{b^6}$

125. $(2x^{-4} y^3)^{-4} = 2^{-4} x^{-4 \cdot -4} y^{3 \cdot -4} = \dfrac{x^{16}}{16y^{12}}$

126. $\dfrac{a^{-3} b^6}{9^{-1} a^{-5} b^{-2}} = 9a^{-3-(-5)} b^{6-(-2)} = 9a^2 b^8$

127. $(6x + 2) + (5x - 7) = 6x + 2 + 5x - 7 = 11x - 5$

128. $(-y^2 - 4) + (3y^2 - 6) = -y^2 - 4 + 3y^2 - 6$
$$= 2x^2 - 10$$

129. $(8y^2 - 3y + 1) - (3y^2 + 2)$
$$= (8y^2 - 3y + 1) + (-3y^2 - 2)$$
$$= 8y^2 - 3y + 1 - 3y^2 - 2$$
$$= 5y^2 - 3y - 1$$

130. $(5x^2 + 2x - 6) - (-x - 4)$
$= (5x^2 + 2x - 6) + (x + 4)$
$= 5x^2 + 2x - 6 + x + 4$
$= 5x^2 + 3x - 2$

131. $4x(7x^2 + 3) = 4x(7x^2) + 4x(3) = 28x^3 + 12x$

132. $(2x + 5)(3x - 2) = 6x^2 - 4x + 15x - 10$
$= 6x^2 + 11x - 10$

133. $(x - 3)(x^2 + 4x - 6)$
$= x(x^2 + 4x - 6) + (-3)(x^2 + 4x - 6)$
$= x^3 + 4x^2 - 6x - 3x^2 - 12x + 18$
$= x^3 + x^2 - 18x + 18$

134. $(7x - 2)(4x - 9) = 28x^2 - 63x - 8x + 18$
$= 28x^2 - 71x + 18$

135. $(5x + 34)^2 = (5x)^2 + 2(5x)(4) + 4^2$
$= 25x^2 + 40x + 16$

136. $(6x + 3)(6x - 3) = (6x)^2 - 3^2 = 36x^2 - 9$

137. $\dfrac{8a^4 - 2a^3 + 4a - 5}{2a^3}$
$= \dfrac{8a^4}{2a^3} - \dfrac{2a^3}{2a^3} + \dfrac{4a}{2a^3} - \dfrac{5}{2a^3}$
$= 4a - 1 + \dfrac{2}{a^2} - \dfrac{5}{2a^3}$

138.
$$\begin{array}{r} x - 3 \\ x + 5 \overline{) x^2 + 2x + 10} \\ \underline{x^2 + 5x} \\ -3x + 10 \\ \underline{-3x - 15} \\ 25 \end{array}$$

$\dfrac{x^2 + 2x + 10}{x + 5} = x - 3 + \dfrac{25}{x + 5}$

139.
$$\begin{array}{r} 2x^2 + 7x + 5 \\ 2x - 3 \overline{) 4x^3 + 8x^2 - 11x + 4} \\ \underline{4x^3 - 6x^2} \\ 14x^2 - 11x \\ \underline{14x^2 - 21x} \\ 10x + 4 \\ \underline{10x - 15} \\ 19 \end{array}$$

$\dfrac{4x^3 + 8x^2 - 11x + 4}{2x - 3} = 2x^2 + 7x + 5 + \dfrac{19}{2x - 3}$

Chapter 3 Test

1. $2^5 = 2 \cdot 2 \cdot 2 \cdot 2 \cdot 2 = 32$

2. $(-3)^4 = (-3)(-3)(-3)(-3) = 81$

3. $-3^4 = -(3 \cdot 3 \cdot 3 \cdot 3) = -81$

4. $4^{-3} = \dfrac{1}{4^3} = \dfrac{1}{64}$

5. $(3x^2)(-5x^9) = 3(-5)(x^2 \cdot x^9)$
$= -15x^{2+9}$
$= -15x^{11}$

6. $\dfrac{y^7}{y^2} = y^{7-2} = y^5$

7. $\dfrac{r^{-8}}{r^{-3}} = r^{-8-(-3)} = r^{-5} = \dfrac{1}{r^5}$

8. $\left(\dfrac{x^2 y^3}{x^3 y^{-4}}\right)^2 = \dfrac{x^{2\cdot 2} y^{3\cdot 2}}{x^{3\cdot 2} y^{-4\cdot 2}}$

$\qquad\qquad = \dfrac{x^4 y^6}{x^6 y^{-8}}$

$\qquad\qquad = x^{4-6} y^{6-(-8)}$

$\qquad\qquad = x^{-2} y^{14}$

$\qquad\qquad = \dfrac{y^{14}}{x^2}$

9. $\dfrac{6^2 x^{-4} y^{-1}}{6^3 x^{-3} y^7} = \dfrac{6^2}{6^3} \cdot \dfrac{x^{-4}}{x^{-3}} \cdot \dfrac{y^{-1}}{y^7}$

$\qquad\qquad = 6^{2-3} x^{-4-(-3)} y^{-1-7}$

$\qquad\qquad = 6^{-1} x^{-1} y^{-8}$

$\qquad\qquad = \dfrac{1}{6} \cdot \dfrac{1}{x} \cdot \dfrac{1}{y^8}$

$\qquad\qquad = \dfrac{1}{6 x y^8}$

10. $563{,}000 = 5.63 \times 10^5$

11. $0.0000863 = 8.63 \times 10^{-5}$

12. $1.5 \times 10^{-3} = 0.0015$

13. $6.23 \times 10^4 = 62{,}300$

14. $(1.2 \times 10^5)(3 \times 10^{-7}) = 3.6 \times 10^{-2} = 0.036$

15. a. $4xy^2 + 7xyz + x^3 y - 2$

Term	Numerical Coefficient	Degree of Terms
$4xy^2$	4	3
$7xyz$	7	3
$x^3 y$	1	4
-2	-2	0

b. The degree of $4xy^2 + 7xyz + x^3 y$ is $3 + 1$ or 4.

16. $5x^2 + 4x - 7x^2 + 11 + 8x = (5 - 7)x^2 + (4 + 8)x + 11$
$$= -2x^2 + 12x + 11$$

17. $(8x^3 + 7x^2 + 4x - 7) + (8x^3 - 7x - 6) = 8x^3 + 7x^2 + 4x - 7 + 8x^3 - 7x - 6$
$$= 8x^3 + 8x^3 + 7x^2 + 4x - 7x - 7 - 6$$
$$= 16x^3 + 7x^2 - 3x - 13$$

18.
$$\begin{array}{r} 5x^3 + x^2 + 5x - 2 \\ - (8x^3 - 4x^2 + x - 7) \\ \hline \end{array} \qquad \begin{array}{r} 5x^3 + x^2 + 5x - 2 \\ -8x^3 + 4x^2 - x + 7 \\ \hline -3x^3 + 5x^2 + 4x + 5 \end{array}$$

19. $[(8x^2 + 7x + 5) + (x^3 - 8)] - (4x + 2) = (8x^2 + 7x + 5) + (x^3 - 8) + (-4x - 2)$
$$= 8x^2 + 7x + 5 + x^3 - 8 - 4x - 2$$
$$= x^3 + 8x^2 + 3x - 5$$

20. $(3x + 7)(x^2 + 5x + 2) = 3x(x^2 + 5x + 2) + 7(x^2 + 5x + 2)$
$$= 3x(x^2) + 3x(5x) + 3x(2) + 7(x^2) + 7(5x) + 7(2)$$
$$= 3x^3 + 15x^2 + 6x + 7x^2 + 35x + 14$$
$$= 3x^3 + 22x^2 + 41x + 14$$

21. $3x^2(2x^2 - 3x + 7) = 3x^2(2x^2) + 3x^2(-3x) + 3x^2(7)$
$$= 6x^4 - 9x^3 + 21x^2$$

22. $(x + 7)(3x - 5) = 3x^2 - 5x + 21x - 35$
$$= 3x^2 + 16x - 35$$

23. $\left(3x - \dfrac{1}{5}\right)\left(3x + \dfrac{1}{5}\right) = (3x)^2 - \left(\dfrac{1}{5}\right)^2 = 9x^2 - \dfrac{1}{25}$

24. $(4x - 2)^2 = (4x)^2 - 2(4x)(2) + 2^2$
$$= 16x^2 - 16x + 4$$

25. $(8x + 3)^2 = (8x)^2 + 2(8x)(3) + (3)^2$
$$= 64x^2 + 48x + 9$$

26. $(x^2 - 9b)(x^2 + 9b) = (x^2)^2 - (9b)^2 = x^4 - 81b^2$

27. $-16t^2 + 1001$

$= -16(0)^2 + 1001 = 1001$ ft

$= -16(1)^2 + 1001 = -16 + 1001$
$= 985$ ft

$= -16(3)^2 + 1001 = -144 + 1001$
$= 857$ ft

$= -16(5)^2 + 1001 = -400 + 1001$
$= 601$ ft

28. $(2x-3)(2x+3) = (2x)^2 - (3)^2$
$$= 4x^2 - 9$$
The area is $(2x-3)(2x+3)$ or $(4x^2 - 9)$ square inches.

29. $\dfrac{4x^2 + 2xy - 7x}{8xy} = \dfrac{4x^2}{8xy} + \dfrac{2xy}{8xy} - \dfrac{7x}{8xy}$
$$= \dfrac{x}{2y} + \dfrac{1}{4} - \dfrac{7}{8y}$$

30.
$$
\begin{array}{r}
x+2 \\
x+5 \overline{\smash{\big)}\ x^2 + 7x + 10} \\
\underline{x^2 + 5x} \\
2x + 10 \\
\underline{2x + 10} \\
0
\end{array}
$$

$\dfrac{x^2 + 7x + 10}{x+5} = x + 2$

31.
$$
\begin{array}{r}
9x^2 - 6x + 4 \\
3x+2 \overline{\smash{\big)}\ 27x^3 + 0x^2 + 0x - 8} \\
\underline{27x^3 + 18x^2} \\
-18x^2 + 0x \\
\underline{-18x^2 - 12x} \\
12x - 8 \\
\underline{12x + 8} \\
-16
\end{array}
$$

$\dfrac{27x^3 - 8}{3x + 2} = 9x^2 - 6x + 4 - \dfrac{16}{3x+2}$

Cumulative Review Chapters 1–3

1. a. 11, 112

   b. 0, 11, 112

   c. $-3, -2, 0, 11, 112$

   d. $-3, -2, 0, \dfrac{1}{4}, 11, 112$

   e. $\sqrt{2}$

   f. $-2, 0, \dfrac{1}{4}, 112, -3, 11, \sqrt{2}$

2. a. $|-7.2| = 7.2$

   b. $|0| = 0$

   c. $\left|-\dfrac{1}{2}\right| = \dfrac{1}{2}$

3. a. $3^2 = 9$

   b. $5^3 = 125$

   c. $2^4 = 16$

   d. $7^1 = 7$

   e. $\left(\dfrac{3}{7}\right)^2 = \dfrac{9}{49}$

4. a. $\dfrac{3}{4} \cdot \dfrac{7}{21} = \dfrac{1}{4} \cdot \dfrac{21}{21} = \dfrac{1}{4}$

   b. $\dfrac{1}{2} \cdot 4\dfrac{5}{6} = \dfrac{1}{2} \cdot \dfrac{29}{6} = \dfrac{29}{12} = 2\dfrac{5}{12}$

5. $\dfrac{3}{2} \cdot \dfrac{1}{2} - \dfrac{1}{2} = \dfrac{3}{4} - \dfrac{1}{2} = \dfrac{3}{4} - \dfrac{2}{4} = \dfrac{1}{4}$

6. $\dfrac{2x-7y}{x^2}$ where $x = 5$ and $y = 1$.

$\dfrac{2(5)-7(1)}{5^2} = \dfrac{10-7}{25} = \dfrac{3}{25}$

7. **a.** $x + 3$

   b. $3x$

   c. $7.3 \div x$ or $\dfrac{7.3}{x}$

   d. $10 - x$

   e. $5x + 7$

8. $8 + 3(2 \cdot 6 - 1) = 8 + 3(12 - 1)$
$$= 8 + 3(11)$$
$$= 8 + 33$$
$$= 41$$

9. $11.4 + (-4.7) = 11.4 - 4.7 = 6.7$

10. $5(1)^2 + 2 = 1 - 8$
$$5 + 2 = 1 - 8$$
$$7 \neq -7; \text{ no}$$

11. **a.** $\dfrac{x-y}{12+x} = \dfrac{2-(-5)}{12+2} = \dfrac{2+5}{12+2} = \dfrac{7}{14} = \dfrac{1}{2}$

   b. $x^2 - y = 2^2 - (-5) = 4 + 5 = 9$

12. **a.** $7 - 40 = -33$

   b. $-5 - (-10) = -5 + 10 = 5$

13. $\dfrac{-30}{-10} = 3$

14. $\dfrac{-48}{6} = -8$

15. $\dfrac{42}{-0.6} = -70$

16. $\dfrac{-30}{-0.2} = 150$

17. $5(3x + 2) = 5(3x) + 5(2) = 15x + 10$

18. $-3(2x - 3) = (-3)(2x) + (-3)(-3) = -6x + 9$

19. $-2(y + 0.3z - 1)$
$$= (-2)(y) + (-2)(0.3z) + (-2)(-1)$$
$$= -2y - 0.6z + 2$$

20. $4x(-x^2 + 6x - 1)$
$$= 4x(-x^2) + 4x(6x) + 4x(-1)$$
$$= -4x^3 + 24x^2 - 4x$$

21. $-(9x + y - 2z + 6) = -9x - y + 2z - 6$

22. $-(-4xy + 6y - 2) = 4xy - 6y + 2$

23. $6(2a - 1) - (11a + 6) = 7$
$$12a - 6 - 11a - 6 = 7$$
$$a - 12 = 7$$
$$a = 19$$

24. $2x + \dfrac{1}{8} = x - \dfrac{3}{8}$

$\underline{-x \qquad\quad -x}$

$x + \dfrac{1}{8} = -\dfrac{3}{8}$

$\underline{-\dfrac{1}{8} \quad\; -\dfrac{1}{8}}$

$x = -\dfrac{4}{8} = -\dfrac{1}{2}$

25. $\dfrac{y}{7} = 20$

$7 \cdot \dfrac{y}{7} = 20 \cdot 7$

$y = 140$

26.
$$10 = 5j - 2$$
$$\underline{+2 \qquad +2}$$
$$12 = 5j$$
$$\frac{12}{5} = j$$

27.
$$0.25x + 0.10(x - 3) = 1.1$$
$$0.25x + 0.10x - 0.30 = 1.1$$
$$0.35x - 0.30 = 1.1$$
$$0.35x = 1.4$$
$$x = 4$$

28.
$$\frac{7x + 5}{3} = x + 3$$
$$7x + 5 = 3(x + 3)$$
$$7x + 5 = 3x + 9$$
$$4x + 5 = 9$$
$$4x = 4$$
$$x = 1$$

29.
$$2(x + 4) = 4x - 12$$
$$2x + 8 = 4x - 12$$
$$8 = 2x - 12$$
$$20 = 2x$$
$$10 = x$$
The number is 10.

30. $(x + 7) - 2x = -x + 7$

31.
$$30 \cdot 2 + 2x = 140$$
$$60 + 2x = 140$$
$$2x = 80$$
$$x = 40$$
The length is 40 feet.

32. $\dfrac{4(-3) + (-8)}{5 + (-5)} = \dfrac{-12 - 8}{5 - 5} = \dfrac{-20}{0} = \text{undefined}$

33.
$$\frac{120}{x} = \frac{15}{100}$$
$$15x = 12,000$$
$$x = 800$$

34.

35.
$$-4x + 7 \geq -9$$
$$-4x \geq -16$$
$$x \leq 4$$
$$\{x \mid x \leq 4\}$$

36. a. $(-5)^2 = (-5)(-5) = 25$

    b. $-5^2 = -(5 \cdot 5) = -25$

    c. $2 \cdot 5^2 = 2 \cdot 25 = 50$

37. a. $x^7 \cdot x^4 = x^{7+4} = x^{11}$

    b. $\left(\dfrac{t}{2}\right)^4 = \dfrac{t^4}{2^4} = \dfrac{t^4}{16}$

    c. $(9y^5)^2 = 9^2 y^{5 \cdot 2} = 81y^{10}$

38.
$$\frac{(z^2)^3 \cdot z^7}{z^9} = \frac{z^6 \cdot z^7}{z^9}$$
$$= \frac{z^{6+7}}{z^9}$$
$$= \frac{z^{13}}{z^9}$$
$$= z^{13-9}$$
$$= z^4$$

39. $\left(\dfrac{3a^2}{b}\right)^{-3} = \dfrac{3^{-3} a^{2 \cdot -3}}{b^{-3}} = \dfrac{3^{-3} a^{-6}}{b^{-3}} = \dfrac{b^3}{27a^6}$

40. $(5x^7)(-3x^9) = -15x^{7+9} = -15x^{16}$

41. $(5y^3)^{-2} = 5^{-2} y^{3 \cdot -2} = 5^{-2} y^{-6} = \dfrac{1}{25y^6}$

42. $(-3)^{-2} = \dfrac{1}{(-3)^2} = \dfrac{1}{9}$

43. $9x^3 + x = (9+1)x^3 = 10x^3$

44. $(5y^2 - 6) - (y^2 + 2) = (5y^2 - 6) + (-y^2 - 2)$
$$= 5y^2 - 6 - y^2 - 2$$
$$= 4y^2 - 8$$

45. $5x^2 + 6x - 9x - 3 = 5x^2 - 3x - 3$

46. $(10x^2 - 3)(10x^2 + 3) = (10x^2)^2 - (3)^2$
$$= 100x^4 - 9$$

47. $7x(x^2 + 2x + 5) = 7x(x^2) + 7x(2x) + 7x(5)$
$$= 7x^3 + 14x^2 + 35x$$

48. $(10x^2 + 3)^2 = (10x^2)^2 + 2(10x^2)(3) + 3^2$
$$= 100x^4 + 60x^2 + 9$$

49. $\dfrac{9x^5 - 12x^2 + 3x}{3x^2} = \dfrac{9x^5}{3x^2} - \dfrac{12x^2}{3x^2} + \dfrac{3x}{3x^2}$
$$= 3x^3 - 4 + \dfrac{1}{x}$$

Chapter 4

Section 4.1 Practice Problems

1. a. $45 = 3 \cdot 3 \cdot 5$
$75 = 3 \cdot 5 \cdot 5$
$\text{GCF} = 3 \cdot 5 = 15$

b. $32 = 2 \cdot 2 \cdot 2 \cdot 2 \cdot 2$
$33 = 3 \cdot 11$
There are no common prime factors;
thus, the GCF is 1.

c. $14 = 2 \cdot 7$
$24 = 2 \cdot 2 \cdot 2 \cdot 3$
$60 = 2 \cdot 2 \cdot 3 \cdot 5$
$\text{GCF} = 2$

2. a. The GCF is y^4, since 4 is the smallest exponent to which y is raised.

b. The GCF is x^1 or x, since 1 is the smallest exponent on x.

3. a. $6x^2 = 2 \cdot 3 \cdot x^2$
$9x^4 = 3 \cdot 3 \cdot x^4$
$-12x^5 = -1 \cdot 2 \cdot 2 \cdot 3 \cdot x^5$
$\text{GCF} = 3 \cdot x^2 = 3x^2$

b. $-16y = -1 \cdot 2 \cdot 2 \cdot 2 \cdot 2 \cdot y$
$-20y^6 = -1 \cdot 2 \cdot 2 \cdot 5 \cdot y^6$
$40y^4 = 2 \cdot 2 \cdot 2 \cdot 5 \cdot y^4$
$\text{GCF} = 2 \cdot 2 \cdot y = 4y$

c. The GCF of a^5, a, and a^3 is a.
The GCF of b^4, b^3, and b^2 is b^2.
Thus, the GCF of $a^5 b^4$, ab^3, and $a^3 b^2$ is ab^2.

4. a. The GCF of terms $10y$ and 25 is 5.
$10y + 25 = 5 \cdot 2y + 5 \cdot 5 = 5(2y + 5)$

b. The GCF of x^4 and x^9 is x^4.
$x^4 - x^9 = x^4(1) - x^4(x^5) = x^4(1 - x^5)$

5. $-10x^3 + 8x^2 - 2x$
$= 2x(-5x^2) + 2x(4x) + 2x(-1)$
$= 2x(-5x^2 + 4x - 1)$

6. $4x^3 + 12x = 4x(x^2 + 3)$

7. $\dfrac{2}{5}a^5 - \dfrac{4}{5}a^3 + \dfrac{1}{5}a^2 = \dfrac{1}{5}a^2(2a^3 - 4a + 1)$

8. $6a^3b + 3a^3b^2 + 9a^2b^4$
$= 3a^2b(2a + ab + 3b^3)$

9. $7(p + 2) + q(p + 2) = (p + 2)(7 + q)$

10. $ab + 7a + 2b + 14 = (ab + 7a) + (2b + 14)$
$= a(b + 7) + 2(b + 7)$
$= (b + 7)(a + 2)$

11. $28x^3 - 7x^2 + 12x - 3$
$= (28x^3 - 7x^2) + (12x - 3)$
$= 7x^2(4x - 1) + 3(4x - 1)$
$= (4x - 1)(7x^2 + 3)$

12. $2xy + 5y^2 - 4x - 10y$
$= (2xy + 5y^2) + (-4x - 10y)$
$= y(2x + 5y) - 2(2x + 5y)$
$= (2x + 5y)(y - 2)$

13. $3x^2 + 4xy + 3x + 4y$
$= (3x^2 + 4xy) + (3x + 4y)$
$= x(3x + 4y) + 1(3x + 4y)$
$= (3x + 4y)(x + 1)$

14. $4x^3 + x - 20x^2 - 5 = x(4x^2 + 1) - 5(4x^2 + 1)$
$= (4x^2 + 1)(x - 5)$

15. $3xy - 4 + x - 12y = (3xy + x) + (-12y - 4)$
$= x(3y + 1) - 4(3y + 1)$
$= (3y + 1)(x - 4)$

16. $2x - 2 + x^3 - 3x^2 = 2(x - 1) + x^2(x - 3)$

There is no common binomial factor that can now be factored out. This polynomial is not factorable by grouping.

Mental Math

1. 2, 16
GCF = 2

2. 3, 18
GCF = 3

3. 6, 7
GCF = 1

4. 9, 11
GCF = 1

5. 14, 35
GCF = 7

6. 33, 55
GCF = 11

Exercise Set 4.1

1. $32 = 2 \cdot 2 \cdot 2 \cdot 2 \cdot 2$
$36 = 2 \cdot 2 \cdot 3 \cdot 3$
GCF $= 2 \cdot 2 = 4$

3. $18 = 2 \cdot 3 \cdot 3$
$42 = 2 \cdot 3 \cdot 7$
$84 = 2 \cdot 2 \cdot 3 \cdot 7$
GCF $= 2 \cdot 3 = 6$

5. $24 = 2 \cdot 2 \cdot 2 \cdot 3$
$14 = 2 \cdot 7$
$21 = 3 \cdot 7$
GCF = 1 since there are no common prime factors.

7. The GCF of y^2, y^4, and y^7 is y^2.

9. The GCF of z^7, z^9, and z^{11} is z^7.

11. The GCF of x^{10}, x, and x^3 is x.
The GCF of y^2, y^2, and y^3 is y^2.
Thus, the GCF of $x^{10}y^2$, xy^2, and x^3y^3 is xy^2.

13. $14x = 2 \cdot 7 \cdot x$
$21 = 3 \cdot 7$
GCF = 7

15. $12y^4 = 2 \cdot 2 \cdot 3 \cdot y^4$
$20y^3 = 2 \cdot 2 \cdot 5 \cdot y^3$
GCF $= 2 \cdot 2 \cdot y^3 = 4y^3$

17. $-10x^2 = -1 \cdot 2 \cdot 5 \cdot x^2$
$15x^3 = 3 \cdot 5 \cdot x^3$
GCF $= 5 \cdot x^2 = 5x^2$

19. $12x^3 = 2 \cdot 2 \cdot 3 \cdot x^3$
$-6x^4 = -1 \cdot 2 \cdot 3 \cdot x^4$
$3x^5 = 3 \cdot x^5$
GCF $= 3 \cdot x^3 = 3x^3$

21. $-18x^2y = -1 \cdot 2 \cdot 3 \cdot 3 \cdot x^2 \cdot y$
$9x^3y^3 = 3 \cdot 3 \cdot x^3 \cdot y^3$
$36x^3y = 2 \cdot 2 \cdot 3 \cdot 3 \cdot x^3 \cdot y$
GCF $= 3 \cdot 3 \cdot x^2 \cdot y = 9x^2y$

23. $20a^6b^2c^8 = 2 \cdot 2 \cdot 5 \cdot a^6 \cdot b^2 \cdot c^8$
$50a^7b = 2 \cdot 5 \cdot 5 \cdot a^7 \cdot b$
GCF $= 2 \cdot 5 \cdot a^6 \cdot b = 10a^6b$

25. $3a + 6 = 3(a + 2)$

27. $30x - 15 = 15(2x - 1)$

29. $x^3 + 5x^2 = x^2(x + 5)$

31. $6y^4 + 2y^3 = 2y^3(3y+1)$

33. $32xy - 18x^2 = 2x(16y - 9x)$

35. $4x - 8y + 4 = 4(x - 2y + 1)$

37. $6x^3 - 9x^2 + 12x = 3x(2x^2 - 3x + 4)$

39. $a^7b^6 - a^3b^2 + a^2b^5 - a^2b^2$
$= a^2b^2(a^5b^4 - a + b^3 - 1)$

41. $5x^3y - 15x^2y + 10xy = 5xy(x^2 - 3x + 2)$

43. $8x^5 + 16x^4 - 20x^3 + 12$
$= 4(2x^5 + 4x^4 - 5x^3 + 3)$

45. $\dfrac{1}{3}x^4 + \dfrac{2}{3}x^3 - \dfrac{4}{3}x^5 + \dfrac{1}{3}x$
$= \dfrac{1}{3}x(x^3 + 2x^2 - 4x^4 + 1)$

47. $y(x^2 + 2) + 3(x^2 + 2) = (x^2 + 2)(y + 3)$

49. $z(y + 4) + 3(y + 4) = (y + 4)(z + 3)$

51. $r(z^2 - 6) + (z - 6) = r(z^2 - 6) + 1(z^2 - 6)$
$= (z^2 - 6)(r + 1)$

53. $-x - 7 = (-1)x + (-1)(7) = -1(x + 7)$

55. $-2 + z = -1(2) + (-1)(-z) = -1(2 - z)$

57. $3a - b + 2 = (-1)(-3a) + (-1)(b) - (-1)(2)$
$= -1(-3a + b - 2)$

59. $x^3 + 2x^2 + 5x + 10 = x^2(x + 2) + 5(x + 2)$
$= (x + 2)(x^2 + 5)$

61. $5x + 15 + xy + 3y = 5(x + 3) + y(x + 3)$
$= (x + 3)(5 + y)$

63. $6x^3 - 4x^2 + 15x - 10$
$= 2x^2(3x - 2) + 5(3x - 2)$
$= (3x - 2)(2x^2 + 5)$

65. $5m^3 + 6mn + 5m^2 + 6n$
$= m(5m^2 + 6n) + 1(5m^2 + 6n)$
$= (5m^2 + 6n)(m + 1)$

67. $2y - 8 + xy - 4x = 2(y - 4) + x(y - 4)$
$= (y - 4)(2 + x)$

69. $2x^3 + x^2 + 8x + 4 = x^2(2x + 1) + 4(2x + 1)$
$= (2x + 1)(x^2 + 4)$

71. $4x^2 - 8xy - 3x + 6y = 4x(x - 2y) - 3(x - 2y)$
$= (x - 2y)(4x - 3)$

73. $5q^2 - 4pq - 5q + 4p$
$= q(5q - 4p) - 1(5q - 4p)$
$= (5q - 4p)(q - 1)$

75. $12x^2y - 42x^2 - 4y + 14$
$= 2(6x^2y - 21x^2 - 2y + 7)$
$= 2[3x^2(2y - 7) - 1(2y - 7)]$
$= 2(2y - 7)(3x^2 - 1)$

77. $(x + 2)(x + 5) = x^2 + 5x + 2x + 10$
$= x^2 + 7x + 10$

79. $(b + 1)(b - 4) = b^2 - 4b + b - 4 = b^2 - 3b - 4$

81. $2 \cdot 6 = 12$
$2 + 6 = 8$
2 and 6 have a product of 12 and a sum of 8.

83. $-1 \cdot (-8) = 8$
$-1 + (-8) = -9$
-1 and -8 have a product of 8 and a sum of -9.

85. $-2 \cdot 5 = -10$
$-2 + 5 = 3$
-2 and 5 have a product of -10 and a sum of 3.

87. $-8 \cdot 3 = -24$
$-8 + 3 = -5$
-8 and 3 have a product of -24 and a sum of -5.

89. $8a - 24 = 8(a - 3)$, which is choice b.

91. $(x + 5)(x + y)$ is factored.

93. Since
$3x(a + 2b) + 2(a + 2b) = (a + 2b)(3x + 2)$,
the given expression is not factored.

95. a. $-13x^2 + 221x + 8476$
$= -13(1)^2 + 221(1) + 8476$
$= -13 + 221 + 8476$
$= 8684$
The average total supply in 2001 was 8684 thousand barrels per day.

b. Let $x = 3$ for 2003.
$-13x^2 + 221x + 8476$
$= -13(3)^2 + 221(3) + 8476$
$= -13(9) + 221(3) + 8476$
$= -117 + 663 + 8476$
$= 9022$
The average total supply in 2003 was 9022 thousand barrels per day.

c. $-13x^2 + 221x + 8476$
$= -13(x^2 - 17x - 652)$
or $13(-x^2 + 17x + 652)$

97. The area of the circle is πx^2. Since the sides of the square have length $2x$, the area of the square is $(2x)^2 = 4x^2$. The shaded region is the region inside the square but outside the circle. The area of the shaded region is
$4x^2 - \pi x^2 = x^2(4 - \pi)$.

99.   Area = width · length
$5x^5 - 5x^2 = 5x^2(x^3 - 1)$
The length is $(x^3 - 1)$ units.

101. answers may vary

103. answers may vary

Section 4.2 Practice Problems

1.

Factors of 20	Sum of Factors
1, 20	21
2, 10	12
4, 5	9

Thus, $x^2 + 12x + 20 = (x + 10)(x + 2)$.

2. a.

Factors of 22	Sum of Factors
$-1, -22$	-23
$-2, -11$	-13

$x^2 - 23x + 22 = (x - 1)(x - 22)$

b.

Factors of 50	Sum of Factors
$-1, -50$	-51
$-2, -25$	-27
$-5, -10$	-15

$x^2 - 27x + 50 = (x - 2)(x - 25)$

3.

Factors of −36	Sum of Factors
−1, 36	35
1, −36	−35
−2, 18	16
2, −18	−16
−3, 12	9
3, −12	−9
−4, 9	5
4, −9	−5
−6, 6	0

$$x^2 + 5x - 36 = (x+9)(x-4)$$

4. a. Two factors of −40 whose sum is −3 are −8 and 5.

$$q^2 - 3q - 40 = (q-8)(q+5)$$

   b. Two factors of −48 whose sum is 2 are 8 and −6.

$$y^2 + 2y - 48 = (y+8)(y-6)$$

5. Since there are no two numbers whose product is 15 and whose sum is 6, the polynomial is prime.

6. a. Two factors of $14y^2$ whose sum is $9y$ are $2y$ and $7y$.

$$x^2 + 9xy + 14y^2 = (x+2y)(x+7y)$$

   b. Two factors of $30b^2$ whose sum is $-13b$ are $-3b$ and $-10b$.

$$a^2 - 13ab + 30b^2 = (a-3b)(a-10b)$$

7. Two factors of 12 whose sum is 8 are 6 and 2.

$$x^4 + 8x^2 + 12 = (x^2+6)(x^2+2)$$

8. $48 - 14x + x^2 = x^2 - 14x + 48$

Two factors of 48 whose sum is −14 are −6 and −8.

$$x^2 - 14x + 48 = (x-6)(x-8)$$

9. a. $4x^2 - 24x + 36 = 4(x^2 - 6x + 9)$

Two factors of 9 whose sum is −6 are −3 and −3.

$$4(x^2 - 6x + 9) = 4(x-3)(x-3) \text{ or}$$
$$4(x-3)^2$$

   b. $x^3 + 3x^2 - 4x = x(x^2 + 3x - 4)$

Two factors of −4 whose sum is 3 are 4 and −1.

$$x(x^2 + 3x - 4) = x(x+4)(x-1)$$

10. $5x^5 - 25x^4 - 30x^3 = 5x^3(x^2 - 5x - 6)$
$$= 5x^3(x+1)(x-6)$$

Mental Math

1. $x^2 + 9x + 20 = (x+4)(x+5)$

2. $x^2 + 12x + 35 = (x+5)(x+7)$

3. $x^2 - 7x + 12 = (x-4)(x-3)$

4. $x^2 - 13x + 22 = (x-2)(x-11)$

5. $x^2 + 4x + 4 = (x+2)(x+2)$

6. $x^2 + 10x + 24 = (x+6)(x+4)$

Exercise Set 4.2

1. Two factors of 6 whose sum is 7 are 6 and 1.

$$x^2 + 7x + 6 = (x+6)(x+1)$$

3. Two factors of 9 whose sum is -10 are -9 and -1.
$$y^2 - 10y + 9 = (y-9)(y-1)$$

5. Two factors of 9 whose sum is -6 are -3 and -3.
$$x^2 - 6x + 9 = (x-3)(x-3) \text{ or } (x-3)^2$$

7. Two factors of -18 whose sum is -3 are -6 and 3.
$$x^2 - 3x - 18 = (x-6)(x+3)$$

9. Two factors of -70 whose sum is 3 are 10 and -7.
$$x^2 + 3x - 70 = (x+10)(x-7)$$

11. Since there are no two numbers whose product is 2 and whose sum is 5, the polynomial is prime.

13. Two factors of $15y^2$ whose sum is $8y$ are $5y$ and $3y$.
$$x^2 + 8xy + 15y^2 = (x+5y)(x+3y)$$

15. Two factors of -15 whose sum is -2 are -5 and 3.
$$a^4 - 2a^2 - 15 = (a^2 - 5)(a^2 + 3)$$

17. $13 + 14m + m^2 = m^2 + 14m + 13$
Two factors of 13 whose sum is 14 are 13 and 1.
$$13 + 14m + m^2 = m^2 + 14m + 13$$
$$= (m+13)(m+1)$$

19. $10t - 24 + t^2 = t^2 + 10t - 24$
Two factors of -24 whose sum is 10 are -2 and 12.
$$10t - 24 + t^2 = t^2 + 10t - 24 = (t-2)(t+12)$$

21. Two factors of $16b^2$ whose sum is $-10b$ are $-2b$ and $-8b$.
$$a^2 - 10ab + 16b^2 = (a-2b)(a-8b)$$

23. $2z^2 + 20z + 32 = 2(z^2 + 10z + 16)$
$$= 2(z+8)(z+2)$$

25. $2x^3 - 18x^2 + 40x = 2x(x^2 - 9x + 20)$
$$= 2x(x-5)(x-4)$$

27. $x^2 - 3xy - 4y^2 = (x-4y)(x+y)$

29. $x^2 + 15x + 36 = (x+12)(x+3)$

31. $x^2 - x - 2 = (x-2)(x+1)$

33. $r^2 - 16r + 48 = (r-12)(r-4)$

35. $x^2 + xy - 2y^2 = (x+2y)(x-y)$

37. $3x^2 + 9x - 30 = 3(x^2 + 3x - 10)$
$$= 3(x+5)(x-2)$$

39. $3x^2 - 60x + 108 = 3(x^2 - 20x + 36)$
$$= 3(x-18)(x-2)$$

41. $x^2 - 18x - 144 = (x-24)(x+6)$

43. $r^2 - 3r + 6$ is prime.

45. $x^2 - 8x + 15 = (x-5)(x-3)$

47. $6x^3 + 54x^2 + 120x = 6x(x^2 + 9x + 20)$
$$= 6x(x+4)(x+5)$$

49. $4x^2y + 4xy - 12y = 4y(x^2 + x - 3)$

51. $x^2 - 4x - 21 = (x-7)(x+3)$

53. $x^2 + 7xy + 10y^2 = (x+5y)(x+2y)$

55. $64 + 24t + 2t^2 = 2t^2 + 24t + 64$
$$= 2(t^2 + 12t + 32)$$
$$= 2(t+8)(t+4)$$

57. $x^3 - 2x^2 - 24x = x(x^2 - 2x - 24)$
$= x(x - 6)(x + 4)$

59. $2t^5 - 14t^4 + 24t^3 = 2t^3(t^2 - 7t + 12)$
$= 2t^3(t - 4)(t - 3)$

61. $5x^3y - 25x^2y^2 - 120xy^3$
$= 5xy(x^2 - 5xy - 24y^2)$
$= 5xy(x - 8y)(x + 3y)$

63. $162 - 45m + 3m^2 = 3m^2 - 45m + 162$
$= 3(m^2 - 15m + 54)$
$= 3(m - 9)(m - 6)$

65. $-x^2 + 12x - 11 = -1(x^2 - 12x + 11)$
$= -1(x - 11)(x - 1)$

67. $\frac{1}{2}y^2 - \frac{9}{2}y - 11 = \frac{1}{2}(y^2 - 9y - 22)$
$= \frac{1}{2}(y - 11)(y + 2)$

69. $x^3y^2 + x^2y - 20x = x(x^2y^2 + xy - 20)$
$= x(xy - 4)(xy + 5)$

71. $(2x + 1)(x + 5) = 2x^2 + 10x + x + 5$
$= 2x^2 + 11x + 5$

73. $(5y - 4)(3y - 1) = 15y^2 - 5y - 12y + 4$
$= 15y^2 - 17y + 4$

75. $(a + 3b)(9a - 4b) = 9a^2 - 4ab + 27ab - 12b^2$
$= 9a^2 + 23ab - 12b^2$

77. $(x - 3)(x + 8) = x^2 + 8x - 3x - 24$
$= x^2 + 5x - 24$

79. answers may vary

81. $P = 2l + 2w$
$= 2(x^2 + 10x) + 2(4x + 33)$
$= 2x^2 + 20x + 8x + 66$
$= 2x^2 + 28x + 66$
$= 2(x^2 + 14x + 33)$
$= 2(x + 3)(x + 11)$

83. $-16t^2 + 64t + 80 = -16(t^2 - 4t - 5)$
$= -16(t - 5)(t + 1)$

85. $x^2 + \frac{1}{2}x + \frac{1}{16} = \left(x + \frac{1}{4}\right)\left(x + \frac{1}{4}\right)$ or
$\left(x + \frac{1}{4}\right)^2$

87. $z^2(x + 1) - 3z(x + 1) - 70(x + 1)$
$= (x + 1)(z^2 - 3z - 70)$
$= (x + 1)(z - 10)(z + 7)$

89. The factors of c must sum to -16. Since c is positive, both factors must have the same sign.
$-1 + (-15) = -16; (-1)(-15) = 15$
$-2 + (-14) = -16; (-2)(-14) = 28$
$-3 + (-13) = -16; (-3)(-13) = 39$
$-4 + (-12) = -16; (-4)(-12) = 48$
$-5 + (-11) = -16; (-5)(-11) = 55$
$-6 + (-10) = -16; (-6)(-10) = 60$
$-7 + (-9) = -16; (-7)(-9) = 63$
$-8 + (-8) = -16; (-8)(-8) = 64$
The possible values of c are 15, 28, 39, 48, 55, 60, 63, and 64.

91. The factors of 20 must sum to b. Since 20 is positive and b is positive, both factors of 20 must be positive.
$1 + 20 = 21; (1)(20) = 20$
$2 + 10 = 12; (2)(10) = 20$
$4 + 5 = 9; (4)(5) = 20$
The possible values of b are 9, 12, and 21.

93. $x^{2n} + 8x^n - 20 = (x^n + 10)(x^n - 2)$

Section 4.3 Practice Problems

1. a. Factors of $5x^2$: $5x^2 = 5x \cdot x$

Factors of 10: $10 = 1 \cdot 10$, $10 = 2 \cdot 5$

$5x^2 + 27x + 10 = (5x + 2)(x + 5)$

b. Factors of $4x^2$: $4x^2 = 4x \cdot x$,

$4x^2 = 2x \cdot 2x$

Factors of 5: $5 = 1 \cdot 5$

$4x^2 + 12x + 5 = (2x + 5)(2x + 1)$

2. a. Factors of $2x^2$: $2x^2 = 2x \cdot x$

Factors of 12: $12 = -1 \cdot -12$,

$12 = -2 \cdot -6$, $12 = -3 \cdot -4$

$2x^2 - 11x + 12 = (2x - 3)(x - 4)$

b. Factors of $6x^2$:

$6x^2 = 6x \cdot x$, $6x^2 = 3x \cdot 2x$

Factors of 1: $1 = -1 \cdot -1$

$6x^2 - 5x + 1 = (3x - 1)(2x - 1)$

3. a. Factors of $3x^2$: $3x^2 = 3x \cdot x$

Factors of -5: $-5 = -5 \cdot 1$, $-5 = 1 \cdot -5$

$3x^2 + 14x - 5 = (3x - 1)(x + 5)$

b. Factors of $35x^2$: $35x^2 = 35x \cdot x$,

$35x^2 = 5x \cdot 7x$

Factors of -4: $-4 = -1 \cdot 4$, $-4 = 1 \cdot -4$,

$-4 = -2 \cdot 2$

$35x^2 + 4x - 4 = (5x + 2)(7x - 2)$

4. a. Factors of $14x^2$: $14x^2 = 14x \cdot x$,

$14x^2 = 7x \cdot 2x$

Factors of $-2y^2$: $-2y^2 = -2y \cdot y$,

$-2y^2 = 2y \cdot -y$

$14x^2 - 3xy - 2y^2 = (7x + 2y)(2x - y)$

b. Factors of $12a^2$: $12a^2 = 12a \cdot a$,

$12a^2 = 6a \cdot 2a$, $12a^2 = 3a \cdot 4a$

Factors of $-3b^2$: $-3b^2 = -3b \cdot b$,

$-3b^2 = 3b \cdot -b$

$12a^2 - 16ab - 3b^2 = (6a + b)(2a - 3b)$

5. Factors of $2x^4$: $2x^4 = 2x^2 \cdot x^2$

Factors of -7: $-7 = -7 \cdot 1$, $-7 = 7 \cdot -1$

$2x^4 - 5x^2 - 7 = (2x^2 - 7)(x^2 + 1)$

6. a. $3x^3 + 17x^2 + 10x = x(3x^2 + 17x + 10)$

Factors of $3x^2$: $3x^2 = 3x \cdot x$

Factors of 10: $10 = 1 \cdot 10$, $10 = 2 \cdot 5$

$x(3x^2 + 17x + 10) = x(3x + 2)(x + 5)$

b. $6xy^2 + 33xy - 18x = 3x(2y^2 + 11y - 6)$

Factors of $2y^2$: $2y^2 = 2y \cdot y$

Factors of -6: $-6 = 1 \cdot -6$, $-6 = -1 \cdot 6$,

$-6 = 2 \cdot -3$, $-6 = -2 \cdot 3$

$3x(2y^2 + 11y - 6) = 3x(2y - 1)(y + 6)$

7. $-5x^2 - 19x + 4 = -1(5x^2 + 19x - 4)$

Factors of $5x^2$: $5x^2 = 5x \cdot x$

Factors of -4: $-4 = -4 \cdot 1$, $-4 = 4 \cdot -1$,

$-4 = 2 \cdot -2$

$-1(5x^2 + 19x - 4) = -1(x + 4)(5x - 1)$

Exercise Set 4.3

1. $5x^2 = 5x \cdot x$

$8 = 2 \cdot 4$

$5x^2 + 22x + 8 = (5x + 2)(x + 4)$

3. $50x^2 = 5x \cdot 10x$

$-2 = 2 \cdot -1$

$50x^2 + 15x - 2 = (5x + 2)(10x - 1)$

5. $20x^2 = 5x \cdot 4x$
$-6 = 2 \cdot -3$
$20x^2 - 7x - 6 = (5x + 2)(4x - 3)$

7. Factors of $2x^2$: $2x^2 = 2x \cdot x$
Factors of 15: $15 = 1 \cdot 15, 15 = 3 \cdot 5$
$2x^2 + 13x + 15 = (2x + 3)(x + 5)$

9. Factors of $8y^2$: $8y^2 = 8y \cdot y, 8y^2 = 4y \cdot 2y.$
Factors of 9: $9 = -1 \cdot -9, -3 \cdot -3.$
$8y^2 - 17y + 9 = (y - 1)(8y - 9)$

11. Factors of $2x^2$: $2x^2 = 2x \cdot x$
Factors of -5: $-5 = 1 \cdot -5, -5 = -1 \cdot 5$
$2x^2 - 9x - 5 = (2x + 1)(x - 5)$

13. Factors of $20r^2$: $20r^2 = 20r \cdot r,$
$20r^2 = 10r \cdot 2r, \ 20r^2 = 5r \cdot 4r.$
Factors of -8: $-8 = -1 \cdot 8, -8 = -2 \cdot 4,$
$-8 = -4 \cdot 2, -8 = -8 \cdot 1$
$20r^2 + 27r - 8 = (4r - 1)(5r + 8)$

15. Factors of $10x^2$:
$10x^2 = 10x \cdot x, 10x^2 = 5x \cdot 2x$
Factors of 3: $3 = 1 \cdot 3$
$10x^2 + 17x + 3 = (5x + 1)(2x + 3)$

17. $x + 3x^2 - 2 = 3x^2 + x - 2$
Factors of $3x^2$: $3x^2 = 3x \cdot x$
Factors of -2: $-2 = -1 \cdot 2, -2 = 2 \cdot -1$
$3x^2 + x - 2 = (3x - 2)(x + 1)$

19. Factors of $6x^2$: $6x^2 = 6x \cdot x, 6x^2 = 3x \cdot 2x$
Factors of $5y^2$: $5y^2 = -5y \cdot -y$
$6x^2 - 13xy + 5y^2 = (3x - 5y)(2x - y)$

21. Factors of $15m^2$: $15m^2 = 15m \cdot m,$
$15m^2 = 5m \cdot 3m.$
Factors of -15: $-15 = -1 \cdot 15, -15 = -3 \cdot 5,$
$-15 = -5 \cdot 3, -15 = -15 \cdot 1$
$15m^2 - 16m - 15 = (3m - 5)(5m + 3)$

23. $-9x + 20 + x^2 = x^2 - 9x + 20$
Factors of x^2: $x^2 = x \cdot x$
Factors of 20: $20 = -1 \cdot -20, 20 = -2 \cdot -10,$
$20 = -4 \cdot -5$
$x^2 - 9x + 20 = (x - 4)(x - 5)$

25. Factors of $2x^2$: $2x^2 = 2x \cdot x$
Factors of -99: $-99 = -1 \cdot 99,$
$-99 = -3 \cdot 33, -99 = -9 \cdot 11, -99 = -11 \cdot 9,$
$-99 = -33 \cdot 3, -99 = -99 \cdot 1$
$2x^2 - 7x - 99 = (2x + 11)(x - 9)$

27. $-27t + 7t^2 - 4 = 7t^2 - 27t - 4$
Factors of $7t^2$: $7t^2 = 7t \cdot t$
Factors of -4: $-4 = -1 \cdot 4, -4 = 1 \cdot -4,$
$-4 = 2 \cdot -2$
$7t^2 - 27t - 4 = (7t + 1)(t - 4)$

29. Factors of $3a^2$: $3a^2 = 3a \cdot a$
Factors of $3b^2$: $3b^2 = b \cdot 3b$
$3a^2 + 10ab + 3b^2 = (3a + b)(a + 3b)$

31. Factors of $49p^2$: $49p^2 = 49p \cdot p,$
$49p^2 = 7p \cdot 7p$
Factors of -2: $-2 = -1 \cdot 2, -2 = 1 \cdot -2$
$49p^2 - 7p - 2 = (7p + 1)(7p - 2)$.

33. Factors of $18x^2$: $18x^2 = 18x \cdot x,$
$18x^2 = 9x \cdot 2x, 18x^2 = 6x \cdot 3x$
Factors of -14: $-14 = -1 \cdot 14, -14 = -2 \cdot 7,$
$-14 = -7 \cdot 2, -14 = -14 \cdot 1$
$18x^2 - 9x - 14 = (6x - 7)(3x + 2)$

35. Factors of $2m^2$: $2m^2 = 2m \cdot m$
Factors of 10: $10 = 1 \cdot 10$, $10 = 2 \cdot 5$
$2m^2 + 17m + 10$ is prime.

37. Factors of $24x^2$: $24x^2 = 24x \cdot x$,
$24x^2 = 12x \cdot 2x$, $24x^2 = 8x \cdot 3x$,
$24x^2 = 6x \cdot 4x$
Factors of 12: $12 = 1 \cdot 12$, $12 = 2 \cdot 6$,
$12 = 3 \cdot 4$
$24x^2 + 41x + 12 = (3x + 4)(8x + 3)$

39. $12x^3 + 11x^2 + 2x = x(12x^2 + 11x + 2)$

Factor of $12x^2$: $12x^2 = 12x \cdot x$,
$12x^2 = 6x \cdot 2x$, $12x^2 = 4x \cdot 3x$
Factors of 2: $2 = 1 \cdot 2$
$12x^3 + 11x^2 + 2x = x(3x + 2)(4x + 1)$

41. $21b^2 - 48b - 45 = 3(7b^2 - 16b - 15)$

Factors of $7b^2$: $7b^2 = 7b \cdot b$
Factors of -15: $-15 = -1 \cdot 15$, $-15 = -3 \cdot 5$,
$-15 = -5 \cdot 3$, $-15 = -15 \cdot 1$
$21b^2 - 48b - 45 = 3(7b + 5)(b - 3)$

43. $7z + 12z^2 - 12 = 12z^2 + 7z - 12$
Factors of $12z^2$: $12z^2 = 12z \cdot z$,
$12z^2 = 6z \cdot 2z$, $12z^2 = 4z \cdot 3z$
Factors of -12: $-12 = -12 \cdot 1$, $-12 = 12 \cdot$
-1, $-12 = -6 \cdot 2$, $-12 = 6 \cdot -2$, $-12 = -4 \cdot 3$,
$-12 = 4 \cdot -3$
$12z^2 + 7z - 12 = (3z + 4)(4z - 3)$

45. $6x^2y^2 - 2xy^2 - 60y^2 = 2y^2(3x^2 - x - 30)$

Factors of $3x^2$: $3x^2 = 3x \cdot x$
Factors of -30: $-30 = -1 \cdot 30$, $-30 = -2 \cdot$
15, $-30 = -3 \cdot 10$, $-30 = -5 \cdot 6$,
$-30 = -6 \cdot 5$, $-30 = -10 \cdot 3$, $-30 = -15 \cdot 2$,
$-30 = -30 \cdot 1$
$6x^2y^2 - 2xy^2 - 60y^2 = 2y^2(3x - 10)(x + 3)$

47. Factors of $4x^2$: $4x^2 = 4x \cdot x$, $4x^2 = 2x \cdot 2x$
Factors of -21: $-21 = -1 \cdot 21$,
$-21 = 1 \cdot -21$, $-21 = -7 \cdot 3$, $-21 = 7 \cdot -3$
$4x^2 - 8x - 21 = (2x - 7)(2x + 3)$

49. $3x^2 - 42x + 63 = 3(x^2 - 14x + 21)$
$x^2 - 14x + 21$ is prime, so $3(x^2 - 14x + 21)$
is a factored form of $3x^2 - 42x + 63$.

51. Factors of $8x^2$: $8x^2 = 8x \cdot x$, $8x^2 = 4x \cdot 2x$
Factors of $-27y^2$: $-27y^2 = -27y \cdot y$,
$-27y^2 = 27y \cdot -y$, $-27y^2 = -9y \cdot 3y$,
$-27y^2 = 9y \cdot -3y$
$8x^2 + 6xy - 27y^2 = (4x + 9y)(2x - 3y)$

53. $-x^2 + 2x + 24 = -1(x^2 - 2x - 24)$
$\qquad\qquad\qquad = -1(x - 6)(x + 4)$

55. $4x^3 - 9x^2 - 9x = x(4x^2 - 9x - 9)$

Factors of $4x^2$: $4x^2 = 4x \cdot x$, $4x^2 = 2x \cdot 2x$
Factors of -9: $-9 = -1 \cdot 9$, $-9 = 1 \cdot -9$,
$-9 = 3 \cdot -3$
$4x^3 - 9x^2 - 9x = x(4x + 3)(x - 3)$

57. Factors of $24x^2$: $24x^2 = 24x \cdot x$,
$24x^2 = 12x \cdot 2x$, $24x^2 = 8x \cdot 3x$,
$24x^2 = 6x \cdot 4x$
Factors of 9: $9 = -1 \cdot -9$, $9 = -3 \cdot -3$
$24x^2 - 58x + 9 = (4x - 9)(6x - 1)$

59. $40a^2b + 9ab - 9b = b(40a^2 + 9a - 9)$
Factors of $40a^2$: $40a^2 = 40a \cdot a$,
$40a^2 = 20a \cdot 2a$, $40a^2 = 10a \cdot 4a$,
$40a^2 = 8a \cdot 5a$

Factors of -9: $-9 = -1 \cdot 9$, $-9 = 1 \cdot -9$,
$-9 = 3 \cdot -3$
$40a^2b + 9ab - 9b = b(8a - 3)(5a + 3)$

61. $30x^3 + 38x^2 + 12x = 2x(15x^2 + 19x + 6)$

Factors of
$15x^2$: $15x^2 = 15x \cdot x$, $15x^2 = 5x \cdot 3x$
Factors of 6: $6 = 1 \cdot 6$, $6 = 2 \cdot 3$
$30x^3 + 38x^2 + 12x = 2x(3x + 2)(5x + 3)$

63. $6y^3 - 8y^2 - 30y = 2y(3y^2 - 4y - 15)$

Factors of $3y^2$: $3y^2 = 3y \cdot y$
Factors of -15: $-15 = -1 \cdot 15$,
$-15 = 1 \cdot -15$, $-15 = -3 \cdot 5$, $-15 = 3 \cdot -5$
$6y^3 - 8y^2 - 30y = 2y(3y + 5)(y - 3)$

65. $10x^4 + 25x^3y - 15x^2y^2$
$= 5x^2(2x^2 + 5xy - 3y^2)$

Factors of $2x^2$: $2x^2 = 2x \cdot x$
Factors of $-3y^2$: $-3y^2 = -y \cdot 3y$,
$-3y^2 = -3y \cdot y$
$10x^4 + 25x^3y - 15x^2y^2$
$= 5x^2(2x - y)(x + 3y)$

67. $-14x^2 + 39x - 10 = -1(14x^2 - 39x + 10)$

Factors of $14x^2$: $14x^2 = 14x \cdot x$,
$14x^2 = 7x \cdot 2x$
Factors of 10: $10 = -1 \cdot -10$, $10 = -2 \cdot -5$
$-14x^2 + 39x - 10 = -1(2x - 5)(7x - 2)$

69. $16p^4 - 40p^3 + 25p^2$
$= p^2(16p^2 - 40p + 25)$

Factors of $16p^2$: $16p^2 = 16p \cdot p$,
$16p^2 = 8p \cdot 2p$, $16p^2 = 4p \cdot 4p$
Factors of 25: $25 = -1 \cdot -25$, $25 = -5 \cdot -5$
$16p^4 - 40p^3 + 25p^2 = p^2(4p - 5)(4p - 5)$
or $p^2(4p - 5)^2$

71. $-2x^2 + 9x + 5 = -1(2x^2 - 9x - 5)$

Factors of $2x^2$: $2x^2 = 2x \cdot x$
Factors of -5: $-5 = -1 \cdot 5$, $-5 = 1 \cdot -5$
$-2x^2 + 9x + 5 = -1(2x + 1)(x - 5)$

73. $-4 + 52x - 48x^2 = -48x^2 + 52x - 4$
$= -4(12x^2 - 13x + 1)$

Factors of $12x^2$: $12x^2 = 12x \cdot x$,
$12x^2 = 6x \cdot 2x$, $12x^2 = 4x \cdot 3x$
Factors of 1: $1 = -1 \cdot -1$
$-4 + 52x - 48x^2 = -4(12x - 1)(x - 1)$

75. Factors of $2t^4$: $2t^4 = 2t^2 \cdot t^2$
Factors of -27: $-27 = -1 \cdot 27$, $-27 = 1 \cdot -27$,
$-27 = -3 \cdot 9$, $-27 = 3 \cdot -9$
$2t^4 + 3t^2 - 27 = (2t^2 + 9)(t^2 - 3)$

77. Factors of $5x^2y^2$: $5x^2y^2 = 5xy \cdot xy$
Factors of 1: $1 = 1 \cdot 1$
There is no combination that gives the
correct middle term, so $5x^2y^2 + 20xy + 1$ is
prime.

79. $6a^5 + 37a^3b^2 + 6ab^4$
$= a(6a^4 + 37a^2b^2 + 6b^4)$

Factors of $6a^4$:
$6a^4 = 6a^2 \cdot a^2$, $6a^4 = 3a^2 \cdot 2a^2$
Factors of $6b^4$: $6b^4 = 6b^2 \cdot b^2$,
$6b^4 = 3b^2 \cdot 2b^2$
$6a^5 + 37a^3b^2 + 6ab^4$
$= a(6a^2 + b^2)(a^2 + 6b^2)$

81. $(x - 4)(x + 4) = (x)^2 - (4)^2 = x^2 - 16$

83. $(x + 2)^2 = x^2 + 2x(2) + 2^2 = x^2 + 4x + 4$

85. $(2x-1)^2 = (2x)^2 - 2(2x)(1) + (1)^2$
$$= 4x^2 - 4x + 1$$

87. No.
$$4x^2 = 2 \cdot 2 \cdot x \cdot x$$
$$19x = 19 \cdot x$$
$$12 = 2 \cdot 2 \cdot 3$$
There is no common factor (other than 1).

89. $(3x^2 + 1) + (6x + 4) + (x^2 + 15x)$
$$= 4x^2 + 21x + 5$$
$$= (4x + 1)(x + 5)$$

91. $4x^2 + 2x + \dfrac{1}{4} = \left(2x + \dfrac{1}{2}\right)\left(2x + \dfrac{1}{2}\right)$ or
$$\left(2x + \dfrac{1}{2}\right)^2$$

93. $4x^2(y-1)^2 + 10x(y-1)^2 + 25(y-1)^2$
$$= (y-1)^2(4x^2 + 10x + 25)$$

95. Factors of $3x^2$: $3x^2 = 3x \cdot x$
Factors of -5: $-5 = -1 \cdot 5, -5 = 1 \cdot -5$
$(3x-1)(x+5) = 3x^2 + 14x - 5$
$(3x+1)(x-5) = 3x^2 - 14x - 5$
$(3x-5)(x+1) = 3x^2 - 2x - 5$
$(3x+5)(x-1) = 3x^2 + 2x - 5$
Since b is positive, the possible values are 2 and 14.

97. Note that $5 + 2 = 7$.
$(5x+2)(x+1) = 5x(x+1) + 2(x+1)$
$$= 5x^2 + 5x + 2x + 2$$
$$= 5x^2 + 7x + 2$$
If $c = 2$, then $5x^2 + 7x + c$ is factorable.

99. answers may vary

Section 4.4 Practice Problems

1. a. $3 \cdot 8 = 24$
$12 \cdot 2 = 24$
$12 + 2 = 14$
$3x^2 + 14x + 8 = 3x^2 + 12x + 2x + 8$
$$= 3x(x+4) + 2(x+4)$$
$$= (x+4)(3x+2)$$

b. $12 \cdot 5 = 60$
$15 \cdot 4 = 60$
$15 + 4 = 19$
$12x^2 + 19x + 5 = 12x^2 + 15x + 4x + 5$
$$= 3x(4x+5) + 1(4x+5)$$
$$= (4x+5)(3x+1)$$

2. a. $30x^2 - 26x + 4 = 2(15x^2 - 13x + 2)$
$15 \cdot 2 = 30$
$-3 \cdot -10 = 30$
$-3 + (-10) = -13$
$30x^2 - 26x + 4$
$$= 2(15x^2 - 13x + 2)$$
$$= 2(15x^2 - 3x - 10x + 2)$$
$$= 2[3x(5x-1) - 2(5x-1)]$$
$$= 2(5x-1)(3x-2)$$

b. $6x^2 y - 7xy - 5y = y(6x^2 - 7x - 5)$
$6 \cdot -5 = -30$
$3 \cdot -10 = -30$
$3 + (-10) = -7$
$6x^2 y - 7xy - 5y$
$$= y(6x^2 - 7x - 5)$$
$$= y(6x^2 + 3x - 10x - 5)$$
$$= y[3x(2x+1) - 5(2x+1)]$$
$$= y(2x+1)(3x-5)$$

3. $12y^5 + 10y^4 - 42y^3 = 2y^3(6y^2 + 5y - 21)$

$6 \cdot -21 = -126$

$14 \cdot -9 = -126$

$14 + (-9) = 5$

$12y^5 + 10y^4 - 42y^3$

$= 2y^3(6y^2 + 5y - 21)$

$= 2y^3(6y^2 + 14y - 9y - 21)$

$= 2y^3[2y(3y + 7) - 3(3y + 7)]$

$= 2y^3(3y + 7)(2y - 3)$

Exercise Set 4.4

1. $x^2 + 3x + 2x + 6 = x(x + 3) + 2(x + 3)$

$\qquad\qquad\qquad\quad = (x + 3)(x + 2)$

3. $y^2 + 8y - 2y - 16 = y(y + 8) - 2(y + 8)$

$\qquad\qquad\qquad\quad = (y + 8)(y - 2)$

5. $8x^2 - 5x - 24x + 15 = x(8x - 5) - 3(8x - 5)$

$\qquad\qquad\qquad\qquad = (8x - 5)(x - 3)$

7. $5x^4 - 3x^2 + 25x^2 - 15$

$= x^2(5x^2 - 3) + 5(5x^2 - 3)$

$= (5x^2 - 3)(x^2 + 5)$

9. a. $9 \cdot 2 = 18$

$\quad 9 + 2 = 11$

9 and 2 are numbers whose product is 18 and whose sum is 11.

b. $11x = 9x + 2x$

c. $6x^2 + 11x + 3 = 6x^2 + 9x + 2x + 3$

$\qquad\qquad\qquad = 3x(2x + 3) + 1(2x + 3)$

$\qquad\qquad\qquad = (2x + 3)(3x + 1)$

11. a. $-3 \cdot -20 = 60$

$\quad -3 + (-20) = -23$

-3 and -20 are numbers whose product is 60 and whose sum is -23.

b. $-23x = -3x - 20x$

c. $15x^2 - 23x + 4 = 15x^2 - 3x - 20x + 4$

$\qquad\qquad\qquad = 3x(5x - 1) - 4(5x - 1)$

$\qquad\qquad\qquad = (5x - 1)(3x - 4)$

13. $21 \cdot 2 = 42$

$14 \cdot 3 = 42$

$14 + 3 = 17$

$21y^2 + 17y + 2 = 21y^2 + 14y + 3y + 2$

$\qquad\qquad\qquad = 7y(3y + 2) + 1(3y + 2)$

$\qquad\qquad\qquad = (3y + 2)(7y + 1)$

15. $7 \cdot -11 = -77$

$-11 \cdot 7 = -77$

$-11 + 7 = -4$

$7x^2 - 4x - 11 = 7x^2 - 11x + 7x - 11$

$\qquad\qquad\qquad = x(7x - 11) + 1(7x - 11)$

$\qquad\qquad\qquad = (7x - 11)(x + 1)$

17. $10 \cdot 2 = 20$

$-4 \cdot -5 = 20$

$-4 + (-5) = -9$

$10x^2 - 9x + 2 = 10x^2 - 4x - 5x + 2$

$\qquad\qquad\qquad = 2x(5x - 2) - 1(5x - 2)$

$\qquad\qquad\qquad = (5x - 2)(2x - 1)$

19. $2 \cdot 5 = 10$

$-5 \cdot -2 = 10$

$-5 + (-2) = -7$

$2x^2 - 7x + 5 = 2x^2 - 5x - 2x + 5$

$\qquad\qquad\qquad = x(2x - 5) - 1(2x - 5)$

$\qquad\qquad\qquad = (2x - 5)(x - 1)$

21. $12x + 4x^2 + 9 = 4x^2 + 12x + 9$

$4 \cdot 9 = 36$

$6 \cdot 6 = 36$

$6 + 6 = 12$

$4x^2 + 12x + 9 = 4x^2 + 6x + 6x + 9$

$\qquad\qquad\qquad = 2x(2x + 3) + 3(2x + 3)$

$\qquad\qquad\qquad = (2x + 3)(2x + 3)$ or $(2x + 3)^2$

23. $4 \cdot -21 = -84$
$6 \cdot -14 = -84$
$6 + (-14) = -8$
$4x^2 - 8x - 21 = 4x^2 + 6x - 14x - 21$
$\qquad = 2x(2x+3) - 7(2x+3)$
$\qquad = (2x+3)(2x-7)$

25. $10 \cdot 12 = 120$
$-8 \cdot -15 = 120$
$-8 + (-15) = -23$
$10x^2 - 23x + 12 = 10x^2 - 8x - 15x + 12$
$\qquad = 2x(5x-4) - 3(5x-4)$
$\qquad = (5x-4)(2x-3)$

27. $2x^3 + 13x^2 + 15x = x(2x^2 + 13x + 15)$
$2 \cdot 15 = 30$
$3 \cdot 10 = 30$
$3 + 10 = 13$
$2x^3 + 13x^2 + 15x = x(2x^2 + 13x + 15)$
$\qquad = x(2x^2 + 3x + 10x + 15)$
$\qquad = x[x(2x+3) + 5(2x+3)]$
$\qquad = x(2x+3)(x+5)$

29. $16y^2 - 34y + 18 = 2(8y^2 - 17y + 9)$
$8 \cdot 9 = 72$
$-9 \cdot -8 = 72$
$-9 + (-8) = -17$
$16y^2 - 34y + 18 = 2(8y^2 - 17y + 9)$
$\qquad = 2(8y^2 - 9y - 8y + 9)$
$\qquad = 2[y(8y-9) - 1(8y-9)]$
$\qquad = 2(8y-9)(y-1)$

31. $-13x + 6 + 6x^2 = 6x^2 - 13x + 6$
$6 \cdot 6 = 36$
$-9 \cdot -4 = 36$
$-9 + (-4) = -13$
$6x^2 - 13x + 6 = 6x^2 - 9x - 4x + 6$
$\qquad = 3x(2x-3) - 2(2x-3)$
$\qquad = (2x-3)(3x-2)$

33. $54a^2 - 9a - 30 = 3(18a^2 - 3a - 10)$
$18 \cdot -10 = -180$
$12 \cdot -15 = -180$
$12 + (-15) = -3$
$54a^2 - 9a - 30 = 3(18a^2 - 3a - 10)$
$\qquad = 3(18a^2 + 12a - 15a - 10)$
$\qquad = 3[6a(3a+2) - 5(3a+2)]$
$\qquad = 3(3a+2)(6a-5)$

35. $20a^3 + 37a^2 + 8a = a(20a^2 + 37a + 8)$
$20 \cdot 8 = 160$
$5 \cdot 32 = 160$
$5 + 32 = 37$
$20a^3 + 37a^2 + 8a = a(20a^2 + 37a + 8)$
$\qquad = a(20a^2 + 5a + 32a + 8)$
$\qquad = a[5a(4a+1) + 8(4a+1)]$
$\qquad = a(4a+1)(5a+8)$

37. $12x^3 - 27x^2 - 27x = 3x(4x^2 - 9x - 9)$
$4 \cdot -9 = -36$
$3 \cdot -12 = -36$
$3 + (-12) = -9$
$12x^3 - 27x^2 - 27x$
$= 3x(4x^2 - 9x - 9)$
$= 3x(4x^2 + 3x - 12x - 9)$
$= 3x[x(4x+3) - 3(4x+3)]$
$= 3x(4x+3)(x-3)$

39. $3x^2y + 4xy^2 + y^3 = y(3x^2 + 4xy + y^2)$
$3 \cdot y^2 = 3y^2$
$y \cdot 3y = 3y^2$
$y + 3y = 4y$
$3x^2y + 4xy^2 + y^3 = y(3x^2 + 4xy + y^2)$
$\qquad = y(3x^2 + xy + 3xy + y^2)$
$\qquad = y[x(3x+y) + y(3x+y)]$
$\qquad = y(3x+y)(x+y)$

41. $20 \cdot 1 = 20$
There are no factors of 20 which sum to 7,
so $20z^2 + 7z + 1$ is prime.

43. $24a^2 - 6ab - 30b^2 = 6(4a^2 - ab - 5b^2)$

$4 \cdot -5b^2 = -20b^2$

$4b \cdot -5b = -20b^2$

$4b + (-5b) = -b$

$24a^2 - 6ab - 30b^2$

$= 6(4a^2 - ab - 5b^2)$

$= 6(4a^2 + 4ab - 5ab - 5b^2)$

$= 6[4a(a+b) - 5b(a+b)]$

$= 6(a+b)(4a - 5b)$

45. $15p^4 + 31p^3q + 2p^2q^2$

$= p^2(15p^2 + 31pq + 2q^2)$

$15 \cdot 2q^2 = 30q^2$

$q \cdot 30q = 30q^2$

$q + 30q = 31q$

$15p^4 + 31p^3q + 2p^2q^2$

$= p^2(15p^2 + 31pq + 2q^2)$

$= p^2(15p^2 + pq + 30pq + 2q^2)$

$= p^2[p(15p+q) + 2q(15p+q)]$

$= p^2(15p+q)(p+2q)$

47. $35 + 12x + x^2 = x^2 + 12x + 35$

$1 \cdot 35 = 35$

$7 \cdot 5 = 35$

$7 + 5 = 12$

$x^2 + 12x + 35$

$= x^2 + 7x + 5x + 35$

$= x(x+7) + 5(x+7)$

$= (x+7)(x+5)$ or $(7+x)(5+x)$

49. $6 - 11x + 5x^2 = 5x^2 - 11x + 6$

$5 \cdot 6 = 30$

$-6 \cdot -5 = 30$

$-6 + (-5) = -11$

$5x^2 - 11x + 6$

$= 5x^2 - 6x - 5x + 6$

$= x(5x - 6) - 1(5x - 6)$

$= (5x - 6)(x - 1)$ or $(6 - 5x)(1 - x)$

51. $(x-2)(x+2) = x^2 - 2^2 = x^2 - 4$

53. $(y+4)(y+4) = (y+4)^2$

$= (y)^2 + 2(y)(4) + (4)^2$

$= y^2 + 8y + 16$

55. $(9z+5)(9z-5) = (9z)^2 - 5^2 = 81z^2 - 25$

57. $(4x-3)^2 = (4x)^2 - 2(4x)(3) + (3)^2$

$= 16x^2 - 24x + 9$

59. $5(2x^2 + 9x + 9) = 10x^2 + 45x + 45$

$2 \cdot 9 = 18$

$3 \cdot 6 = 18$

$3 + 6 = 9$

$5(2x^2 + 9x + 9) = 5(2x^2 + 3x + 6x + 9)$

$= 5[x(2x+3) + 3(2x+3)]$

$= 5(2x+3)(x+3)$

The perimeter is $10x^2 + 45x + 45$ or
$5(2x+3)(x+3)$.

61. $x^{2n} + 2x^n + 3x^n + 6 = x^n(x^n + 2) + 3(x^n + 2)$

$= (x^n + 2)(x^n + 3)$

63. $3 \cdot -35 = -105$

$-5 \cdot 21 = -105$

$-5 + 21 = 16$

$3x^{2n} + 16x^n - 35 = 3x^{2n} - 5x^n + 21x^n - 35$

$= x^n(3x^n - 5) + 7(3x^n - 5)$

$= (3x^n - 5)(x^n + 7)$

65. answers may vary

Section 4.5 Practice Problems

1. a. Since $36 = 6^2$ and $12x = 2 \cdot 6 \cdot x$,
$x^2 + 12x + 36$ is a perfect square
trinomial.

b. Since $100 = 10^2$ and $20x = 2 \cdot 10 \cdot x$,

$x^2 + 20x + 100$ is a perfect square trinomial.

2. a. Since $9x^2 = (3x)^2$ and $25 = 5^2$, but $20x \neq 2 \cdot 3x \cdot 5$, the polynomial is not a perfect square trinomial.

b. Since $4x^2 = (2x)^2$, but 11 is not a perfect square, the polynomial is not a perfect square trinomial.

3. a. Since $25x^2 = (5x)^2$ and $1 = 1^2$, and $2 \cdot 5x \cdot 1 = 10x$ which is the opposite of $-10x$, the trinomial is a perfect square trinomial.

b. Since $9x^2 = (3x)^2$ and $49 = 7^2$, and $2 \cdot 3x \cdot 7 = 42x$ which is the opposite of $-42x$, the trinomial is a perfect square trinomial.

4. $x^2 + 16x + 64 = (x)^2 + 2 \cdot x \cdot 8 + 8^2$

$\qquad\qquad\qquad = (x+8)^2$

5. $9r^2 + 24rs + 16s^2 = (3r)^2 + 2 \cdot 3r \cdot 4s + (4s)^2$

$\qquad\qquad\qquad\qquad = (3r + 4s)^2$

6. $9n^4 - 6n^2 + 1 = (3n^2)^2 - 2 \cdot 3n^2 \cdot 1 + 1^2$

$\qquad\qquad\qquad = (3n^2 - 1)^2$

7. Notice that this trinomial is not a perfect square trinomial.

$9x^2 = (3x)^2$ and $4 = 2^2$, but $2 \cdot 3x \cdot 2 = 12x$ and $12x$ is not the middle term $15x$.

Factor by grouping.

$9 \cdot 4 = 36$

$3 \cdot 12 = 36$

$3 + 12 = 15$

$9x^2 + 15x + 4 = 9x^2 + 3x + 12x + 4$

$\qquad\qquad\qquad = 3x(3x+1) + 4(3x+1)$

$\qquad\qquad\qquad = (3x+1)(3x+4)$

8. a. $8n^2 + 40n + 50$

$= 2(4n^2 + 20n + 25)$

$= 2[(2n)^2 + 2 \cdot 2n \cdot 5 + 5^2]$

$= 2(2n + 5)^2$

b. $12x^3 - 84x^2 + 147x$

$= 3x(4x^2 - 28x + 49)$

$= 3x[(2x)^2 - 2 \cdot 2x \cdot 7 + 7^2]$

$= 3x(2x - 7)^2$

9. $x^2 - 9 = x^2 - 3^2 = (x-3)(x+3)$

10. $a^2 - 16 = a^2 - 4^2 = (a-4)(a+4)$

11. $c^2 - \dfrac{9}{25} = c^2 - \left(\dfrac{3}{5}\right)^2 = \left(c - \dfrac{3}{5}\right)\left(c + \dfrac{3}{5}\right)$

12. $s^2 + 9$ is prime since it is the sum of two squares.

13. $9s^2 - 1 = (3s)^2 - 1^2 = (3s-1)(3s+1)$

14. $16x^2 - 49y^2 = (4x)^2 - (7y)^2$

$\qquad\qquad\qquad = (4x - 7y)(4x + 7y)$

15. $p^4 - 81 = (p^2)^2 - 9^2$

$\qquad\qquad = (p^2 + 9)(p^2 - 9)$

$\qquad\qquad = (p^2 + 9)(p + 3)(p - 3)$

16. $9x^3 - 25x = x(9x^2 - 25)$

$\qquad\qquad\quad = x[(3x)^2 - 5^2]$

$\qquad\qquad\quad = x(3x - 5)(3x + 5)$

17. $48x^4 - 3 = 3(16x^4 - 1)$

$\qquad\qquad\quad = 3[(4x^2)^2 - 1^2]$

$\qquad\qquad\quad = 3(4x^2 + 1)(4x^2 - 1)$

$\qquad\qquad\quad = 3(4x^2 + 1)(2x + 1)(2x - 1)$

18. $-9x^2 + 100 = -1(9x^2 - 100)$
$$= -1[(3x)^2 - 10^2]$$
$$= -1(3x - 10)(3x + 10)$$

19. $121 - m^2 = 11^2 - m^2 = (11 + m)(11 - m)$
or
$$121 - m^2 = -m^2 + 121$$
$$= -1(m^2 - 121)$$
$$= -1(m^2 - 11^2)$$
$$= -1(m + 11)(m - 11)$$

Calculator Explorations

	$x^2 - 2x + 1$	$x^2 - 2x - 1$	$(x-1)^2$
$x = 5$	16	14	16
$x = -3$	16	14	16
$x = 2.7$	2.89	0.89	2.89
$x = -12.1$	171.61	169.61	171.61
$x = 0$	1	1	1

Mental Math

1. $1 = 1^2$

2. $25 = 5^2$

3. $81 = 9^2$

4. $64 = 8^2$

5. $9 = 3^2$

6. $100 = 10^2$

7. $9x^2 = (3x)^2$

8. $16y^2 = (4y)^2$

9. $25a^2 = (5a)^2$

10. $81b^2 = (9b)^2$

11. $36p^4 = (6p^2)^2$

12. $4q^4 = (2q^2)^2$

Exercise Set 4.5

1. Since $64 = 8^2$ and $16x = 2 \cdot 8 \cdot x$,
$x^2 + 16x + 64$ is a perfect square trinomial.

3. Since $25 = 5^2$ but $5y \neq 2 \cdot 5 \cdot y$,
$y^2 + 5y + 25$ is not a perfect square
trinomial.

5. Since $1 = 1^2$ and $-2m = -2 \cdot 1 \cdot m$,
$m^2 - 2m + 1$ is a perfect square trinomial.

7. Since $49 = 7^2$ but $16a \neq 2 \cdot 7 \cdot a$,
$a^2 - 16a + 49$ is not a perfect square
trinomial.

9. $4x^2 = (2x)^2$ but $8y^2$ is not a perfect square,
so $4x^2 + 12xy + 8y^2$ is not a perfect square
trinomial.

11. $25a^2 = (5a)^2$, $16b^2 = (4b)^2$, and
$40ab = 2 \cdot 5a \cdot 4b$, so $25a^2 - 40ab + 16b^2$ is
a perfect square trinomial.

13. $x^2 + 22x + 121 = x^2 + 2 \cdot x \cdot 11 + 11^2$
$$= (x + 11)^2$$

15. $x^2 - 16x + 64 = x^2 - 2 \cdot x \cdot 8 + 8^2 = (x - 8)^2$

17. $16a^2 - 24a + 9 = (4a)^2 - 2 \cdot 4a \cdot 3 + 3^2$
$$= (4a - 3)^2$$

19. $x^4 + 4x^2 + 4 = (x^2)^2 + 2 \cdot x^2 \cdot 2 + 2^2$
$$= (x^2 + 2)^2$$

21. $2n^2 - 28n + 98 = 2(n^2 - 14n + 49)$
$$= 2[n^2 - 2 \cdot n \cdot 7 + 7^2]$$
$$= 2(n-7)^2$$

23. $16y^2 + 40y + 25 = (4y)^2 + 2 \cdot 4y \cdot 5 + 5^2$
$$= (4y + 5)^2$$

25. $x^2 y^2 - 10xy + 25 = (xy)^2 - 2 \cdot xy \cdot 5 + 5^2$
$$= (xy - 5)^2$$

27. $m^3 + 18m^2 + 81m = m(m^2 + 18m + 81)$
$$= m(m^2 + 2 \cdot m \cdot 9 + 9^2)$$
$$= m(m+9)^2$$

29. Since $1 = 1^2$ and $x^4 = (x^2)^2$, but

$6x^2 \neq 2 \cdot 1 \cdot x^2$, the polynomial is not a perfect square trinomial. Since there are no factors of $1 \cdot 1 = 1$ that sum to 6, the trinomial is prime.

31. $9x^2 - 24xy + 16y^2$
$$= (3x)^2 - 2 \cdot 3x \cdot 4y + (4y)^2$$
$$= (3x - 4y)^2$$

33. $x^2 - 4 = x^2 - (2)^2 = (x+2)(x-2)$

35. $81 - p^2 = 9^2 - p^2 = (9+p)(9-p)$
or
$81 - p^2 = -p^2 + 81$
$$= -1(p^2 - 81)$$
$$= -1(p - 9^2)$$
$$= -1(p+9)(p-9)$$

37. $-4r^2 + 1 = -1(4r^2 - 1)$
$$= -1[(2r)^2 - 1^2]$$
$$= -1(2r+1)(2r-1)$$

39. $9x^2 - 16 = (3x)^2 - 4^2 = (3x+4)(3x-4)$

41. $16r^2 + 1$ is the sum of two squares, which is prime.

43. $-36 + x^2 = x^2 - 36$
$$= x^2 - 6^2$$
$$= (x-6)(x+6) \text{ or } (-6+x)(6+x)$$
or
$-36 + x^2 = -1(36 - x^2)$
$$= -1(6^2 - x^2)$$
$$= -1(6+x)(6-x)$$

45. $m^4 - 1 = (m^2)^2 - 1^2$
$$= (m^2 + 1)(m^2 - 1)$$
$$= (m^2 + 1)(m^2 - 1^2)$$
$$= (m^2 + 1)(m+1)(m-1)$$

47. $x^2 - 169y^2 = x^2 - (13y)^2$
$$= (x^2 + 13y)(x - 13y)$$

49. $18r^2 - 8 = 2(9r^2 - 4)$
$$= 2[(3r)^2 - 2^2]$$
$$= 2(3r+2)(3r-2)$$

51. $9xy^2 - 4x = x(9y^2 - 4)$
$$= x[(3y)^2 - 2^2]$$
$$= x(3y+2)(3y-2)$$

53. $16x^4 - 64x^2 = 16x^2(x^2 - 4)$
$$= 16x^2(x^2 - 2^2)$$
$$= 16x^2(x+2)(x-2)$$

55. $xy^3 - 9xyz^2 = xy(y^2 - 9z^2)$
$$= xy[y^2 - (3z)^2]$$
$$= xy(y - 3z)(y + 3z)$$

57. $36x^2 - 64y^2 = 4(9x^2 - 16y^2)$
$$= 4[(3x)^2 - (4y)^2]$$
$$= 4(3x+4y)(3x-4y)$$

59. $144 - 81x^2 = 9(16 - 9x^2)$
$$= 9[4^2 - (3x)^2]$$
$$= 9(4 - 3x)(4 + 3x)$$

61. $25y^2 - 9 = (5y)^2 - 3^2 = (5y + 3)(5y - 3)$

63. $121m^2 - 100n^2 = (11m)^2 - (10n)^2$
$$= (11m + 10n)(11m - 10n)$$

65. $x^2 y^2 - 1 = (xy)^2 - 1^2 = (xy + 1)(xy - 1)$

67. $x^2 - \dfrac{1}{4} = x^2 - \left(\dfrac{1}{2}\right)^2 = \left(x - \dfrac{1}{2}\right)\left(x + \dfrac{1}{2}\right)$

69. $49 - \dfrac{9}{25}m^2 = 7^2 - \left(\dfrac{3}{5}m\right)^2$
$$= \left(7 + \dfrac{3}{5}m\right)\left(7 - \dfrac{3}{5}m\right)$$

71. $81a^2 - 25b^2 = (9a)^2 - (5b)^2$
$$= (9a + 5b)(9a - 5b)$$

73. $x^2 + 14xy + 49y^2 = x^2 + 2 \cdot x \cdot 7y + (7y)^2$
$$= (x + 7y)^2$$

75. $32n^4 - 112n^2 + 98$
$$= 2(16n^4 - 56n^2 + 49)$$
$$= 2[(4n^2)^2 - 2 \cdot 4n^2 \cdot 7 + 7^2]$$
$$= 2(4n^2 - 7)^2$$

77. $x^6 - 81x^2 = x^2(x^4 - 81)$
$$= x^2[(x^2)^2 - 9^2]$$
$$= x^2(x^2 + 9)(x^2 - 9)$$
$$= x^2(x^2 + 9)[(x)^2 - 3^2]$$
$$= x^2(x^2 + 9)(x + 3)(x - 3)$$

79. $64p^3 q - 81pq^3 = pq(64p^2 - 81q^2)$
$$= pq[(8p)^2 - (9q)^2]$$
$$= pq(8p - 9q)(8p + 9q)$$

81. $x - 6 = 0$
$$x - 6 + 6 = 0 + 6$$
$$x = 6$$

83. $2m + 4 = 0$
$$2m + 4 - 4 = 0 - 4$$
$$2m = -4$$
$$\dfrac{2m}{2} = \dfrac{-4}{2}$$
$$m = -2$$

85. $5z - 1 = 0$
$$5z - 1 + 1 = 0 + 1$$
$$5z = 1$$
$$\dfrac{5z}{5} = \dfrac{1}{5}$$
$$z = \dfrac{1}{5}$$

87. $x^2 - \dfrac{2}{3}x + \dfrac{1}{9} = x^2 - 2 \cdot x \cdot \dfrac{1}{3} + \left(\dfrac{1}{3}\right)^2$
$$= \left(x - \dfrac{1}{3}\right)^2$$

89. $(x + 2)^2 - y^2 = [(x + 2) + y][(x + 2) - y]$
$$= (x + 2 + y)(x + 2 - y)$$

91. $a^2(b - 4) - 16(b - 4) = (b - 4)(a^2 - 16)$
$$= (b - 4)(a^2 - 4^2)$$
$$= (b - 4)(a - 4)(a + 4)$$

93. $(x^2 + 6x + 9) - 4y^2$
$$= (x^2 + 2 \cdot x \cdot 3 + 3^2) - 4y^2$$
$$= (x + 3)^2 - 4y^2$$
$$= (x + 3)^2 - (2y)^2$$
$$= [(x + 3) + 2y][(x + 3) - 2y]$$
$$= (x + 3 + 2y)(x + 3 - 2y)$$

95. $x^{2n} - 100 = (x^n)^2 - 10^2 = (x^n + 10)(x^n - 10)$

97. $x^2 = x^2$ and $16 = 4^2$.

$(x+4)^2 = x^2 + 2 \cdot x \cdot 4 + 4^2 = x^2 + 8x + 16$,

so the number 8 makes the given expression a perfect square trinomial.

99. answers may vary

101. The difference of two squares is the result of multiplying binomials of the form $(x + a)$ and $(x - a)$, so multiplying $(x - 6)$ by $(x + 6)$ results in the difference of two squares.

103. $(a+b)^2 = a^2 + 2 \cdot a \cdot b + b^2 = a^2 + 2ab + b^2$

105. a. $841 - 16t^2 = 841 - 16(2)^2$
$$= 841 - 16 \cdot 4$$
$$= 841 - 64$$
$$= 777$$
After 2 seconds, the height of the object is 777 feet.

b. $841 - 16t^2 = 841 - 16(5)^2$
$$= 841 - 16 \cdot 25$$
$$= 841 - 400$$
$$= 441$$
After 5 seconds, the height of the object is 441 feet.

c. The object hits the ground when its height is 0 feet.
$$841 - 16t^2 = 0$$
$$(29 + 4t)(29 - 4t) = 0$$
$$29 + 4t = 0 \quad \text{or} \quad 29 - 4t = 0$$
$$4t = -29 \qquad\qquad -4t = -29$$
$$t = -\frac{29}{4} \qquad\qquad t = \frac{29}{4}$$
Discard $t = -\dfrac{29}{4}$ since time cannot be negative. The object hits the ground after $\dfrac{29}{4} \approx 7$ seconds.

d. $841 - 16t^2 = (29)^2 - (4t)^2$
$$= (29 + 4t)(29 - 4t)$$

107. a. $1600 - 16t^2 = 1600 - 16(3)^2$
$$= 1600 - 16 \cdot 9$$
$$= 1600 - 144$$
$$= 1456$$
After 3 seconds, the height of the bolt is 1456 feet.

b. $1600 - 16t^2 = 1600 - 16(7)^2$
$$= 1600 - 16(49)$$
$$= 1600 - 784$$
$$= 816$$
After 7 seconds, the height of the bolt is 816 feet.

c. The bolt hits the ground when its height is 0 feet.
$$1600 - 16t^2 = 0$$
$$(40 - 4t)(40 + 4t) = 0$$
$$40 - 4t = 0 \quad \text{or} \quad 40 + 4t = 0$$
$$-4t = -40 \qquad\qquad 4t = -40$$
$$t = 10 \qquad\qquad t = -10$$
Discard $t = -10$ since time cannot be negative. The bolt hits the ground after 10 seconds.

d. $1600 - 16t^2 = 16(100 - t^2)$
$$= 16(10^2 - t^2)$$
$$= 16(10 + t)(10 - t)$$

Integrated Review

1–74. Factoring methods may vary.

1. $x^2 + x - 12 = (x - 3)(x + 4)$

2. $x^2 - 10x + 16 = (x - 8)(x - 2)$

3. $x^2 - x - 6 = (x + 2)(x - 3)$

4. $x^2 + 2x + 1 = x^2 + 2 \cdot x \cdot 1 + 1^2$
$$= (x + 1)^2$$

5. $x^2 - 6x + 9 = x^2 - 2 \cdot x \cdot 3 + 3^2$
$$= (x - 3)^2$$

6. $x^2 + x - 2 = (x + 2)(x - 1)$

7. $x^2 + x - 6 = (x + 3)(x - 2)$

8. $x^2 + 7x + 12 = (x + 3)(x + 4)$

9. $x^2 - 7x + 10 = (x - 5)(x - 2)$

10. $x^2 - x - 30 = (x - 6)(x + 5)$

11. $2x^2 - 98 = 2(x^2 - 49)$
$$= 2(x^2 - 7^2)$$
$$= 2(x - 7)(x + 7)$$

12. $3x^2 - 75 = 3(x^2 - 25)$
$$= 3(x^2 - 5^2)$$
$$= 3(x - 5)(x + 5)$$

13. $x^2 + 3x + 5x + 15 = x(x + 3) + 5(x + 3)$
$$= (x + 3)(x + 5)$$

14. $3y - 21 + xy - 7x = 3(y - 7) + x(y - 7)$
$$= (y - 7)(3 + x)$$

15. $x^2 + 6x - 16 = (x + 8)(x - 2)$

16. $x^2 - 3x - 28 = (x - 7)(x + 4)$

17. $4x^3 + 20x^2 - 56x = 4x(x^2 + 5x - 14)$
$$= 4x(x + 7)(x - 2)$$

18. $6x^3 - 6x^2 - 120x = 6x(x^2 - x - 20)$
$$= 6x(x - 5)(x + 4)$$

19. $12x^2 + 34x + 24 = 2(6x^2 + 17x + 12)$
$$= 2(3x + 4)(2x + 3)$$

20. $8a^2 + 6ab - 5b^2 = (2a - b)(4a + 5b)$

21. $4a^2 - b^2 = (2a)^2 - b^2 = (2a + b)(2a - b)$

22. $x^2 - 25y^2 = x^2 - (5y)^2 = (x + 5y)(x - 5y)$

23. $28 - 13x - 6x^2 = (4 - 3x)(7 + 2x)$

24. $20 - 3x - 2x^2 = (5 - 2x)(4 + x)$

25. $4 - 2x + x^2$ is prime.

26. $a + a^2 - 3$ is prime.

27. $6y^2 + y - 15 = (3y + 5)(2y - 3)$

28. $4x^2 - x - 5 = (4x - 5)(x + 1)$

29. $18x^3 - 63x^2 + 9x = 9x(2x^2 - 7x + 1)$

30. $12a^3 - 24a^2 + 4a = 4a(3a^2 - 6a + 1)$

31. $16a^2 - 56a + 49 = (4a)^2 - 2 \cdot 4a \cdot 7 + 7^2$
$$= (4a - 7)^2$$

32. $25p^2 - 70p + 49 = (5p)^2 - 2 \cdot 5p \cdot 7 + 7^2$
$$= (5p - 7)^2$$

33. $14 + 5x - x^2 = (7 - x)(2 + x)$

34. $3 - 2x - x^2 = (3 + x)(1 - x)$

35. $3x^4y + 6x^3y - 72x^2y = 3x^2y(x^2 + 2x - 24)$
$$= 3x^2y(x + 6)(x - 4)$$

36. $2x^3y + 8x^2y^2 - 10xy^3$
$$= 2xy(x^2 + 4xy - 5y^2)$$
$$= 2xy(x + 5y)(x - y)$$

37. $12x^3y + 243xy = 3xy(4x^2 + 81)$

38. $6x^3y^2 + 8xy^2 = 2xy^2(3x^2 + 4)$

39. $2xy - 72x^3y = 2xy(1 - 36x^2)$
$$= 2xy[1 - (6x)^2]$$
$$= 2xy(1 - 6x)(1 + 6x)$$

40. $2x^3 - 18x = 2x(x^2 - 9)$
$$= 2x(x^2 - 3^2)$$
$$= 2x(x - 3)(x + 3)$$

41. $x^3 + 6x^2 - 4x - 24 = x^2(x + 6) - 4(x + 6)$
$$= (x + 6)(x^2 - 4)$$
$$= (x + 6)(x + 2)(x - 2)$$

42. $x^3 - 2x^2 - 36x + 72 = x^2(x - 2) - 36(x - 2)$
$$= (x - 2)(x^2 - 36)$$
$$= (x - 2)(x - 6)(x + 6)$$

43. $6a^3 + 10a^2 = 2a^2(3a + 5)$

44. $4n^2 - 6n = 2n(2n - 3)$

45. $3x^3 - x^2 + 12x - 4 = x^2(3x - 1) + 4(3x - 1)$
$$= (3x - 1)(x^2 + 4)$$

46. $x^3 - 2x^2 + 3x - 6 = x^2(x - 2) + 3(x - 2)$
$$= (x - 2)(x^2 + 3)$$

47. $6x^2 + 18xy + 12y^2 = 6(x^2 + 3xy + 2y^2)$
$$= 6(x + 2y)(x + y)$$

48. $12x^2 + 46xy - 8y^2 = 2(6x^2 + 23xy - 4y^2)$
$$= 2(x + 4y)(6x - y)$$

49. $5(x + y) + x(x + y) = (x + y)(5 + x)$

50. $7(x - y) + y(x - y) = (x - y)(7 + y)$

51. $14t^2 - 9t + 1 = (7t - 1)(2t - 1)$

52. $3t^2 - 5t + 1$ is prime.

53. $-3x^2 - 2x + 5 = -1(3x^2 + 2x - 5)$
$$= -1(3x + 5)(x - 1)$$

54. $-7x^2 - 19x + 6 = -1(7x^2 + 19x - 6)$
$$= -1(7x - 2)(x + 3)$$

55. $1 - 8a - 20a^2 = (1 - 10a)(1 + 2a)$

56. $1 - 7a - 60a^2 = (1 + 5a)(1 - 12a)$

57. $x^4 - 10x^2 + 9 = (x^2 - 9)(x^2 - 1)$
$$= (x - 3)(x + 3)(x - 1)(x + 1)$$

58. $x^4 - 13x^2 + 36 = (x^2 - 9)(x^2 - 4)$
$$= (x - 3)(x + 3)(x - 2)(x + 2)$$

59. $x^2 - 23x + 120 = (x - 15)(x - 8)$

60. $y^2 + 22y + 96 = (y + 16)(y + 6)$

61. $x^2 - 14x - 48$ is prime.

62. $16a^2 - 56ab + 49b^2$
$$= (4a)^2 - 2 \cdot 4a \cdot 7b + (7b)^2$$
$$= (4a - 7b)^2$$

63. $25p^2 - 70pq + 49q^2$
$$= (5p)^2 - 2 \cdot 5p \cdot 7q + (7q)^2$$
$$= (5p - 7q)^2$$

64. $7x^2 + 24xy + 9y^2 = (7x + 3y)(x + 3y)$

65. $-x^2 - x + 30 = -1(x^2 + x - 30)$
$$= -1(x - 5)(x + 6)$$

66. $-x^2 + 6x - 8 = -1(x^2 - 6x + 8)$
$= -1(x - 2)(x - 4)$

67. $3rs - s + 12r - 4 = s(3r - 1) + 4(3r - 1)$
$= (3r - 1)(s + 4)$

68. $x^3 - 2x^2 + x - 2 = x^2(x - 2) + 1(x - 2)$
$= (x - 2)(x^2 + 1)$

69. $4x^2 - 8xy - 3x + 6y = 4x(x - 2y) - 3(x - 2y)$
$= (x - 2y)(4x - 3)$

70. $4x^2 - 2xy - 7yz + 14xz$
$= 2x(2x - y) - 7z(y - 2x)$
$= 2x(2x - y) + 7z(2x - y)$
$= (2x - y)(2x + 7z)$

71. $x^2 + 9xy - 36y^2 = (x + 12y)(x - 3y)$

72. $3x^2 + 10xy - 8y^2 = (3x - 2y)(x + 4y)$

73. $x^4 - 14x^2 - 32 = (x^2 + 2)(x^2 - 16)$
$= (x^2 + 2)(x + 4)(x - 4)$

74. $x^4 - 22x^2 - 75 = (x^2 + 3)(x^2 - 25)$
$= (x^2 + 3)(x + 5)(x - 5)$

75. answers may vary

76. Yes; $9x^2 + 81y^2 = 9(x^2 + 9y^2)$

Section 4.6 Practice Problems

1. $(x - 7)(x + 2) = 0$
$x - 7 = 0$ or $x + 2 = 0$
$x = 7$ $x = -2$
The solutions are 7 and −2.

2. $(x - 10)(3x + 1) = 0$
$x - 10 = 0$ or $3x + 1 = 0$
$x = 10$ $3x = -1$
$x = -\dfrac{1}{3}$
The solutions are 10 and $-\dfrac{1}{3}$.

3. a. $y(y + 3) = 0$
$y = 0$ or $y + 3 = 0$
$y = -3$
The solutions are 0 and −3.

b. $x(4x - 3) = 0$
$x = 0$ or $4x - 3 = 0$
$4x = 3$
$x = \dfrac{3}{4}$
The solutions are 0 and $\dfrac{3}{4}$.

4. $x^2 - 3x - 18 = 0$
$(x - 6)(x + 3) = 0$
$x - 6 = 0$ or $x + 3 = 0$
$x = 6$ $x = -3$
The solutions are 6 and −3.

5. $9x^2 - 24x = -16$
$9x^2 - 24x + 16 = 0$
$(3x - 4)(3x - 4) = 0$
$3x - 4 = 0$ or $3x - 4 = 0$
$3x = 4$ $3x = 4$
$x = \dfrac{4}{3}$ $x = \dfrac{4}{3}$
The solution is $\dfrac{4}{3}$.

6. a.
$$x(x-4) = 5$$
$$x^2 - 4x = 5$$
$$x^2 - 4x - 5 = 0$$
$$(x-5)(x+1) = 0$$
$$x - 5 = 0 \quad \text{or} \quad x + 1 = 0$$
$$x = 5 \qquad\qquad x = -1$$
The solutions are 5 and −1.

b.
$$x(3x+7) = 6$$
$$3x^2 + 7x = 6$$
$$3x^2 + 7x - 6 = 0$$
$$(3x-2)(x+3) = 0$$
$$3x - 2 = 0 \quad \text{or} \quad x + 3 = 0$$
$$3x = 2 \qquad\qquad x = -3$$
$$x = \frac{2}{3}$$

The solutions are $\frac{2}{3}$ and −3.

7.
$$2x^3 - 18x = 0$$
$$2x(x^2 - 9) = 0$$
$$2x(x-3)(x+3) = 0$$
$$2x = 0 \quad \text{or} \quad x - 3 = 0 \quad \text{or} \quad x + 3 = 0$$
$$x = 0 \qquad\qquad x = 3 \qquad\qquad x = -3$$
The solutions are 0, 3, and −3.

8. $(x+3)(3x^2 - 20x - 7) = 0$

$$x + 3 = 0 \quad \text{or} \quad 3x^2 - 20x - 7 = 0$$
$$x = -3 \qquad\qquad (3x+1)(x-7) = 0$$
$$3x + 1 = 0 \quad \text{or} \quad x - 7 = 0$$
$$3x = -1 \qquad\qquad x = 7$$
$$x = -\frac{1}{3}$$

The solutions are −3, $-\frac{1}{3}$, and 7.

Mental Math

1. $(a-3)(a-7) = 0$
$a = 3 \quad \text{or} \quad a = 7$
The solutions are 3 and 7.

2. $(a-5)(a-2) = 0$
$a = 5 \quad \text{or} \quad a = 2$
The solutions are 5 and 2.

3. $(x+8)(x+6) = 0$
$x = -8 \quad \text{or} \quad x = -6$
The solutions are −8 and −6.

4. $(x+2)(x+3) = 0$
$x = -2 \quad \text{or} \quad x = -3$
The solutions are −2 and −3.

5. $(x+1)(x-3) = 0$
$x = -1 \quad \text{or} \quad x = 3$
The solutions are −1 and 3.

6. $(x-1)(x+2) = 0$
$x = 1 \quad \text{or} \quad x = -2$
The solutions are 1 and −2.

Exercise Set 4.6

1. $(x-2)(x+1) = 0$
$$x - 2 = 0 \quad \text{or} \quad x + 1 = 0$$
$$x = 2 \qquad\qquad x = -1$$
The solutions are 2 and −1.

3. $(x-6)(x-7) = 0$
$$x - 6 = 0 \quad \text{or} \quad x - 7 = 0$$
$$x = 6 \qquad\qquad x = 7$$
The solutions are 6 and 7.

5. $(x+9)(x+17) = 0$
$$x + 9 = 0 \quad \text{or} \quad x + 17 = 0$$
$$x = -9 \qquad\qquad x = -17$$
The solutions are −9 and −17.

7. $x(x+6) = 0$
$$x = 0 \quad \text{or} \quad x + 6 = 0$$
$$x = -6$$
The solutions are 0 and −6.

9. $3x(x-8) = 0$
$$3x = 0 \quad \text{or} \quad x - 8 = 0$$
$$x = 0 \qquad\qquad x = 8$$
The solutions are 0 and 8.

11. $(2x + 3)(4x - 5) = 0$

$2x + 3 = 0$ or $4x - 5 = 0$

$2x = -3$ $4x = 5$

$x = -\dfrac{3}{2}$ $x = \dfrac{5}{4}$

The solutions are $-\dfrac{3}{2}$ and $\dfrac{5}{4}$.

13. $(2x - 7)(7x + 2) = 0$

$2x - 7 = 0$ or $7x + 2 = 0$

$2x = 7$ $7x = -2$

$x = \dfrac{7}{2}$ $x = -\dfrac{2}{7}$

The solutions are $\dfrac{7}{2}$ and $-\dfrac{2}{7}$.

15. $\left(x - \dfrac{1}{2}\right)\left(x + \dfrac{1}{3}\right) = 0$

$x - \dfrac{1}{2} = 0$ or $x + \dfrac{1}{3} = 0$

$x = \dfrac{1}{2}$ $x = -\dfrac{1}{3}$

The solutions are $\dfrac{1}{2}$ and $-\dfrac{1}{3}$.

17. $(x + 0.2)(x + 1.5) = 0$

$x + 0.2 = 0$ or $x + 1.5 = 0$

$x = -0.2$ $x = -1.5$

The solutions are -0.2 and -1.5.

19. $x^2 - 13x + 36 = 0$

$(x - 9)(x - 4) = 0$

$x - 9 = 0$ or $x - 4 = 0$

$x = 9$ $x = 4$

The solutions are 9 and 4.

21. $x^2 + 2x - 8 = 0$

$(x + 4)(x - 2) = 0$

$x + 4 = 0$ or $x - 2 = 0$

$x = -4$ $x = 2$

The solutions are -4 and 2.

23. $x^2 - 7x = 0$

$x(x - 7) = 0$

$x = 0$ or $x - 7 = 0$

$x = 7$

The solutions are 0 and 7.

25. $x^2 + 20x = 0$

$x(x + 20) = 0$

$x = 0$ or $x + 20 = 0$

$x = -20$

The solutions are 0 and -20.

27. $x^2 = 16$

$x^2 - 16 = 0$

$(x - 4)(x + 4) = 0$

$x - 4 = 0$ or $x + 4 = 0$

$x = 4$ $x = -4$

The solutions are 4 and -4.

29. $x^2 - 4x = 32$

$x^2 - 4x - 32 = 0$

$(x - 8)(x + 4) = 0$

$x - 8 = 0$ or $x + 4 = 0$

$x = 8$ $x = -4$

The solutions are 8 and -4.

31. $(x + 4)(x - 9) = 4x$

$x^2 - 5x - 36 = 4x$

$x^2 - 9x - 36 = 0$

$(x + 3)(x - 12) = 0$

$x + 3 = 0$ or $x - 12 = 0$

$x = -3$ $x = 12$

The solutions are -3 and 12.

33.
$$x(3x-1) = 14$$
$$3x^2 - x = 14$$
$$3x^2 - x - 14 = 0$$
$$3x^2 - 7x + 6x - 14 = 0$$
$$x(3x-7) + 2(3x-7) = 0$$
$$(3x-7)(x+2) = 0$$
$$3x-7 = 0 \quad \text{or} \quad x+2 = 0$$
$$3x = 7 \qquad\qquad x = -2$$
$$x = \frac{7}{3}$$

The solutions are $\dfrac{7}{3}$ and -2.

35. $3x^2 + 19x - 72 = 0$
$$(3x-8)(x+9) = 0$$
$$3x-8 = 0 \quad \text{or} \quad x+9 = 0$$
$$3x = 8 \qquad\qquad x = -9$$
$$x = \frac{8}{3}$$

The solutions are $\dfrac{8}{3}$ and -9.

37.
$$4x^3 - x = 0$$
$$x(4x^2 - 1) = 0$$
$$x[(2x)^2 - (1)^2] = 0$$
$$x(2x+1)(2x-1) = 0$$
$$x = 0 \quad \text{or} \quad 2x+1 = 0 \quad \text{or} \quad 2x-1 = 0$$
$$2x = -1 \qquad\qquad 2x = 1$$
$$x = -\frac{1}{2} \qquad\qquad x = \frac{1}{2}$$

The solutions are 0, $-\dfrac{1}{2}$, and $\dfrac{1}{2}$.

39. $4(x-7) = 6$
$$4x - 28 = 6$$
$$4x = 34$$
$$x = \frac{34}{4}$$
$$x = \frac{17}{2}$$

The solution is $\dfrac{17}{2}$.

41.
$$(4x-3)(16x^2 - 24x + 9) = 0$$
$$(4x-3)[(4x)^2 - 2 \cdot 4x \cdot 3 + (3)^2] = 0$$
$$(4x-3)(4x-3)^2 = 0$$
$$(4x-3)(4x-3)(4x-3) = 0$$
$$4x-3 = 0$$
$$4x = 3$$
$$x = \frac{3}{4}$$

The solution is $\dfrac{3}{4}$.

43.
$$4y^2 - 1 = 0$$
$$(2y-1)(2y+1) = 0$$
$$2y-1 = 0 \quad \text{or} \quad 2y+1 = 0$$
$$2y = 1 \qquad\qquad 2y = -1$$
$$y = \frac{1}{2} \qquad\qquad y = -\frac{1}{2}$$

The solutions are $\dfrac{1}{2}$ and $-\dfrac{1}{2}$.

45.
$$(2x+3)(2x^2 - 5x - 3) = 0$$
$$(2x+3)(2x^2 + x - 6x - 3) = 0$$
$$(2x+3)[x(2x+1) - 3(2x+1)] = 0$$
$$(2x+3)(2x+1)(x-3) = 0$$
$$2x+3 = 0 \quad \text{or} \quad 2x+1 = 0 \quad \text{or} \quad x-3 = 0$$
$$2x = -3 \qquad\qquad 2x = -1 \qquad\qquad x = 3$$
$$x = -\frac{3}{2} \qquad\qquad x = -\frac{1}{2}$$

The solutions are $-\dfrac{3}{2}$, $-\dfrac{1}{2}$, and 3.

47. $x^2 - 15 = -2x$

$x^2 + 2x - 15 = 0$

$(x+5)(x-3) = 0$

$x + 5 = 0$ or $x - 3 = 0$

$x = -5$ $x = 3$

The solutions are -5 and 3.

49. $30x^2 - 11x = 30$

$30x^2 - 11x - 30 = 0$

$30x^2 + 25x - 36x - 30 = 0$

$5x(6x+5) - 6(6x+5) = 0$

$(6x+5)(5x-6) = 0$

$6x + 5 = 0$ or $5x - 6 = 0$

$6x = -5$ $5x = 6$

$x = -\dfrac{5}{6}$ $x = \dfrac{6}{5}$

The solutions are $-\dfrac{5}{6}$ and $\dfrac{6}{5}$.

51. $5x^2 - 6x - 8 = 0$

$(5x+4)(x-2) = 0$

$5x + 4 = 0$ or $x - 2 = 0$

$5x = -4$ $x = 2$

$x = -\dfrac{4}{5}$

The solutions are $-\dfrac{4}{5}$ and 2.

53. $6y^2 - 22y - 40 = 0$

$2(3y^2 - 11y - 20) = 0$

$2(3y^2 + 4y - 15y - 20) = 0$

$2[y(3y+4) - 5(3y+4)] = 0$

$2(3y+4)(y-5) = 0$

$3y + 4 = 0$ or $y - 5 = 0$

$3y = -4$ $y = 5$

$y = -\dfrac{4}{3}$

The solutions are $-\dfrac{4}{3}$ and 5.

55. $(y-2)(y+3) = 6$

$y^2 + y - 6 = 6$

$y^2 + y - 12 = 0$

$(y+4)(y-3) = 0$

$y + 4 = 0$ or $y - 3 = 0$

$y = -4$ $y = 3$

The solutions are -4 and 3.

57. $x^3 - 12x^2 + 32x = 0$

$x(x^2 - 12x + 32) = 0$

$x(x-8)(x-4) = 0$

$x = 0$ or $x - 8 = 0$ or $x - 4 = 0$

$x = 8$ $x = 4$

The solutions are 0, 8, and 4.

59. $x^2 + 14x + 49 = 0$

$(x+7)(x+7) = 0$

$x + 7 = 0$

$x = -7$

The solution is -7.

61. $12y = 8y^2$

$0 = 8y^2 - 12y$

$0 = 4y(2y-3)$

$4y = 0$ or $2y - 3 = 0$

$y = 0$ $2y = 3$

$y = \dfrac{3}{2}$

The solutions are 0 and $\dfrac{3}{2}$.

63. $7x^3 - 7x = 0$

$7x(x^2 - 1) = 0$

$7x(x-1)(x+1) = 0$

$7x = 0$ or $x - 1 = 0$ or $x + 1 = 0$

$x = 0$ $x = 1$ $x = -1$

The solutions are 0, 1, and -1.

65.
$$3x^2 + 8x - 11 = 13 - 6x$$
$$3x^2 + 14x - 11 = 13$$
$$3x^2 + 14x - 24 = 0$$
$$3x^2 + 18x - 4x - 24 = 0$$
$$3x(x+6) - 4(x+6) = 0$$
$$(x+6)(3x-4) = 0$$
$$x + 6 = 0 \quad \text{or} \quad 3x - 4 = 0$$
$$x = -6 \qquad\qquad 3x = 4$$
$$x = \frac{4}{3}$$

The solutions are -6 and $\frac{4}{3}$.

67.
$$3x^2 - 20x = -4x^2 - 7x - 6$$
$$7x^2 - 13x + 6 = 0$$
$$(7x-6)(x-1) = 0$$
$$7x - 6 = 0 \quad \text{or} \quad x - 1 = 0$$
$$7x = 6 \qquad\qquad x = 1$$
$$x = \frac{6}{7}$$

The solutions are $\frac{6}{7}$ and 1.

69. $\dfrac{3}{5} + \dfrac{4}{9} = \dfrac{3 \cdot 9}{5 \cdot 9} + \dfrac{4 \cdot 5}{9 \cdot 5} = \dfrac{27}{45} + \dfrac{20}{45} = \dfrac{47}{45}$

71. $\dfrac{7}{10} - \dfrac{5}{12} = \dfrac{7 \cdot 6}{10 \cdot 6} - \dfrac{5 \cdot 5}{12 \cdot 5} = \dfrac{42}{60} - \dfrac{25}{60} = \dfrac{17}{60}$

73. $\dfrac{4}{5} \cdot \dfrac{7}{8} = \dfrac{4 \cdot 7}{5 \cdot 8} = \dfrac{4 \cdot 7}{5 \cdot 4 \cdot 2} = \dfrac{7}{5 \cdot 2} = \dfrac{7}{10}$

75. The equation must first be written in standard form.
$$x(x-2) = 8$$
$$x^2 - 2x = 8$$
$$x^2 - 2x - 8 = 0$$
$$(x-4)(x+2) = 0$$
$$x - 4 = 0 \quad \text{or} \quad x + 2 = 0$$
$$x = 4 \qquad\qquad x = -2$$
The solutions are 4 and -2.

77. answers may vary, for example
$$(x-6)(x+1) = 0$$

79. answers may vary, for example
$$(x-5)(x-7) = 0$$
$$x^2 - 7x - 5x + 35 = 0$$
$$x^2 - 12x + 35 = 0$$

81. a.

Time, x (seconds)	Height, y (feet)
0	$-16(0)^2 + 20(0) + 300 = 300$
1	$-16(1)^2 + 20(1) + 300 = 304$
2	$-16(2)^2 + 20(2) + 300 = 276$
3	$-16(3)^2 + 20(3) + 300 = 216$
4	$-16(4)^2 + 20(4) + 300 = 124$
5	$-16(5)^2 + 20(5) + 300 = 0$
6	$-16(6)^2 + 20(6) + 300 = -156$

b. When the compass strikes the ground, its height is 0 feet, so it strikes the ground after 5 seconds.

c. The maximum height listed in the table is 304 feet, after 1 second.

83.
$$(x-3)(3x+4) = (x+2)(x-6)$$
$$3x^2 + 4x - 9x - 12 = x^2 - 6x + 2x - 12$$
$$3x^2 - 5x - 12 = x^2 - 4x - 12$$
$$2x^2 - x = 0$$
$$x(2x-1) = 0$$
$$x = 0 \quad \text{or} \quad 2x - 1 = 0$$
$$x = \frac{1}{2}$$

The solutions are 0 and $\frac{1}{2}$.

85. $(2x-3)(x+8) = (x-6)(x+4)$
$2x^2 + 16x - 3x - 24 = x^2 + 4x - 6x - 24$
$2x^2 + 13x - 24 = x^2 - 2x - 24$
$x^2 + 15x = 0$
$x(x+15) = 0$
$x = 0$　or　$x + 15 = 0$
$x = -15$
The solutions are 0 and -15.

The Bigger Picture

1. $-7 + (-27) = -34$

2. $\dfrac{(x^3)^4}{(x^{-2})^5} = \dfrac{x^{12}}{x^{-10}} = x^{12-(-10)} = x^{12+10} = x^{22}$

3. $(x^3 - 6x^2 + 2) - (5x^3 - 6)$
$= x^3 - 6x^2 + 2 - 5x^3 + 6$
$= x^3 - 5x^3 - 6x^2 + 2 + 6$
$= -4x^3 - 6x^2 + 8$

4. $\dfrac{3y^3 - 3y^2 + 9}{3y^2} = \dfrac{3y^3}{3y^2} - \dfrac{3y^2}{3y^2} + \dfrac{9}{3y^2}$
$= y - 1 + \dfrac{3}{y^2}$

5. $10x^3 - 250x = 10x(x^2 - 25)$
$= 10x(x^2 - 5^2)$
$= 10x(x+5)(x-5)$

6. $x^2 - 36x + 35 = (x-1)(x-35)$

7. $6xy + 15x - 6y - 15 = 3(2xy + 5x - 2y - 5)$
$= 3[x(2y+5) - 1(2y+5)]$
$= 3(2y+5)(x-1)$

8. $5xy^2 - 2xy - 7x = x(5y^2 - 2y - 7)$
$= x(5y-7)(y+1)$

9. $(x-5)(2x+1) = 0$
$x - 5 = 0$　or　$2x + 1 = 0$
$x = 5$　　　　　$2x = -1$
$x = -\dfrac{1}{2}$
The solutions are 5 and $-\dfrac{1}{2}$.

10. $5x - 5 = 0$
$5(x-1) = 0$
$x - 1 = 0$
$x = 1$
The solution is 1.

11. $x(x-12) = 28$
$x^2 - 12x = 28$
$x^2 - 12x - 28 = 0$
$(x+2)(x-14) = 0$
$x + 2 = 0$　or　$x - 14 = 0$
$x = -2$　　　　　$x = 14$
The solutions are -2 and 14.

12. $7(x-3) + 2(5x+1) = 14$
$7x - 21 + 10x + 2 = 14$
$17x - 19 = 14$
$17x = 33$
$\dfrac{17x}{17} = \dfrac{33}{17}$
$x = \dfrac{33}{17}$
The solution is $\dfrac{33}{17}$.

Section 4.7 Practice Problems

1. The diver will be at a height of 0 when he or she reaches the pool.
$h = -16t^2 + 64$
$0 = -16t^2 + 64$
$0 = -16(t^2 - 4)$
$0 = -16(t-2)(t+2)$
$t - 2 = 0$　or　$t + 2 = 0$
$t = 2$　　　　　$t = -2$
It takes the diver 2 seconds to reach the pool.

2. Let x be the number.

$$x^2 - 2x = 63$$
$$x^2 - 2x - 63 = 0$$
$$(x-9)(x+7) = 0$$
$$x - 9 = 0 \quad \text{or} \quad x + 7 = 0$$
$$x = 9 \qquad\qquad x = -7$$

The numbers are 9 and −7.

3. Let x be the width. Then $x + 5$ is the length.

Area = length · width
$$176 = (x+5)\cdot x$$
$$176 = x^2 + 5x$$
$$0 = x^2 + 5x - 176$$
$$0 = (x+16)(x-11)$$
$$x + 16 = 0 \quad \text{or} \quad x - 11 = 0$$
$$x = -16 \qquad\qquad x = 11$$

The width is 11 feet and the length is $x + 5 = 11 + 5 = 16$ feet.

4. Let x be the first odd integer. Then $x + 2$ is the next consecutive odd integer.

$$x(x+2) = x + (x+2) + 23$$
$$x^2 + 2x = 2x + 25$$
$$x^2 = 25$$
$$x^2 - 25 = 0$$
$$(x+5)(x-5) = 0$$
$$x + 5 = 0 \quad \text{or} \quad x - 5 = 0$$
$$x = -5 \qquad\qquad x = 5$$
$$x + 2 = -3 \qquad x + 2 = 7$$

The two consecutive odd integers are −5 and −3 or 5 and 7.

5. Let x be the length of one leg. Then $x - 7$ is the length of the other leg.

$$x^2 + (x-7)^2 = 13^2$$
$$x^2 + x^2 - 14x + 49 = 169$$
$$2x^2 - 14x - 120 = 0$$
$$2(x^2 - 7x - 60) = 0$$
$$2(x+5)(x-12) = 0$$
$$x + 5 = 0 \quad \text{or} \quad x - 12 = 0$$
$$x = -5 \qquad\qquad x = 12$$

Discard $x = -5$ since length cannot be negative. If $x = 12$, then $x - 7 = 5$. The lengths of the legs are 5 meters and 12 meters.

Exercise Set 4.7

1. Let x be the width of the rectangle, then the length is $x + 4$.

3. Let x be the first odd integer, then the next consecutive odd integer is $x + 2$.

5. Let x be the base of the triangle, then the height is $4x + 1$.

7. area = (side)^2

$$121 = x^2$$
$$0 = x^2 - 121$$
$$0 = (x-11)(x+11)$$
$$x - 11 = 0 \quad \text{or} \quad x + 11 = 0$$
$$x = 11 \qquad\qquad x = -11$$

Discard $x = -11$ since length cannot be negative. The length of its sides are 11 units.

9. The perimeter is the sum of the lengths of the sides.

$$(x+5) + (x^2 - 3x) + (3x-8) + (x+3) = 120$$
$$x^2 + 2x = 120$$
$$x^2 + 2x - 120 = 0$$
$$(x+12)(x-10) = 0$$
$$x + 12 = 0 \quad \text{or} \quad x - 10 = 0$$
$$x = -12 \qquad\qquad x = 10$$

For $x = -12$, $x + 5$, $x + 3$, and $3x - 8$ are negative. Since lengths cannot be negative, $x = 10$ is the only solution that works in this context.
$$x + 3 = 10 + 3 = 13$$
$$x + 5 = 10 + 5 = 15$$
$$x^2 - 3x = (10)^2 - 3(10) = 100 - 30 = 70$$
$$3x - 8 = 3(10) - 8 = 30 - 8 = 22$$
The sides have lengths 13 cm, 15 cm, 70 cm, and 22 cm.

11. Area = base · height
$$96 = (x+5)(x-5)$$
$$96 = x^2 - 25$$
$$0 = x^2 - 121$$
$$0 = (x+11))(x-11)$$
$$x+11 = 0 \quad \text{or} \quad x-11 = 0$$
$$x = -11 \qquad\qquad x = 11$$
For $x = -11$, $x + 5$ and $x - 5$ are negative. Since lengths cannot be negative, $x = 11$ is the only solution that works in this context.
$$x + 5 = 11 + 5 = 16$$
$$x - 5 = 11 - 5 = 6$$
The base is 16 miles and the height is 6 miles.

13. The object will hit the ground when its height is 0.
$$0 = -16t^2 + 64t + 80$$
$$0 = -16(t^2 - 4t - 5)$$
$$0 = -16(t-5)(t+1)$$
$$0 = t-5 \quad \text{or} \quad 0 = t+1$$
$$5 = t \qquad\qquad -1 = t$$
Since the time t cannot be negative, we discard $t = -1$. The object hits the ground after 5 seconds.

15. Let x be the width. Then $2x - 7$ is the length.
Area = length · width
$$30 = (2x-7) \cdot (x)$$
$$30 = 2x^2 - 7x$$
$$0 = 2x^2 - 7x - 30$$
$$0 = (2x+5)(x-6)$$
$$2x+5 = 0 \quad \text{or} \quad x-6 = 0$$
$$x = -\frac{5}{2} \qquad\qquad x = 6$$
Discard $x = -\dfrac{5}{2}$ since length cannot be negative. The width is 6 cm and the length is $2x - 7 = 2(6) - 7 = 12 - 7 = 5$ cm.

17. $D = \dfrac{1}{2}n(n-3)$
$$= \frac{1}{2}(12)(12-3)$$
$$= \frac{1}{2}(12)(9)$$
$$= 54$$
A polygon with 12 sides has 54 diagonals.

19. $D = \dfrac{1}{2}n(n-3)$
$$35 = \frac{1}{2}n(n-3)$$
$$2 \cdot 35 = 2 \cdot \frac{1}{2}n(n-3)$$
$$70 = n(n-3)$$
$$70 = n^2 - 3n$$
$$0 = n^2 - 3n - 70$$
$$0 = (n-10)(n+7)$$
$$n-10 = 0 \quad \text{or} \quad n+7 = 0$$
$$n = 10 \qquad\qquad n = -7$$
Discard $n = -7$. The polygon has 10 sides.

21. Let x be the number.
$$x + x^2 = 132$$
$$x^2 + x - 132 = 0$$
$$(x+12)(x-11) = 0$$
$$x+12 = 0 \quad \text{or} \quad x-11 = 0$$
$$x = -12 \qquad\qquad x = 11$$
The number is -12 or 11.

23. Let x be the first number. Then $x + 1$ is the next consecutive number.
$$x(x+1) = 210$$
$$x^2 + x = 210$$
$$x^2 + x - 210 = 0$$
$$(x-14)(x+15) = 0$$
$$x-14 = 0 \quad \text{or} \quad x+15 = 0$$
$$x = 14 \qquad\qquad x = -15$$
Discard $x = -15$. The room numbers are 14 and $x + 1 = 14 + 1 = 15$.

25. Use the Pythagorean theorem where
x = hypotenuse, $(x - 1)$ = one leg, and
5 = other leg.

$$x^2 = (x-1)^2 + 5^2$$
$$x^2 = x^2 - 2x + 1 + 25$$
$$x^2 = x^2 - 2x + 26$$
$$0 = -2x + 26$$
$$2x = 26$$
$$x = 13$$

The length of the ladder is 13 feet.

27.
$$(x+3)^2 = 64$$
$$x^2 + 6x + 9 = 64$$
$$x^2 + 6x - 55 = 0$$
$$(x-5)(x+11) = 0$$
$$x - 5 = 0 \quad \text{or} \quad x + 11 = 0$$
$$x = 5 \qquad\qquad x = -11$$

Discard $x = -11$. The original square had
sides of 5 inches.

29. Let x be the length of the shorter leg. Then
the length of the other leg is $x + 4$ and the
length of the hypotenuse is $x + 8$.

$$(x+8)^2 = x^2 + (x+4)^2$$
$$x^2 + 16x + 64 = x^2 + x^2 + 8x + 16$$
$$0 = x^2 - 8x - 48$$
$$0 = (x-12)(x+4)$$
$$0 = x - 12 \quad \text{or} \quad 0 = x + 4$$
$$12 = x \qquad\qquad -4 = x$$

Discard $x = -4$ since length cannot be
negative.

$x + 4 = 12 + 4 = 16$
$x + 8 = 12 + 8 = 20$

The lengths of the sides are 12 mm, 16 mm,
and 20 mm.

31. $\text{area} = \dfrac{1}{2} \cdot \text{base} \cdot \text{height}$

$$100 = \frac{1}{2} \cdot 2x \cdot x$$
$$100 = x^2$$
$$0 = x^2 - 100$$
$$0 = (x-10)(x+10)$$
$$x - 10 = 0 \quad \text{or} \quad x + 10 = 0$$
$$x = 10 \qquad\qquad x = -10$$

Discard $x = -10$. The height is
10 kilometers.

33. Let x be the length of the shorter leg. Then
$x + 12$ is the length of the longer leg and
$2x - 12$ is the length of the hypotenuse.

$$(2x-12)^2 = x^2 + (x+12)^2$$
$$4x^2 - 48x + 144 = x^2 + x^2 + 24x + 144$$
$$2x^2 - 72x = 0$$
$$2x(x-36) = 0$$
$$2x = 0 \quad \text{or} \quad x - 36 = 0$$
$$x = 0 \qquad\qquad x = 36$$

Discard $x = 0$ since the length must be
positive. The shorter leg of the triangle has
length 36 feet.

35. When the object reaches the ground, the
height is 0.

$$h = -16t^2 + 1444$$
$$0 = -16t^2 + 1444$$
$$0 = -4(4t^2 - 361)$$
$$0 = -4(2t-19)(2t+19)$$
$$2t - 19 = 0 \quad \text{or} \quad 2t + 19 = 0$$
$$t = \frac{19}{2} \qquad\qquad t = -\frac{19}{2}$$

Discard $t = -\dfrac{19}{2}$ since time must be
positive. The object reaches the ground after
$\dfrac{19}{2}$ or 9.5 seconds.

37. Use $A = P(1+r)^2$ when $P = 100$ and $A = 144$.
$$144 = 100(1+r)^2$$
$$144 = 100(1+2r+r^2)$$
$$144 = 100 + 200r + 100r^2$$
$$0 = 100r^2 + 200r - 44$$
$$0 = 4(25r^2 + 50r - 11)$$
$$0 = 4(5r - 1)(5r + 11)$$
$$0 = 5r - 1 \quad \text{or} \quad 0 = 5r + 11$$
$$1 = 5r \qquad\qquad -11 = 5r$$
$$\frac{1}{5} = r \qquad\qquad -\frac{11}{5} = r$$

Discard $r = -\dfrac{11}{5}$ since the interest rate must

be positive. The interest rate is
$\dfrac{1}{5} = 0.20 = 20\%$.

39. Let x be the length. Then $x - 7$ is the width.
area = length · width
$$120 = x(x - 7)$$
$$0 = x^2 - 7x - 120$$
$$0 = (x - 15)(x + 8)$$
$$x - 15 = 0 \quad \text{or} \quad x + 8 = 0$$
$$x = 15 \qquad\qquad x = -8$$
Discard $x = -8$ since length is positive. The
length is 15 miles and the width is
$x - 7 = 15 - 7 = 8$ miles.

41. Let $C = 9500$ in $C = x^2 - 15x + 50$.
$$9500 = x^2 - 15x + 50$$
$$0 = x^2 - 15x - 9450$$
$$0 = (x + 90)(x - 105)$$
$$0 = x + 90 \quad \text{or} \quad 0 = x - 105$$
$$-90 = x \qquad\qquad 105 = x$$
Discard $x = -90$ since the number of units
manufactured cannot be negative.
105 units are manufactured at a cost of
$9500.

43. From the graph, there were approximately
11,250 thousand acres of farmland in
Georgia in 1997.

45. From the graph, there were approximately
10,750 thousand acres of farmland in
Georgia in 1987.

47. From the graph, the lines intersect at
approximately 1995.

49. answers may vary

51. Since x is the width of the rectangle, the
length is $x + 6$.
$$x^2 + x(x + 6) = 176$$
$$x^2 + x^2 + 6x = 176$$
$$2x^2 + 6x - 176 = 0$$
$$2(x^2 + 3x - 88) = 0$$
$$2(x - 8)(x + 11) = 0$$
$$x - 8 = 0 \quad \text{or} \quad x + 11 = 0$$
$$x = 8 \qquad\qquad x = -11$$
Discard $x = -11$ since length cannot be
negative. The side of the square is 8 meters.

53. Let x be one of the numbers. Since the
numbers sum to 25, the other number is
$25 - x$.
$$x^2 + (25 - x)^2 = 325$$
$$x^2 + 625 - 50x + x^2 = 325$$
$$2x^2 - 50x + 300 = 0$$
$$2(x^2 - 25x + 150) = 0$$
$$2(x - 10)(x - 15) = 0$$
$$x - 10 = 0 \quad \text{or} \quad x - 15 = 0$$
$$x = 10 \qquad\qquad x = 15$$
If $x = 10$, then $25 - x = 25 - 10 = 15$.
If $x = 15$, then $25 - x = 25 - 15 = 10$.
The numbers are 10 and 15.

55. Dimensions of total area:
length $= x + 6 + 4 + 4 = x + 14$
width $= x + 4 + 4 = x + 8$
     Total area = pool area $+ 576$
$(x+14)(x+8) = x(x+6) + 576$
$x^2 + 22x + 112 = x^2 + 6x + 576$
    $22x + 112 = 6x + 576$
        $16x = 464$
          $x = 29$
$x + 6 = 29 + 6 = 35$
The pool is 29 meters by 35 meters.

Chapter 4 Vocabulary Check

1. An equation that can be written in the form $ax^2 + bx + c = 0$ (with a not 0) is called a <u>quadratic equation</u>.

2. <u>Factoring</u> is the process of writing an expression as a product.

3. The <u>greatest common factor</u> of a list of terms is the product of all common factors.

4. A trinomial that is the square of some binomial is called a <u>perfect square trinomial</u>.

Chapter 4 Review

1. $6x^2 - 15x = 3x \cdot 2x - 3x \cdot 5 = 3x(2x - 5)$

2. $4x^5 + 2x - 10x^4 = 2x \cdot 2x^4 + 2x \cdot 1 - 2x \cdot 5x^3$
                $= 2x(2x^4 + 1 - 5x^3)$

3. $5m + 30 = 5 \cdot m + 5 \cdot 6 = 5(m + 6)$

4. $20x^3 + 12x^2 + 24x$
   $= 4x \cdot 5x^2 + 4x \cdot 3x + 4x \cdot 6$
   $= 4x(5x^2 + 3x + 6)$

5. $3x(2x + 3) - 5(2x + 3) = (2x + 3)(3x - 5)$

6. $5x(x+1) - (x+1) = 5x(x+1) - 1 \cdot (x+1)$
                   $= (x+1)(5x - 1)$

7. $3x^2 - 3x + 2x - 2 = 3x(x-1) + 2(x-1)$
                    $= (x-1)(3x + 2)$

8. $6x^2 + 10x - 3x - 5 = 2x(3x + 5) - 1(3x + 5)$
                    $= (3x + 5)(2x - 1)$

9. $3a^2 + 9ab + 3b^2 + ab$
   $= 3a(a + 3b) + b(3b + a)$
   $= (a + 3b)(3a + b)$

10. $x^2 + 6x + 8 = (x+4)(x+2)$

11. $x^2 - 11x + 24 = (x-8)(x-3)$

12. $x^2 + x + 2$ is prime.

13. $x^2 - 5x - 6 = (x-6)(x+1)$

14. $x^2 + 2x - 8 = (x+4)(x-2)$

15. $x^2 + 4xy - 12y^2 = (x+6y)(x-2y)$

16. $x^2 + 8xy + 15y^2 = (x+5y)(x+3y)$

17. $72 - 18x - 2x^2 = 2(36 - 9x - x^2)$
                     $= 2(3 - x)(12 + x)$
or
$72 - 18x - 2x^2 = -2x^2 - 18x + 72$
                  $= -2(x^2 + 9x - 36)$
                  $= -2(x - 3)(x + 12)$

18. $32 + 12x - 4x^2 = 4(8 + 3x - x^2)$
or
$32 + 12x - 4x^2 = -4x^2 + 12x + 32$
                  $= -4(x^2 - 3x - 8)$

19. $5y^3 - 50y^2 + 120y = 5y(y^2 - 10y + 24)$
                     $= 5y(y - 6)(y - 4)$

20. To factor $x^2 + 2x - 48$, think of two numbers whose product is -48 and whose sum is 2.

21. The first step to factor $3x^2 + 15x + 30$ is to factor out the GCF, 3.

22. Factors of $2x^2$: $2x \cdot x$
Factors of 6: $6 = 1 \cdot 6, 6 = 2 \cdot 3$
$2x^2 + 13x + 6 = (2x + 1)(x + 6)$

23. Factors of $4x^2$: $4x^2 = 4x \cdot x, 4x^2 = 2x \cdot 2x$
Factors of -3: $-3 = -1 \cdot 3, -3 = 1 \cdot -3$
$4x^2 + 4x - 3 = (2x + 3)(2x - 1)$

24. Factors of $6x^2$: $6x^2 = 6x \cdot x, 6x^2 = 3x \cdot 2x$
Factors of $-4y^2$: $-4y^2 = -4y \cdot y$,
$-4y^2 = 4y \cdot -y, -4y^2 = -2y \cdot 2y$
$6x^2 + 5xy - 4y^2 = (3x + 4y)(2x - y)$

25. $x^2 - x + 2$ is prime.

26. $2 \cdot -39 = -78$
$3 \cdot -26 = -78$
$3 + (-26) = -23$
$2x^2 - 23x - 39 = 2x^2 + 3x - 26x - 39$
$\qquad = x(2x + 3) - 13(2x + 3)$
$\qquad = (2x + 3)(x - 13)$

27. $18 \cdot -20y^2 = -360y^2$
$15y \cdot -24y = -360y^2$
$15y + (-24y) = -9y$
$18x^2 - 9xy - 20y^2$
$= 18x^2 + 15xy - 24xy - 20y^2$
$= 3x(6x + 5y) - 4y(6x + 5y)$
$= (6x + 5y)(3x - 4y)$

28. $10y^3 + 25y^2 - 60y = 5y(2y^2 + 5y - 12)$
$2 \cdot -12 = -24$
$-3 \cdot 8 = -24$
$-3 + 8 = 5$
$10y^3 + 25y^2 - 60y$
$= 5y(2y^2 + 5y - 12)$
$= 5y(2y^2 - 3y + 8y - 12)$
$= 5y[y(2y - 3) + 4(2y - 3)]$
$= 5y(2y - 3)(y + 4)$

29. $(x^2 - 2) + (x^2 - 4x) + (3x^2 - 5x)$
$= x^2 - 2 + x^2 - 4x + 3x^2 - 5x$
$= 5x^2 - 9x - 2$
$5 \cdot -2 = -10$
$1 \cdot -10 = -10$
$1 + (-10) = -9$
$5x^2 - 9x - 2 = 5x^2 + x - 10x - 2$
$\qquad = x(5x + 1) - 2(5x + 1)$
$\qquad = (5x + 1)(x - 2)$
The perimeter is $5x^2 - 9x - 2$ or $(5x + 1)(x - 2)$.

30. $2(2x^2 + 3) + 2(6x^2 - 14x)$
$= 4x^2 + 6 + 12x^2 - 28x$
$= 16x^2 - 28x + 6$
$= 2(8x^2 - 14x + 3)$
$8 \cdot 3 = 24$
$-2 \cdot -12 = 24$
$-2 + (-12) = -14$
$2(8x^2 - 14x + 3) = 2(8x^2 - 2x - 12x + 3)$
$\qquad = 2[2x(4x - 1) - 3(4x - 1)]$
$\qquad = 2(4x - 1)(2x - 3)$
The perimeter is $16x^2 - 28x + 6$ or $2(4x - 1)(2x - 3)$.

31. Since $9 = 3^2$ and $6x = 2 \cdot 3 \cdot x$, $x^2 + 6x + 9$ is a perfect square trinomial.

32. Since $64 = 8^2$, but $8x \neq 2 \cdot 8 \cdot x$,

$x^2 + 8x + 64$ is not a perfect square trinomial.

33. $9m^2 = (3m)^2$ and $16 = 4^2$, but

$2 \cdot 3m \cdot 4 \neq 12m$, so $9m^2 - 12m + 16$ is not a perfect square trinomial.

34. Since $4y^2 = (2y)^2$ and $49 = 7^2$ and

$2 \cdot 2y \cdot 7 = 28y$, $4y^2 - 28y + 49$ is a perfect square trinomial.

35. Yes; $x^2 - 9$ or $x^2 - 3^2$ is the difference of two squares.

36. No; $x^2 + 16$ or $x^2 + 4^2$ is the sum of two squares.

37. Yes; $4x^2 - 25y^2$ or $(2x)^2 - (5y)^2$ is the difference of two squares.

38. No; $9a^3 - 1$ is not the difference of two squares [note: $9a^2 - 1$ or $(3a)^2 - 1^2$ is the difference of two squares.]

39. $x^2 - 81 = x^2 - 9^2 = (x - 9)(x + 9)$

40. $x^2 + 12x + 36 = x^2 + 2 \cdot x \cdot 6 + 6^2 = (x + 6)^2$

41. $4x^2 - 9 = (2x)^2 - 3^2 = (2x - 3)(2x + 3)$

42. $9t^2 - 25s^2 = (3t)^2 - (5s)^2$
$= (3t - 5s)(3t + 5s)$

43. $16x^2 = (4x)^2$ and y^2 are both perfect squares, but they are added instead of subtracted, so the polynomial is prime.

44. $n^2 - 18n + 81 = n^2 - 2 \cdot n \cdot 9 + 9^2 = (n - 9)^2$

45. $3r^2 + 36r + 108 = 3(r^2 + 12 + 36)$
$= 3(r^2 + 2 \cdot r \cdot 6 + 6^2)$
$= 3(r + 6)^2$

46. $9y^2 - 42y + 49 = (3y)^2 - 2 \cdot 3y \cdot 7 + 7^2$
$= (3y - 7)^2$

47. $5m^8 - 5m^6 = 5m^6(m^2 - 1)$
$= 5m^6(m^2 - 1^2)$
$= 5m^6(m + 1)(m - 1)$

48. $4x^2 - 28xy + 49y^2$
$= (2x)^2 - 2 \cdot 2x \cdot 7y + (7y)^2$
$= (2x - 7y)^2$

49. $3x^2y + 6xy^2 + 3y^3 = 3y(x^2 + 2xy + y^2)$
$= 3y(x + y)^2$

50. $16x^4 - 1 = (4x^2)^2 - 1^2$
$= (4x^2 - 1)(4x^2 + 1)$
$= [(2x)^2 - 1^2](4x^2 + 1)$
$= (2x - 1)(2x + 1)(4x^2 + 1)$

51. $(x + 6)(x - 2) = 0$
$x + 6 = 0$ or $x - 2 = 0$
$x = -6$ $x = 2$
The solutions are -6 and 2.

52. $3x(x + 1)(7x - 2) = 0$
$3x = 0$ or $x + 1 = 0$ or $7x - 2 = 0$
$x = 0$ $x = -1$ $x = \dfrac{2}{7}$

The solutions are 0, -1, and $\dfrac{2}{7}$.

53. $4(5x+1)(x+3) = 0$
$\quad 5x+1 = 0 \quad$ or $\quad x+3 = 0$
$\qquad x = -\dfrac{1}{5} \qquad\qquad x = -3$

The solutions are $-\dfrac{1}{5}$ and -3.

54. $x^2 + 8x + 7 = 0$
$\quad (x+7)(x+1) = 0$
$\quad x+7 = 0 \quad$ or $\quad x+1 = 0$
$\qquad x = -7 \qquad\qquad x = -1$
The solutions are -7 and -1.

55. $x^2 - 2x - 24 = 0$
$\quad (x+4)(x-6) = 0$
$\quad x+4 = 0 \quad$ or $\quad x-6 = 0$
$\qquad x = -4 \qquad\qquad x = 6$
The solutions are -4 and 6.

56. $x^2 + 10x = -25$
$\quad x^2 + 10x + 25 = 0$
$\qquad (x+5)^2 = 0$
$\qquad x+5 = 0$
$\qquad\quad x = -5$
The solution is -5.

57. $x(x-10) = -16$
$\quad x^2 - 10x = -16$
$\quad x^2 - 10x + 16 = 0$
$\quad (x-2)(x-8) = 0$
$\quad x-2 = 0 \quad$ or $\quad x-8 = 0$
$\qquad x = 2 \qquad\qquad x = 8$
The solutions are 2 and 8.

58. $(3x-1)(9x^2 + 3x + 1) = 0$
$\quad 3x-1 = 0 \quad$ or $\quad 9x^2 + 3x + 1 = 0$
$\qquad x = \dfrac{1}{3} \qquad\qquad 9x^2 + 3x + 1 > 0$

The solution is $\dfrac{1}{3}$.

59. $56x^2 - 5x - 6 = 0$
$\quad (7x+2)(8x-3) = 0$
$\quad 7x+2 = 0 \quad$ or $\quad 8x-3 = 0$
$\qquad x = -\dfrac{2}{7} \qquad\qquad x = \dfrac{3}{8}$

The solutions are $-\dfrac{2}{7}$ and $\dfrac{3}{8}$.

60. $m^2 = 6m$
$\quad m^2 - 6m = 0$
$\quad m(m-6) = 0$
$\quad m = 0 \quad$ or $\quad m-6 = 0$
$\qquad\qquad\qquad\quad m = 6$
The solutions are 0 and 6.

61. $r^2 = 25$
$\quad r^2 - 25 = 0$
$\quad (r-5)(r+5) = 0$
$\quad r-5 = 0 \quad$ or $\quad r+5 = 0$
$\qquad r = 5 \qquad\qquad r = -5$
The solutions are 5 and -5.

62. $(x-4)(x-5) = 0$
$\quad x^2 - 9x + 20 = 0$

63. Let x be the width. Then $2x$ is the length.
perimeter $= 2 \cdot$ length $+ 2 \cdot$ width
$\qquad 24 = 2(2x) + 2x$
$\qquad 24 = 4x + 2x$
$\qquad 24 = 6x$
$\qquad\; 4 = x$
$8 = 2x$
The dimensions are 4 inches by 8 inches.
The choice is c.

64. Let x be the width. Then $3x + 1$ is the length.

area = length $\cdot$ width

$$80 = (3x+1) \cdot x$$
$$80 = 3x^2 + x$$
$$0 = 3x^2 + x - 80$$
$$0 = (3x+16)(x-5)$$
$$3x+16 = 0 \quad \text{or} \quad x-5 = 0$$
$$x = -\frac{16}{3} \qquad\qquad x = 5$$

Discard $x = -\dfrac{16}{3}$ since length cannot be negative. The dimensions are 5 meters by $3x + 1 = 3(5) + 1 = 15 + 1 = 16$ meters. The choice is d.

65. area = side^2

$$81 = x^2$$
$$0 = x^2 - 81$$
$$0 = (x-9)(x+9)$$
$$x-9 = 0 \quad \text{or} \quad x+9 = 0$$
$$x = 9 \qquad\qquad x = -9$$

Discard $x = -9$ since length cannot be negative. The length of each side is 9 units.

66. The perimeter is the sum of the sides.

$$(2x+3) + (3x+1) + (x^2 - 3x) + (x+3) = 47$$
$$x^2 + 3x + 7 = 47$$
$$x^2 + 3x - 40 = 0$$
$$(x-5)(x+8) = 0$$
$$x-5 = 0 \quad \text{or} \quad x+8 = 0$$
$$x = 5 \qquad\qquad x = -8$$

Discard $x = -8$ since, for example, $x + 3$ would be negative.
$2x + 3 = 2(5) + 3 = 10 + 3 = 13$
$3x + 1 = 3(5) + 1 = 15 + 1 = 16$
$x^2 - 3x = 5^2 - 3(5) = 25 - 15 = 10$
$x + 3 = 5 + 3 = 8$
The lengths of the sides are 13 units, 16 units, 10 units, and 8 units.

67. Let x be the width. Then $2x - 15$ is the length.

area = length $\cdot$ width

$$500 = (2x-15) \cdot x$$
$$500 = 2x^2 - 15x$$
$$0 = 2x^2 - 15x - 500$$
$$0 = (2x+25)(x-20)$$
$$2x+25 = 0 \quad \text{or} \quad x-20 = 0$$
$$x = -\frac{25}{2} \qquad\qquad x = 20$$

Discard $x = -\dfrac{25}{2}$ since length cannot be negative. The width is 20 inches and the length is $2 \cdot 20 - 15 = 40 - 15 = 25$ inches.

68. Let x be the height. Then $4x$ is the base.

area $= \dfrac{1}{2} \cdot$ base $\cdot$ height

$$162 = \frac{1}{2} \cdot 4x \cdot x$$
$$162 = 2x^2$$
$$0 = 2x^2 - 162$$
$$0 = 2(x^2 - 81)$$
$$0 = 2(x-9)(x+9)$$
$$x-9 = 0 \quad \text{or} \quad x+9 = 0$$
$$x = 9 \qquad\qquad x = -9$$

Discard $x = -9$ since length cannot be negative. The height is 9 yards and the base is $4x = 4(9) = 36$ yards.

69. Let x be the first positive integer. Then $x + 1$ is the next consecutive integer.

$$x(x+1) = 380$$
$$x^2 + x = 380$$
$$x^2 + x - 380 = 0$$
$$(x+20)(x-19) = 0$$
$$x+20 = 0 \quad \text{or} \quad x-19 = 0$$
$$x = -20 \qquad\qquad x = 19$$

Discard $x = -20$ since it is not positive. The integers are 19 and $19 + 1 = 20$.

70. a.

$$h = -16t^2 + 440t$$
$$2800 = -16t^2 + 440t$$
$$16t^2 - 440t + 2800 = 0$$
$$8(2t^2 - 55t + 350) = 0$$
$$8(2t - 35)(t - 10) = 0$$

$$2t - 35 = 0 \quad \text{or} \quad t - 10 = 0$$
$$t = \frac{35}{2} \text{ or } 17.5 \qquad t = 10$$

The rocket reaches a height of 2800 feet at 10 seconds on the way up and at 17.5 seconds on the way down.

b. The height is 0 when the rocket reaches the ground.

$$h = -16t^2 + 440t$$
$$0 = -16t^2 + 440t$$
$$0 = -8t(2t - 55)$$

$$-8t = 0 \quad \text{or} \quad 2t - 55 = 0$$
$$t = 0 \qquad\qquad 2t = 55$$
$$\qquad\qquad t = 27.5$$

The rocket reaches the ground again after 27.5 seconds.

71. Let x be the length of the longer leg. Then $x + 8$ is the length of the hypotenuse and $x - 8$ is the length of the shorter leg.

$$(x+8)^2 = (x-8)^2 + x^2$$
$$x^2 + 16x + 64 = x^2 - 16x + 64 + x^2$$
$$0 = x^2 - 32x$$
$$0 = x(x - 32)$$

$$x = 0 \quad \text{or} \quad x - 32 = 0$$
$$x = 32$$

The longer leg is 32 centimeters.

72. $6x + 24 = 6 \cdot x + 6 \cdot 4 = 6(x + 4)$

73. $7x - 63 = 7 \cdot x - 7 \cdot 9 = 7(x - 9)$

74. $11x(4x - 3) - 6(4x - 3) = (4x - 3)(11x - 6)$

75. $2x(x-5) - (x-5) = 2x(x-5) - 1(x-5)$
$$= (x-5)(2x-1)$$

76. $3x^3 - 4x^2 + 6x - 8 = x^2(3x-4) + 2(3x-4)$
$$= (3x-4)(x^2+2)$$

77. $xy + 2x - y - 2 = x(y+2) - 1(y+2)$
$$= (y+2)(x-1)$$

78. $2x^2 + 2x - 24 = 2(x^2 + x - 12)$
$$= 2(x+4)(x-3)$$

79. $3x^3 - 30x^2 + 27x = 3x(x^2 - 10x + 9)$
$$= 3x(x-9)(x-1)$$

80. $4x^2 - 81 = (2x)^2 - 9^2 = (2x+9)(2x-9)$

81. $2x^2 - 18 = 2(x^2 - 9)$
$$= 2(x^2 - 3^2)$$
$$= 2(x-3)(x+3)$$

82. $16x^2 - 24x + 9 = (4x)^2 - 2 \cdot 4x \cdot 3 + 3^2$
$$= (4x-3)^2$$

83. $5x^2 + 20x + 20 = 5(x^2 + 4x + 4)$
$$= 5(x^2 + 2 \cdot x \cdot 2 + 2^2)$$
$$= 5(x+2)^2$$

84.

$$2x^2 - x - 28 = 0$$
$$(2x+7)(x-4) = 0$$
$$2x + 7 = 0 \quad \text{or} \quad x - 4 = 0$$
$$x = -\frac{7}{2} \qquad\qquad x = 4$$

The solutions are $-\dfrac{7}{2}$ and 4.

85.

$$x^2 - 2x = 15$$
$$x^2 - 2x - 15 = 0$$
$$(x+3)(x-5) = 0$$
$$x + 3 = 0 \quad \text{or} \quad x - 5 = 0$$
$$x = -3 \qquad\qquad x = 5$$

The solutions are -3 and 5.

86. $2x(x+7)(x+4) = 0$

$2x = 0$ or $x+7 = 0$ or $x+4 = 0$

$x = 0$ $x = -7$ $x = -4$

The solutions are 0, –7, and –4.

87. $x(x-5) = -6$

$x^2 - 5x = -6$

$x^2 - 5x + 6 = 0$

$(x-3)(x-2) = 0$

$x-3 = 0$ or $x-2 = 0$

$x = 3$ $x = 2$

The solutions are 3 and 2.

88. $x^2 = 16x$

$x^2 - 16x = 0$

$x(x-16) = 0$

$x = 0$ or $x-16 = 0$

$x = 16$

The solutions are 0 and 16.

89. The perimeter is the sum of the sides.

$(x^2+3) + 2x + (4x+5) = 48$

$x^2 + 6x + 8 = 48$

$x^2 + 6x - 40 = 0$

$(x-4)(x+10) = 0$

$x-4 = 0$ or $x+10 = 0$

$x = 4$ $x = -10$

Discard $x = -10$ since length, such as $2x = 2(-10) = -20$ cannot be negative.

$x^2 + 3 = 4^2 + 3 = 16 + 3 = 19$

$2x = 2 \cdot 4 = 8$

$4x + 5 = 4 \cdot 4 + 5 = 16 + 5 = 21$

The lengths are 19 inches, 8 inches, and 21 inches.

90. Let x be the length. Then $x - 4$ is the width.

area = length $\cdot$ width

$12 = x(x-4)$

$12 = x^2 - 4x$

$0 = x^2 - 4x - 12$

$0 = (x-6)(x+2)$

$x-6 = 0$ or $x+2 = 0$

$x = 6$ $x = -2$

Discard $x = -2$ since length cannot be negative. The length is 6 inches and the width is $x - 4 = 6 - 4 = 2$ inches.

Chapter 4 Test

1. $9x^2 - 3x = 3x(3x-1)$

2. $x^2 + 11x + 28 = (x+7)(x+4)$

3. $49 - m^2 = 7^2 - m^2 = (7-m)(7+m)$

4. $y^2 + 22y + 121 = y^2 + 2 \cdot y \cdot 11 + 11^2$

$= (y+11)^2$

5. $x^4 - 16 = (x^2)^2 - (4)^2$

$= (x^2+4)(x^2-4)$

$= (x^2+4)[(x)^2-(2)^2]$

$= (x^2+4)(x+2)(x-2)$

6. $4(a+3) - y(a+3) = (a+3)(4-y)$

7. $x^2 + 4$ is prime.

8. $y^2 - 8y - 48 = (y-12)(y+4)$

9. $3a^2 + 3ab - 7a - 7b = 3a(a+b) - 7(a+b)$

$= (a+b)(3a-7)$

10. $3x^2 - 5x + 2 = (3x-2)(x-1)$

11. $180 - 5x^2 = 5(36-x^2)$

$= 5(6^2-x^2)$

$= 5(6-x)(6+x)$

12. $3x^3 - 21x^2 + 30x = 3x(x^2 - 7x + 10)$

$= 3x(x-5)(x-2)$

13. $6t^2 - t - 5 = 6t^2 + 5t - 6t - 5$

$= t(6t+5) - 1(6t+5)$

$= (6t+5)(t-1)$

14. $xy^2 - 7y^2 - 4x + 28 = y^2(x-7) - 4(x-7)$
$$= (x-7)(y^2-4)$$
$$= (x-7)(y-2)(y+2)$$

15. $x - x^5 = x(1-x^4)$
$$= x[1^2 - (x^2)^2]$$
$$= x(1+x^2)(1-x^2)$$
$$= x(1+x^2)(1+x)(1-x)$$

16. $x^2 + 14xy + 24y^2 = x^2 + 12xy + 2xy + 24y^2$
$$= x(x+12y) + 2y(x+12y)$$
$$= (x+12y)(x+2y)$$

17. $(x-3)(x+9) = 0$
$x-3=0$ or $x+9=0$
$x=3$ $x=-9$
The solutions are 3 and −9.

18. $x^2 + 5x = 14$
$x^2 + 5x - 14 = 0$
$(x+7)(x-2) = 0$
$x+7=0$ or $x-2=0$
$x=-7$ $x=2$
The solutions are −7 and 2.

19. $x(x+6) = 7$
$x^2 + 6x = 7$
$x^2 + 6x - 7 = 0$
$(x+7)(x-1) = 0$
$x+7=0$ or $x-1=0$
$x=-7$ $x=1$
The solutions are −7 and 1.

20. $3x(2x-3)(3x+4) = 0$
$3x=0$ or $2x-3=0$ or $3x+4=0$
$x=0$ $x=\dfrac{3}{2}$ $x=-\dfrac{4}{3}$
The solutions are 0, $\dfrac{3}{2}$, and $-\dfrac{4}{3}$.

21. $5t^3 - 45t = 0$
$5t(t^2-9) = 0$
$5t[(t)^2 - (3)^2] = 0$
$5t(t+3)(t-3) = 0$
$5t=0$ or $t+3=0$ or $t-3=0$
$t=0$ $t=-3$ $t=3$
The solutions are 0, −3, and 3.

22. $t^2 - 2t - 15 = 0$
$(t+3)(t-5) = 0$
$t+3=0$ or $t-5=0$
$t=-3$ $t=5$
The solutions are −3 and 5.

23. $6x^2 = 15x$
$6x^2 - 15x = 0$
$3x(2x-5) = 0$
$3x=0$ or $2x-5=0$
$x=0$ $x=\dfrac{5}{2}$
The solutions are 0 and $\dfrac{5}{2}$.

24. Let x be the height. Then the base is $x+9$.
$$\text{area} = \frac{1}{2}\cdot \text{base}\cdot \text{height}$$
$$68 = \frac{1}{2}\cdot (x+9)\cdot x$$
$$2\cdot 68 = 2\cdot \frac{1}{2}\cdot (x+9)\cdot x$$
$$136 = x(x+9)$$
$$136 = x^2 + 9x$$
$$0 = x^2 + 9x - 136$$
$$0 = (x+17)(x-8)$$
$x+17=0$ or $x-8=0$
$x=-17$ $x=8$
Discard $x = -17$ since length cannot be negative. The height is 8 feet and the base is $x + 9 = 8 + 9 = 17$ feet.

25. $(x-1)(x+2) = 54$

$x^2 + 2x - x - 2 = 54$

$x^2 + x - 56 = 0$

$(x+8)(x-7) = 0$

$x + 8 = 0$ or $x - 7 = 0$

$x = -8$ $x = 7$

Discard $x = -8$ since length cannot be negative.

$x - 1 = 7 - 1 = 6$

$x + 2 = 7 + 2 = 9$

The width of the rectangle is 6 units and the length is 9 units.

26. The object is at a height of 0 when it reaches the ground.

$h = -16t^2 + 784$

$0 = -16t^2 + 784$

$0 = -16(t^2 - 49)$

$0 = -16(t-7)(t+7)$

$t - 7 = 0$ or $t + 7 = 0$

$t = 7$ $t = -7$

Discard $t = -7$ since time cannot be negative. The object reaches the ground after 7 seconds.

27. Let x be the length of the shorter leg. Then $x + 10$ is the length of the hypotenuse and $x + 10 - 5 = x + 5$ is the length of the longer leg.

$$x^2 + (x+5)^2 = (x+10)^2$$

$$x^2 + x^2 + 10x + 25 = x^2 + 20x + 100$$

$$x^2 - 10x - 75 = 0$$

$$(x-15)(x+5) = 0$$

$x - 15 = 0$ or $x + 5 = 0$

$x = 15$ $x = -5$

Discard $x = -5$ since length cannot be negative. The shorter leg is 15 cm, the longer leg is $15 + 5 = 20$ cm, and the hypotenuse is $15 + 10 = 25$ cm.

28. The height of the object is 0 when it reaches the ground.

$h = -16t^2 + 1089$

$0 = -16t^2 + 1089$

$0 = -1(16t^2 - 1089)$

$0 = -1[(4t)^2 - 33^2]$

$0 = -1(4t - 33)(4t + 33)$

$4t - 33 = 0$ or $4t + 33 = 0$

$t = \dfrac{33}{4}$ or 8.25 $t = -\dfrac{33}{4}$

Discard $t = -\dfrac{33}{4}$ since time cannot be negative. The object reaches the ground after 8.25 seconds.

Cumulative Review Chapters 1–4

1. a. Nine is less than or equal to eleven is translated as $9 \le 11$.

b. Eight is greater than one is translated as $8 > 1$.

c. Three is not equal to four is translated as $3 \ne 4$.

2. a. $|-5| = 5$

$|-3| = 3$

$|-5| > |-3|$ since $5 > 3$

b. $|0| = 0$

$|-2| = 2$

$|0| < |-2|$ since $0 < 2$

3. Replace x with 2.

$3x + 10 = 8x$

$3(2) + 10 \stackrel{?}{=} 8(2)$

$6 + 10 \stackrel{?}{=} 16$

$16 = 16$ True

Since a true statement results, 2 is a solution of the equation.

4. Replace x with 20 and y with 10.

$$\frac{x}{y} + 5x = \frac{20}{10} + 5(20) = 2 + 100 = 102$$

5. $-4 - 8 = -4 + (-8) = -12$

6. Replace x with -20 and y with 10.

$$\frac{x}{y} + 5x = \frac{-20}{10} + 5(-20)$$
$$= -2 + (-100)$$
$$= -102$$

7. Replace x with -2 and y with -4.

a. $\dfrac{3x}{2y} = \dfrac{3(-2)}{2(-4)} = \dfrac{-6}{-8} = \dfrac{-2 \cdot 3}{-2 \cdot 4} = \dfrac{3}{4}$

b. $x^3 - y^2 = (-2)^3 - (-4)^2$
$$= -8 - 16$$
$$= -8 + (-16)$$
$$= -24$$

8. Replace x with -20 and y with -10.

$$\frac{x}{y} + 5x = \frac{-20}{-10} + 5(-20) = 2 + (-100) = -98$$

9. $2x + 3x + 5 + 2 = (2x + 3x) + (5 + 2)$
$$= 5x + 7$$

10. $5 - 2(3x - 7) = 5 - 2 \cdot 3x - 2(-7)$
$$= 5 - 6x + 14$$
$$= 19 - 6x$$

11. $-5a - 3 + a + 2 = -5a + a - 3 + 2 = -4a - 1$

12. $5(x - 6) + 9(-2x + 1)$
$$= 5 \cdot x - 5 \cdot 6 + 9 \cdot (-2x) + 9 \cdot 1$$
$$= 5x - 30 - 18x + 9$$
$$= -13x - 21$$

13. $2.3x + 5x - 6 = 7.3x - 6$

14. $\quad 0.8y + 0.2(y - 1) = 1.8$
$$10[0.8y + 0.2(y - 1)] = 10(1.8)$$
$$8y + 2(y - 1) = 18$$
$$8y + 2y - 2 = 18$$
$$10y - 2 = 18$$
$$10y - 2 + 2 = 18 + 2$$
$$10y = 20$$
$$\frac{10y}{10} = \frac{20}{10}$$
$$y = 2$$

The solution is 2.

15. $-3x = 33$
$$\frac{-3x}{-3} = \frac{33}{-3}$$
$$x = -11$$

The solution is -11.

16. $\quad \dfrac{x}{-7} = -4$
$$-7 \cdot \frac{x}{-7} = -7 \cdot -4$$
$$x = 28$$

The solution is 28.

17. $\quad 3(x - 4) = 3x - 12$
$$3x - 12 = 3x - 12$$
$$-3x + 3x - 12 = -3x + 3x - 12$$
$$-12 = -12$$

Since this results in a true statement, the solution is every real number.

18. $\quad -\dfrac{2}{3}x = -22$
$$-\frac{3}{2} \cdot -\frac{2}{3}x = -\frac{3}{2} \cdot -22$$
$$x = 33$$

The solution is 33.

19. $V = lwh$
$$\frac{V}{wh} = \frac{lwh}{wh}$$
$$\frac{V}{wh} = l \text{ or } l = \frac{V}{wh}$$

20.
$$3x + 2y = -7$$
$$-3x + 3x + 2y = -3x - 7$$
$$2y = -3x - 7$$
$$\frac{2y}{2} = \frac{-3x - 7}{2}$$
$$y = -\frac{3}{2}x - \frac{7}{2}$$

21. $(5^3)^6 = 5^{3 \cdot 6} = 5^{18}$

22. $5^2 + 5^1 = 25 + 5 = 30$

23. $(y^8)^2 = y^{8 \cdot 2} = y^{16}$

24. $y^8 \cdot y^2 = y^{8+2} = y^{10}$

25.
$$\frac{(x^3)^4 x}{x^7} = \frac{x^{3 \cdot 4} \cdot x^1}{x^7}$$
$$= \frac{x^{12} \cdot x^1}{x^7}$$
$$= \frac{x^{12+1}}{x^7}$$
$$= \frac{x^{13}}{x^7}$$
$$= x^{13-7}$$
$$= x^6$$

26. $3^{-2} = \frac{1}{3^2} = \frac{1}{9}$

27. $(y^{-3}z^6)^{-6} = y^{-3 \cdot -6}z^{6 \cdot -6} = y^{18}z^{-36} = \frac{y^{18}}{z^{36}}$

28. $\frac{x^{-3}}{x^{-7}} = x^{-3-(-7)} = x^{-3+7} = x^4$

29.
$$\frac{x^{-7}}{(x^4)^3} = \frac{x^{-7}}{x^{4 \cdot 3}}$$
$$= \frac{x^{-7}}{x^{12}}$$
$$= \frac{1}{x^{12-(-7)}}$$
$$= \frac{1}{x^{12+7}}$$
$$= \frac{1}{x^{19}}$$

30. $\frac{(5a^7)^2}{a^5} = \frac{5^2 a^{7 \cdot 2}}{a^5} = \frac{25a^{14}}{a^5} = 25a^{14-5} = 25a^9$

31. $-3x + 7x = (-3 + 7)x = 4x$

32.
$$\frac{2}{3}x + 23 + \frac{1}{6}x - 100$$
$$= \left(\frac{2}{3}x + \frac{1}{6}x\right) + (23 - 100)$$
$$= \left(\frac{4}{6}x + \frac{1}{6}x\right) + (23 - 100)$$
$$= \frac{5}{6}x - 77$$

33. $11x^2 + 5 + 2x^2 - 7 = 11x^2 + 2x^2 + 5 - 7$
$$= 13x^2 - 2$$

34. $0.2x - 1.1 + 2.3 - 0.7x$
$$= 0.2x - 0.7x - 1.1 + 2.3$$
$$= -0.5x + 1.2$$

35. $(2x - y)^2$
$$= (2x - y)(2x - y)$$
$$= 2x(2x) + 2x(-y) + (-y)(2x) + (-y)(-y)$$
$$= 4x^2 - 2xy - 2xy + y^2$$
$$= 4x^2 - 4xy + y^2$$

36. $(3x - 7y)^2 = (3x - 7y)(3x - 7y)$
$$= 3x(3x) + 3x(-7y) + (-7y)(3x) + (-7y)(-7y)$$
$$= 9x^2 - 21xy - 21xy + 49y^2$$
$$= 9x^2 - 42xy + 49y^2$$

37. $(t + 2)^2 = t^2 + 2(t)(2) + 2^2 = t^2 + 4t + 4$

38. $(x - 13)^2 = x^2 - 2(x)(13) + 13^2$
$$= x^2 - 26x + 169$$

39. $(x^2 - 7y)^2 = (x^2)^2 - 2(x^2)(7y) + (7y)^2$
$$= x^4 - 14x^2 y + 49y^2$$

40. $(7x + y)^2 = (7x)^2 + 2(7x)(y) + (y)^2$
$$= 49x^2 + 14xy + y^2$$

41. $\dfrac{8x^2 y^2 - 16xy + 2x}{4xy} = \dfrac{8x^2 y^2}{4xy} - \dfrac{16xy}{4xy} + \dfrac{2x}{4xy}$
$$= 2xy - 4 + \dfrac{1}{2y}$$

42. $z^3 + 7z + z^2 + 7 = z(z^2 + 7) + 1(z^2 + 7)$
$$= (z^2 + 7)(z + 1)$$

43. $5(x + 3) + y(x + 3) = (x + 3)(5 + y)$

44. $2x^3 + 2x^2 - 84x = 2x(x^2 + x - 42)$
$$= 2x(x^2 + 7x - 6x - 42)$$
$$= 2x[x(x + 7) - 6(x + 7)]$$
$$= 2x(x + 7)(x - 6)$$

45. $x^4 + 5x^2 + 6 = x^4 + 2x^2 + 3x^2 + 6$
$$= x^2(x^2 + 2) + 3(x^2 + 2)$$
$$= (x^2 + 2)(x^2 + 3)$$

46.

$$-4x^2 - 23x + 6 = -4x^2 + x - 24x + 6$$
$$= x(-4x+1) + 6(-4x+1)$$
$$= (-4x+1)(x+6)$$

or

$$-4x^2 - 23x + 6 = -1(4x^2 + 23x - 6)$$
$$= -1(4x^2 - x + 24x - 6)$$
$$= -1[x(4x-1) + 6(4x-1)]$$
$$= -1(4x-1)(x+6)$$

47.

$$6x^2 - 2x - 20 = 2(3x^2 - x - 10)$$
$$= 2(3x^2 - 6x + 5x - 10)$$
$$= 2[3x(x-2) + 5(x-2)]$$
$$= 2(x-2)(3x+5)$$

48.

$$9xy^2 - 16x = x(9y^2 - 16)$$
$$= x[(3y)^2 - 4^2]$$
$$= x(3y+4)(3y-4)$$

49. The height is 0 when the diver reaches the ocean.

$$h = -16t^2 + 144$$
$$0 = -16t^2 + 144$$
$$0 = -16(t^2 - 9)$$
$$0 = -16(t-3)(t+3)$$
$$t - 3 = 0 \quad \text{or} \quad t + 3 = 0$$
$$t = 3 \qquad\qquad t = -3$$

Discard $t = -3$ since time cannot be negative. The diver reaches the ocean after 3 seconds.

50.

$$x^2 - 13x = -36$$
$$x^2 - 13x + 36 = 0$$
$$x^2 - 9x - 4x + 36 = 0$$
$$x(x-9) - 4(x-9) = 0$$
$$(x-9)(x-4) = 0$$
$$x - 9 = 0 \quad \text{or} \quad x - 4 = 0$$
$$x = 9 \qquad\qquad x = 4$$

The solutions are 9 and 4.

Chapter 5

Section 5.1 Practice Problems

1. a. $\dfrac{x-3}{5x+1} = \dfrac{4-3}{5(4)+1} = \dfrac{1}{20+1} = \dfrac{1}{21}$

b. $\dfrac{x-3}{5x+1} = \dfrac{(-3)-3}{5(-3)+1} = \dfrac{-6}{-15+1} = \dfrac{-6}{-14} = \dfrac{3}{7}$

2. a. $x+8 = 0$
$x = -8$

When $x = -8$, the expression $\dfrac{x}{x+8}$ is undefined.

b. $x^2 + 5x + 4 = 0$
$(x+4)(x+1) = 0$
$x+4 = 0 \quad$ or $\quad x+1 = 0$
$x = -4 \qquad\qquad x = -1$
When $x = -4$ or $x = -1$, the expression

$\dfrac{x-3}{x^2+5x+4}$ is undefined.

c. The denominator of $\dfrac{x^2-3x+2}{5}$ is

never 0, so there are no values of x for which this expression is undefined.

3. $\dfrac{x^4+x^3}{5x+5} = \dfrac{x^3(x+1)}{5(x+1)} = \dfrac{x^3}{5}$

4. $\dfrac{x^2+11x+18}{x^2+x-2} = \dfrac{(x+9)(x+2)}{(x-1)(x+2)} = \dfrac{x+9}{x-1}$

5. $\dfrac{x^2+10x+25}{x^2+5x} = \dfrac{(x+5)(x+5)}{x(x+5)} = \dfrac{x+5}{x}$

6. $\dfrac{x+5}{x^2-25} = \dfrac{x+5}{(x+5)(x-5)} = \dfrac{1}{x-5}$

7. a. $\dfrac{x+4}{4+x} = \dfrac{x+4}{x+4} = 1$

b. $\dfrac{x-4}{4-x} = \dfrac{x-4}{(-1)(x-4)} = \dfrac{1}{-1} = -1$

8. $-\dfrac{3x+7}{x-6} = \dfrac{-(3x+7)}{x-6} = \dfrac{-3x-7}{x-6}$ or

$\dfrac{3x+7}{-(x-6)} = \dfrac{3x+7}{-x+6}$

Mental Math

1. When $x = 0$, the expression $\dfrac{x+5}{x}$ is undefined.

2. When $x = 3$, the expression $\dfrac{x^2-5x}{x-3}$ is undefined.

3. When $x = 0$ or $x = 1$, the expression $\dfrac{x^2+4x-2}{x(x-1)}$ is undefined.

4. When $x = 5$ or $x = 6$, the expression $\dfrac{x+2}{(x-5)(x-6)}$ is undefined.

Exercise Set 5.1

1. $\dfrac{x+5}{x+2} = \dfrac{2+5}{2+2} = \dfrac{7}{4}$

3. $\dfrac{y^3}{y^2-1} = \dfrac{(-2)^3}{(-2)^2-1} = \dfrac{-8}{4-1} = -\dfrac{8}{3}$

212

5. $\dfrac{x^2+8x+2}{x^2-x-6}=\dfrac{2^2+8(2)+2}{2^2-2-6}$

$\qquad\qquad\quad =\dfrac{4+16+2}{4-2-6}$

$\qquad\qquad\quad =\dfrac{22}{-4}$

$\qquad\qquad\quad =-\dfrac{11}{2}$

7. a. $A=\dfrac{3x+400}{x}=\dfrac{3(1)+400}{1}=403$

The cost of producing 1 DVD is $403.

b. $A=\dfrac{3(100)+400}{100}$

$\qquad =\dfrac{300+400}{100}$

$\qquad =\dfrac{700}{100}$

$\quad t=7$

The average cost for producing 100 DVDs is $7.

c. decrease; answers may vary

9. $2x=0$

$\quad x=0$

$\dfrac{7}{2x}$ is undefined for $x=0$.

11. $x+2=0$

$\qquad x=-2$

$\dfrac{x+3}{x+2}$ is undefined for $x=-2$.

13. $2x-5=0$

$\qquad 2x=5$

$\qquad\ x=\dfrac{5}{2}$

$\dfrac{x-4}{2x-5}$ is undefined for $x=\dfrac{5}{2}$.

15. $15x^2+30x=0$

$\quad 15x(x+2)=0$

$\quad 15x=0\quad$ or $\quad x+2=0$

$\qquad x=0\qquad\qquad x=-2$

$\dfrac{9x^3+4}{15x^2+30x}$ is undefined for $x=0$ and $x=-2$.

17. Since $4\neq0$, there are no values of x for which $\dfrac{x^2-5x-2}{4}$ is undefined.

19. $x^2-5x-6=0$

$\quad (x-6)(x+1)=0$

$\quad x-6=0\quad$ or $\quad x+1=0$

$\qquad x=6\qquad\qquad x=-1$

$\dfrac{3x^2+9}{x^2-5x-6}$ is undefined for $x=6$ and $x=-1$.

21. $3x^2+13x+14=0$

$\quad (x+2)(3x+7)=0$

$\quad x+2=0\quad$ or $\quad 3x+7=0$

$\qquad x=-2\qquad\qquad 3x=-7$

$\qquad\qquad\qquad\qquad\qquad x=-\dfrac{7}{3}$

$\dfrac{x}{3x^2+13x+14}$ is undefined for $x=-2$ and $x=-\dfrac{7}{3}$.

23. $\dfrac{x+7}{7+x}=\dfrac{x+7}{x+7}=1$

25. $\dfrac{x-7}{7-x}=\dfrac{-1(-x+7)}{7-x}=\dfrac{-1(7-x)}{7-x}=-1$

27. $\dfrac{2}{8x+16}=\dfrac{2}{8(x+2)}=\dfrac{1}{4(x+2)}$

29. $\dfrac{x-2}{x^2-4} = \dfrac{x-2}{(x+2)(x-2)} = \dfrac{1}{x+2}$

31. $\dfrac{2x-10}{3x-30} = \dfrac{2(x-5)}{3(x-10)}$ can't be simplified.

33. $\dfrac{-5a-5b}{a+b} = \dfrac{-5(a+b)}{a+b} = -5$

35. $\dfrac{7x+35}{x^2+5x} = \dfrac{7(x+5)}{x(x+5)} = \dfrac{7}{x}$

37. $\dfrac{x+5}{x^2-4x-45} = \dfrac{x+5}{(x-9)(x+5)} = \dfrac{1}{x-9}$

39. $\dfrac{5x^2+11x+2}{x+2} = \dfrac{(5x+1)(x+2)}{x+2} = 5x+1$

41. $\dfrac{x^3+7x^2}{x^2+5x-14} = \dfrac{x^2(x+7)}{(x-2)(x+7)} = \dfrac{x^2}{x-2}$

43. $\dfrac{14x^2-21x}{2x-3} = \dfrac{7x(2x-3)}{2x-3} = 7x$

45. $\dfrac{x^2+7x+10}{x^2-3x-10} = \dfrac{(x+5)(x+2)}{(x+2)(x-5)} = \dfrac{x+5}{x-5}$

47. $\dfrac{3x^2+7x+2}{3x^2+13x+4} = \dfrac{(3x+1)(x+2)}{(3x+1)(x+4)} = \dfrac{x+2}{x+4}$

49. $\dfrac{2x^2-8}{4x-8} = \dfrac{2(x^2-4)}{4(x-2)}$
$= \dfrac{2(x+2)(x-2)}{2\cdot 2(x-2)}$
$= \dfrac{x+2}{2}$

51. $\dfrac{4-x^2}{x-2} = \dfrac{-(x^2-4)}{x-2}$
$= \dfrac{-(x+2)(x-2)}{x-2}$
$= -(x+2)$

53. $\dfrac{x^2-1}{x^2-2x+1} = \dfrac{(x+1)(x-1)}{(x-1)^2} = \dfrac{x+1}{x-1}$

55. $\dfrac{x^2+xy+2x+2y}{x+2} = \dfrac{x(x+y)+2(x+y)}{x+2}$
$= \dfrac{(x+2)(x+y)}{x+2}$
$= x+y$

57. $\dfrac{5x+15-xy-3y}{2x+6} = \dfrac{5(x+3)-y(x+3)}{2(x+3)}$
$= \dfrac{(x+3)(5-y)}{2(x+3)}$
$= \dfrac{5-y}{2}$

59. $\dfrac{2xy+5x-2y-5}{3xy+4x-3y-4} = \dfrac{x(2y+5)-1(2y+5)}{x(3y+4)-1(3y+4)}$
$= \dfrac{(2y+5)(x-1)}{(3y+4)(x-1)}$
$= \dfrac{2y+5}{3y+4}$

61. $-\dfrac{x-10}{x+8} = \dfrac{-(x-10)}{x+8}$
$= \dfrac{-x+10}{x+8}$
$= \dfrac{x-10}{-(x+8)}$
$= \dfrac{x-10}{-x-8}$

63.
$$-\frac{5y-3}{y-12} = \frac{-(5y-3)}{y-12}$$
$$= \frac{-5y+3}{y-12}$$
$$= \frac{5y-3}{-(y-12)}$$
$$= \frac{5y-3}{-y+12}$$

65.
$$\frac{9-x^2}{x-3} = \frac{9-x^2}{-(3-x)}$$
$$= \frac{(3+x)(3-x)}{-(3-x)}$$
$$= \frac{3+x}{-1}$$
$$= -3-x$$

The given answer is correct.

67.
$$\frac{7-34x-5x^2}{25x^2-1} = \frac{(7+x)(1-5x)}{(5x+1)(5x-1)}$$
$$= \frac{(7+x)(1-5x)}{-(5x+1)(1-5x)}$$
$$= \frac{7+x}{-5x-1}$$

The given answer is correct.

69.
$$\frac{1}{3}\cdot\frac{9}{11} = \frac{1\cdot 9}{3\cdot 11} = \frac{1\cdot 3\cdot 3}{3\cdot 11} = \frac{3}{11}$$

71.
$$\frac{1}{3}\div\frac{1}{4} = \frac{1}{3}\cdot\frac{4}{1} = \frac{1\cdot 4}{3\cdot 1} = \frac{4}{3}$$

73.
$$\frac{13}{20}\div\frac{2}{9} = \frac{13}{20}\cdot\frac{9}{2} = \frac{13\cdot 9}{20\cdot 2} = \frac{117}{40}$$

75.
$$\frac{5a-15}{5} = \frac{5(a-3)}{5} = a-3$$

The given answer is correct.

77.
$$\frac{1+2}{1+3} = \frac{3}{4} \neq \frac{2}{3}$$

The given answer is incorrect.

79. answers may vary

80. answers may vary

81. answers may vary

83. Use $C = \dfrac{DA}{A+12}$ with $A = 8$ and $D = 1000$.

$$C = \frac{DA}{A+12} = \frac{1000 \cdot 8}{8+12} = \frac{8000}{20} = 400$$

The child should receive a dose of 400 milligrams.

85. Use $S = \dfrac{h+d+2t+3r}{b}$ with $h = 262$, $d = 24$, $t = 5$, $r = 8$, and $b = 704$.

$$S = \frac{262+24+2(5)+3(8)}{704} = \frac{320}{704} \approx 0.4545$$

Suzuki's slugging percentage was about 45.5%.

87. Use $\left[\dfrac{20C+0.5A+Y+20T-100I}{A}\right]\left(\dfrac{25}{6}\right)$ with $A = 527$, $C = 317$, $Y = 3620$, $T = 23$, and $I = 12$.

$$\left[\frac{20C+0.5A+Y+80T-100I}{A}\right]\left(\frac{25}{6}\right) = \left[\frac{20(317)+0.5(527)+3620+80(23)-100(12)}{527}\right]\left(\frac{25}{6}\right)$$

$$= \left[\frac{6340+263.5+3620+1840-1200}{527}\right]\left(\frac{25}{6}\right)$$

$$= \left(\frac{10{,}863.5}{527}\right)\left(\frac{25}{6}\right)$$

$$\approx 85.9$$

Brady's quarterback rating for the 2004–2005 season is 85.9.

Section 5.2 Practice Problems

1. a. $\dfrac{16y}{3} \cdot \dfrac{1}{x^2} = \dfrac{16y \cdot 1}{3 \cdot x^2} = \dfrac{16y}{3x^2}$

b. $\dfrac{-5a^3}{3b^3} \cdot \dfrac{2b^2}{15a} = \dfrac{-5a^3 \cdot 2b^2}{3b^3 \cdot 15a}$

$$= \frac{-5 \cdot a^3 \cdot 2 \cdot b^2}{3 \cdot b^3 \cdot 3 \cdot 5 \cdot a}$$

$$= -\frac{2a^2}{9b}$$

2. $\dfrac{3x+6}{14} \cdot \dfrac{7x^2}{x^3+2x^2} = \dfrac{3(x+2) \cdot 7x^2}{2 \cdot 7 \cdot x^2 (x+2)} = \dfrac{3}{2}$

3. $\dfrac{4x+8}{7x^2-14x} \cdot \dfrac{3x^2-5x-2}{9x^2-1}$

$= \dfrac{4(x+2)}{7x(x-2)} \cdot \dfrac{(3x+1)(x-2)}{(3x+1)(3x-1)}$

$= \dfrac{4(x+2)(3x+1)(x-2)}{7x(x-2)(3x+1)(3x-1)}$

$= \dfrac{4(x+2)}{7x(3x-1)}$

4. $\dfrac{7x^2}{6} \div \dfrac{x}{2y} = \dfrac{7x^2}{6} \cdot \dfrac{2y}{x} = \dfrac{7 \cdot x \cdot x \cdot 2 \cdot y}{2 \cdot 3 \cdot x} = \dfrac{7xy}{3}$

5. $\dfrac{(x-4)^2}{6} \div \dfrac{3x-12}{2} = \dfrac{(x-4)^2}{6} \cdot \dfrac{2}{3x-12}$

$= \dfrac{(x-4)(x-4) \cdot 2}{2 \cdot 3 \cdot 3 \cdot (x-4)}$

$= \dfrac{x-4}{9}$

6. $\dfrac{10x+4}{x^2-4} \div \dfrac{5x^3+2x^2}{x+2}$

$= \dfrac{10x+4}{x^2-4} \cdot \dfrac{x+2}{5x^3+2x^2}$

$= \dfrac{2(5x+2)(x+2)}{(x-2)(x+2) \cdot x^2(5x+2)}$

$= \dfrac{2}{x^2(x-2)}$

7. $\dfrac{3x^2-10x+8}{7x-14} \div \dfrac{9x-12}{21}$

$= \dfrac{3x^2-10x+8}{7x-14} \cdot \dfrac{21}{9x-12}$

$= \dfrac{(3x-4)(x-2) \cdot 3 \cdot 7}{7(x-2) \cdot 3(3x-4)}$

$= 1$

8. a. $\dfrac{x+3}{x} \cdot \dfrac{7}{x+3} = \dfrac{(x+3) \cdot 7}{x \cdot (x+3)} = \dfrac{7}{x}$

b. $\dfrac{x+3}{x} \div \dfrac{7}{x+3} = \dfrac{x+3}{x} \cdot \dfrac{x+3}{7} = \dfrac{(x+3)^2}{7x}$

c. $\dfrac{3-x}{x^2+6x+5} \cdot \dfrac{2x+10}{x^2-7x+12}$

$= \dfrac{(3-x) \cdot 2(x+5)}{(x+5)(x+1) \cdot (x-4)(x-3)}$

$= \dfrac{-1(x-3) \cdot 2}{(x+1)(x-4)(x-3)}$

$= -\dfrac{2}{(x+1)(x-4)}$

9. $288 \text{ sq in.} = \dfrac{288 \text{ sq in.}}{1} \cdot \dfrac{1 \text{ sq ft}}{144 \text{ sq in.}} = 2 \text{ sq ft}$

10. $3.5 \text{ sq ft} = \dfrac{3.5 \text{ sq ft}}{1} \cdot \dfrac{144 \text{ sq in.}}{1 \text{ sq ft}} = 504 \text{ sq in.}$

11. $35,000 \text{ sq yd} = 35,000 \text{ sq yd} \cdot \dfrac{9 \text{ sq ft}}{1 \text{ sq yd}}$

$= 315,000 \text{ sq ft}$

12. 102.7 feet/second

$= \dfrac{102.7 \text{ feet}}{1 \text{ second}} \cdot \dfrac{3600 \text{ seconds}}{1 \text{ hour}} \cdot \dfrac{1 \text{ mile}}{5280 \text{ feet}}$

$= \dfrac{102.7 \cdot 3600}{5280} \text{ miles/hour}$

$\approx 70.0 \text{ miles/hour}$

Mental Math

1. $\dfrac{2}{y} \cdot \dfrac{x}{3} = \dfrac{2x}{3y}$

2. $\dfrac{3x}{4} \cdot \dfrac{1}{y} = \dfrac{3x}{4y}$

3. $\dfrac{5}{7} \cdot \dfrac{y^2}{x^2} = \dfrac{5y^2}{7x^2}$

4. $\dfrac{x^5}{11} \cdot \dfrac{4}{z^3} = \dfrac{4x^5}{11z^3}$

5. $\dfrac{9}{x} \cdot \dfrac{x}{5} = \dfrac{9}{5}$

6. $\dfrac{y}{7} \cdot \dfrac{3}{y} = \dfrac{3}{7}$

Exercise Set 5.2

1. $\dfrac{3x}{y^2} \cdot \dfrac{7y}{4x} = \dfrac{3 \cdot 7 \cdot x \cdot y}{4 \cdot x \cdot y \cdot y} = \dfrac{21}{4y}$

3. $\dfrac{8x}{2} \cdot \dfrac{x^5}{4x^2} = \dfrac{2 \cdot 4 \cdot x \cdot x \cdot x \cdot x \cdot x \cdot x}{2 \cdot 4 \cdot x \cdot x} = x^4$

5. $-\dfrac{5a^2b}{30a^2b^2} \cdot b^3 = -\dfrac{5a^2b}{30a^2b^2} \cdot \dfrac{b^3}{1}$

$\qquad = -\dfrac{5 \cdot a^2b \cdot b \cdot b^2}{5 \cdot 6 \cdot a^2 \cdot b^2}$

$\qquad = -\dfrac{b \cdot b}{6}$

$\qquad = -\dfrac{b^2}{6}$

7. $\dfrac{x}{2x-14} \cdot \dfrac{x^2-7x}{5} = \dfrac{x}{2(x-7)} \cdot \dfrac{x(x-7)}{5}$

$\qquad = \dfrac{x \cdot x}{2 \cdot 5}$

$\qquad = \dfrac{x^2}{10}$

9. $\dfrac{6x+6}{5} \cdot \dfrac{10}{36x+36} = \dfrac{6(x+1)}{5} \cdot \dfrac{10}{36(x+1)}$

$\qquad = \dfrac{6 \cdot 10}{5 \cdot 36}$

$\qquad = \dfrac{6 \cdot 2 \cdot 5}{5 \cdot 6 \cdot 2 \cdot 3}$

$\qquad = \dfrac{1}{3}$

11. $\dfrac{(m+n)^2}{m-n} \cdot \dfrac{m}{m^2+mn}$

$\qquad = \dfrac{(m+n)(m+n)}{m-n} \cdot \dfrac{m}{m(m+n)}$

$\qquad = \dfrac{m+n}{m-n}$

13. $\dfrac{x^2-25}{x^2-3x-10} \cdot \dfrac{x+2}{x} = \dfrac{(x+5)(x-5)}{(x+2)(x-5)} \cdot \dfrac{x+2}{x}$

$\qquad = \dfrac{x+5}{x}$

15. $\dfrac{x^2+6x+8}{x^2+x-20} \cdot \dfrac{x^2+2x-15}{x^2+8x+16}$

$\qquad = \dfrac{(x+4)(x+2)}{(x+5)(x-4)} \cdot \dfrac{(x+5)(x-3)}{(x+4)(x+4)}$

$\qquad = \dfrac{(x+2)(x-3)}{(x-4)(x+4)}$

17. $\dfrac{5x^7}{2x^5} \div \dfrac{15x}{4x^3} = \dfrac{5x^7}{2x^5} \cdot \dfrac{4x^3}{15x}$

$\qquad = \dfrac{5 \cdot x^5 \cdot x \cdot x}{2 \cdot x^5} \cdot \dfrac{2 \cdot 2 \cdot x^3}{5 \cdot 3 \cdot x}$

$\qquad = \dfrac{2 \cdot x \cdot x^3}{3}$

$\qquad = \dfrac{2x^4}{3}$

19. $\dfrac{8x^2}{y^3} \div \dfrac{4x^2y^3}{6} = \dfrac{8x^2}{y^3} \cdot \dfrac{6}{4x^2y^3}$

$\qquad = \dfrac{2 \cdot 4 \cdot 6 \cdot x \cdot x}{4 \cdot x \cdot x \cdot y \cdot y \cdot y \cdot y \cdot y \cdot y}$

$\qquad = \dfrac{2 \cdot 6}{y \cdot y \cdot y \cdot y \cdot y \cdot y}$

$\qquad = \dfrac{12}{y^6}$

21. $\dfrac{(x-6)(x+4)}{4x} \div \dfrac{2x-12}{8x^2}$

$\qquad = \dfrac{(x-6)(x+4)}{4x} \cdot \dfrac{8x^2}{2x-12}$

$\qquad = \dfrac{(x-6)(x+4)}{4x} \cdot \dfrac{4x \cdot 2 \cdot x}{2(x-6)}$

$\qquad = \dfrac{x(x+4)}{1}$

$\qquad = x(x+4)$

23. $\dfrac{3x^2}{x^2-1} \div \dfrac{x^5}{(x+1)^2}$

$= \dfrac{3x^2}{x^2-1} \cdot \dfrac{(x+1)^2}{x^5}$

$= \dfrac{3x^2}{(x-1)(x+1)} \cdot \dfrac{(x+1)(x+1)}{x^2 \cdot x^3}$

$= \dfrac{3(x+1)}{x^3(x-1)}$

25. $\dfrac{m^2-n^2}{m+n} \div \dfrac{m}{m^2+nm}$

$= \dfrac{m^2-n^2}{m+n} \cdot \dfrac{m^2+nm}{m}$

$= \dfrac{(m+n)(m-n)}{m+n} \cdot \dfrac{m(m+n)}{m}$

$= (m-n)(m+n)$

$= m^2-n^2$

27. $\dfrac{x+2}{7-x} \div \dfrac{x^2-5x+6}{x^2-9x+14}$

$= \dfrac{x+2}{7-x} \cdot \dfrac{x^2-9x+14}{x^2-5x+6}$

$= \dfrac{x+2}{7-x} \cdot \dfrac{(x-7)(x-2)}{(x-3)(x-2)}$

$= \dfrac{x+2}{-(x-7)} \cdot \dfrac{(x-7)(x-2)}{(x-3)(x-2)}$

$= -\dfrac{x+2}{x-3}$

29. $\dfrac{x^2+7x+10}{x-1} \div \dfrac{x^2+2x-15}{x-1}$

$= \dfrac{x^2+7x+10}{x-1} \cdot \dfrac{x-1}{x^2+2x-15}$

$= \dfrac{(x+5)(x+2)}{x-1} \cdot \dfrac{x-1}{(x+5)(x-3)}$

$= \dfrac{x+2}{x-3}$

31. $\dfrac{5x-10}{12} \div \dfrac{4x-8}{8} = \dfrac{5x-10}{12} \cdot \dfrac{8}{4x-8}$

$= \dfrac{5(x-2)}{2\cdot6} \cdot \dfrac{4\cdot2}{4(x-2)}$

$= \dfrac{5}{6}$

33. $\dfrac{x^2+5x}{8} \cdot \dfrac{9}{3x+15} = \dfrac{x(x+5)}{8} \cdot \dfrac{3\cdot3}{3(x+5)} = \dfrac{3x}{8}$

35. $\dfrac{7}{6p^2+q} \div \dfrac{14}{18p^2+3q} = \dfrac{7}{6p^2+q} \cdot \dfrac{18p^2+3q}{14}$

$= \dfrac{7}{6p^2+q} \cdot \dfrac{3(6p^2+q)}{7\cdot2}$

$= \dfrac{3}{2}$

37. $\dfrac{3x+4y}{x^2+4xy+4y^2} \cdot \dfrac{x+2y}{2} = \dfrac{3x+4y}{(x+2y)^2} \cdot \dfrac{x+2y}{2}$

$= \dfrac{3x+4y}{2(x+2y)}$

39. $\dfrac{(x+2)^2}{x-2} \div \dfrac{x^2-4}{2x-4}$

$= \dfrac{(x+2)^2}{x-2} \cdot \dfrac{2x-4}{x^2-4}$

$= \dfrac{(x+2)(x+2)}{x-2} \cdot \dfrac{2(x-2)}{(x-2)(x+2)}$

$= \dfrac{2(x+2)}{x-2}$

41. $\dfrac{x^2-4}{24x} \div \dfrac{2-x}{6xy} = \dfrac{x^2-4}{24x} \cdot \dfrac{6xy}{2-x}$

$= \dfrac{(x+2)(x-2)}{6\cdot4\cdot x} \cdot \dfrac{6\cdot x\cdot y}{-1(x-2)}$

$= \dfrac{y(x+2)}{-4}$

$= -\dfrac{y(x+2)}{4}$

43. $\dfrac{a^2+7a+12}{a^2+5a+6}\cdot\dfrac{a^2+8a+15}{a^2+5a+4}$

$=\dfrac{(a+4)(a+3)}{(a+3)(a+2)}\cdot\dfrac{(a+5)(a+3)}{(a+4)(a+1)}$

$=\dfrac{(a+5)(a+3)}{(a+2)(a+1)}$

45. $\dfrac{5x-20}{3x^2+x}\cdot\dfrac{3x^2+13x+4}{x^2-16}$

$=\dfrac{5(x-4)}{x(3x+1)}\cdot\dfrac{(x+4)(3x+1)}{(x+4)(x-4)}$

$=\dfrac{5}{x}$

47. $\dfrac{8n^2-18}{2n^2-5n+3}\div\dfrac{6n^2+7n-3}{n^2-9n+8}$

$=\dfrac{8n^2-18}{2n^2-5n+3}\cdot\dfrac{n^2-9n+8}{6n^2+7n-3}$

$=\dfrac{2(4n^2-9)}{(2n-3)(n-1)}\cdot\dfrac{(n-8)(n-1)}{(3n-1)(2n+3)}$

$=\dfrac{2(2n-3)(2n+3)}{(2n-3)(n-1)}\cdot\dfrac{(n-8)(n-1)}{(3n-1)(2n+3)}$

$=\dfrac{2(n-8)}{3n-1}$

49. 10 square feet

$=\dfrac{10\text{ square feet}}{1}\cdot\dfrac{144\text{ square inches}}{1\text{ square foot}}$

$=1440$ square inches

51. 45 square feet

$=\dfrac{45\text{ square feet}}{1}\cdot\dfrac{1\text{ square yard}}{9\text{ square feet}}$

$=5$ square yards

53. 3 cubic yards $=\dfrac{3\text{ cubic yards}}{1}\cdot\dfrac{27\text{ cubic feet}}{1\text{ cubic yard}}$

$=81$ cubic feet

55. $\dfrac{50\text{ miles}}{1\text{ hour}}$

$=\dfrac{50\text{ miles}}{1\text{ hour}}\cdot\dfrac{5280\text{ feet}}{1\text{ mile}}\cdot\dfrac{1\text{ hour}}{3600\text{ seconds}}$

≈ 73 feet per second

57. 6.3 square yards

$=\dfrac{6.3\text{ square yards}}{1}\cdot\dfrac{9\text{ square feet}}{1\text{ square yard}}$

$=56.7$ square feet

59. 3,705,793 square feet

$=\dfrac{3,705,793\text{ square feet}}{1}\cdot\dfrac{1\text{ square yard}}{9\text{ square feet}}$

$\approx 411,755$ square yards

61. 930 miles per hour

$=\dfrac{930\text{ miles}}{1\text{ hour}}\cdot\dfrac{5280\text{ feet}}{1\text{ mile}}\cdot\dfrac{1\text{ hour}}{3600\text{ seconds}}$

$=\dfrac{930\cdot 5280\text{ feet}}{3600\text{ seconds}}$

$=1364$ feet per second

63. $\dfrac{1}{5}+\dfrac{4}{5}=\dfrac{1+4}{5}=\dfrac{5}{5}=1$

65. $\dfrac{9}{9}-\dfrac{19}{9}=\dfrac{9-19}{9}=\dfrac{-10}{9}=-\dfrac{10}{9}$

67. $\dfrac{6}{5}+\left(\dfrac{1}{5}-\dfrac{8}{5}\right)=\dfrac{6}{5}+\left(-\dfrac{7}{5}\right)=-\dfrac{1}{5}$

69. $\dfrac{4}{a}\cdot\dfrac{1}{b}=\dfrac{4\cdot 1}{a\cdot b}=\dfrac{4}{ab}$

The statement is true.

71. $\dfrac{x}{5}\cdot\dfrac{x+3}{4}=\dfrac{x^2+3x}{20}\neq\dfrac{2x+3}{20}$

The statement is false.

73. $\dfrac{2x}{x^2-25} \cdot \dfrac{x+5}{9x} = \dfrac{2 \cdot x}{(x+5)(x-5)} \cdot \dfrac{x+5}{9 \cdot x}$

$= \dfrac{2}{9(x-5)}$

The area is $\dfrac{2}{9(x-5)}$ square feet.

75. $\left(\dfrac{x^2-y^2}{x^2+y^2} \div \dfrac{x^2-y^2}{3x} \right) \cdot \dfrac{x^2+y^2}{6}$

$= \dfrac{x^2-y^2}{x^2+y^2} \cdot \dfrac{3x}{x^2-y^2} \cdot \dfrac{x^2+y^2}{6}$

$= \dfrac{3x}{6}$

$= \dfrac{x}{2}$

77. $\left(\dfrac{2a+b}{b^2} \cdot \dfrac{3a^2-2ab}{ab+2b^2} \right) \div \dfrac{a^2-3ab+2b^2}{5ab-10b^2}$

$= \dfrac{2a+b}{b^2} \cdot \dfrac{3a^2-2ab}{ab+2b^2} \cdot \dfrac{5ab-10b^2}{a^2-3ab+2b^2}$

$= \dfrac{2a+b}{b^2} \cdot \dfrac{a(3a-2b)}{b(a+2b)} \cdot \dfrac{5 \cdot b(a-2b)}{(a-b)(a-2b)}$

$= \dfrac{5a(2a+b)(3a-2b)}{b^2(a-b)(a+2b)}$

79. answers may vary

81. $\$2000 \text{ US} = \dfrac{\$2000 \text{ US}}{1} \cdot \dfrac{1 \text{ euro}}{\$1.2955 \text{ US}}$

$= \dfrac{2000}{1.2955} \text{ euros}$

$\approx 1543.805 \text{ euros}$

On that day, $2000 US was worth 1543.81 euros.

Section 5.3 Practice Problems

1. $\dfrac{8x}{3y} + \dfrac{x}{3y} = \dfrac{8x+x}{3y} = \dfrac{9x}{3y} = \dfrac{3x}{y}$

2. $\dfrac{3x}{3x-7} - \dfrac{7}{3x-7} = \dfrac{3x-7}{3x-7} = \dfrac{1}{1}$ or 1

3. $\dfrac{2x^2+5x}{x+2} - \dfrac{4x+6}{x+2} = \dfrac{2x^2+5x-(4x+6)}{x+2}$

$= \dfrac{2x^2+5x-4x-6}{x+2}$

$= \dfrac{2x^2+x-6}{x+2}$

$= \dfrac{(2x-3)(x+2)}{x+2}$

$= 2x-3$

4. a. $\dfrac{2}{9}, \dfrac{7}{15}$

$9 = 3^2$ and $15 = 3 \cdot 5$

$\text{LCD} = 3^2 \cdot 5 = 9 \cdot 5 = 45$

b. $\dfrac{5}{6x^3}, \dfrac{11}{8x^5}$

$6x^3 = 2 \cdot 3 \cdot x^3$ and $8x^5 = 2^3 \cdot x^5$

$\text{LCD} = 3 \cdot 2^3 \cdot x^5 = 3 \cdot 8 \cdot x^5 = 24x^5$

5. $\dfrac{3a}{a+5}, \dfrac{7a}{a-5}$

$\text{LCD} = (a+5)(a-5)$

6. $\dfrac{7x^2}{(x-4)^2}, \dfrac{5x}{3x-12}$

$(x-4)^2 = (x-4)(x-4)$

and $3x - 12 = 3(x-4)$

$\text{LCD} = 3(x-4)(x-4) = 3(x-4)^2$

7. $\dfrac{y+5}{y^2+2y-3}, \dfrac{y+4}{y^2-3y+2}$

$y^2+2y-3 = (y+3)(y-1)$

$y^2-3y+2 = (y-2)(y-1)$

$\text{LCD} = (y+3)(y-2)(y-1)$

8. $\dfrac{6}{x-4}, \dfrac{9}{4-x}$

$(4-x) = -(x-4)$

$\text{LCD} = (x-4)$ or $(4-x)$

9. $\dfrac{2x}{5y} = \dfrac{2x}{5y} \cdot 1 = \dfrac{2x}{5y} \cdot \dfrac{4x^2 y}{4x^2 y} = \dfrac{8x^3 y}{20x^2 y^2}$

10. $\dfrac{3}{x^2 - 25} = \dfrac{3}{(x-5)(x+5)}$

$\qquad = \dfrac{3}{(x-5)(x+5)} \cdot \dfrac{x-3}{x-3}$

$\qquad = \dfrac{3(x-3)}{(x-5)(x+5)(x-3)}$

$\qquad = \dfrac{3x-9}{(x-5)(x+5)(x-3)}$

Mental Math

1. $\dfrac{2}{3} + \dfrac{1}{3} = \dfrac{3}{3} = 1$

2. $\dfrac{5}{11} + \dfrac{1}{11} = \dfrac{6}{11}$

3. $\dfrac{3x}{9} + \dfrac{4x}{9} = \dfrac{7x}{9}$

4. $\dfrac{3y}{8} + \dfrac{2y}{8} = \dfrac{5y}{8}$

5. $\dfrac{8}{9} - \dfrac{7}{9} = \dfrac{1}{9}$

6. $\dfrac{14}{12} - \dfrac{3}{12} = \dfrac{11}{12}$

7. $\dfrac{7y}{5} + \dfrac{10y}{5} = \dfrac{17y}{5}$

8. $\dfrac{12x}{7} - \dfrac{4x}{7} = \dfrac{8x}{7}$

Exercise Set 5.3

1. $\dfrac{a}{13} + \dfrac{9}{13} = \dfrac{a+9}{13}$

3. $\dfrac{4m}{3n} + \dfrac{5m}{3n} = \dfrac{4m+5m}{3n} = \dfrac{9m}{3n} = \dfrac{3m}{n}$

5. $\dfrac{4m}{m-6} - \dfrac{24}{m-6} = \dfrac{4m-24}{m-6} = \dfrac{4(m-6)}{m-6} = 4$

7. $\dfrac{9}{3+y} + \dfrac{y+1}{3+y} = \dfrac{9+y+1}{3+y} = \dfrac{y+10}{3+y}$

9. $\dfrac{5x^2 + 4x}{x-1} - \dfrac{2x+3}{x-1} = \dfrac{5x^2 + 4x - (2x+3)}{x-1}$

$\qquad = \dfrac{5x^2 + 2x - 3}{x-1}$

$\qquad = \dfrac{(5x-3)(x+1)}{x-1}$

11. $\dfrac{4a}{a^2 + 2a - 15} - \dfrac{12}{a^2 + 2a - 15} = \dfrac{4a-12}{a^2 + 2a - 15}$

$\qquad = \dfrac{4(a-3)}{(a+5)(a-3)}$

$\qquad = \dfrac{4}{a+5}$

13. $\dfrac{2x+3}{x^2 - x - 30} - \dfrac{x-2}{x^2 - x - 30} = \dfrac{2x+3-(x-2)}{x^2 - x - 30}$

$\qquad = \dfrac{x+5}{(x-6)(x+5)}$

$\qquad = \dfrac{1}{x-6}$

15. $2x = 2 \cdot x$

$\qquad 4x^3 = 2^2 \cdot x^3$

$\qquad \text{LCD} = 2^2 \cdot x^3 = 4x^3$

17. $8x = 2 \cdot 2 \cdot 2 \cdot x$

$\qquad 2x + 4 = 2(x+2)$

$\qquad \text{LCD} = 2 \cdot 2 \cdot 2 \cdot x(x+2) = 8x(x+2)$

19. $x + 3 = x + 3$

$\qquad x - 2 = x - 2$

$\qquad \text{LCD} = (x+3)(x-2)$

21. $x + 6 = x + 6$
$3x + 18 = 3(x + 6)$
LCD $= 3(x + 6)$

23. $(x - 6)^2 = (x - 6)(x - 6)$
$5x - 30 = 5(x - 6)$
LCD $= 5(x - 6)(x - 6) = 5(x - 6)^2$

25. $3x + 3 = 3(x + 1)$
$2x^2 + 4x + 2 = 2(x^2 + 2x + 1) = 2(x + 1)^2$
LCD $= 2 \cdot 3 \cdot (x + 1)^2 = 6(x + 1)^2$

27. $8 - x = -(x - 8)$
LCD $= x - 8$ or $8 - x$

29. $x^2 + 3x - 4 = (x - 1)(x + 4)$
$x^2 + 2x - 3 = (x - 1)(x + 3)$
LCD $= (x - 1)(x + 4)(x + 3)$

31. $3x^2 + 4x + 1 = (3x + 1)(x + 1)$
$2x^2 - x - 1 = (2x + 1)(x - 1)$
LCD $= (3x + 1)(x + 1)(2x + 1)(x - 1)$

33. $x^2 - 16 = (x + 4)(x - 4)$
$2x^3 - 8x^2 = 2x^2(x - 4)$
LCD $= 2x^2(x + 4)(x - 4)$

35. $\dfrac{3}{2x} = \dfrac{3 \cdot 2x}{2x \cdot 2x} = \dfrac{6x}{4x^2}$

37. $\dfrac{6}{3a} = \dfrac{6 \cdot 4b^2}{3a \cdot 4b^2} = \dfrac{24b^2}{12ab^2}$

39. $\dfrac{9}{2x + 6} = \dfrac{9}{2(x + 3)} = \dfrac{9y}{2(x + 3)y} = \dfrac{9y}{2y(x + 3)}$

41. $\dfrac{9a + 2}{5a + 10} = \dfrac{9a + 2}{5(a + 2)} = \dfrac{(9a + 2) \cdot b}{5(a + 2) \cdot b} = \dfrac{9ab + 2b}{5b(a + 2)}$

43. $\dfrac{x}{x^3 + 6x^2 + 8x} = \dfrac{x}{x(x^2 + 6x + 8)}$
$= \dfrac{x}{x(x + 4)(x + 2)}$
$= \dfrac{x(x + 1)}{x(x + 4)(x + 2)(x + 1)}$
$= \dfrac{x^2 + x}{x(x + 4)(x + 2)(x + 1)}$

45. $\dfrac{9y - 1}{15x^2 - 30} = \dfrac{(9y - 1) \cdot 2}{(15x^2 - 30) \cdot 2} = \dfrac{18y - 2}{30x^2 - 60}$

47. $\dfrac{5x}{7} + \dfrac{9x}{7} = \dfrac{5x + 9x}{7} = \dfrac{14x}{7} = 2x$

49. $\dfrac{x + 3}{4} \div \dfrac{2x - 1}{4} = \dfrac{x + 3}{4} \cdot \dfrac{4}{2x - 1} = \dfrac{x + 3}{2x - 1}$

51. $\dfrac{x^2}{x - 6} - \dfrac{5x + 6}{x - 6} = \dfrac{x^2 - (5x + 6)}{x - 6}$
$= \dfrac{x^2 - 5x - 6}{x - 6}$
$= \dfrac{(x - 6)(x + 1)}{x - 6}$
$= x + 1$

53. $\dfrac{-2x}{x^3 - 8x} + \dfrac{3x}{x^3 - 8x} = \dfrac{-2x + 3x}{x^3 - 8x}$
$= \dfrac{x}{x(x^2 - 8)}$
$= \dfrac{1}{x^2 - 8}$

55. $\dfrac{12x - 6}{x^2 + 3x} \cdot \dfrac{4x^2 + 13x + 3}{4x^2 - 1}$
$= \dfrac{6(2x - 1)}{x(x + 3)} \cdot \dfrac{(4x + 1)(x + 3)}{(2x - 1)(2x + 1)}$
$= \dfrac{6(4x + 1)}{x(2x + 1)}$

57. $\dfrac{2}{3}+\dfrac{5}{7}=\dfrac{2\cdot 7}{3\cdot 7}+\dfrac{5\cdot 3}{7\cdot 3}=\dfrac{14}{21}+\dfrac{15}{21}=\dfrac{29}{21}$

59. $\dfrac{2}{6}-\dfrac{3}{4}=\dfrac{2\cdot 2}{6\cdot 2}-\dfrac{3\cdot 3}{4\cdot 3}=\dfrac{4}{12}-\dfrac{9}{12}=-\dfrac{5}{12}$

61. $\begin{aligned}\dfrac{1}{12}+\dfrac{3}{20}&=\dfrac{1\cdot 5}{12\cdot 5}+\dfrac{3\cdot 3}{20\cdot 3}\\[4pt]&=\dfrac{5}{60}+\dfrac{9}{60}\\[4pt]&=\dfrac{14}{60}\\[4pt]&=\dfrac{7}{30}\end{aligned}$

63. $4a-20=4(a-5)$

$(a-5)^2=(a-5)(a-5)$

$\text{LCD}=4(a-5)(a-5)=4(a-5)^2;\ \text{d}$

65. The perimeter of a square is 4 times the side length.

$4\cdot\dfrac{5}{x-2}=\dfrac{4}{1}\cdot\dfrac{5}{x-2}=\dfrac{20}{x-2}$

The perimeter is $\dfrac{20}{x-2}$ meters.

67. answers may vary

69. $8=2\cdot 2\cdot 2$

$12=2\cdot 2\cdot 3$

$\text{LCD}=2\cdot 2\cdot 2\cdot 3=24$

$24=8\cdot 3$

$24=12\cdot 2$

You should buy 3 packages of hot dogs and 2 packages of buns.

71. answers may vary

73. answers may vary

Section 5.4 Practice Problems

1. a. $\dfrac{y}{5}-\dfrac{3y}{15}=\dfrac{y\cdot 3}{5\cdot 3}-\dfrac{3y}{15}=\dfrac{3y-3y}{15}=\dfrac{0}{15}=0$

b. $\begin{aligned}\dfrac{5}{8x}+\dfrac{11}{10x^2}&=\dfrac{5\cdot 5x}{8x\cdot 5x}+\dfrac{11\cdot 4}{10x^2\cdot 4}\\[4pt]&=\dfrac{25x}{40x^2}+\dfrac{44}{40x^2}\\[4pt]&=\dfrac{25x+44}{40x^2}\end{aligned}$

2. $\begin{aligned}\dfrac{10x}{x^2-9}&-\dfrac{5}{x+3}\\[4pt]&=\dfrac{10x}{(x-3)(x+3)}-\dfrac{5(x-3)}{(x+3)(x-3)}\\[4pt]&=\dfrac{10x-5(x-3)}{(x+3)(x-3)}\\[4pt]&=\dfrac{10x-5x+15}{(x+3)(x-3)}\\[4pt]&=\dfrac{5x+15}{(x+3)(x-3)}\\[4pt]&=\dfrac{5(x+3)}{(x+3)(x-3)}\\[4pt]&=\dfrac{5}{x-3}\end{aligned}$

3. $\begin{aligned}\dfrac{5}{7x}+\dfrac{2}{x+1}&=\dfrac{5(x+1)}{7x(x+1)}+\dfrac{2(7x)}{(x+1)(7x)}\\[4pt]&=\dfrac{5(x+1)+2(7x)}{7x(x+1)}\\[4pt]&=\dfrac{5x+5+14x}{7x(x+1)}\\[4pt]&=\dfrac{19x+5}{7x(x+1)}\end{aligned}$

4. $\begin{aligned}\dfrac{10}{x-6}-\dfrac{15}{6-x}&=\dfrac{10}{x-6}-\dfrac{15}{-(x-6)}\\[4pt]&=\dfrac{10}{x-6}-\dfrac{-15}{x-6}\\[4pt]&=\dfrac{10-(-15)}{x-6}\\[4pt]&=\dfrac{25}{x-6}\end{aligned}$

5. $2 + \dfrac{x}{x+5} = \dfrac{2}{1} + \dfrac{x}{x+5}$

$\qquad = \dfrac{2(x+5)}{1(x+5)} + \dfrac{x}{x+5}$

$\qquad = \dfrac{2x+10+x}{x+5}$

$\qquad = \dfrac{3x+10}{x+5}$

6. $\dfrac{4}{3x^2+2x} - \dfrac{3x}{12x+8}$

$\qquad = \dfrac{4}{x(3x+2)} - \dfrac{3x}{4(3x+2)}$

$\qquad = \dfrac{4(4)}{x(3x+2)(4)} - \dfrac{3x(x)}{4(3x+2)(x)}$

$\qquad = \dfrac{16-3x^2}{4x(3x+2)}$

7. $\dfrac{6x}{x^2+4x+4} + \dfrac{x}{x^2-4}$

$\qquad = \dfrac{6x}{(x+2)(x+2)} + \dfrac{x}{(x+2)(x-2)}$

$\qquad = \dfrac{6x(x-2)}{(x+2)(x+2)(x-2)} + \dfrac{x(x+2)}{(x+2)(x-2)(x+2)}$

$\qquad = \dfrac{6x(x-2)+x(x+2)}{(x+2)^2(x-2)}$

$\qquad = \dfrac{6x^2-12x+x^2+2x}{(x+2)^2(x-2)}$

$\qquad = \dfrac{7x^2-10x}{(x+2)^2(x-2)}$

$\qquad = \dfrac{x(7x-10)}{(x+2)^2(x-2)}$

Exercise Set 5.4

1. $\dfrac{4}{2x} + \dfrac{9}{3x} = \dfrac{4 \cdot 3}{2x \cdot 3} + \dfrac{9 \cdot 2}{3x \cdot 2} = \dfrac{12}{6x} + \dfrac{18}{6x} = \dfrac{30}{6x} = \dfrac{5}{x}$

3. $\dfrac{15a}{b} + \dfrac{6b}{5} = \dfrac{15a \cdot 5}{b \cdot 5} + \dfrac{6b \cdot b}{5 \cdot b}$

$\qquad = \dfrac{75a}{5b} + \dfrac{6b^2}{5b}$

$\qquad = \dfrac{75a+6b^2}{5b}$

5. $\dfrac{3}{x} + \dfrac{5}{2x^2} = \dfrac{3 \cdot 2x}{x \cdot 2x} + \dfrac{5}{2x^2}$

$\qquad = \dfrac{6x}{2x^2} + \dfrac{5}{2x^2}$

$\qquad = \dfrac{6x+5}{2x^2}$

7. $\dfrac{6}{x+1} + \dfrac{10}{2x+2} = \dfrac{6}{x+1} + \dfrac{10}{2(x+1)}$

$\qquad = \dfrac{6}{x+1} + \dfrac{5}{x+1}$

$\qquad = \dfrac{11}{x+1}$

9. $\dfrac{3}{x+2} - \dfrac{2x}{x^2-4}$

$\qquad = \dfrac{3(x-2)}{(x+2)(x-2)} - \dfrac{2x}{(x+2)(x-2)}$

$\qquad = \dfrac{3x-6-2x}{(x+2)(x-2)}$

$\qquad = \dfrac{x-6}{(x+2)(x-2)}$

11. $\dfrac{3}{4x} + \dfrac{8}{x-2} = \dfrac{3(x-2)}{4x(x-2)} + \dfrac{8(4x)}{(x-2)(4x)}$

$\qquad = \dfrac{3x-6+32x}{4x(x-2)}$

$\qquad = \dfrac{35x-6}{4x(x-2)}$

13. $\dfrac{6}{x-3}+\dfrac{8}{3-x}=\dfrac{6}{x-3}+\dfrac{8}{-(x-3)}$

$\qquad = \dfrac{6}{x-3}+\dfrac{-8}{x-3}$

$\qquad = \dfrac{6+(-8)}{x-3}$

$\qquad = \dfrac{-2}{x-3}$

$\qquad = -\dfrac{2}{x-3}$

15. $\dfrac{9}{x-3}+\dfrac{9}{3-x}=\dfrac{9}{x-3}+\dfrac{9}{-(x-3)}$

$\qquad = \dfrac{9}{x-3}+\dfrac{-9}{x-3}$

$\qquad = \dfrac{0}{x-3}$

$\qquad = 0$

17. $\dfrac{-8}{x^2-1}-\dfrac{7}{1-x^2}=\dfrac{-8}{x^2-1}-\dfrac{7}{-(x^2-1)}$

$\qquad = \dfrac{-8}{x^2-1}-\dfrac{-7}{x^2-1}$

$\qquad = \dfrac{-8-(-7)}{x^2-1}$

$\qquad = \dfrac{-1}{x^2-1}$

$\qquad = -\dfrac{1}{x^2-1}$

19. $\dfrac{5}{x}+2=\dfrac{5}{x}+\dfrac{2}{1}=\dfrac{5}{x}+\dfrac{2\cdot x}{1\cdot x}=\dfrac{5+2x}{x}$

21. $\dfrac{5}{x-2}+6=\dfrac{5}{x-2}+\dfrac{6}{1}$

$\qquad = \dfrac{5}{x-2}+\dfrac{6(x-2)}{1(x-2)}$

$\qquad = \dfrac{5+6x-12}{x-2}$

$\qquad = \dfrac{6x-7}{x-2}$

23. $\dfrac{y+2}{y+3}-2=\dfrac{y+2}{y+3}-\dfrac{2}{1}$

$\qquad = \dfrac{y+2}{y+3}-\dfrac{2(y+3)}{1(y+3)}$

$\qquad = \dfrac{y+2}{y+3}-\dfrac{2y+6}{y+3}$

$\qquad = \dfrac{y+2-(2y+6)}{y+3}$

$\qquad = \dfrac{y+2-2y-6}{y+3}$

$\qquad = \dfrac{-y-4}{y+3}$

$\qquad = \dfrac{-(y+4)}{y+3}$

$\qquad = -\dfrac{y+4}{y+3}$

25. $\dfrac{-x+2}{x}-\dfrac{x-6}{4x}=\dfrac{4(-x+2)}{4x}-\dfrac{x-6}{4x}$

$\qquad = \dfrac{-4x+8-(x-6)}{4x}$

$\qquad = \dfrac{-4x+8-x+6}{4x}$

$\qquad = \dfrac{-5x+14}{4x} \text{ or } -\dfrac{5x-14}{4x}$

27. $\dfrac{5x}{x+2}-\dfrac{3x-4}{x+2}=\dfrac{5x-(3x-4)}{x+2}$

$\qquad = \dfrac{5x-3x+4}{x+2}$

$\qquad = \dfrac{2x+4}{x+2}$

$\qquad = \dfrac{2(x+2)}{x+2}$

$\qquad = 2$

29. $\dfrac{3x^4}{7}-\dfrac{4x^2}{21}=\dfrac{3x^4\cdot 3}{7\cdot 3}-\dfrac{4x^2}{21}$

$\qquad = \dfrac{9x^4}{21}-\dfrac{4x^2}{21}$

$\qquad = \dfrac{9x^4-4x^2}{21}$

31. $\dfrac{1}{x+3} - \dfrac{1}{(x+3)^2}$

$= \dfrac{1 \cdot (x+3)}{(x+3)(x+3)} - \dfrac{1}{(x+3)(x+3)}$

$= \dfrac{x+3-1}{(x+3)^2}$

$= \dfrac{x+2}{(x+3)^2}$

33. $\dfrac{4}{5b} + \dfrac{1}{b-1} = \dfrac{4(b-1)}{5b(b-1)} + \dfrac{1 \cdot 5b}{(b-1)(5b)}$

$= \dfrac{4b-4}{5b(b-1)} + \dfrac{5b}{5b(b-1)}$

$= \dfrac{4b-4+5b}{5b(b-1)}$

$= \dfrac{9b-4}{5b(b-1)}$

35. $\dfrac{2}{m} + 1 = \dfrac{2}{m} + \dfrac{m}{m} = \dfrac{2+m}{m}$

37. $\dfrac{2x}{x-7} - \dfrac{x}{x-2}$

$= \dfrac{2x(x-2)}{(x-7)(x-2)} - \dfrac{x(x-7)}{(x-2)(x-7)}$

$= \dfrac{2x^2 - 4x - (x^2 - 7x)}{(x-7)(x-2)}$

$= \dfrac{2x^2 - 4x - x^2 + 7x}{(x-7)(x-2)}$

$= \dfrac{x^2 + 3x}{(x-7)(x-2)}$ or $\dfrac{x(x+3)}{(x-7)(x-2)}$

39. $\dfrac{6}{1-2x} - \dfrac{4}{2x-1} = \dfrac{6}{1-2x} - \dfrac{4}{-(1-2x)}$

$= \dfrac{6}{1-2x} - \dfrac{-4}{1-2x}$

$= \dfrac{6+4}{1-2x}$

$= \dfrac{10}{1-2x}$

41. $\dfrac{7}{(x+1)(x-1)} + \dfrac{8}{(x+1)^2}$

$= \dfrac{7(x+1)}{(x+1)(x-1)(x+1)} + \dfrac{8(x-1)}{(x+1)^2(x-1)}$

$= \dfrac{7x+7+8x-8}{(x+1)^2(x-1)}$

$= \dfrac{15x-1}{(x+1)^2(x-1)}$

43. $\dfrac{x}{x^2-1} - \dfrac{2}{x^2-2x+1}$

$= \dfrac{x(x-1)}{(x-1)(x+1)(x-1)} - \dfrac{2(x+1)}{(x-1)^2(x+1)}$

$= \dfrac{x^2 - x - 2x - 2}{(x-1)^2(x+1)}$

$= \dfrac{x^2 - 3x - 2}{(x-1)^2(x+1)}$

45. $\dfrac{3a}{2a+6} - \dfrac{a-1}{a+3} = \dfrac{3a}{2(a+3)} - \dfrac{(a-1)(2)}{(a+3)(2)}$

$= \dfrac{3a - (2a-2)}{2(a+3)}$

$= \dfrac{3a - 2a + 2}{2(a+3)}$

$= \dfrac{a+2}{2(a+3)}$

47. $\dfrac{y-1}{2y+3} + \dfrac{3}{(2y+3)^2}$

$= \dfrac{(y-1)(2y+3)}{(2y+3)^2} + \dfrac{3}{(2y+3)^2}$

$= \dfrac{2y^2 + 3y - 2y - 3 + 3}{(2y+3)^2}$

$= \dfrac{2y^2 + y}{(2y+3)^2}$

$= \dfrac{y(2y+1)}{(2y+3)^2}$

49. $\dfrac{5}{2-x} + \dfrac{x}{2x-4} = \dfrac{5}{-(x-2)} + \dfrac{x}{2(x-2)}$

$= \dfrac{-5(2)}{(x-2)(2)} + \dfrac{x}{2(x-2)}$

$= \dfrac{-10+x}{2(x-2)}$

$= \dfrac{x-10}{2(x-2)}$

51. $\dfrac{15}{x^2+6x-19} + \dfrac{2}{x+3}$

$= \dfrac{15}{(x+3)^2} + \dfrac{2(x+3)}{(x+3)(x+3)}$

$= \dfrac{15+2x+6}{(x+3)^2}$

$= \dfrac{2x+21}{(x+3)^2}$

53. $\dfrac{13}{x^2-5x+6} - \dfrac{5}{x-3}$

$= \dfrac{13}{(x-3)(x-2)} - \dfrac{5(x-2)}{(x-3)(x-2)}$

$= \dfrac{13-(5x-10)}{(x-3)(x-2)}$

$= \dfrac{13-5x+10}{(x-3)(x-2)}$

$= \dfrac{-5x+23}{(x-3)(x-2)}$

55. $\dfrac{70}{m^2-100} + \dfrac{7}{2(m+10)}$

$= \dfrac{70\cdot 2}{(m-10)(m+10)2} + \dfrac{7(m-10)}{2(m+10)(m-10)}$

$= \dfrac{140+7m-70}{2(m-10)(m+10)}$

$= \dfrac{7m+70}{2(m-10)(m+10)}$

$= \dfrac{7(m+10)}{2(m-10)(m+10)}$

$= \dfrac{7}{2(m-10)}$

57. $\dfrac{x+8}{x^2-5x-6} + \dfrac{x+1}{x^2-4x-5}$

$= \dfrac{x+8}{(x+1)(x-6)} + \dfrac{x+1}{(x+1)(x-5)}$

$= \dfrac{(x+8)(x-5)+(x+1)(x-6)}{(x+1)(x-6)(x-5)}$

$= \dfrac{x^2+3x-40+x^2-5x-6}{(x+1)(x-6)(x-5)}$

$= \dfrac{2x^2-2x-46}{(x+1)(x-6)(x-5)}$

59. $\dfrac{5}{4n^2-12n+8} - \dfrac{3}{3n^2-6n}$

$= \dfrac{5\cdot 3n}{4(n-2)(n-1)3n} - \dfrac{3(n-1)\cdot 4}{3n(n-2)(n-1)4}$

$= \dfrac{15n-12n+12}{12n(n-2)(n-1)}$

$= \dfrac{3n+12}{12n(n-2)(n-1)}$

$= \dfrac{3(n+4)}{12n(n-2)(n-1)}$

$= \dfrac{n+4}{4n(n-2)(n-1)}$

61. $\dfrac{15x}{x+8} \cdot \dfrac{2x+16}{3x} = \dfrac{5\cdot 3x}{x+8} \cdot \dfrac{2(x+8)}{3x} = \dfrac{5\cdot 2}{1} = 10$

63. $\dfrac{8x+7}{3x+5} - \dfrac{2x-3}{3x+5} = \dfrac{8x+7-(2x-3)}{3x+5}$

$= \dfrac{8x+7-2x+3}{3x+5}$

$= \dfrac{6x+10}{3x+5}$

$= \dfrac{2(3x+5)}{3x+5}$

$= 2$

65.

$$\frac{5a+10}{18} \div \frac{a^2-4}{10a} = \frac{5a+10}{18} \cdot \frac{10a}{a^2-4}$$

$$= \frac{5(a+2)}{9\cdot 2} \cdot \frac{2\cdot 5a}{(a+2)(a-2)}$$

$$= \frac{5\cdot 5a}{9(a-2)}$$

$$= \frac{25a}{9(a-2)}$$

67.

$$\frac{5}{x^2-3x+2} + \frac{1}{x-2} = \frac{5}{(x-2)(x-1)} + \frac{1(x-1)}{(x-2)(x-1)}$$

$$= \frac{5+x-1}{(x-2)(x-1)}$$

$$= \frac{x+4}{(x-2)(x-1)}$$

69. $3x+5=7$

$$3x=2$$

$$x=\frac{2}{3}$$

71. $\quad 2x^2-x-1=0$

$$(2x+1)(x-1)=0$$

$$2x+1=0 \quad \text{or} \quad x-1=0$$

$$2x=-1 \qquad\qquad x=1$$

$$x=-\frac{1}{2}$$

73. $4(x+6)+3=-3$

$$4x+24+3=-3$$

$$4x+27=-3$$

$$4x=-30$$

$$x=\frac{-30}{4}$$

$$x=-\frac{15}{2}$$

75. $\dfrac{3}{x} - \dfrac{2x}{x^2-1} + \dfrac{5}{x+1} = \dfrac{3}{x} - \dfrac{2x}{(x-1)(x+1)} + \dfrac{5}{x+1}$

$$= \frac{3(x-1)(x+1)}{x(x-1)(x+1)} - \frac{2x \cdot x}{(x-1)(x+1)x} + \frac{5 \cdot x(x-1)}{(x+1)x(x-1)}$$

$$= \frac{3x^2 - 3 - 2x^2 + 5x^2 - 5x}{x(x-1)(x+1)}$$

$$= \frac{6x^2 - 5x - 3}{x(x-1)(x+1)}$$

77. $\dfrac{5}{x^2-4} + \dfrac{2}{x^2-4x+4} - \dfrac{3}{x^2-x-6} = \dfrac{5}{(x+2)(x-2)} + \dfrac{2}{(x-2)^2} - \dfrac{3}{(x+2)(x-3)}$

$$= \frac{5(x-2)(x-3) + 2(x+2)(x-3) - 3(x-2)^2}{(x+2)(x-2)^2(x-3)}$$

$$= \frac{5(x^2-5x+6) + 2(x^2-x-6) - 3(x^2-4x+4)}{(x-2)^2(x+2)(x-3)}$$

$$= \frac{5x^2 - 25x + 30 + 2x^2 - 2x - 12 - 3x^2 + 12x - 12}{(x-2)^2(x+2)(x-3)}$$

$$= \frac{4x^2 - 15x + 6}{(x-2)^2(x+2)(x-3)}$$

79. $\dfrac{9}{x^2+9x+14} - \dfrac{3x}{x^2+10x+21} + \dfrac{x+4}{x^2+5x+6}$

$$= \frac{9}{(x+2)(x+7)} - \frac{3x}{(x+3)(x+7)} + \frac{x+4}{(x+2)(x+3)}$$

$$= \frac{9(x+3)}{(x+2)(x+7)(x+3)} - \frac{3x(x+2)}{(x+3)(x+7)(x+2)} + \frac{(x+4)(x+7)}{(x+2)(x+3)(x+7)}$$

$$= \frac{9x + 27 - 3x^2 - 6x + x^2 + 7x + 4x + 28}{(x+2)(x+7)(x+3)}$$

$$= \frac{-2x^2 + 14x + 55}{(x+2)(x+7)(x+3)}$$

81. $\dfrac{3}{x+4} - \dfrac{1}{x-4} = \dfrac{3(x-4) - 1(x+4)}{(x+4)(x-4)}$

$$= \frac{3x - 12 - x - 4}{(x+4)(x-4)}$$

$$= \frac{2x - 16}{(x+4)(x-4)}$$

The other piece of the board measures $\dfrac{2x-16}{(x+4)(x-4)}$ inches.

83. $1 - \dfrac{G}{P} = \dfrac{P}{P} - \dfrac{G}{P} = \dfrac{P-G}{P}$

85. answers may vary

87. $90 - \dfrac{40}{x} = \dfrac{90}{1} - \dfrac{40}{x}$

$= \dfrac{90x}{x} - \dfrac{40}{x}$

$= \dfrac{90x - 40}{x}$

The complement measures $\left(\dfrac{90x - 40}{x}\right)^{\circ}$.

89. answers may vary

The Bigger Picture

1. $-8.6 + (-9.1) = -17.7$

2. $(-8.6)(-9.1) = 78.26$

3. $14 - (-14) = 14 + 14 = 28$

4. $3x^4 - 7 + x^4 - x^2 - 10 = 4x^4 - x^2 - 17$

5. $\dfrac{5x^2 - 5}{25x + 25} = \dfrac{5(x^2 - 1)}{25(x+1)} = \dfrac{5(x-1)(x+1)}{5 \cdot 5(x+1)} = \dfrac{x-1}{5}$

6. $\dfrac{7x}{x^2 + 4x + 3} \div \dfrac{x}{2x + 6}$

$= \dfrac{7x}{x^2 + 4x + 3} \cdot \dfrac{2x + 6}{x}$

$= \dfrac{7x}{(x+3)(x+1)} \cdot \dfrac{2(x+3)}{x}$

$= \dfrac{14}{x+1}$

7. $\dfrac{2}{9} - \dfrac{5}{6} = \dfrac{2 \cdot 2}{9 \cdot 2} - \dfrac{5 \cdot 3}{6 \cdot 3} = \dfrac{4 - 15}{18} = -\dfrac{11}{18}$

8. $\dfrac{x}{9} - \dfrac{x+3}{5} = \dfrac{x \cdot 5}{9 \cdot 5} - \dfrac{(x+3) \cdot 9}{5 \cdot 9}$

$= \dfrac{5x}{45} - \dfrac{9x + 27}{45}$

$= \dfrac{5x - 9x - 27}{45}$

$= \dfrac{-4x - 27}{45}$ or $-\dfrac{4x + 27}{45}$

9. $9x^3 - 2x^2 - 11x = x(9x^2 - 2x - 11)$
$= x(9x - 11)(x + 1)$

10. $12xy - 21x + 4y - 7 = 3x(4y - 7) + 1(4y - 7)$
$= (4y - 7)(3x + 1)$

11. $7x - 14 = 5x + 10$
$7x = 5x + 24$
$2x = 24$
$x = 12$

12. $\dfrac{-x + 2}{5} < \dfrac{3}{10}$

$-x + 2 < \dfrac{3}{2}$

$-x < -\dfrac{1}{2}$

$x > \dfrac{1}{2}$

$\left\{ x \middle| x > \dfrac{1}{2} \right\}$

13. $1 + 4(x + 4) = 3^2 + x$
$1 + 4x + 16 = 9 + x$
$4x + 17 = 9 + x$
$3x + 17 = 9$
$3x = -8$
$x = -\dfrac{8}{3}$

14. $x(x-2) = 24$

$x^2 - 2x - 24 = 0$

$(x-6)(x+4) = 0$

$x-6 = 0$ or $x+4 = 0$

     $x = 6$ $x = -4$

Section 5.5 Practice Problems

1. The LCD is 20.

$$\frac{x}{4} + \frac{4}{5} = \frac{1}{20}$$

$$20\left(\frac{x}{4} + \frac{4}{5}\right) = 20\left(\frac{1}{20}\right)$$

$$20\left(\frac{x}{4}\right) + 20\left(\frac{4}{5}\right) = 20\left(\frac{1}{20}\right)$$

$$5x + 16 = 1$$

$$5x = -15$$

$$x = -3$$

The solution is -3.

2. The LCD is 15.

$$\frac{x+2}{3} - \frac{x-1}{5} = \frac{1}{15}$$

$$15\left(\frac{x+2}{3} - \frac{x-1}{5}\right) = 15\left(\frac{1}{15}\right)$$

$$15\left(\frac{x+2}{3}\right) - 15\left(\frac{x-1}{5}\right) = 15\left(\frac{1}{15}\right)$$

$$5(x+2) - 3(x-1) = 1$$

$$5x + 10 - 3x + 3 = 1$$

$$2x + 13 = 1$$

$$2x = -12$$

$$x = -6$$

The solution is -6.

3. The LCD is x.

$$2 + \frac{6}{x} = x + 7$$

$$x\left(2 + \frac{6}{x}\right) = x(x + 7)$$

$$x(2) + x\left(\frac{6}{x}\right) = x \cdot x + x \cdot 7$$

$$2x + 6 = x^2 + 7x$$

$$0 = x^2 + 5x - 6$$

$$0 = (x + 6)(x - 1)$$

$$x + 6 = 0 \quad \text{or} \quad x - 1 = 0$$

$$x = -6 \qquad \qquad x = 1$$

Neither -6 nor 1 makes the denominator 0 in the original equation. Both -6 and 1 are solutions.

4. The LCD is $(x + 3)(x - 3)$.

$$\frac{2}{x+3} + \frac{3}{x-3} = \frac{-2}{x^2 - 9}$$

$$(x+3)(x-3)\left(\frac{2}{x+3} + \frac{3}{x-3}\right) = (x+3)(x-3)\left(\frac{-2}{x^2-9}\right)$$

$$(x+3)(x-3) \cdot \frac{2}{x+3} + (x+3)(x-3) \cdot \frac{3}{x-3} = (x+3)(x-3)\left(\frac{-2}{x^2-9}\right)$$

$$2(x-3) + 3(x+3) = -2$$

$$2x - 6 + 3x + 9 = -2$$

$$5x + 3 = -2$$

$$5x = -5$$

$$x = -1$$

The solution is -1.

5. The LCD is $x - 1$.

$$\frac{5x}{x-1} = \frac{5}{x-1} + 3$$

$$(x-1)\left(\frac{5x}{x-1}\right) = (x-1)\left(\frac{5}{x-1} + 3\right)$$

$$(x-1) \cdot \frac{5x}{x-1} = (x-1) \cdot \frac{5}{x-1} + (x-1) \cdot 3$$

$$5x = 5 + 3(x-1)$$

$$5x = 5 + 3x - 3$$

$$2x = 2$$

$$x = 1$$

1 makes the denominator 0 in the original equation. Therefore 1 is *not* a solution and this equation has no solution.

6. The LCD is $x + 3$.

$$x - \frac{6}{x+3} = \frac{2x}{x+3} + 2$$

$$(x+3)\left(x - \frac{6}{x+3}\right) = (x+3)\left(\frac{2x}{x+3} + 2\right)$$

$$(x+3)(x) - (x+3)\left(\frac{6}{x+3}\right) = (x+3)\left(\frac{2x}{x+3}\right) + (x+3)(2)$$

$$(x+3)(x) - 6 = 2x + 2(x+3)$$

$$x^2 + 3x - 6 = 2x + 2x + 6$$

$$x^2 - x - 12 = 0$$

$$(x-4)(x+3) = 0$$

$$x - 4 = 0 \quad \text{or} \quad x + 3 = 0$$
$$x = 4 \qquad\qquad x = 3$$

$x = 3$ can't be a solution of the original equation. The only solution is 4.

7. The LCD is abx.

$$\frac{1}{a} + \frac{1}{b} = \frac{1}{x}$$

$$abx\left(\frac{1}{a} + \frac{1}{b}\right) = abx\left(\frac{1}{x}\right)$$

$$abx\left(\frac{1}{a}\right) + abx\left(\frac{1}{b}\right) = abx\left(\frac{1}{x}\right)$$

$$bx + ax = ab$$

$$bx = ab - ax$$

$$bx = a(b - x)$$

$$\frac{bx}{b-x} = \frac{a(b-x)}{b-x}$$

$$\frac{bx}{b-x} = a$$

Mental Math

1. $\dfrac{x}{5} = 2$

$$5\left(\frac{x}{5}\right) = 5(2)$$

$$x = 10$$

2. $\dfrac{x}{8} = 4$

$$8\left(\frac{x}{8}\right) = 8(4)$$

$$x = 32$$

3. $\dfrac{z}{6} = 6$

$$6\left(\dfrac{z}{6}\right) = 6(6)$$
$$z = 36$$

4. $\dfrac{y}{7} = 8$

$$7\left(\dfrac{y}{7}\right) = 7(8)$$
$$y = 56$$

Exercise Set 5.5

1. The LCD is 5.

$$\dfrac{x}{5} + 3 = 9$$
$$5\left(\dfrac{x}{5} + 3\right) = 5(9)$$
$$x + 15 = 45$$
$$x = 30$$

Check: $\dfrac{x}{5} + 3 = 9$

$$\dfrac{30}{5} + 3 \stackrel{?}{=} 9$$
$$6 + 3 \stackrel{?}{=} 9$$
$$9 = 9 \quad \text{True}$$

The solution is 30.

3. The LCD is 12.

$$\dfrac{x}{2} + \dfrac{5x}{4} = \dfrac{x}{12}$$
$$12\left(\dfrac{x}{2} + \dfrac{5x}{4}\right) = 12\left(\dfrac{x}{12}\right)$$
$$6x + 15x = x$$
$$21x = x$$
$$20x = 0$$
$$x = 0$$

Check: $\dfrac{x}{2} + \dfrac{5x}{4} = \dfrac{x}{12}$

$$\dfrac{0}{2} + \dfrac{5(0)}{4} \stackrel{?}{=} \dfrac{0}{12}$$
$$0 + 0 \stackrel{?}{=} 0$$
$$0 = 0 \quad \text{True}$$

The solution is 0.

5. The LCD is x.

$$2 - \dfrac{8}{x} = 6$$
$$x\left(2 - \dfrac{8}{x}\right) = x(6)$$
$$2x - 8 = 6x$$
$$-8 = 4x$$
$$-2 = x$$

Check: $2 - \dfrac{8}{x} = 6$

$$2 - \dfrac{8}{-2} \stackrel{?}{=} 6$$
$$2 - (-4) \stackrel{?}{=} 6$$
$$2 + 4 = 6$$
$$6 = 6 \quad \text{True}$$

The solution is -2.

7. The LCD is x.

$$2 + \dfrac{10}{x} = x + 5$$
$$x\left(2 + \dfrac{10}{x}\right) = x(x + 5)$$
$$2x + 10 = x^2 + 5x$$
$$0 = x^2 + 3x - 10$$
$$0 = (x + 5)(x - 2)$$
$$x + 5 = 0 \quad \text{or} \quad x - 2 = 0$$
$$x = -5 \qquad\qquad x = 2$$

Check -5: $2 + \dfrac{10}{x} = x + 5$

$$2 + \dfrac{10}{-5} \stackrel{?}{=} -5 + 5$$
$$2 + (-2) \stackrel{?}{=} -5 + 5$$
$$0 = 0 \quad \text{True}$$

Check 2: $2 + \dfrac{10}{x} = x + 5$

$$2 + \dfrac{10}{2} \stackrel{?}{=} 2 + 5$$
$$2 + 5 \stackrel{?}{=} 2 + 5$$
$$7 = 7 \quad \text{True}$$

The solutions are -5 and 2.

9. The LCD is $5 \cdot 2 = 10$.

$$\frac{a}{5} = \frac{a-3}{2}$$

$$10\left(\frac{a}{5}\right) = 10\left(\frac{a-3}{2}\right)$$

$$2a = 5(a-3)$$

$$2a = 5a-15$$

$$-3a = -15$$

$$a = 5$$

Check: $\dfrac{a}{5} = \dfrac{a-3}{2}$

$$\frac{5}{5} \stackrel{?}{=} \frac{5-3}{2}$$

$$1 \stackrel{?}{=} \frac{2}{2}$$

$$1 = 1 \quad \text{True}$$

The solution is 5.

11. The LCD is $5 \cdot 2 = 10$.

$$\frac{x-3}{5} + \frac{x-2}{2} = \frac{1}{2}$$

$$10\left(\frac{x-3}{5} + \frac{x-2}{2}\right) = 10\left(\frac{1}{2}\right)$$

$$2(x-3) + 5(x-2) = \frac{10}{2}$$

$$2x-6+5x-10 = 5$$

$$7x-16 = 5$$

$$7x = 21$$

$$x = 3$$

Check: $\dfrac{x-3}{5} + \dfrac{x-2}{2} = \dfrac{1}{2}$

$$\frac{3-3}{5} + \frac{3-2}{2} \stackrel{?}{=} \frac{1}{2}$$

$$\frac{0}{5} + \frac{1}{2} \stackrel{?}{=} \frac{1}{2}$$

$$\frac{1}{2} = \frac{1}{2} \quad \text{True}$$

The solution is 3.

13. The LCD is $2a - 5$.

$$\frac{3}{2a-5} = -1$$

$$(2a-5)\left(\frac{3}{2a-5}\right) = (2a-5)(-1)$$

$$3 = -2a+5$$

$$-2 = -2a$$

$$1 = a$$

Check: $\dfrac{3}{2a-5} = -1$

$$\frac{3}{2(1)-5} \stackrel{?}{=} -1$$

$$\frac{3}{2-5} \stackrel{?}{=} -1$$

$$\frac{3}{-3} \stackrel{?}{=} -1$$

$$-1 = -1 \quad \text{True}$$

The solution is 1.

15. The LCD is $y - 4$.

$$\frac{4y}{y-4} + 5 = \frac{5y}{y-4}$$

$$(y-4)\left(\frac{4y}{y-4} + 5\right) = (y-4)\left(\frac{5y}{y-4}\right)$$

$$4y+5(y-4) = 5y$$

$$4y+5y-20 = 5y$$

$$9y-20 = 5y$$

$$-20 = -4y$$

$$5 = y$$

Check: $\dfrac{4y}{y-4} + 5 = \dfrac{5y}{y-4}$

$$\frac{4 \cdot 5}{5-4} + 5 \stackrel{?}{=} \frac{5 \cdot 5}{5-4}$$

$$\frac{20}{1} + 5 \stackrel{?}{=} \frac{25}{1}$$

$$25 = 25 \quad \text{True}$$

The solution is 5.

17. The LCD is $a - 3$.

$$2 + \frac{3}{a-3} = \frac{a}{a-3}$$

$$(a-3)\left(2 + \frac{3}{a-3}\right) = (a-3)\frac{a}{a-3}$$

$$2(a-3) + 3 = a$$

$$2a - 6 + 3 = a$$

$$2a - 3 = a$$

$$-3 = -a$$

$$3 = a$$

Check: $\ 2 + \dfrac{3}{a-3} = \dfrac{a}{a-3}$

$$2 + \frac{3}{3-3} \stackrel{?}{=} \frac{3}{3-3}$$

$$2 + \frac{3}{0} \stackrel{?}{=} \frac{3}{0}$$

Since $\dfrac{3}{0}$ is undefined, $a = 3$ does not check

and the equation has no solution.

19. The LCD is $(x+3)(x-3) = x^2 - 9$.

$$\frac{1}{x+3} + \frac{6}{x^2 - 9} = 1$$

$$(x^2 - 9)\left(\frac{1}{x+3} + \frac{6}{x^2-9}\right) = 1(x^2 - 9)$$

$$(x-3) + 6 = x^2 - 9$$

$$x + 3 = x^2 - 9$$

$$0 = x^2 - x - 12$$

$$0 = (x-4)(x+3)$$

$$x - 4 = 0 \quad \text{or} \quad x + 3 = 0$$

$$x = 4 \qquad\qquad x = -3$$

Check 4: $\ \dfrac{1}{x+3} + \dfrac{6}{x^2 - 9} = 1$

$$\frac{1}{4+3} + \frac{6}{4^2 - 9} \stackrel{?}{=} 1$$

$$\frac{1}{7} + \frac{6}{16 - 9} \stackrel{?}{=} 1$$

$$\frac{1}{7} + \frac{6}{7} \stackrel{?}{=} 1$$

$$\frac{7}{7} = 1 \quad \text{True}$$

Check -3: $\qquad \dfrac{1}{x+3} + \dfrac{6}{x^2-9} = 1$

$$\frac{1}{-3+3} + \frac{6}{(-3)^2 - 9} \stackrel{?}{=} 1$$

$$\frac{1}{0} + \frac{6}{0} \stackrel{?}{=} 1$$

Since $\dfrac{1}{0}$ and $\dfrac{6}{0}$ are undefined, $x = -3$ does

not check. The solution is 4.

21. The LCD is $y + 4$.

$$\frac{2y}{y+4} + \frac{4}{y+4} = 3$$

$$(y+4)\left(\frac{2y}{y+4} + \frac{4}{y+4}\right) = (y+4)(3)$$

$$2y + 4 = 3y + 12$$

$$4 = y + 12$$

$$-8 = y$$

Check: $\qquad \dfrac{2y}{y+4} + \dfrac{4}{y+4} = 3$

$$\frac{2(-8)}{-8+4} + \frac{4}{-8+4} \stackrel{?}{=} 3$$

$$\frac{-16}{-4} + \frac{4}{-4} \stackrel{?}{=} 3$$

$$4 - 1 \stackrel{?}{=} 3$$

$$3 = 3 \quad \text{True}$$

The solution is -8.

23. The LCD is $(x+2)(x-2) = x^2 - 4$.

$$\frac{2x}{x+2} - 2 = \frac{x-8}{x-2}$$

$$(x^2 - 4)\left(\frac{2x}{x+2} - 2\right) = (x^2 - 4)\left(\frac{x-8}{x-2}\right)$$

$$2x(x-2) - 2(x^2 - 4) = (x+2)(x-8)$$

$$2x^2 - 4x - 2x^2 + 8 = x^2 - 8x + 2x - 16$$

$$-4x + 8 = x^2 - 6x - 16$$

$$0 = x^2 - 2x - 24$$

$$0 = (x-6)(x+4)$$

$$x - 6 = 0 \quad \text{or} \quad x + 4 = 0$$

$$x = 6 \qquad\qquad x = -4$$

Check 6: $\dfrac{2x}{x+2} - 2 = \dfrac{x-8}{x-2}$

$\dfrac{2(6)}{6+2} - 2 \stackrel{?}{=} \dfrac{6-8}{6-2}$

$\dfrac{12}{8} - 2 \stackrel{?}{=} \dfrac{-2}{4}$

$\dfrac{3}{2} - \dfrac{4}{2} \stackrel{?}{=} -\dfrac{1}{2}$

$-\dfrac{1}{2} = -\dfrac{1}{2}$ True

Check -4: $\dfrac{2x}{x+2} - 2 = \dfrac{x-8}{x-2}$

$\dfrac{2(-4)}{-4+2} - 2 \stackrel{?}{=} \dfrac{-4-8}{-4-2}$

$\dfrac{-8}{-2} - 2 \stackrel{?}{=} \dfrac{-12}{-6}$

$4 - 2 \stackrel{?}{=} 2$

$2 = 2$ True

The solutions are 6 and -4.

25. The LCD is $2y$.

$\dfrac{2}{y} + \dfrac{1}{2} = \dfrac{5}{2y}$

$2y\left(\dfrac{2}{y} + \dfrac{1}{2}\right) = 2y\left(\dfrac{5}{2y}\right)$

$2(2) + y(1) = 5$

$4 + y = 5$

$y = 1$

The solution $y = 1$ checks.

27. The LCD is $(a-6)(a-1)$.

$\dfrac{a}{a-6} = \dfrac{-2}{a-1}$

$(a-6)(a-1)\left(\dfrac{a}{a-6}\right) = (a-6)(a-1)\left(\dfrac{-2}{a-1}\right)$

$a(a-1) = -2(a-6)$

$a^2 - a = -2a + 12$

$a^2 + a - 12 = 0$

$(a+4)(a-3) = 0$

$a+4 = 0$ or $a-3 = 0$

$a = -4$ $a = 3$

The solutions $a = -4$ and $a = 3$ check.

29. The LCD is $2x \cdot 3 = 6x$.

$\dfrac{11}{2x} + \dfrac{2}{3} = \dfrac{7}{2x}$

$6x\left(\dfrac{11}{2x} + \dfrac{2}{3}\right) = 6x\left(\dfrac{7}{2x}\right)$

$3(11) + 2x(2) = 3(7)$

$33 + 4x = 21$

$4x = -12$

$x = -3$

The solution $x = -3$ checks.

31. The LCD is $(x-2)(x+2)$.

$\dfrac{2}{x-2} + 1 = \dfrac{x}{x+2}$

$(x-2)(x+2)\left(\dfrac{2}{x-2} + 1\right) = (x-2)(x+2)\left(\dfrac{x}{x+2}\right)$

$2(x+2) + 1(x-2)(x+2) = x(x-2)$

$2x + 4 + x^2 - 4 = x^2 - 2x$

$4x = 0$

$x = 0$

The solution $x = 0$ checks.

33. The LCD is 6.

$\dfrac{x+1}{3} - \dfrac{x-1}{6} = \dfrac{1}{6}$

$6\left(\dfrac{x+1}{3} - \dfrac{x-1}{6}\right) = 6\left(\dfrac{1}{6}\right)$

$2(x+1) - (x-1) = 1$

$2x + 2 - x + 1 = 1$

$x + 3 = 1$

$x = -2$

The solution $x = -2$ checks.

35. The LCD is $6(t-4)$.

$\dfrac{t}{t-4} = \dfrac{t+4}{6}$

$6(t-4)\left(\dfrac{t}{t-4}\right) = 6(t-4)\left(\dfrac{t+4}{6}\right)$

$6t = (t-4)(t+4)$

$6t = t^2 - 16$

$0 = t^2 - 6t - 16$

$0 = (t-8)(t+2)$

$$t - 8 = 0 \quad \text{or} \quad t + 2 = 0$$
$$t = 8 \qquad\qquad t = -2$$

The solutions $t = 8$ and $t = -2$ check.

37. $2y + 2 = 2(y + 1)$

$4y + 4 = 4(y + 1) = 2 \cdot 2(y + 1)$

The LCD is $4(y + 1)$.

$$\frac{y}{2y+2} + \frac{2y-16}{4y+4} = \frac{2y-3}{y+1}$$

$$4(y+1)\left(\frac{y}{2(y+1)} + \frac{2y-16}{4(y+1)} \right) = 4(y+1)\left(\frac{2y-3}{y+1} \right)$$

$$2y + 2y - 16 = 4(2y - 3)$$
$$4y - 16 = 8y - 12$$
$$-16 = 4y - 12$$
$$-4 = 4y$$
$$-1 = y$$

The solution $y = -1$ makes the denominators $2y + 2$, $4y + 4$, and $y + 1$ zero, so the equation has no solution.

39. $r^2 + 5r - 14 = (r + 7)(r - 2)$

The LCD is $(r + 7)(r - 2)$.

$$\frac{4r - 4}{r^2 + 5r - 14} + \frac{2}{r+7} = \frac{1}{r-2}$$

$$(r+7)(r-2)\left(\frac{4r-4}{(r+7)(r-2)} + \frac{2}{r+7} \right) = (r+7)(r-2)\left(\frac{1}{r-2} \right)$$

$$4r - 4 + 2(r - 2) = 1(r + 7)$$
$$4r - 4 + 2r - 4 = r + 7$$
$$6r - 8 = r + 7$$
$$5r = 15$$
$$r = 3$$

The solution $r = 3$ checks.

41. $x^2 + x - 6 = (x+3)(x-2)$

The LCD is $(x+3)(x-2)$.

$$\frac{x+1}{x+3} = \frac{x^2 - 11x}{x^2 + x - 6} - \frac{x-3}{x-2}$$

$$(x+3)(x-2)\left(\frac{x+1}{x+3}\right) = (x+3)(x-2)\left(\frac{x^2-11x}{x^2+x-6} - \frac{x-3}{x-2}\right)$$

$$(x-2)(x+1) = x^2 - 11x - (x+3)(x-3)$$

$$x^2 - x - 2 = x^2 - 11x - (x^2 - 9)$$

$$x^2 - x - 2 = x^2 - 11x - x^2 + 9$$

$$x^2 - x - 2 = -11x + 9$$

$$x^2 + 10x - 11 = 0$$

$$(x+11)(x-1) = 0$$

$$x + 11 = 0 \quad \text{or} \quad x - 1 = 0$$

$$x = -11 \qquad \qquad x = 1$$

The solutions $x = -11$ and $x = 1$ check.

43. The LCD is I.

$$R = \frac{E}{I}$$

$$I(R) = I\left(\frac{E}{I}\right)$$

$$IR = E$$

$$I = \frac{E}{R}$$

45. The LCD is $B + E$.

$$T = \frac{2U}{B+E}$$

$$(B+E)T = (B+E)\left(\frac{2U}{B+E}\right)$$

$$BT + ET = 2U$$

$$BT = 2U - ET$$

$$B = \frac{2U - ET}{T}$$

47. The LCD is h^2.

$$B = \frac{705w}{h^2}$$

$$h^2(B) = h^2\left(\frac{705w}{h^2}\right)$$

$$Bh^2 = 705w$$

$$\frac{Bh^2}{705} = w$$

49. The LCD is G.

$$N = R + \frac{V}{G}$$

$$G(N) = G\left(R + \frac{V}{G}\right)$$

$$GN = GR + V$$

$$GN - GR = V$$

$$G(N - R) = V$$

$$G = \frac{V}{N - R}$$

51. The LCD is πr.

$$\frac{C}{\pi r} = 2$$

$$\pi r\left(\frac{C}{\pi r}\right) = \pi r(2)$$

$$C = 2\pi r$$

$$\frac{C}{2\pi} = r$$

53. The LCD is $3xy$.

$$\frac{1}{y} + \frac{1}{3} = \frac{1}{x}$$

$$3xy\left(\frac{1}{y} + \frac{1}{3}\right) = 3xy\left(\frac{1}{x}\right)$$

$$3x + xy = 3y$$

$$x(3 + y) = 3y$$

$$x = \frac{3y}{3 + y}$$

55. The reciprocal of x is $\dfrac{1}{x}$.

57. The reciprocal of x, added to the reciprocal of 2 is $\dfrac{1}{x}+\dfrac{1}{2}$.

59. $\dfrac{1}{3}$ of the tank is filled in 1 hour.

61. $a^2+4a+3=(a+3)(a+1)$

$a^2+a-6=(a+3)(a-2)$

$a^2-a-2=(a+1)(a-2)$

The LCD is $(a+3)(a+1)(a-2)$.

$$\frac{4}{a^2+4a+3}+\frac{2}{a^2+a-6}-\frac{3}{a^2-a-2}=0$$

$$(a+3)(a+1)(a-2)\left(\frac{4}{a^2+4a+3}+\frac{2}{a^2+a-6}-\frac{3}{a^2-a-2}\right)=(a+3)(a+1)(a-2)(0)$$

$$4(a-2)+2(a+1)-3(a+3)=0$$

$$4a-8+2a+2-3a-9=0$$

$$3a-15=0$$

$$3a=15$$

$$a=5$$

The solution $a = 5$ checks.

63. $\dfrac{20x}{3}+\dfrac{32x}{6}=180$

The LCD is 6.

$$6\left(\frac{20x}{3}+\frac{32x}{6}\right)=6(180)$$

$$2(20x)+32x=1080$$

$$40x+32x=1080$$

$$72x=1080$$

$$x=15$$

$$\frac{20x}{3}=\frac{20(15)}{3}=100°$$

$$\frac{32x}{6}=\frac{32(15)}{6}=80°$$

The angles measure 100° and 80°.

65. $\dfrac{450}{x}+\dfrac{150}{x}=90$

The LCD is x.

$x\left(\dfrac{450}{x}+\dfrac{150}{x}\right)=x(90)$

$450+150=90x$

$600=90x$

$\dfrac{600}{90}=x$

$\dfrac{20}{3}=x$

$\dfrac{450}{x}=450\div\dfrac{20}{3}=\dfrac{450}{1}\cdot\dfrac{3}{20}=\dfrac{1350}{20}=67.5$

$\dfrac{150}{x}=150\div\dfrac{20}{3}=\dfrac{150}{1}\cdot\dfrac{3}{20}=\dfrac{450}{20}=22.5$

The angles measure $22.5°$ and $67.5°$.

67. No; multiplying both terms in the expressions by 4 changes the value of the original expressions.

Integrated Review

1. $\dfrac{1}{x}+\dfrac{2}{3}=\dfrac{1\cdot3}{x\cdot3}+\dfrac{2\cdot x}{3\cdot x}=\dfrac{3}{3x}+\dfrac{2x}{3x}=\dfrac{3+2x}{3x}$

This is an expression.

2. $\dfrac{3}{a}+\dfrac{5}{6}=\dfrac{3\cdot6}{a\cdot6}+\dfrac{5\cdot a}{6\cdot a}=\dfrac{18}{6a}+\dfrac{5a}{6a}=\dfrac{18+5a}{6a}$

This is an expression.

3. $\dfrac{1}{x}+\dfrac{2}{3}=\dfrac{3}{x}$

$3x\left(\dfrac{1}{x}+\dfrac{2}{3}\right)=3x\left(\dfrac{3}{x}\right)$

$3+2x=9$

$2x=6$

$x=3$

This is an equation.

4. $\dfrac{3}{a}+\dfrac{5}{6}=1$

$6a\left(\dfrac{3}{a}+\dfrac{5}{6}\right)=6a(1)$

$18+5a=6a$

$18=a$

This is an equation.

5. $\dfrac{2}{x+1}-\dfrac{1}{x}=\dfrac{2\cdot x}{(x+1)\cdot x}-\dfrac{1\cdot(x+1)}{x\cdot(x+1)}$

$=\dfrac{2x}{x(x+1)}-\dfrac{x+1}{x(x+1)}$

$=\dfrac{2x-x-1}{x(x+1)}$

$=\dfrac{x-1}{x(x+1)}$

This is an expression.

6. $\dfrac{4}{x-3}-\dfrac{1}{x}=\dfrac{4\cdot x}{(x-3)\cdot x}-\dfrac{1\cdot(x-3)}{x\cdot(x-3)}$

$=\dfrac{4x}{x(x-3)}-\dfrac{x-3}{x(x-3)}$

$=\dfrac{4x-x+3}{x(x-3)}$

$=\dfrac{3x+3}{x(x-3)}$

$=\dfrac{3(x+1)}{x(x-3)}$

This is an expression.

7. $\dfrac{2}{x+1}-\dfrac{1}{x}=1$

$x(x+1)\left(\dfrac{2}{x+1}-\dfrac{1}{x}\right)=x(x+1)(1)$

$2x-(x+1)=x(x+1)$

$2x-x-1=x^2+x$

$x-1=x^2+x$

$0=x^2+1$

There are no solutions. This is an equation.

8.
$$\frac{4}{x-3} - \frac{1}{x} = \frac{6}{x(x-3)}$$
$$x(x-3)\left(\frac{4}{x-3} - \frac{1}{x}\right) = x(x-3)\left(\frac{6}{x(x-3)}\right)$$
$$4x - 1(x-3) = 6$$
$$4x - x + 3 = 6$$
$$3x = 3$$
$$x = 1$$
This is an equation.

9.
$$\frac{15x}{x+8} \cdot \frac{2x+16}{3x} = \frac{3 \cdot 5x}{x+8} \cdot \frac{2(x+8)}{3x} = 5 \cdot 2 = 10$$
This is an expression.

10.
$$\frac{9z+5}{15} \cdot \frac{5z}{81z^2 - 25} = \frac{9z+5}{3 \cdot 5} \cdot \frac{5z}{(9z+5)(9z-5)}$$
$$= \frac{z}{3(9z-5)}$$

This is an expression.

11.
$$\frac{2x+3}{x-3} + \frac{3x+6}{x-3} = \frac{2x+1+3x+6}{x-3} = \frac{5x+7}{x-3}$$
This is an expression.

12.
$$\frac{4p-3}{2p+7} + \frac{3p+8}{2p+7} = \frac{4p-3+3p+8}{2p+7} = \frac{7p+5}{2p+7}$$
This is an expression.

13.
$$\frac{x+5}{7} = \frac{8}{2}$$
$$14\left(\frac{x+5}{7}\right) = 14\left(\frac{8}{2}\right)$$
$$2(x+5) = 7(8)$$
$$2x + 10 = 56$$
$$2x = 46$$
$$x = 23$$
This is an equation.

14.
$$\frac{1}{2} = \frac{x+1}{8}$$
$$8\left(\frac{1}{2}\right) = 8\left(\frac{x+1}{8}\right)$$
$$4 = x + 1$$
$$3 = x$$
This is an equation.

15.
$$\frac{5a+10}{18} \div \frac{a^2 - 4}{10a} = \frac{5a+10}{18} \cdot \frac{10a}{a^2 - 4}$$
$$= \frac{5(a+2)}{2 \cdot 9} \cdot \frac{2 \cdot 5a}{(a+2)(a-2)}$$
$$= \frac{5 \cdot 5a}{9(a-2)}$$
$$= \frac{25a}{9(a-2)}$$

This is an expression.

16.
$$\frac{9}{x^2 - 1} \div \frac{12}{3x+3} = \frac{9}{x^2 - 1} \cdot \frac{3x+3}{12}$$
$$= \frac{9}{(x-1)(x+1)} \cdot \frac{3(x+1)}{3 \cdot 4}$$
$$= \frac{9}{4(x-1)}$$

17.
$$\frac{x+2}{3x-1} + \frac{5}{(3x-1)^2}$$
$$= \frac{(x+2)(3x-1)}{(3x-1)(3x-1)} + \frac{5}{(3x-1)^2}$$
$$= \frac{(x+2)(3x-1)+5}{(3x-1)^2}$$
$$= \frac{3x^2 - x + 6x - 2 + 5}{(3x-1)^2}$$
$$= \frac{3x^2 + 5x + 3}{(3x-1)^2}$$
This is an expression.

18. $\dfrac{4}{(2x-5)^2} + \dfrac{x+1}{2x-5}$

$= \dfrac{4}{(2x-5)^2} + \dfrac{(x+1)(2x-5)}{(2x-5)(2x-5)}$

$= \dfrac{4+(x+1)(2x-5)}{(2x-5)^2}$

$= \dfrac{4+2x^2-5x+2x-5}{(2x-5)^2}$

$= \dfrac{2x^2-3x-1}{(2x-5)^2}$

This is an expression.

19. $\dfrac{x-7}{x} - \dfrac{x+2}{5x} = \dfrac{(x-7)\cdot 5}{x\cdot 5} - \dfrac{x+2}{5x}$

$= \dfrac{5x-35-x-2}{5x}$

$= \dfrac{4x-37}{5x}$

This is an expression.

20. $\dfrac{9}{x^2-4} + \dfrac{2}{x+2} = \dfrac{-1}{x-2}$

$(x^2-4)\left(\dfrac{9}{x^2-4} + \dfrac{2}{x+2}\right) = (x^2-4)\left(\dfrac{-1}{x-2}\right)$

$9 + 2(x-2) = -1(x+2)$

$9 + 2x - 4 = -x - 2$

$2x + 5 = -x - 2$

$3x = -7$

$x = -\dfrac{7}{3}$

This is an equation.

21. $\dfrac{3}{x+3} = \dfrac{5}{x^2-9} - \dfrac{2}{x-3}$

$(x^2-9)\left(\dfrac{3}{x+3}\right) = (x^2-9)\left(\dfrac{5}{x^2-9} - \dfrac{2}{x-3}\right)$

$3(x-3) = 5 - 2(x+3)$

$3x - 9 = 5 - 2x - 6$

$3x - 9 = -2x - 1$

$5x = 8$

$x = \dfrac{8}{5}$

This is an equation.

22. $\dfrac{10x-9}{x} - \dfrac{x-4}{3x} = \dfrac{(10x-9)\cdot 3}{x\cdot 3} - \dfrac{x-4}{3x}$

$= \dfrac{3(10x-9)-x+4}{3x}$

$= \dfrac{30x-27-x+4}{3x}$

$= \dfrac{29x-23}{3x}$

This is an expression.

23. answers may vary

24. answers may vary

Section 5.6 Practice Problems

1. $\dfrac{3}{8} = \dfrac{63}{x}$

$3 \cdot x = 63 \cdot 8$

$3x = 504$

$x = 168$

2. $\dfrac{2x+1}{7} = \dfrac{x-3}{5}$

$5(2x+1) = 7(x-3)$

$10x + 5 = 7x - 21$

$3x = -26$

$x = -\dfrac{26}{3}$

3.
$$\frac{26}{250} = \frac{x}{50,000}$$
$$26(50,000) = 250x$$
$$1,300,000 = 250x$$
$$5200 = x$$
5200 people are expected to have a flu shot.

4.
$$\frac{12}{x} = \frac{9}{15}$$
$$12 \cdot 15 = 9 \cdot x$$
$$180 = 9x$$
$$20 = x$$
The missing length is 20 units.

5.
$$\frac{x}{2} - \frac{1}{3} = \frac{x}{6}$$
$$6\left(\frac{x}{2} - \frac{1}{3}\right) = 6\left(\frac{x}{6}\right)$$
$$3x - 2 = x$$
$$2x = 2$$
$$x = 1$$
The number is 1.

6.

	Hours to Complete Total Job	Part of Job Completed in 1 Hour
Andrew	2	$\frac{1}{2}$
Tim	3	$\frac{1}{3}$
Together	x	$\frac{1}{x}$

$$\frac{1}{2} + \frac{1}{3} = \frac{1}{x}$$
$$6x\left(\frac{1}{2}\right) + 6x\left(\frac{1}{3}\right) = 6x\left(\frac{1}{x}\right)$$
$$3x + 2x = 6$$
$$5x = 6$$
$$x = \frac{6}{5} = 1\frac{1}{5} \text{ hr}$$

Together they can sort one batch in $1\frac{1}{5}$ hours.

7.

	distance	=	rate	·	time
Car	600		$x + 15$		$\frac{600}{x+15}$
Motorcycle	450		x		$\frac{450}{x}$

$$\frac{600}{x+15} = \frac{450}{x}$$
$$600x = 450(x+15)$$
$$600x = 450x + 6750$$
$$150x = 6750$$
$$x = 45$$
$$x + 15 = 45 + 15 = 60$$

The speed of the motorcycle is 45 mph. The speed of the car is 60 mph.

Mental Math

1. The time to complete the job working together will be less than both of the individual times. The answer is c.

2. The time to fill the pond with both pipes on at the same time will be less than both of the individual times. The answer is a.

Exercise Set 5.6

1. The LCD is 6.
$$\frac{2}{3} = \frac{x}{6}$$
$$6\left(\frac{2}{3}\right) = 6\left(\frac{x}{6}\right)$$
$$4 = x$$

3. $\dfrac{x}{10} = \dfrac{5}{9}$
$$9x = 50$$
$$x = \frac{50}{9}$$

5. $\dfrac{x+1}{2x+3} = \dfrac{2}{3}$
$$3(x+1) = 2(2x+3)$$
$$3x+3 = 4x+6$$
$$3 = x+6$$
$$-3 = x$$

7. $\dfrac{9}{5} = \dfrac{12}{3x+2}$
$$9(3x+2) = 12 \cdot 5$$
$$27x+18 = 60$$
$$27x = 42$$
$$x = \dfrac{42}{27} = \dfrac{14}{9}$$

9. Let x be the elephant's weight on Pluto.
$$\dfrac{100}{3} = \dfrac{4100}{x}$$
$$100x = 3(4100)$$
$$100x = 12{,}300$$
$$x = 123$$
The elephant weighs 123 pounds on Pluto.

11. Let y be the number of calories in 42.6 grams.
$$\dfrac{110}{28.4} = \dfrac{y}{42.6}$$
$$42.6(110) = 28.4y$$
$$4686 = 28.4y$$
$$165 = y$$
There are 165 calories in 42.6 grams of Crispy Rice.

13. $\dfrac{16}{10} = \dfrac{34}{y}$
$$16y = 34(10)$$
$$16y = 340$$
$$y = \dfrac{340}{16}$$
$$y = 21.25$$

15. $\dfrac{y}{8} = \dfrac{20}{28}$
$$28y = 20(8)$$
$$28y = 160$$
$$y = \dfrac{160}{28} = \dfrac{40}{7} = 5\dfrac{5}{7} \text{ ft}$$

17. Let x be the number.
$$3\left(\dfrac{1}{x}\right) = 9\left(\dfrac{1}{6}\right)$$
$$\dfrac{3}{x} = \dfrac{9}{6}$$
$$3(6) = 9x$$
$$18 = 9x$$
$$2 = x$$
The number is 2.

19. Let x be the number.
$$\dfrac{2x+3}{x+1} = \dfrac{3}{2}$$
$$2(2x+3) = 3(x+1)$$
$$4x+6 = 3x+3$$
$$x = -3$$
The number is -3.

21. Let x be the time in hours that it takes them to complete the job working together.

The experienced surveyor completes $\dfrac{1}{4}$ of the job in 1 hour. The apprentice surveyor completes $\dfrac{1}{5}$ of the job in 1 hour. Together, they complete $\dfrac{1}{x}$ of the job in 1 hour.

$$\dfrac{1}{4} + \dfrac{1}{5} = \dfrac{1}{x}$$
The LCD is $4 \cdot 5 \cdot x = 20x$.
$$20x\left(\dfrac{1}{4} + \dfrac{1}{5}\right) = 20x\left(\dfrac{1}{x}\right)$$
$$5x + 4x = 20$$
$$9x = 20$$
$$x = \dfrac{20}{9}$$

It takes them $\dfrac{20}{9} = 2\dfrac{2}{9}$ hours to survey the roadbed together.

23. Let x be the time in minutes that it takes the two belts to complete the job working together. The first belt completes $\frac{1}{2}$ of the job in 1 minute. The smaller belt completes $\frac{1}{6}$ of the job in 1 minute. Together, they complete $\frac{1}{x}$ of the job in 1 minute.

$$\frac{1}{2}+\frac{1}{6}=\frac{1}{x}$$

The LCD is $6x$.

$$6x\left(\frac{1}{2}+\frac{1}{6}\right)=6x\left(\frac{1}{x}\right)$$
$$3x+x=6$$
$$4x=6$$
$$x=\frac{6}{4}=\frac{3}{2}=1\frac{1}{2}$$

It will take $1\frac{1}{2}$ minutes to move the cans to the storage when both belts are used.

25. Let r be her jogging rate.

	distance	=	rate	·	time
Trip to park	12		r		$\frac{12}{r}$
Return trip	18		r		$\frac{18}{r}$

Since the return trip took 1 hour longer, $\frac{18}{r}=1+\frac{12}{r}$. The LCD is r.

$$r\left(\frac{18}{r}\right)=r\left(1+\frac{12}{r}\right)$$
$$18=r+12$$
$$6=r$$

Her jogging rate is 6 miles per hour.

27. Let x be the speed for the first portion.

	distance	=	rate	·	time
1st portion	20		x		$\frac{20}{x}$
2nd portion	16		$x-2$		$\frac{16}{x-2}$

$$\frac{20}{x}=\frac{16}{x-2}$$

The LCD is $x(x-2)$.

$$x(x-2)\left(\frac{20}{x}\right) = x(x-2)\left(\frac{16}{x-2}\right)$$

$$20(x-2) = 16x$$
$$20x - 40 = 16x$$
$$4x = 40$$
$$x = 10$$

The cyclist's speed for first portion is 10 mph and the speed for the second portion is 8 mph.

29. $40 \text{ students} \cdot \dfrac{9 \text{ square feet}}{1 \text{ student}} = 360 \text{ square feet}$

40 students need a minimum of 360 square feet.

31. Let n be the number.

$$\frac{1}{4} = \frac{n}{8}$$

$$8\left(\frac{1}{4}\right) = 8\left(\frac{n}{8}\right)$$

$$2 = n$$

The number is 2.

33. Let x be the time in hours that it takes Marcus and Tony to do the job working together.

Marcus lays $\dfrac{1}{6}$ of a slab in 1 hour. Tony lays $\dfrac{1}{4}$ of a slab in 1 hour. Together, they lay $\dfrac{1}{x}$ of the slab in 1 hour.

$$\frac{1}{6} + \frac{1}{4} = \frac{1}{x}$$

The LCD is $12x$.

$$12x\left(\frac{1}{6} + \frac{1}{4}\right) = 12x\left(\frac{1}{x}\right)$$

$$2x + 3x = 12$$
$$5x = 12$$
$$x = \frac{12}{5}$$

It will take Tony and Marcus $\dfrac{12}{5}$ hours to lay the slab, so the labor estimate should be $\dfrac{12}{5}(\$45) = \$108$.

35. Let w be the speed of the wind.

	distance	=	rate	·	time
With wind	400		$230 + w$		$\dfrac{400}{230+w}$
Against wind	336		$230 - w$		$\dfrac{336}{230-w}$

$$\frac{400}{230+w} = \frac{336}{230-w}$$
$$400(230-w) = 336(230+w)$$
$$92{,}000 - 400w = 77{,}280 + 336w$$
$$14{,}720 = 736w$$
$$20 = w$$

The speed of the wind is 20 mph.

37. $\dfrac{2}{3} = \dfrac{25}{y}$

$2y = 3(25)$

$2y = 75$

$y = \dfrac{75}{2}$

The unknown length is $y = \dfrac{75}{2} = 37.5$ feet.

39. $\dfrac{4045}{12} \approx 337$

His average is approximately 337 yards per game.

41. Let x be the number.

$$\frac{2}{x-3} - \frac{4}{x+3} = 8 \cdot \frac{1}{x^2-9}$$

The LCD is $x^2 - 9 = (x+3)(x-3)$.

$$(x+3)(x-3)\left(\frac{2}{x-3} - \frac{4}{x+3}\right) = (x^2-9)\left(\frac{8}{x^2-9}\right)$$
$$2(x+3) - 4(x-3) = 8$$
$$2x + 6 - 4x + 12 = 8$$
$$-2x + 18 = 8$$
$$-2x = -10$$
$$x = 5$$

The solution $x = 5$ checks, so the number is 5.

43. Let x be the rate in still air.

	distance	=	rate	$\cdot$	time
With wind	630		$x + 35$		$\frac{630}{x+35}$
Against wind	450		$x - 35$		$\frac{450}{x-35}$

$$\frac{630}{x+35} = \frac{455}{x-35}$$
$$630(x-35) = 455(x+35)$$
$$630x - 22,050 = 455x + 15,925$$
$$175x = 37,975$$
$$x = 217$$

The plane flies at a rate of 217 mph in still air.

45. Let x be the number of gallons of water needed.

$$\frac{8 \text{ tsp}}{2 \text{ gal}} = \frac{36 \text{ tsp}}{x \text{ gal}}$$
$$\frac{8}{2} = \frac{36}{x}$$
$$8x = 36(2)$$
$$8x = 72$$
$$x = 9$$

9 gallons of water are needed to mix with a box of weed killer.

47. Let x be the rate of the wind.

	distance	=	rate	·	time
With wind	48		$16 + x$		$\frac{48}{16+x}$
Into wind	16		$16 - x$		$\frac{16}{16-x}$

$$\frac{48}{16+x} = \frac{16}{16-x}$$
$$42(16-x) = 16(16+x)$$
$$768 - 48x = 256 + 16x$$
$$512 = 64x$$
$$8 = x$$

The rate of the wind is 8 mph.

49. Let x be the rate of the slower hiker. Then the rate of the faster hiker is $x + 1.1$. In 2 hours, the slower hiker walks $2x$ miles, while the faster hiker walks $2(x + 1.1)$ miles.

$$2x + 2(x+1.1) = 11$$
$$2x + 2x + 2.2 = 11$$
$$4x + 2.2 = 11$$
$$4x = 8.8$$
$$x = 2.2$$
$$x + 1.1 = 2.2 + 1.1 = 3.3$$

The hikers walk 2.2 miles per hour and 3.3 miles per hour.

51. Let x be the time it takes for the second worker to do the same job alone.

$$\frac{1}{3} + \frac{1}{x} = \frac{1}{\frac{3}{2}}$$

$$\frac{1}{3} + \frac{1}{x} = \frac{2}{3}$$

$$3x\left(\frac{1}{3} + \frac{1}{x}\right) = 3x\left(\frac{2}{3}\right)$$

$$x + 3 = 2x$$

$$3 = x$$

It will take the second worker 3 hours to get the job done.

53. $$\frac{20 \text{ feet}}{6 \text{ inches}} = \frac{x \text{ feet}}{8 \text{ inches}}$$

$$\frac{20}{6} = \frac{x}{8}$$

$$8(20) = 6x$$

$$160 = 6x$$

$$\frac{160}{6} = x$$

$$\frac{80}{3} = x$$

The missing dimension is $\frac{80}{3} = 26\frac{2}{3}$ feet.

55. Let x be the number of other nuts.

$$\frac{3}{2} = \frac{324}{x}$$

$$3x = 2(324)$$

$$3x = 648$$

$$x = 216$$

There are 216 other nuts in the can.

57. Let t be the time in hours that the jet plane travels.

	distance	=	rate	·	time
jet plane	500t		500		t
propeller plane	200(t + 2)		200		t + 2

$500t = 200(t + 2)$

$500t = 200t + 400$

$300t = 400$

$t = \dfrac{400}{300}$

$t = \dfrac{4}{3}$

distance $= 500t = 500\left(\dfrac{4}{3}\right) = 666\dfrac{2}{3}$

The planes are $666\dfrac{2}{3}$ miles from the starting point.

59. Let x be the time that it takes the third pipe to fill the pool alone.

$$\dfrac{1}{20} + \dfrac{1}{15} + \dfrac{1}{x} = \dfrac{1}{6}$$

$$60x\left(\dfrac{1}{20} + \dfrac{1}{15} + \dfrac{1}{x}\right) = 60x\left(\dfrac{1}{6}\right)$$

$$3x + 4x + 60 = 10x$$

$$7x + 60 = 10x$$

$$60 = 3x$$

$$20 = x$$

It will take the third pump 20 hours to do the job alone.

61. Let r be the motorcycle's speed.

	distance	=	rate	·	time
car	280		$r + 10$		$\frac{280}{r+10}$
motorcycle	240		r		$\frac{240}{r}$

$\dfrac{280}{r+10} = \dfrac{240}{r}$

$280r = 240(r + 10)$

$280r = 240r + 2400$

$40r = 2400$

$r = 60$

$r + 10 = 60 + 10 = 70$

The motorcycle's speed was 60 miles per hour and the car's speed was 70 miles per hour.

63. Let x be the time for the third cook to prepare the same number of pies.

$$\frac{1}{6}+\frac{1}{7}+\frac{1}{x}=\frac{1}{2}$$

$$42x\left(\frac{1}{6}+\frac{1}{7}+\frac{1}{x}\right)=42x\left(\frac{1}{2}\right)$$

$$7x+6x+42=21x$$

$$13x+42=21x$$

$$42=8x$$

$$\frac{42}{8}=x$$

$$\frac{42}{8}=\frac{21}{4}=5\frac{1}{4}$$

It will take the third cook $5\frac{1}{4}$ hours to prepare the pies working alone.

65. Let x be the time in minutes that it takes for the faster pump to fill the tank, so it fills $\frac{1}{x}$ of the tank in 1 minute. It takes the slower pump $3x$ minutes to fill the tank, so the slower pump fills $\frac{1}{3x}$ of the tank in 1 minute. Together, the pumps fill $\frac{1}{21}$ of the tank in 1 minute.

$$\frac{1}{x}+\frac{1}{3x}=\frac{1}{21}$$

$$21x\left(\frac{1}{x}+\frac{1}{3x}\right)=21x\left(\frac{1}{21}\right)$$

$$21+7=x$$

$$28=x$$

$3x=3(28)=84$

The faster pump fills the tank in 28 minutes, while the slower pump takes 84 minutes.

67. $\dfrac{x}{4}=\dfrac{14}{7}$

$7x=4(14)$

$7x=56$

$x=8$

The missing length is $x=8$.

69. $\dfrac{14}{7}=\dfrac{7}{y}$

$14y=7(7)$

$14y=49$

$y=\dfrac{49}{14}$

$y=\dfrac{7}{2}$

The missing length is $y=\dfrac{7}{2}=3.5$.

71. $\dfrac{\frac{9}{5}+\frac{6}{5}}{\frac{17}{6}+\frac{7}{6}}=\dfrac{\frac{15}{5}}{\frac{24}{6}}=\dfrac{3}{4}$

73. $\dfrac{\frac{1}{4}+\frac{5}{4}}{\frac{3}{8}+\frac{7}{8}}=\dfrac{\frac{6}{4}}{\frac{3}{8}+\frac{7}{8}}$

$$=\dfrac{\frac{6}{4}}{\frac{10}{8}}$$

$$=\frac{6}{4}\div\frac{10}{8}$$

$$=\frac{6}{4}\cdot\frac{8}{10}$$

$$=\frac{2\cdot3\cdot2\cdot4}{4\cdot2\cdot5}$$

$$=\frac{6}{5}$$

75. $\dfrac{2-x}{5}=\dfrac{1+x}{3}$

$3(2-x)=5(1+x)$

$6-3x=5+5x$

$1=8x$

$\dfrac{1}{8}=x$

The answer is a.

77. $D=RT$

$\dfrac{D}{T}=\dfrac{RT}{T}$

$\dfrac{D}{T}=R$ or $R=\dfrac{D}{T}$

79. Let t be the time it takes for the hyena to overtake the giraffe.

$$0.5 + 32t = 40t$$
$$0.5 = 8t$$
$$0.0625 = t$$

$$0.0625 \text{ hr} \cdot \frac{60 \text{ min}}{1 \text{ hr}} = 3.75 \text{ min}$$

It will take the hyena 3.75 minutes to overtake the giraffe.

The Bigger Picture

1. $(3x-2)(4x^2 - x - 5)$

$= 3x(4x^2 - x - 5) + (-2)(4x^2 - x - 5)$

$= 12x^3 - 3x^2 - 15x - 8x^2 + 2x + 10$

$= 12x^3 - 11x^2 - 13x + 10$

2. $(2x - y)^2 = (2x)^2 - 2(2x)(y) + (y)^2$

$= 4x^2 - 4xy + y^2$

3. $8y^3 - 20y^5 = 4y^3(2 - 5y^2)$

4. $9m^2 - 11mn + 2n^2 = (9m - 2n)(m - n)$

5.
$$\frac{7}{x} = \frac{9}{x-10}$$
$$7(x-10) = 9x$$
$$7x - 70 = 9x$$
$$-70 = 2x$$
$$-35 = x$$

6. The LCD is $x(x - 10)$.

$$\frac{7}{x} + \frac{9}{x-10} = \frac{7(x-10)}{x(x-10)} + \frac{9x}{(x-10)x}$$

$$= \frac{7(x-10) + 9x}{x(x-10)}$$

$$= \frac{7x - 70 + 9x}{x(x-10)}$$

$$= \frac{16x - 70}{x(x-10)}$$

$$= \frac{2(8x - 35)}{x(x-10)}$$

7. $(-3x^5)\left(\frac{1}{2}x^7\right)(8x) = -12x^{13}$

8. $5x - 1 = |-4| + |-5|$

$5x - 1 = 4 + 5$

$5x - 1 = 9$

$5x = 10$

$x = 2$

9. $\dfrac{8-12}{12 \div 3 \cdot 2} = \dfrac{8-12}{4 \cdot 2} = \dfrac{-4}{8} = -\dfrac{1}{2}$

10. $-2(3y - 4) \le 5y - 7 - 7y - 1$

$-6y + 8 \le -2y - 8$

$16 \le 4y$

$4 \le y$

$\{y | y \ge 4\}$

11. $\dfrac{7}{x} + \dfrac{5}{2x+3} = \dfrac{-2}{x}$

The LCD is $x(2x + 3)$.

$$x(2x+3)\left(\frac{7}{x} + \frac{5}{2x+3}\right) = x(2x+3)\left(\frac{-2}{x}\right)$$

$$7(2x+3) + 5x = -2(2x+3)$$
$$14x + 21 + 5x = -4x - 6$$
$$19x + 21 = -4x - 6$$
$$23x = -27$$
$$x = -\frac{27}{23}$$

12. $\dfrac{(a^{-3}b^2)^{-5}}{ab^4} = \dfrac{a^{15}b^{-10}}{ab^4} = \dfrac{a^{15}a^{-1}}{b^{10}b^4} = \dfrac{a^{14}}{b^{14}}$

Section 5.7 Practice Problems

1. $\dfrac{\frac{3}{7}}{\frac{5}{9}} = \dfrac{3}{7} \div \dfrac{5}{9} = \dfrac{3}{7} \cdot \dfrac{9}{5} = \dfrac{27}{35}$

2. $\dfrac{\frac{3}{4}-\frac{2}{3}}{\frac{1}{2}+\frac{3}{8}} = \dfrac{\frac{3(3)}{4(3)}-\frac{2(4)}{3(4)}}{\frac{1(4)}{2(4)}+\frac{3}{8}}$

$= \dfrac{\frac{9}{12}-\frac{8}{12}}{\frac{4}{8}+\frac{3}{8}}$

$= \dfrac{\frac{1}{12}}{\frac{7}{8}}$

$= \dfrac{1}{12}\cdot\dfrac{8}{7}$

$= \dfrac{1\cdot2\cdot4}{3\cdot4\cdot7}$

$= \dfrac{2}{21}$

3. $\dfrac{\frac{2}{5}-\frac{1}{x}}{\frac{2x}{15}-\frac{1}{3}} = \dfrac{\frac{2(x)}{5(x)}-\frac{1(5)}{x(5)}}{\frac{2x}{15}-\frac{1(5)}{3(5)}}$

$= \dfrac{\frac{2x}{5x}-\frac{5}{5x}}{\frac{2x}{15}-\frac{5}{15}}$

$= \dfrac{\frac{2x-5}{5x}}{\frac{2x-5}{15}}$

$= \dfrac{2x-5}{5x}\cdot\dfrac{15}{2x-5}$

$= \dfrac{2x-15}{5x}\cdot\dfrac{3\cdot5}{2x-15}$

$= \dfrac{3}{x}$

4. The LCD is 24.

$\dfrac{\frac{3}{4}-\frac{2}{3}}{\frac{1}{2}+\frac{3}{8}} = \dfrac{24\left(\frac{3}{4}-\frac{2}{3}\right)}{24\left(\frac{1}{2}+\frac{3}{8}\right)}$

$= \dfrac{24\left(\frac{3}{4}\right)-24\left(\frac{2}{3}\right)}{24\left(\frac{1}{2}\right)+24\left(\frac{3}{8}\right)}$

$= \dfrac{18-16}{12+9}$

$= \dfrac{2}{21}$

5. The LCD is y.

$\dfrac{1+\frac{x}{y}}{\frac{2x+1}{y}} = \dfrac{y\left(1+\frac{x}{y}\right)}{y\left(\frac{2x+1}{y}\right)} = \dfrac{y(1)+y\left(\frac{x}{y}\right)}{y\left(\frac{2x+1}{y}\right)} = \dfrac{y+x}{2x+1}$

6. The LCD is $6xy$.

$\dfrac{\frac{5}{6y}+\frac{y}{x}}{\frac{y}{3}-x} = \dfrac{6xy\left(\frac{5}{6y}+\frac{y}{x}\right)}{6xy\left(\frac{y}{3}-x\right)}$

$= \dfrac{6xy\left(\frac{5}{6y}\right)+6xy\left(\frac{y}{x}\right)}{6xy\left(\frac{y}{3}\right)-6xy(x)}$

$= \dfrac{5x+6y^2}{2xy^2-6x^2y}$

$= \dfrac{5x+6y^2}{2xy(y-3x)}$

Exercise Set 5.7

1. $\dfrac{\frac{1}{2}}{\frac{3}{4}} = \dfrac{1}{2}\div\dfrac{3}{4} = \dfrac{1}{2}\cdot\dfrac{4}{3} = \dfrac{1\cdot4}{2\cdot3} = \dfrac{2}{3}$

3. $\dfrac{-\frac{4x}{9}}{-\frac{2x}{3}} = -\dfrac{4x}{9}\div-\dfrac{2x}{3} = -\dfrac{4x}{9}\cdot\dfrac{3}{-2x} = \dfrac{4x\cdot3}{9\cdot2x} = \dfrac{2}{3}$

5. $\dfrac{\frac{1+x}{6}}{\frac{1+x}{3}} = \dfrac{1+x}{6}\div\dfrac{1+x}{3}$

$= \dfrac{1+x}{6}\cdot\dfrac{3}{1+x}$

$= \dfrac{3(1+x)}{6(1+x)}$

$= \dfrac{1}{2}$

7. $\dfrac{\frac{1}{2}+\frac{2}{3}}{\frac{5}{9}-\frac{5}{6}} = \dfrac{18\left(\frac{1}{2}+\frac{2}{3}\right)}{18\left(\frac{5}{9}-\frac{5}{6}\right)} = \dfrac{9+12}{10-15} = \dfrac{21}{-5} = -\dfrac{21}{5}$

9. $\dfrac{2+\frac{7}{10}}{1+\frac{3}{5}} = \dfrac{10\left(2+\frac{7}{10}\right)}{10\left(1+\frac{3}{5}\right)} = \dfrac{20+7}{10+6} = \dfrac{27}{16}$

11. $\dfrac{\frac{1}{3}}{\frac{1}{2}-\frac{1}{4}} = \dfrac{12\left(\frac{1}{3}\right)}{12\left(\frac{1}{2}-\frac{1}{4}\right)} = \dfrac{4}{6-3} = \dfrac{4}{3}$

13. $\dfrac{-\frac{2}{9}}{-\frac{14}{3}} = -\dfrac{2}{9} \div \left(-\dfrac{14}{3}\right)$

$\qquad = -\dfrac{2}{9} \cdot \left(-\dfrac{3}{14}\right)$

$\qquad = \dfrac{2 \cdot 3}{9 \cdot 14}$

$\qquad = \dfrac{1}{21}$

15. $\dfrac{-\frac{5}{12x^2}}{\frac{25}{16x^3}} = \dfrac{-5}{12x^2} \div \dfrac{25}{16x^3}$

$\qquad = -\dfrac{5}{12x^2} \cdot \dfrac{16x^3}{25}$

$\qquad = -\dfrac{5 \cdot 16x^3}{12x^2 \cdot 25}$

$\qquad = -\dfrac{4x}{15}$

17. $\dfrac{\frac{m}{n}-1}{\frac{m}{n}+1} = \dfrac{n\left(\frac{m}{n}-1\right)}{n\left(\frac{m}{n}+1\right)} = \dfrac{m-n}{m+n}$

19. $\dfrac{\frac{1}{5}-\frac{1}{x}}{\frac{7}{10}+\frac{1}{x^2}} = \dfrac{10x^2\left(\frac{1}{5}-\frac{1}{x}\right)}{10x^2\left(\frac{7}{10}+\frac{1}{x^2}\right)}$

$\qquad = \dfrac{2x^2-10x}{7x^2+10}$

$\qquad = \dfrac{2x(x-5)}{7x^2+10}$

21. $\dfrac{1+\frac{1}{y-2}}{y+\frac{1}{y-2}} = \dfrac{(y-2)\left(1+\frac{1}{y-2}\right)}{(y-2)\left(y+\frac{1}{y-2}\right)}$

$\qquad = \dfrac{y-2+1}{(y-2)y+1}$

$\qquad = \dfrac{y-1}{y^2-2y+1}$

$\qquad = \dfrac{y-1}{(y-1)^2}$

$\qquad = \dfrac{1}{y-1}$

23. $\dfrac{\frac{4y-8}{16}}{\frac{6y-12}{4}} = \dfrac{16\left(\frac{4y-8}{16}\right)}{16\left(\frac{6y-12}{4}\right)}$

$\qquad = \dfrac{4y-8}{24y-48}$

$\qquad = \dfrac{4(y-2)}{24(y-2)}$

$\qquad = \dfrac{1}{6}$

25. $\dfrac{\frac{x}{y}+1}{\frac{x}{y}-1} = \dfrac{y\left(\frac{x}{y}+1\right)}{y\left(\frac{x}{y}-1\right)} = \dfrac{x+y}{x-y}$

27. $\dfrac{1}{2+\frac{1}{3}} = \dfrac{3(1)}{3\left(2+\frac{1}{3}\right)} = \dfrac{3}{6+1} = \dfrac{3}{7}$

29. $\dfrac{\frac{ax+ab}{x^2-b^2}}{\frac{x+b}{x-b}} = \dfrac{ax+ab}{x^2-b^2} \div \dfrac{x+b}{x-b}$

$\qquad = \dfrac{ax+ab}{x^2-b^2} \cdot \dfrac{x-b}{x+b}$

$\qquad = \dfrac{a(x+b)}{(x+b)(x-b)} \cdot \dfrac{x-b}{x+b}$

$\qquad = \dfrac{a}{x+b}$

31.
$$\frac{-\dfrac{3+y}{4}}{\dfrac{8+y}{28}} = \frac{28\left(\dfrac{-3+y}{4}\right)}{28\left(\dfrac{8+y}{28}\right)}$$

$$= \frac{-21+7y}{8+y}$$

$$= \frac{7y-21}{8+y}$$

$$= \frac{7(y-3)}{8+y}$$

33.
$$\frac{3+\dfrac{12}{x}}{1-\dfrac{16}{x^2}} = \frac{x^2\left(3+\dfrac{12}{x}\right)}{x^2\left(1-\dfrac{16}{x^2}\right)}$$

$$= \frac{3x^2+12x}{x^2-16}$$

$$= \frac{3x(x+4)}{(x+4)(x-4)}$$

$$= \frac{3x}{x-4}$$

35.
$$\frac{\dfrac{8}{x+4}+2}{\dfrac{12}{x+4}-2} = \frac{(x+4)\left(\dfrac{8}{x+4}+2\right)}{(x+4)\left(\dfrac{12}{x+4}-2\right)}$$

$$= \frac{8+2x+8}{12-2x-8}$$

$$= \frac{2x+16}{-2x+4}$$

$$= \frac{2(x+8)}{2(-x+2)}$$

$$= -\frac{x+8}{x-2}$$

37.
$$\frac{\dfrac{s}{r}+\dfrac{r}{s}}{\dfrac{s}{r}-\dfrac{r}{s}} = \frac{rs\left(\dfrac{s}{r}+\dfrac{r}{s}\right)}{rs\left(\dfrac{s}{r}-\dfrac{r}{s}\right)} = \frac{s^2+r^2}{s^2-r^2}$$

39.
$$\frac{\dfrac{6}{x-5}+\dfrac{x}{x-2}}{\dfrac{3}{x-6}-\dfrac{2}{x-5}} = \frac{\dfrac{6(x-2)}{(x-5)(x-2)}+\dfrac{x(x-5)}{(x-2)(x-5)}}{\dfrac{3(x-5)}{(x-6)(x-5)}-\dfrac{2(x-6)}{(x-5)(x-6)}}$$

$$= \frac{\dfrac{6x-12+x^2-5x}{(x-5)(x-2)}}{\dfrac{3x-15-2x+12}{(x-6)(x-5)}}$$

$$= \frac{\dfrac{x^2+x-12}{(x-5)(x-2)}}{\dfrac{x-3}{(x-6)(x-5)}}$$

$$= \frac{(x+4)(x-3)}{(x-5)(x-2)} \cdot \frac{(x-6)(x-5)}{x-3}$$

$$= \frac{(x-6)(x+4)}{x-2}$$

41. The longest bar corresponds to Steffi Graf, so Steffi Graf has won the most prize money in her career.

43. Martina Navratilova and Steffi Graf are shown to have earned over $20 million in prize money over their careers.

45. answers may vary

47.
$$\frac{\dfrac{1}{3}+\dfrac{3}{4}}{2} = \frac{\dfrac{1\cdot4}{3\cdot4}+\dfrac{3\cdot3}{4\cdot3}}{2} = \frac{\dfrac{4}{12}+\dfrac{9}{12}}{2} = \frac{\dfrac{13}{12}}{2} = \frac{13}{12}\cdot\frac{1}{2} = \frac{13}{24}$$

49.
$$\frac{1}{\dfrac{1}{R_1}+\dfrac{1}{R_2}} = \frac{R_1R_2(1)}{R_1R_2\left(\dfrac{1}{R_1}+\dfrac{1}{R_2}\right)} = \frac{R_1R_2}{R_2+R_1}$$

51.
$$\frac{x^{-1}+2^{-1}}{x^{-2}-4^{-1}} = \frac{\frac{1}{x}+\frac{1}{2}}{\frac{1}{x^2}-\frac{1}{4}}$$

$$= \frac{\frac{1\cdot2}{x\cdot2}+\frac{1\cdot x}{2\cdot x}}{\frac{1\cdot4}{x^2\cdot4}-\frac{1\cdot x^2}{4\cdot x^2}}$$

$$= \frac{\frac{2+x}{2x}}{\frac{4-x^2}{4x^2}}$$

$$= \frac{2+x}{2x}\cdot\frac{4x^2}{4-x^2}$$

$$= \frac{2+x}{2x}\cdot\frac{4x^2}{(2-x)(2+x)}$$

$$= \frac{2x}{2-x}$$

53.
$$\frac{y^{-2}}{1-y^{-2}} = \frac{\frac{1}{y^2}}{1-\frac{1}{y^2}} = \frac{y^2\left(\frac{1}{y^2}\right)}{y^2\left(1-\frac{1}{y^2}\right)} = \frac{1}{y^2-1}$$

55.
$$t = \frac{d}{r} = \frac{\frac{20x}{3}}{\frac{5x}{9}} = \frac{20x}{3}\cdot\frac{9}{5x} = 12$$

The time is 12 hours.

Chapter 5 Vocabulary Check

1. A <u>ratio</u> is the quotient of two numbers.

2. $\dfrac{x}{2}=\dfrac{7}{16}$ is an example of a <u>proportion</u>.

3. If $\dfrac{a}{b}=\dfrac{c}{d}$, the ad and bc are called <u>cross products</u>.

4. A <u>rational expression</u> is an expression that can be written in the form $\dfrac{P}{Q}$, where P and Q are polynomials and Q is not 0.

5. In a <u>complex fraction</u>, the numerator or denominator or both may contain fractions.

6. A <u>rate</u> is a special type of ratio where different measurements are used.

Chapter 5 Review

1.
$$x^2-4=0$$
$$(x-2)(x+2)=0$$
$$x-2=0 \quad\text{or}\quad x+2=0$$
$$x=2 \qquad\qquad x=-2$$

The expression $\dfrac{x+5}{x^2-4}$ is undefined for $x=2$ and $x=-2$.

2.
$$4x^2-4x-15=0$$
$$(2x-5)(2x+3)=0$$
$$2x-5=0 \quad\text{or}\quad 2x+3=0$$
$$2x=5 \qquad\qquad 2x=3$$
$$x=\frac{5}{2} \qquad\qquad x=-\frac{3}{2}$$

The expression $\dfrac{5x+9}{4x^2-4x-15}$ is undefined for $x=\dfrac{5}{2}$ and $x=-\dfrac{3}{2}$.

3. Replace z with -2.
$$\frac{2-z}{z+5}=\frac{2-(-2)}{-2+5}=\frac{4}{3}$$

4. Replace x with 5 and y with 7.
$$\frac{x^2+xy-y^2}{x+y}=\frac{(5)^2+(5)(7)-(7)^2}{5+7}$$
$$=\frac{25+35-49}{12}$$
$$=\frac{11}{12}$$

5. $\dfrac{2x+6}{x^2+3x}=\dfrac{2(x+3)}{x(x+3)}=\dfrac{2}{x}$

6. $\dfrac{3x-12}{x^2-4x}=\dfrac{3(x-4)}{x(x-4)}=\dfrac{3}{x}$

7. $\dfrac{x+2}{x^2-3x-10} = \dfrac{x+2}{(x-5)(x+2)} = \dfrac{1}{x-5}$

8. $\dfrac{x+4}{x^2+5x+4} = \dfrac{x+4}{(x+4)(x+1)} = \dfrac{1}{x+1}$

9. $\dfrac{x^3-4x}{x^2+3x+2} = \dfrac{x(x^2-4)}{(x+2)(x+1)}$
$= \dfrac{x(x-2)(x+2)}{(x+2)(x+1)}$
$= \dfrac{x(x-2)}{x+1}$

10. $\dfrac{5x^2-125}{x^2+2x-15} = \dfrac{5(x^2-25)}{(x+5)(x-3)}$
$= \dfrac{5(x-5)(x+5)}{(x+5)(x-3)}$
$= \dfrac{5(x-5)}{x-3}$

11. $\dfrac{x^2-x-6}{x^2-3x-10} = \dfrac{(x-3)(x+2)}{(x-5)(x+2)} = \dfrac{x-3}{x-5}$

12. $\dfrac{x^2-2x}{x^2+2x-8} = \dfrac{x(x-2)}{(x+4)(x-2)} = \dfrac{x}{x+4}$

13. $\dfrac{x^2+xa+xb+ab}{x^2-xc+bx-bc} = \dfrac{(x^2+xa)+(xb+ab)}{(x^2-xc)+(bx-bc)}$
$= \dfrac{x(x+a)+b(x+a)}{x(x-c)+b(x-c)}$
$= \dfrac{(x+b)(x+a)}{(x+b)(x-c)}$
$= \dfrac{x+a}{x-c}$

14. $\dfrac{x^2+5x-2x-10}{x^2-3x-2x+6} = \dfrac{(x^2+5x)+(-2x-10)}{(x^2-3x)+(-2x+6)}$
$= \dfrac{x(x+5)-2(x+5)}{x(x-3)-2(x-3)}$
$= \dfrac{(x-2)(x+5)}{(x-2)(x-3)}$
$= \dfrac{x+5}{x-3}$

15. $\dfrac{15x^3y^2}{z} \cdot \dfrac{z}{5xy^3} = \dfrac{3\cdot5\cdot x\cdot x^2\cdot y^2\cdot z}{z\cdot5\cdot x\cdot y\cdot y^2} = \dfrac{3x^2}{y}$

16. $\dfrac{-y^3}{8} \cdot \dfrac{9x^2}{y^3} = -\dfrac{y^3\cdot9x^2}{8\cdot y^3} = -\dfrac{9x^2}{8}$

17. $\dfrac{x^2-9}{x^2-4} \cdot \dfrac{x-2}{x+3} = \dfrac{(x-3)(x+3)}{(x-2)(x+2)} \cdot \dfrac{x-2}{x+3} = \dfrac{x-3}{x+2}$

18. $\dfrac{2x+5}{x-6} \cdot \dfrac{2x}{-x+6} = \dfrac{2x+5}{x-6} \cdot \dfrac{2x}{-(x-6)}$
$= \dfrac{-2x(2x+5)}{(x-6)^2}$

19. $\dfrac{x^2-5x-24}{x^2-x-12} \div \dfrac{x^2-10x+16}{x^2+x-6}$
$= \dfrac{x^2-5x-24}{x^2-x-12} \cdot \dfrac{x^2+x-6}{x^2-10x+16}$
$= \dfrac{(x-8)(x+3)}{(x-4)(x+3)} \cdot \dfrac{(x+3)(x-2)}{(x-8)(x-2)}$
$= \dfrac{x+3}{x-4}$

20. $\dfrac{4x+4y}{xy^2} \div \dfrac{3x+3y}{x^2y} = \dfrac{4x+4y}{xy^2} \cdot \dfrac{x^2y}{3x+3y}$
$= \dfrac{4(x+y)}{xy^2} \cdot \dfrac{x^2y}{3(x+y)}$
$= \dfrac{4x}{3y}$

21. $\dfrac{x^2 + x - 42}{x - 3} \cdot \dfrac{(x-3)^2}{x+7}$

$= \dfrac{(x+7)(x-6)}{x-3} \cdot \dfrac{(x-3)^2}{x+7}$

$= (x-6)(x-3)$

22. $\dfrac{2a+2b}{3} \cdot \dfrac{a-b}{a^2-b^2} = \dfrac{2(a+b)}{3} \cdot \dfrac{a-b}{(a-b)(a+b)}$

$= \dfrac{2}{3}$

23. $\dfrac{2x^2 - 9x + 9}{8x - 12} \div \dfrac{x^2 - 3x}{2x}$

$= \dfrac{2x^2 - 9x + 9}{8x - 12} \cdot \dfrac{2x}{x^2 - 3x}$

$= \dfrac{(2x-3)(x-3)}{4(2x-3)} \cdot \dfrac{2x}{x(x-3)}$

$= \dfrac{1}{2}$

24. $\dfrac{x^2 - y^2}{x^2 + xy} \div \dfrac{3x^2 - 2xy - y^2}{3x^2 + 6x}$

$= \dfrac{x^2 - y^2}{x^2 + xy} \cdot \dfrac{3x^2 + 6x}{3x^2 - 2xy - y^2}$

$= \dfrac{(x-y)(x+y)}{x(x+y)} \cdot \dfrac{3x(x+2)}{(3x+y)(x-y)}$

$= \dfrac{3(x+2)}{3x+y}$

25. $\dfrac{x}{x^2 + 9x + 14} + \dfrac{7}{x^2 + 9x + 14} = \dfrac{x+7}{x^2 + 9x + 14}$

$= \dfrac{x+7}{(x+7)(x+2)}$

$= \dfrac{1}{x+2}$

26. $\dfrac{x}{x^2 + 2x - 15} + \dfrac{5}{x^2 + 2x - 15} = \dfrac{x+5}{x^2 + 2x - 15}$

$= \dfrac{x+5}{(x+5)(x-3)}$

$= \dfrac{1}{x-3}$

27. $\dfrac{4x-5}{3x^2} - \dfrac{2x+5}{3x^2} = \dfrac{4x-5-2x-5}{3x^2} = \dfrac{2x-10}{3x^2}$

28. $\dfrac{9x+7}{6x^2} - \dfrac{3x+4}{6x^2} = \dfrac{9x+7-3x-4}{6x^2}$

$= \dfrac{6x+3}{6x^2}$

$= \dfrac{3(2x+1)}{6x^2}$

$= \dfrac{2x+1}{2x^2}$

29. The LCD is $2 \cdot 7 \cdot x$ or $14x$.

30. $x^2 - 5x - 24 = (x-8)(x+3)$

$x^2 + 11x + 24 = (x+8)(x+3)$

The LCD is $(x+3)(x+8)(x-8)$.

31. $\dfrac{5}{7x} = \dfrac{5 \cdot 2x^2 y}{7x \cdot 2x^2 y} = \dfrac{10x^2 y}{14x^3 y}$

32. $\dfrac{9}{4y} = \dfrac{9 \cdot 4y^2 x}{4y \cdot 4y^2 x} = \dfrac{36y^2 x}{16y^3 x}$

33. $\dfrac{x+2}{x^2 + 11x + 18} = \dfrac{x+2}{(x+2)(x+9)}$

$= \dfrac{(x+2)(x-5)}{(x+2)(x+9)(x-5)}$

$= \dfrac{x^2 - 3x - 10}{(x+2)(x-5)(x+9)}$

34.
$$\frac{3x-5}{x^2+4x+4} = \frac{3x-5}{(x+2)^2}$$
$$= \frac{(3x-5)(x+3)}{(x+2)^2(x+3)}$$
$$= \frac{3x^2+4x-15}{(x+2)^2(x+3)}$$

35. $\dfrac{4}{5x^2} - \dfrac{6}{y} = \dfrac{4y}{5x^2 y} - \dfrac{6\cdot 5x^2}{y\cdot 5x^2} = \dfrac{4y-30x^2}{5x^2 y}$

36.
$$\frac{2}{x-3} - \frac{4}{x-1} = \frac{2(x-1)-4(x-3)}{(x-3)(x-1)}$$
$$= \frac{2x-2-4x+12}{(x-3)(x-1)}$$
$$= \frac{-2x+10}{(x-3)(x-1)}$$

37.
$$\frac{4}{x+3} - 2 = \frac{4-2(x+3)}{x+3}$$
$$= \frac{4-2x-6}{x+3}$$
$$= \frac{-2x-2}{x+3}$$

38.
$$\frac{3}{x^2+2x-8} + \frac{2}{x^2-3x+2}$$
$$= \frac{3}{(x+4)(x-2)} + \frac{2}{(x-2)(x-1)}$$
$$= \frac{3(x-1)+2(x+4)}{(x+4)(x-2)(x-1)}$$
$$= \frac{3x-3+2x+8}{(x+4)(x-2)(x-1)}$$
$$= \frac{5x+5}{(x+4)(x-2)(x-1)}$$

39.
$$\frac{2x-5}{6x+9} - \frac{4}{2x^2+3x} = \frac{2x-5}{3(2x+3)} - \frac{4}{x(2x+3)}$$
$$= \frac{x(2x-5)-4(3)}{3x(2x+3)}$$
$$= \frac{2x^2-5x-12}{3x(2x+3)}$$
$$= \frac{(2x+3)(x-4)}{3x(2x+3)}$$
$$= \frac{x-4}{3x}$$

40.
$$\frac{x-1}{x^2-2x+1} - \frac{x+1}{x-1} = \frac{x-1}{(x-1)(x-1)} - \frac{x+1}{x-1}$$
$$= \frac{1}{x-1} - \frac{x+1}{x-1}$$
$$= \frac{1-x-1}{x-1}$$
$$= -\frac{x}{x-1}$$

41.
$$\frac{n}{10} = 9 - \frac{n}{5}$$
$$10\left(\frac{n}{10}\right) = 10\left(9 - \frac{n}{5}\right)$$
$$n = 90 - 2n$$
$$3n = 90$$
$$n = 30$$
The solution is 30.

42.
$$\frac{2}{x+1} - \frac{1}{x-2} = -\frac{1}{2}$$
$$2(x+1)(x-2)\left(\frac{2}{x+1} - \frac{1}{x-2}\right) = 2(x+1)(x-2)\left(-\frac{1}{2}\right)$$
$$2 \cdot 2(x-2) - 2(x+1) = -1(x+1)(x-2)$$
$$4x - 8 - 2x - 2 = -(x^2 - x - 2)$$
$$2x - 10 = -x^2 + x + 2$$
$$x^2 + x - 12 = 0$$
$$(x+4)(x-3) = 0$$
$$x+4 = 0 \quad \text{or} \quad x-3 = 0$$
$$x = -4 \qquad\qquad x = 3$$
The solutions are −4 and 3.

43.
$$\frac{y}{2y+2} + \frac{2y-16}{4y+4} = \frac{y-3}{y+1}$$
$$\frac{y}{2(y+1)} + \frac{2y-16}{4(y+1)} = \frac{y-3}{y+1}$$
$$4(y+1)\left(\frac{y}{2(y+1)} + \frac{2y-16}{4(y+1)}\right) = 4(y+1)\left(\frac{y-3}{y+1}\right)$$
$$2y + 2y - 16 = 4(y-3)$$
$$4y - 16 = 4y - 12$$
$$0y = 4$$
$$0 = 4; \text{ no solution}$$

44.
$$\frac{2}{x-3} - \frac{4}{x+3} = \frac{8}{x^2 - 9}$$
$$(x^2 - 9)\left(\frac{2}{x-3} - \frac{4}{x+3}\right) = (x^2 - 9)\left(\frac{8}{x^2 - 9}\right)$$
$$2(x+3) - 4(x-3) = 8$$
$$2x + 6 - 4x + 12 = 8$$
$$-2x + 18 = 8$$
$$-2x = -10$$
$$x = 5$$
The solution is 5.

45.

$$\frac{x-3}{x+1} - \frac{x-6}{x+5} = 0$$

$$(x+1)(x+5)\left(\frac{x-3}{x+1} - \frac{x-6}{x+5}\right) = (x+1)(x+5)0$$

$$(x+5)(x-3) - (x+1)(x-6) = 0$$

$$x^2 + 2x - 15 - x^2 + 5x + 6 = 0$$

$$7x - 9 = 0$$

$$7x = 9$$

$$x = \frac{9}{7}$$

The solution is $\frac{9}{7}$.

46.

$$x + 5 = \frac{6}{x}$$

$$x(x+5) = x\left(\frac{6}{x}\right)$$

$$x^2 + 5x = 6$$

$$x^2 + 5x - 6 = 0$$

$$(x+6)(x-1) = 0$$

$$x + 6 = 0 \quad \text{or} \quad x - 1 = 0$$

$$x = -6 \qquad\qquad x = 1$$

The solutions are −6 and 1.

47.

$$\frac{2}{x-1} = \frac{3}{x+3}$$

$$2(x+3) = 3(x-1)$$

$$2x + 6 = 3x - 3$$

$$9 = x$$

The solution is 9.

48.

$$\frac{4}{y-3} = \frac{2}{y-3}$$

$$4(y-3) = 2(y-3)$$

$$4y - 12 = 2y - 6$$

$$2y = 6$$

$$y = 3$$

$y = 3$ does not check. There is no solution.

49. $\dfrac{300}{20} = \dfrac{x}{45}$

$300(45) = 20x$

$13{,}500 = 20x$

$675 = x$

The machine will process 675 parts in 45 minutes.

50. $\dfrac{90}{8} = \dfrac{x}{3}$

$90(3) = 8x$

$270 = 8x$

$33.75 = x$

Mr. Visconti charges \$33.75 for 3 hours of consulting.

51. Let n be the number.

$$5\left(\dfrac{1}{n}\right) = \dfrac{3}{2}\left(\dfrac{1}{n}\right) + \dfrac{7}{6}$$

$$\dfrac{5}{n} = \dfrac{3}{2n} + \dfrac{7}{6}$$

$$6n\left(\dfrac{5}{n}\right) = 6n\left(\dfrac{3}{2n} + \dfrac{7}{6}\right)$$

$$30 = 9 + 7n$$

$$21 = 7n$$

$$3 = n$$

The number is 3.

52. Let n be the number.

$$\dfrac{1}{n} = \dfrac{1}{4-n}$$

$$n(4-n)\left(\dfrac{1}{n}\right) = n(4-n)\left(\dfrac{1}{4-n}\right)$$

$$4-n = n$$

$$4 = 2n$$

$$2 = n$$

The number is 2.

53. Let x be the speed of the faster car.

	distance	=	rate	·	time
Faster car	90		x		$\dfrac{90}{x}$
Slower car	60		$x - 10$		$\dfrac{60}{x-10}$

The times are equal.

$$\frac{90}{x} = \frac{60}{x-10}$$
$$90(x-10) = 60x$$
$$90x - 900 = 60x$$
$$30x = 900$$
$$x = 30$$
$$x - 10 = 20$$

The faster car is traveling at 30 mph and the slower car is traveling at 20 mph.

54. Let x be the speed of the boat in still water.

	distance	=	rate	·	time
Upstream	48		$x - 4$		$\frac{48}{x-4}$
Downstream	72		$x + 4$		$\frac{72}{x+4}$

The times are equal.
$$\frac{48}{x-4} = \frac{72}{x+4}$$
$$48(x+4) = 72(x-4)$$
$$48x + 192 = 72x - 288$$
$$480 = 24x$$
$$20 = x$$

The speed of the boat in still water is 20 mph.

55. Let x be the time for Maria alone. Together, Mark and Maria complete $\frac{1}{5}$ of the job in 1 hour. Individually, they complete $\frac{1}{7}$ and $\frac{1}{x}$ of the job in 1 hour.

$$\frac{1}{5} = \frac{1}{7} + \frac{1}{x}$$
$$35x\left(\frac{1}{5}\right) = 35x\left(\frac{1}{7} + \frac{1}{x}\right)$$
$$7x = 5x + 35$$
$$2x = 35$$
$$x = \frac{35}{2} = 17\frac{1}{2}$$

It will take Maria $17\frac{1}{2}$ hours to manicure Mr. Sturgeon's lawn alone.

56. Let x be the time for the pipes to fill the pond working together. Pipe A fills $\frac{1}{20}$ of the pond in 1 day, pipe B fills $\frac{1}{15}$ of the pond in 1 day, and together they fill $\frac{1}{x}$ of the pond in one day.

$$\frac{1}{20} + \frac{1}{15} = \frac{1}{x}$$
$$60x\left(\frac{1}{20} + \frac{1}{15}\right) = 60x\left(\frac{1}{x}\right)$$
$$3x + 4x = 60$$
$$7x = 60$$
$$x = \frac{60}{7} = 8\frac{4}{7}$$

It takes $8\frac{4}{7}$ days to fill the pond using both pipes.

57.
$$\frac{10}{2} = \frac{x}{3}$$
$$10(3) = 2x$$
$$30 = 2x$$
$$15 = x$$
The missing length is $x = 15$.

58.
$$\frac{18}{12} = \frac{x}{4}$$
$$18(4) = 12x$$
$$72 = 12x$$
$$6 = x$$
The missing length is $x = 6$.

59.
$$\frac{\frac{5x}{27}}{-\frac{10xy}{21}} = \frac{5x}{27} \div -\frac{10xy}{21}$$
$$= \frac{5x}{27} \cdot \frac{-21}{10xy}$$
$$= -\frac{5 \cdot x \cdot 3 \cdot 7}{3 \cdot 9 \cdot 2 \cdot 5 \cdot x \cdot y}$$
$$= -\frac{7}{18y}$$

60.
$$\frac{\frac{3}{5} + \frac{2}{7}}{\frac{1}{5} + \frac{5}{6}} = \frac{\frac{3 \cdot 7}{5 \cdot 7} + \frac{2 \cdot 5}{7 \cdot 5}}{\frac{1 \cdot 6}{5 \cdot 6} + \frac{5 \cdot 5}{6 \cdot 5}}$$
$$= \frac{\frac{21 + 10}{35}}{\frac{6 + 25}{30}}$$
$$= \frac{\frac{31}{35}}{\frac{31}{30}}$$
$$= \frac{31}{35} \div \frac{31}{30}$$
$$= \frac{31}{35} \cdot \frac{30}{31}$$
$$= \frac{31 \cdot 5 \cdot 6}{5 \cdot 7 \cdot 31}$$
$$= \frac{6}{7}$$

61.
$$\frac{3 - \frac{1}{y}}{2 - \frac{1}{y}} = \frac{y\left(3 - \frac{1}{y}\right)}{y\left(2 - \frac{1}{y}\right)} = \frac{3y - 1}{2y - 1}$$

62.
$$\frac{\frac{6}{x+2} + 4}{\frac{8}{x+2} - 4} = \frac{(x+2)\left(\frac{6}{x+2} + 4\right)}{(x+2)\left(\frac{8}{x+2} - 4\right)}$$
$$= \frac{6 + 4(x+2)}{8 - 4(x+2)}$$
$$= \frac{6 + 4x + 8}{8 - 4x - 8}$$
$$= \frac{4x + 14}{-4x}$$
$$= -\frac{2(2x+7)}{2 \cdot 2x}$$
$$= -\frac{2x + 7}{2x}$$

63.
$$\frac{4x + 12}{8x^2 + 24x} = \frac{4(x+3)}{8x(x+3)} = \frac{1}{2x}$$

64.
$$\begin{aligned} \frac{x^3 - 6x^2 + 9x}{x^2 + 4x - 21} &= \frac{x(x^2 - 6x + 9)}{(x+7)(x-3)} \\ &= \frac{x(x-3)(x-3)}{(x+7)(x-3)} \\ &= \frac{x(x-3)}{x+7} \end{aligned}$$

65.
$$\begin{aligned} \frac{x^2 + 9x + 20}{x^2 - 25} \cdot \frac{x^2 - 9x + 20}{x^2 + 8x + 16} &= \frac{(x+4)(x+5)}{(x-5)(x+5)} \cdot \frac{(x-4)(x-5)}{(x+4)(x+4)} \\ &= \frac{x-4}{x+4} \end{aligned}$$

66.
$$\begin{aligned} \frac{x^2 - x - 72}{x^2 - x - 30} \div \frac{x^2 + 6x - 27}{x^2 - 9x + 18} &= \frac{x^2 - x - 72}{x^2 - x - 30} \cdot \frac{x^2 - 9x + 18}{x^2 + 6x - 27} \\ &= \frac{(x-9)(x+8)}{(x-6)(x+5)} \cdot \frac{(x-6)(x-3)}{(x+9)(x-3)} \\ &= \frac{(x-9)(x+8)}{(x+5)(x+9)} \end{aligned}$$

67.
$$\begin{aligned} \frac{x}{x^2 - 36} + \frac{6}{x^2 - 36} &= \frac{x+6}{x^2 - 36} \\ &= \frac{x+6}{(x+6)(x-6)} \\ &= \frac{1}{x-6} \end{aligned}$$

68.
$$\frac{5x - 1}{4x} - \frac{3x - 2}{4x} = \frac{5x - 1 - 3x + 2}{4x} = \frac{2x + 1}{4x}$$

69.
$$\begin{aligned} \frac{4}{3x^2 + 8x - 3} + \frac{2}{3x^2 - 7x + 2} &= \frac{4}{(3x-1)(x+3)} + \frac{2}{(3x-1)(x-2)} \\ &= \frac{4(x-2)}{(3x-1)(x+3)(x-2)} + \frac{2(x+3)}{(3x-1)(x-2)(x+3)} \\ &= \frac{4x - 8 + 2x + 6}{(3x-1)(x+3)(x-2)} \\ &= \frac{6x - 2}{(3x-1)(x+3)(x-2)} \\ &= \frac{2(3x-1)}{(3x-1)(x+3)(x-2)} \\ &= \frac{2}{(x+3)(x-2)} \end{aligned}$$

70. $\dfrac{3x}{x^2+9x+14}-\dfrac{6x}{x^2+4x-21}$

$=\dfrac{3x}{(x+7)(x+2)}-\dfrac{6x}{(x+7)(x-3)}$

$=\dfrac{3x(x-3)}{(x+7)(x+2)(x-3)}-\dfrac{6x(x+2)}{(x+7)(x-3)(x+2)}$

$=\dfrac{3x^2-9x-6x^2-12x}{(x+7)(x+2)(x-3)}$

$=\dfrac{-3x^2-21x}{(x+7)(x+2)(x-3)}$

$=\dfrac{-3x(x+7)}{(x+7)(x+2)(x-3)}$

$=-\dfrac{3x}{(x+2)(x-3)}$

71. $\dfrac{4}{a-1}+2=\dfrac{3}{a-1}$

$(a-1)\left(\dfrac{4}{a-1}+2\right)=(a-1)\left(\dfrac{3}{a-1}\right)$

$4+2(a-1)=3$

$4+2a-2=3$

$2+2a=3$

$2a=1$

$a=\dfrac{1}{2}$

The solution is $\dfrac{1}{2}$.

72. $\dfrac{x}{x+3}+4=\dfrac{x}{x+3}$

$(x+3)\left(\dfrac{x}{x+3}+4\right)=(x+3)\left(\dfrac{x}{x+3}\right)$

$x+4(x+3)=x$

$x+4x+12=x$

$5x+12=x$

$4x+12=0$

$4x=-12$

$x=-3$

-3 makes the denominator $x+3$ zero.
Therefore there is no solution.

73. Let n be the number.

$\dfrac{2n}{3}-\dfrac{1}{6}=\dfrac{n}{2}$

$6\left(\dfrac{2n}{3}-\dfrac{1}{6}\right)=6\left(\dfrac{n}{2}\right)$

$4n-1=3n$

$n-1=0$

$n=1$

The number is 1.

74. Let x be the time to paint the shed together.

Mr. Crocker can paint $\dfrac{1}{3}$ of the shed in one

day, and his son can paint $\dfrac{1}{4}$. Together,

they can paint $\dfrac{1}{x}$.

$\dfrac{1}{3}+\dfrac{1}{4}=\dfrac{1}{x}$

$12x\left(\dfrac{1}{3}+\dfrac{1}{4}\right)=12x\left(\dfrac{1}{x}\right)$

$4x+3x=12$

$7x=12$

$x=\dfrac{12}{7}$ or $1\dfrac{5}{7}$

It will take $1\dfrac{5}{7}$ days to paint the shed when

they work together.

75. $\dfrac{x}{10}=\dfrac{3}{5}$

$5x=3(10)$

$5x=30$

$x=6$

The missing length is $x=6$.

76. $\dfrac{x}{4}=\dfrac{18}{6}$

$6x=4(18)$

$6x=72$

$x=12$

The missing length is $x=12$.

77. $\dfrac{\frac{1}{4}}{\frac{1}{3}+\frac{1}{2}} = \dfrac{12\left(\frac{1}{4}\right)}{12\left(\frac{1}{3}+\frac{1}{2}\right)} = \dfrac{3}{4+6} = \dfrac{3}{10}$

78. $\dfrac{4+\frac{2}{x}}{6+\frac{3}{x}} = \dfrac{x\left(4+\frac{2}{x}\right)}{x\left(6+\frac{3}{x}\right)} = \dfrac{4x+2}{6x+3} = \dfrac{2(2x+1)}{3(2x+1)} = \dfrac{2}{3}$

Chapter 5 Test

1. $x^2 + 4x + 3 = 0$
$(x+1)(x+3) = 0$
$x+1 = 0 \quad \text{or} \quad x+3 = 0$
$\quad x = -1 \qquad\qquad x = -3$

The expression $\dfrac{x+5}{x^2+4x+3}$ is undefined for
$x = -1$ and $x = -3$.

2. a. $C = \dfrac{100x+3000}{x}$
$= \dfrac{100(200)+3000}{200}$
$= \$115$

   b. $C = \dfrac{100x+3000}{x}$
$= \dfrac{100(1000)+3000}{1000}$
$= \$103$

3. $\dfrac{3x-6}{5x-10} = \dfrac{3(x-2)}{5(x-2)} = \dfrac{3}{5}$

4. $\dfrac{x+6}{x^2+12x+36} = \dfrac{x+6}{(x+6)(x+6)} = \dfrac{1}{x+6}$

5. $\dfrac{7-x}{x-7} = \dfrac{-(x-7)}{x-7} = -1$

6. $\dfrac{y-x}{x^2-y^2} = \dfrac{-(x-y)}{(x-y)(x+y)} = -\dfrac{1}{x+y}$

7. $\dfrac{2m^3-2m^2-12m}{m^2-5m+6} = \dfrac{2m(m^2-m-6)}{(m-3)(m-2)}$
$= \dfrac{2m(m-3)(m+2)}{(m-3)(m-2)}$
$= \dfrac{2m(m+2)}{m-2}$

8. $\dfrac{ay+3a+2y+6}{ay+3a+5y+15} = \dfrac{(ay+3a)+(2y+6)}{(ay+3a)+(5y+15)}$
$= \dfrac{a(y+3)+2(y+3)}{a(y+3)+5(y+3)}$
$= \dfrac{(a+2)(y+3)}{(a+5)(y+3)}$
$= \dfrac{a+2}{a+5}$

9. $\dfrac{x^2-13x+42}{x^2+10x+21} \div \dfrac{x^2-4}{x^2+x-6}$
$= \dfrac{x^2-13x+42}{x^2+10x+21} \cdot \dfrac{x^2+x-6}{x^2-4}$
$= \dfrac{(x-6)(x-7)}{(x+3)(x+7)} \cdot \dfrac{(x+3)(x-2)}{(x+2)(x-2)}$
$= \dfrac{(x-6)(x-7)}{(x+7)(x+2)}$

10. $\dfrac{3}{x-1} \cdot (5x-5) = \dfrac{3}{x-1} \cdot 5(x-1) = 15$

11. $\dfrac{y^2-5y+6}{2y+4} \cdot \dfrac{y+2}{2y-6}$
$= \dfrac{(y-3)(y-2)}{2(y+2)} \cdot \dfrac{y+2}{2(y-3)}$
$= \dfrac{y-2}{4}$

12. $\dfrac{5}{2x+5} - \dfrac{6}{2x+5} = \dfrac{5-6}{2x+5} = -\dfrac{1}{2x+5}$

13. $\dfrac{5a}{a^2-a-6}-\dfrac{2}{a-3}$

$=\dfrac{5a}{(a-3)(a+2)}-\dfrac{2}{a-3}$

$=\dfrac{5a}{(a-3)(a+2)}-\dfrac{2(a+2)}{(a-3)(a+2)}$

$=\dfrac{5a-2(a+2)}{(a-3)(a+2)}$

$=\dfrac{5a-2a-4}{(a-3)(a+2)}$

$=\dfrac{3a-4}{(a-3)(a+2)}$

14. $\dfrac{6}{x^2-1}+\dfrac{3}{x+1}=\dfrac{6}{(x-1)(x+1)}+\dfrac{3}{x+1}$

$=\dfrac{6}{(x-1)(x+1)}+\dfrac{3(x-1)}{(x+1)(x-1)}$

$=\dfrac{6+3x-3}{(x+1)(x-1)}$

$=\dfrac{3x+3}{(x+1)(x-1)}$

$=\dfrac{3(x+1)}{(x+1)(x-1)}$

$=\dfrac{3}{x-1}$

15. $\dfrac{x^2-9}{x^2-3x}\div\dfrac{x^2+4x+1}{2x+10}$

$=\dfrac{x^2-9}{x^2-3x}\cdot\dfrac{2x+10}{x^2+4x+1}$

$=\dfrac{(x-3)(x+3)}{x(x-3)}\cdot\dfrac{2(x+5)}{x^2+4x+1}$

$=\dfrac{2(x+3)(x+5)}{x(x^2+4x+1)}$

16. $\dfrac{x+2}{x^2+11x+18}+\dfrac{5}{x^2-3x-10}$

$=\dfrac{x+2}{(x+9)(x+2)}+\dfrac{5}{(x-5)(x+2)}$

$=\dfrac{(x+2)(x-5)}{(x+9)(x+2)(x-5)}+\dfrac{5(x+9)}{(x-5)(x+2)(x+9)}$

$=\dfrac{x^2-3x-10+5x+45}{(x+9)(x+2)(x-5)}$

$=\dfrac{x^2+2x+35}{(x+9)(x+2)(x-5)}$

17. $\dfrac{4y}{y^2+6y+5}-\dfrac{3}{y^2+5y+4}$

$=\dfrac{4y}{(y+5)(y+1)}-\dfrac{3}{(y+1)(y+4)}$

$=\dfrac{4y(y+4)-3(y+5)}{(y+1)(y+5)(y+4)}$

$=\dfrac{4y^2+16y-3y-15}{(y+1)(y+5)(y+4)}$

$=\dfrac{4y^2+13y-15}{(y+1)(y+5)(y+4)}$

18. The LCD is $3\cdot5\cdot y=15y$.

$\dfrac{4}{y}-\dfrac{5}{3}=-\dfrac{1}{5}$

$15y\left(\dfrac{4}{y}-\dfrac{5}{3}\right)=15y\left(-\dfrac{1}{5}\right)$

$60-25y=-3y$

$60=22y$

$\dfrac{60}{22}=y$

$\dfrac{30}{11}=y$

The solution is $\dfrac{30}{11}$.

19.
$$\frac{5}{y+1} = \frac{4}{y+2}$$
$$5(y+2) = 4(y+1)$$
$$5y+10 = 4y+4$$
$$y = -6$$
The solution is −6.

20. The LCD is $2(a-3)$.
$$\frac{a}{a-3} = \frac{3}{a-3} - \frac{3}{2}$$
$$2(a-3)\left(\frac{a}{a-3}\right) = 2(a-3)\left(\frac{3}{a-3} - \frac{3}{2}\right)$$
$$2a = 6 - 3(a-3)$$
$$2a = 6 - 3a + 9$$
$$5a = 15$$
$$a = 3$$
Since $a = 3$ causes the denominator $a - 3$ to be 0, the equation has no solution.

21. The LCD is $x^2 - 25 = (x+5)(x-5)$.
$$\frac{10}{x^2-25} = \frac{3}{x+5} + \frac{1}{x-5}$$
$$(x^2-25)\left(\frac{10}{x^2-25}\right) = (x+5)(x-5)\left(\frac{3}{x+5} + \frac{1}{x-5}\right)$$
$$10 = 3(x-5) + 1(x+5)$$
$$10 = 3x - 15 + x + 5$$
$$10 = 4x - 10$$
$$20 = 4x$$
$$5 = x$$
Since $x = 5$ causes the denominators $x^2 - 25$ and $x - 5$ to be 0, the equation has no solution.

22.
$$x - \frac{14}{x-1} = 4 - \frac{2x}{x-1}$$
$$(x-1)\left(x - \frac{14}{x-1}\right) = (x-1)\left(4 - \frac{2x}{x-1}\right)$$
$$x(x-1) - 14 = 4(x-1) - 2x$$
$$x^2 - x - 14 = 4x - 4 - 2x$$
$$x^2 - x - 14 = 2x - 4$$
$$x^2 - 3x - 10 = 0$$
$$(x-5)(x+2) = 0$$
$$x - 5 = 0 \quad \text{or} \quad x + 2 = 0$$
$$x = 5 \qquad \qquad x = -2$$
The solutions are 5 and −2.

23.
$$\frac{\frac{5x^2}{yz^2}}{\frac{10x}{z^3}} = \frac{5x^2}{yz^2} \div \frac{10x}{z^3}$$
$$= \frac{5x^2}{yz^2} \cdot \frac{z^3}{10x}$$
$$= \frac{5 \cdot x \cdot x \cdot z \cdot z^2}{y \cdot z^2 \cdot 2 \cdot 5 \cdot x}$$
$$= \frac{xz}{2y}$$

24.
$$\frac{\frac{b}{a} - \frac{a}{b}}{\frac{1}{b} + \frac{1}{a}} = \frac{\left(\frac{b}{a} - \frac{a}{b}\right)ab}{\left(\frac{1}{b} + \frac{1}{a}\right)ab}$$
$$= \frac{b^2 - a^2}{a + b}$$
$$= \frac{(b-a)(b+a)}{a+b}$$
$$= b - a$$

25.
$$\frac{5 - \frac{1}{y^2}}{\frac{1}{y} + \frac{2}{y^2}} = \frac{y^2\left(5 - \frac{1}{y^2}\right)}{y^2\left(\frac{1}{y} + \frac{2}{y^2}\right)} = \frac{5y^2 - 1}{y + 2}$$

26.
$$\frac{8}{x} = \frac{10}{15}$$
$$8(15) = 10x$$
$$120 = 10x$$
$$12 = x$$

27. Let n be the number.

$$n + 5\left(\frac{1}{n}\right) = 6$$

$$n + \frac{5}{n} = 6$$

$$n\left(n + \frac{5}{n}\right) = n(6)$$

$$n^2 + 5 = 6n$$

$$n^2 - 6n + 5 = 0$$

$$(n-5)(n-1) = 0$$

$$n - 5 = 0 \quad \text{or} \quad n - 1 = 0$$

$$n = 5 \qquad\qquad n = 1$$

The number is 1 or 5.

28. Let x be the speed of the boat in still water.

Let $x + 2$ be the speed of the boat going downstream. Let $x - 2$ be the speed of the boat going upstream.

	distance	=	rate	·	time
Upstream	14		$x - 2$		$\frac{14}{x-2}$
Downstream	16		$x + 2$		$\frac{16}{x+2}$

$$\frac{14}{x-2} = \frac{16}{x+2}$$

$$14(x+2) = 16(x-2)$$

$$14x + 28 = 16x - 32$$

$$60 = 2x$$

$$30 = x$$

The speed of the boat in still water is 30 mph.

29. Let x be the time in hours that it takes for both inlet pipes together to fill the tank.

The first pipe fills $\frac{1}{12}$ of the tank in 1 hour, the second pipe fills $\frac{1}{15}$ of the tank in 1 hour, and the two pipes fill $\frac{1}{x}$ of the tank in 1 hour.

$$\frac{1}{12} + \frac{1}{15} = \frac{1}{x}$$

The LCD is $60x$.

$$60x\left(\frac{1}{12} + \frac{1}{15}\right) = 60x\left(\frac{1}{x}\right)$$
$$5x + 4x = 60$$
$$9x = 60$$
$$x = \frac{60}{9}$$
$$x = \frac{20}{3}$$

It takes both pipes $\frac{20}{3} = 6\frac{2}{3}$ hours to fill the tank.

30.
$$\frac{3}{85} = \frac{x}{510}$$
$$3(510) = 85x$$
$$1530 = 85x$$
$$18 = x$$

18 defective bulbs should be found in 510 bulbs.

Cumulative Review Chapters 1–5

1. a. The quotient of 15 and a number is 4 is written as $\frac{15}{x} = 4$.

    b. Three subtracted from 12 is a number is written as $12 - 3 = x$.

    c. 17 added to four times a number is 21 is written as $4x + 17 = 21$.

2. a. The difference of 12 and a number is −45 is written as $12 - x = -45$.

    b. The product of 12 and a number is −45 is written as $12x = -45$.

    c. A number less 10 is twice the number is written as $x - 10 = 2x$.

3. a. $3 + (-7) + (-8) = (-4) + (-8) = -12$

    b. $[7 + (-10)] + [-2 + (-4)] = (-3) + (-6)$
$$= -9$$

4. a. $28 - 6 - 30 = 22 - 30 = -8$

    b. $7 - 2 - 22 = 5 - 22 = -17$

5. $3(x + y) = 3 \cdot x + 3 \cdot y$ illustrates the distributive property.

6. $3 + y = y + 3$ illustrates the commutative property of addition.

7. $(x + 7) + 9 = x + (7 + 9)$ illustrates the associative property of addition.

8. $(x \cdot 7) \cdot 9 = x \cdot (7 \cdot 9)$ illustrates the associative property of multiplication.

9.
$$3 - x = 7$$
$$2 - x - 3 = 7 - 3$$
$$-x = 4$$
$$(-1)(-x) = (-1)4$$
$$x = -4$$
The solution is −4.

10.
$$7x - 6 = 6x - 6$$
$$7x - 6 + 6 = 6x - 6 + 6$$
$$7x = 6x$$
$$7x - 6x = 6x - 6x$$
$$x = 0$$
The solution is 0.

11. Let x be the length of the shorter piece. Then $4x$ is the length of the longer piece.
$$x + 4x = 10$$
$$5x = 10$$
$$x = 2$$
$$4x = 4(2) = 8$$
The shorter piece is 2 feet. The longer piece is 8 feet.

12. Let n be the first even integer. Then $n + 2$ is the second.
$$n + (n + 2) = 382$$
$$2n + 2 = 382$$
$$2n = 380$$
$$n = 190$$
$$n + 2 = 192$$
The two consecutive even integers are 190 and 192.

13.
$$y = mx + b$$
$$y - b = mx$$
$$\frac{y - b}{m} = x$$

14. $3x - 2y = 6$
$$3x = 6 + 2y$$
$$x = \frac{6 + 2y}{3}$$

15.
$$x + 4 \leq -6$$
$$x + 4 - 4 \leq -6 - 4$$
$$x \leq -10$$

16. $-3x + 7 > -x + 9$
$$7 > 2x + 9$$
$$-2 > 2x$$
$$-1 > x$$
$$\{x | x < -1\}$$

17. $\dfrac{x^5}{x^2} = x^{5-2} = x^3$

18. $\dfrac{y^{14}}{y^{14}} = 1$

19. $\dfrac{4^7}{4^3} = 4^{7-3} = 4^4 = 256$

20. $(x^5 y^2)^3 = x^{5 \cdot 3} y^{2 \cdot 3} = x^{15} y^6$

21. $\dfrac{(-3)^5}{(-3)^2} = (-3)^{5-2} = (-3)^3 = -27$

22. $\dfrac{x^{19} y^5}{xy} = \dfrac{x^{19}}{x} \cdot \dfrac{y^5}{y} = x^{19-1} \cdot y^{5-1} = x^{18} y^4$

23. $\dfrac{2x^5 y^2}{xy} = \dfrac{2}{1} \cdot \dfrac{x^5}{x} \cdot \dfrac{y^2}{y} = 2x^{5-1} y^{2-1} = 2x^4 y$

24. $(-3a^2 b)(5a^3 b) = -3 \cdot 5 \cdot a^2 \cdot a^3 \cdot b \cdot b$
$$= -15a^{2+3} b^{1+1}$$
$$= -15a^5 b^2$$

25. $2x^{-3} = 2 \cdot \dfrac{1}{x^3} = \dfrac{2}{x^3}$

26. $7^{-2} = \dfrac{1}{7^2} = \dfrac{1}{49}$

27. $(-2)^{-4} = \dfrac{1}{(-2)^4} = \dfrac{1}{16}$

28. $5z^{-7} = 5 \cdot \dfrac{1}{z^7} = \dfrac{5}{z^7}$

29. $5x(2x^3 + 6) = 5x(2x^3) + 5x(6) = 10x^4 + 30x$

30. $(x + 9)^2 = (x)^2 + 2(x)(9) + (9)^2$
$$= x^2 + 18x + 81$$

31. $-3x^2(5x^2 + 6x - 1)$
$$= -3x^2(5x^2) + (-3x^2)(6x) - (-3x^2)(1)$$
$$= -15x^4 - 18x^3 + 3x^2$$

32. $(2x + 1)(2x - 1) = (2x)^2 - (1)^2 = 4x^2 - 1$

33. $\dfrac{4x^2 + 7 + 8x^3}{2x + 3} = 4x^2 - 4x + 6 - \dfrac{11}{2x + 3}$

$$
\require{enclose}
\begin{array}{r}
4x^2 - 4x + 6 \\
2x+3 \enclose{longdiv}{8x^3 + 4x^2 + 0x^1 + 7} \\
\underline{8x^3 + 12x^2} \\
-8x^2 + 0x^1 \\
\underline{-8x^2 - 12x} \\
12x + 7 \\
\underline{12x + 18} \\
-11
\end{array}
$$

34. $\dfrac{4x^3 - 9x + 2}{x - 4} = 4x^2 + 16x + 55 + \dfrac{222}{x - 4}$

$$
\begin{array}{r}
4x^2 + 16x + 55 \\
x-4\overline{\smash{)}4x^3 -\ 0x^2 - 9x +\ \ \ 2} \\
\underline{4x^3 - 16x^2} \\
16x^2 -\ 9x \\
\underline{16x^2 - 64x} \\
55x +\ \ 2 \\
\underline{55x - 220} \\
222
\end{array}
$$

35. $x^2 + 7x + 12 = (x + 4)(x + 3)$

36. $-2a^2 + 10a + 12 = -2(a^2 - 5a - 6)$
$\qquad\qquad\qquad\quad = -2(a - 6)(a + 1)$

37. $25x^2 + 20xy + 4y^2 = (5x + 2y)(5x + 2y)$
$\qquad\qquad\qquad\qquad\ = (5x + 2y)^2$

38. $x^2 - 4 = (x - 2)(x + 2)$

39. $x^2 - 9x - 22 = 0$
$\quad (x - 11)(x + 2) = 0$
$\quad x - 11 = 0 \quad$ or $\quad x + 2 = 0$
$\qquad\ x = 11 \qquad\qquad\ x = -2$
The solutions are 11 and -2.

40. $\qquad 3x^2 + 5x = 2$
$\qquad 3x^2 + 5x - 2 = 0$
$\qquad (3x - 1)(x + 2) = 0$
$\quad 3x - 1 = 0 \quad$ or $\quad x + 2 = 0$
$\qquad\ 3x = 1 \qquad\qquad\ x = -2$
$\qquad\ \ x = \dfrac{1}{3}$

The solutions are $\dfrac{1}{3}$ and -2.

41. $\dfrac{x^2 + x}{3x} \cdot \dfrac{6}{5x + 5} = \dfrac{x(x+1)}{3x} \cdot \dfrac{2 \cdot 3}{5(x+1)} = \dfrac{2}{5}$

42. $\dfrac{2x^2 - 50}{4x^4 - 20x^3} = \dfrac{2(x^2 - 25)}{4x^3(x - 5)}$
$\qquad\qquad\quad = \dfrac{2(x - 5)(x + 5)}{2 \cdot 2x^3(x - 5)}$
$\qquad\qquad\quad = \dfrac{x + 5}{2x^3}$

43. $\dfrac{3x^2 + 2x}{x - 1} - \dfrac{10x - 5}{x - 1} = \dfrac{3x^2 + 2x - 10x + 5}{x - 1}$
$\qquad\qquad\qquad\qquad\ = \dfrac{3x^2 - 8x + 5}{x - 1}$
$\qquad\qquad\qquad\qquad\ = \dfrac{(3x - 5)(x - 1)}{x - 1}$
$\qquad\qquad\qquad\qquad\ = 3x - 5$

44. $7x^6 - 7x^5 + 7x^4 = 7x^4(x^2 - x + 1)$

$\dfrac{6x}{x^2 - 4} - \dfrac{3}{x + 2}$
$= \dfrac{6x}{(x + 2)(x - 2)} - \dfrac{3(x - 2)}{(x + 2)(x - 2)}$
$= \dfrac{6x - 3x + 6}{(x + 2)(x - 2)}$
$= \dfrac{3x + 6}{(x + 2)(x - 2)}$
$= \dfrac{3(x + 2)}{(x + 2)(x - 2)}$

45. $= \dfrac{3}{x - 2}$

46. $4x^2 + 12x + 9 = (2x + 3)(2x + 3) = (2x + 3)^2$

47. $\dfrac{t - 4}{2} - \dfrac{t - 3}{9} = \dfrac{5}{18}$
$18\left(\dfrac{t - 4}{2} - \dfrac{t - 3}{9}\right) = 18\left(\dfrac{5}{18}\right)$
$9(t - 4) - 2(t - 3) = 5$
$9t - 36 - 2t + 6 = 5$
$7t - 30 = 5$
$7t = 35$
$t = 5$
The solution is 5.

48. $\dfrac{6x^2-18x}{3x^2-2x}\cdot\dfrac{15x-10}{x^2-9}=\dfrac{6x(x-3)}{x(3x-2)}\cdot\dfrac{5(3x-2)}{(x-3)(x+3)}$

$$=\dfrac{30}{x+3}$$

49. Let x be the time in hours for Sam and Frank to complete the tour together. Sam completes $\dfrac{1}{3}$ of a tour in 1 hour, and Frank completes $\dfrac{1}{7}$ of a tour in 1 hour. Together, they complete $\dfrac{1}{x}$ of a tour in 1 hour.

$$\dfrac{1}{3}+\dfrac{1}{7}=\dfrac{1}{x}$$
$$21x\left(\dfrac{1}{3}+\dfrac{1}{7}\right)=21x\left(\dfrac{1}{x}\right)$$
$$7x+3x=21$$
$$10x=21$$
$$x=\dfrac{21}{10}=2\dfrac{1}{10}$$

Sam and Frank can complete a quality control tour together in $2\dfrac{1}{10}$ hours.

50. $\dfrac{\frac{m}{3}+\frac{n}{6}}{\frac{m+n}{12}}=\dfrac{12\left(\frac{m}{3}+\frac{n}{6}\right)}{12\left(\frac{m+n}{12}\right)}=\dfrac{4m+2n}{m+n}=\dfrac{2(2m+n)}{m+n}$

Chapter 6

Section 6.1 Practice Problems

1. a. Look for the shortest bar, which is the bar representing the Africa/Middle East region. To approximate the number, move from the right edge of this bar vertically downward. This region has approximately 11 million Internet users.

b. The Asia/Pacific region has approximately 187 million Internet users. The Latin America region has approximately 33 million Internet users. Subtract, $187 - 33 = 154$, to find that approximately 154 million more Internet users are in the Asia/Pacific region.

2. a. Locate 40 along the time axis and move vertically upward to the line. Then move horizontally to the left until the pulse rate axis is reached. The pulse rate is 70 beats per minute 40 minutes after lighting a cigarette.

b. Read the pulse rate for time = 0. The pulse rate is 60 beats per minute when the cigarette in being lit.

c. Read the time when the pulse rate is the highest, i.e., when the line is at the peak. The pulse rate is the highest 5 minutes after lighting the cigarette.

3. a. Point (4, 2) lies in quadrant I.

b. Point (−1, −3) lies in quadrant III.

c. Point (2, −2) lies in quadrant IV.

d. Point (−5, 1) lies in quadrant II.

e. Point (0, 3) lies on the y-axis.

f. Point (3, 0) lies on the x-axis.

g. Point (0, −4) lies on the y-axis.

h. Point $\left(-2\frac{1}{2}, 0\right)$ lies on the x-axis.

i. Point $\left(1, -3\frac{3}{4}\right)$ lies in quadrant IV.

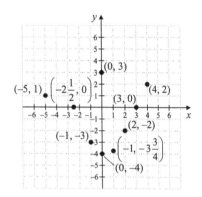

4. a. (1998, 1424), (1999, 1343), (2000, 997), (2001, 1216), (2002, 941), (2003, 1376)

b.

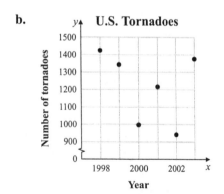

c. The number of tornadoes varies greatly from year to year.

5. a. $x + 2y = 8$
In (0,   ), the x-coordinate is 0.
$0 + 2y = 8$
$2y = 8$
$y = 4$
The ordered pair is (0, 4).

b. $x + 2y = 8$
In (, 3) the y-coordinate is 3.
$x + 2(3) = 8$
$x + 6 = 8$
$x = 2$
The ordered pair is (2, 3).

c. $x + 2y = 8$
In (−4,), the x-coordinate is −4.
$-4 + 2y = 8$
$2y = 12$
$y = 6$
The ordered pair is (−4, 6).

6. a. Let $x = -3$
$y = -2x$
$y = -2(-3)$
$y = 6$

b. Let $y = 0$.
$y = -2x$
$0 = -2x$
$0 = x$

c. Let $y = 10$.
$y = -2x$
$10 = -2x$
$-5 = x$

The ordered pairs are (−3, 6), (0, 0), and (−5, 10).

	x	y
a.	−3	6
b.	0	0
c.	−5	10

7. a. Let $x = -3$.
$y = \dfrac{1}{3}x - 1$
$y = \dfrac{1}{3}(-3) - 1$
$y = -1 - 1$
$y = -2$

b. Let $x = 0$.
$y = \dfrac{1}{3}x - 1$
$y = \dfrac{1}{3}(0) - 1$
$y = 0 - 1$
$y = -1$

c. Let $y = 0$.
$0 = \dfrac{1}{3}x - 1$
$1 = \dfrac{1}{3}x$
$3(1) = x$
$3 = x$

The ordered pairs are (−3, −2), (0, −1), and (3, 0).

	x	y
a.	−3	−2
b.	0	−1
c.	3	0

8. When $x = 1$,
$y = -50x + 400$
$y = -50(1) + 400$
$y = -50 + 400$
$y = 350$

When $x = 2$,
$y = -50x + 400$
$y = -50(2) + 400$
$y = -100 + 400$
$y = 300$

When $x = 3$,
$y = -50x + 400$
$y = -50(3) + 400$
$y = -150 + 400$
$y = 250$

When $x = 4$,
$y = -50x + 400$
$y = -50(4) + 400$
$y = -200 + 400$
$y = 200$

When $x = 5$,
$y = -50x + 400$
$y = -50(5) + 400$
$y = -250 + 400$
$y = 150$

When $x = 6$,
$y = -50x + 400$
$y = -50(6) + 400$
$y = -300 + 400$
$y = 100$

When $x = 7$,
$y = -50x + 400$
$y = -50(7) + 400$
$y = -350 + 400$
$y = 50$

The completed table is shown.

x	1	2	3	4	5	6	7
y	350	300	250	200	150	100	50

Mental Math

1. $x + y = 10$
   Answers may vary.
   Example: $(5, 5)$, $(7, 3)$

2. $x + y = 6$
   Answers may vary.
   Example: $(0, 6)$, $(6, 0)$

Exercise Set 6.1

1. The tallest bar corresponds to France, so France is the most popular tourist destination.

3. Find the bars that have heights greater than 40. France, U.S., and Spain have more than 40 million tourists per year.

5. The height of the bar is near 40, so approximately 40 million tourists go to Italy each year.

7. The line is at about 72,600 for year 2000. Thus, the attendance is 72,600 in 2000.

9. The highest point corresponds to 1999, and is at a height of about 74,800.

11. The line is at about 50 for year 1986. Thus, there were approximately 50 students per computer in 1986.

13. The greatest decrease was from 1984 to 1986.

15. The line falls below 15 in 1994.

17.

$(1, 5)$ and $(3.7, 2.2)$ are in quadrant I.

$\left(-1, 4\frac{1}{2}\right)$ is in quadrant II.

$(-5, -2)$ is in quadrant III.

$(2, -4)$ and $\left(\frac{1}{2}, -3\right)$ are in quadrant IV.

$(-3, 0)$ lies on the *x*-axis.
$(0, -1)$ lies on the *y*-axis.

19. Point *A* is at the origin, so its coordinates are $(0, 0)$.

21. Point *C* is 3 units right and 2 units up from the origin, so its coordinates are $(3, 2)$.

23. Point *E* is 2 units left and 2 units down from the origin, so its coordinates are $(-2, -2)$.

25. Point *G* is 2 units right and 1 unit down from the origin, so its coordinates are $(2, -1)$.

27. Point *B* is on the *y*-axis and 3 units down from the origin, so its coordinates are $(0, -3)$.

29. Point *D* is 1 unit right and 3 units up from the origin, so its coordinates are $(1, 3)$.

31. Point *F* is 3 units left and 1 unit down from the origin, so its coordinates are $(-3, -1)$.

33. a. The ordered pairs are $(2001, 28.5)$, $(2002, 29.5)$, $(2003, 32.4)$, and $(2004, 34.3)$.

b. The ordered pair $(2004, 34.3)$ indicates that in 2004, \$34.3 billion was spent on pet-related expenditure.

c.

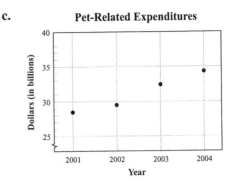

d. The scatter diagram shows that pet-related expenditures increased every year.

35. a. The ordered pairs are $(0.50, 10)$, $(0.75, 12)$, $(1.00, 15)$, $(1.25, 16)$, $(1.50, 18)$, $(1.50, 19)$, $(1.75, 19)$, and $(2.00, 20)$.

b. The ordered pair $(1.25, 16)$ indicates that when Minh studied 1.25 hours, her quiz score was 16.

c.

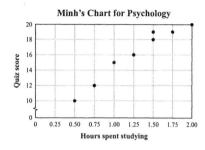

Minh's Chart for Psychology

d. answers may vary

37. $x - 4y = 4$

In (, -2), the y-coordinate is -2.

$x - 4(-2) = 4$

$\quad x + 8 = 4$

$\quad\quad x = -4$

In (4,), the x-coordinate is 4.

$4 - 4y = 4$

$\quad -4y = 0$

$\quad\quad y = 0$

The completed coordinates are $(-4, -2)$ and $(4, 0)$.

39. $y = \dfrac{1}{4}x - 3$

In (-8,), the x-coordinate is -8.

$y = \dfrac{1}{4}(-8) - 3 = -2 - 3 = -5$

In (, 1), the y-coordinate is 1.

$1 = \dfrac{1}{4}x - 3$

$4 = \dfrac{1}{4}x$

$16 = x$

The completed coordinates are $(-8, -5)$ and $(16, 1)$.

41. $y = -7x$

$-\dfrac{1}{7}y = x$

$x = -\dfrac{1}{7}y$	$y = -7x$
0	$-7(0) = 0$
-1	$-7(-1) = 7$
$-\dfrac{1}{7}(2) = -\dfrac{2}{7}$	2

43. $y = -x + 2$

$y - 2 = -x$

$-y + 2 = x$

$x = -y + 2$	$y = -x + 2$
0	$-0 + 2 = 2$
$-0 + 2 = 2$	0
-3	$-(-3) + 2 = 3 + 2 = 5$

45. $y = \dfrac{1}{2}x$

$2y = x$

$x = 2y$	$y = \frac{1}{2}x$
0	$\frac{1}{2}(0) = 0$
-6	$\frac{1}{2}(-6) = -3$
$2(1) = 2$	1

47. $x + 3y = 6$ $3y = -x + 6$

$\quad x = -3y + 6$ $y = -\dfrac{1}{3}x + 2$

$x = -3y + 6$	$y = -\frac{1}{3}x + 2$
0	$-\frac{1}{3}(0) + 2 = 0 + 2 = 2$
$-3(0) + 6 = 0 + 6$ $= 6$	0
$-3(1) + 6 = -3 + 6$ $= 3$	1

49.
$$y = 2x - 12$$
$$y + 12 = 2x$$
$$\frac{1}{2}(y+12) = x$$

$x = \frac{1}{2}(y+12)$	$y = 2x - 12$
0	$2(0) - 12 = 0 - 12 = -12$
$\frac{1}{2}(-2+12) = \frac{1}{2}(10) = 5$	-2
3	$2(3) - 12 = 6 - 12 = -6$

51.
$$2x + 7y = 5 \qquad\qquad 7y = -2x + 5$$
$$2x = -7y + 5 \qquad\qquad y = \frac{-2x+5}{7}$$
$$x = \frac{-7y+5}{2}$$

$x = \frac{-7y+5}{2}$	$y = \frac{-2x+5}{7}$
0	$\frac{-2(0)+5}{7} = \frac{5}{7}$
$\frac{-7(0)+5}{2} = \frac{5}{2}$	0
$\frac{-7(1)+5}{2} = \frac{-2}{2} = -1$	1

53. $x = -5y$
$y = 0: x = -5(0) = 0$
$y = 1: x = -5(1) = -5$
$x = 10: \ 10 = -5y$
$\qquad\qquad -2 = y$

The ordered pairs are $(0, 0)$, $(-5, 1)$, and $(10, -2)$.

x	y
0	0
-5	1
10	-2

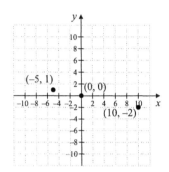

55. $y = \dfrac{1}{3}x + 2$

$x = 0$: $\quad y = \dfrac{1}{3}(0) + 2 = 0 + 2 = 2$

$x = -3$: $\quad y = \dfrac{1}{3}(-3) + 2 = -1 + 2 = 1$

$y = 0$: $\quad 0 = \dfrac{1}{3}x + 2$

$\qquad\qquad -2 = \dfrac{1}{3}x$

$\qquad\qquad -6 = x$

The ordered pairs are $(0, 2)$, $(-3, 1)$, and $(-6, 0)$.

x	y
0	2
−3	1
−6	0

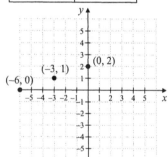

57. a.

x	100	200	300
$y = 80x + 5000$	$80(100) + 5000$ $= 8000 + 5000$ $= 13{,}000$	$80(200) + 5000$ $= 16{,}000 + 5000$ $= 21{,}000$	$80(300) + 5000$ $= 24{,}000 + 5000$ $= 29{,}000$

b. Find x when $y = 8600$.

$$8600 = 80x + 5000$$
$$3600 = 80x$$
$$45 = x$$

45 computer desks can be produced for $8600.

59. a.

x	1	3	5
$y = -2.4x + 13$	$-2.4(1) + 13$ $= -2.4 + 13$ $= 10.6$	$-2.4(3) + 13$ $= -7.2 + 13$ $= 5.8$	$-2.4(5) + 13$ $= -12.0 + 13$ $= 1$

b. Find x when $y = 3$.

$$3 = -2.4x + 13$$
$$-10 = -2.4x$$
$$4 = x$$
$$1998 + 4 = 2002$$

Approximately 3% were cassettes in 2002.

61. $x + y = 5$
$$y = 5 - x$$

63. $2x + 4y = 5$
$$4y = 5 - 2x$$
$$y = \frac{5 - 2x}{4}$$

65. $10x = -5y$
$$-2x = y$$
$$y = -2x$$

67. False; point $(-1, 5)$ lies in quadrant II.

69. True

71. Points in quadrant III are to the left and down from the origin, so the x- and y-coordinates are negative. (negative, negative) corresponds to quadrant III.

73. Points in quadrant IV are to the right and down from the origin, so the x-coordinate is positive and the y-coordinate is negative. (positive, negative) corresponds to quadrant IV.

75. The origin corresponds to $(0, 0)$.

77. If the x-coordinate of a point is 0, the point is neither to the left nor to the right of the origin, so it is on the y-axis.

79. no; answers may vary

81. answers may vary

83. A point four units to right of the y-axis and seven units below the x-axis has coordinates $(4, -7)$.

85. The length of the rectangle is $3 - (-1) = 4$ and the width of the rectangle is $5 - (-4) = 9$.
$$\begin{aligned} \text{Perimeter} &= 2(\text{length}) + 2(\text{width}) \\ &= 2(4) + 2(9) \\ &= 8 + 18 \\ &= 26 \end{aligned}$$
The perimeter is 26 units.

87. The revenues are approximately $26 billion, $27 billion, $26 billion, and $28 billion.

89. answers may vary

Section 6.2 Practice Problems

1. $x + 3y = 6$
$$\begin{aligned} x = 0:\ 0 + 3y &= 6 \\ 3y &= 6 \\ y &= 2 \end{aligned}$$
$$\begin{aligned} x = 3:\ 3 + 3y &= 6 \\ 3y &= 3 \\ y &= 1 \end{aligned}$$
$$\begin{aligned} y = 0:\ x + 3(0) &= 6 \\ x + 0 &= 6 \\ x &= 6 \end{aligned}$$
The ordered pairs are $(0, 2)$, $(3, 1)$, and $(6, 0)$.

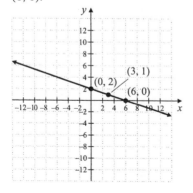

2. $-2x + 4y = 8$
$$\begin{aligned} x = -2:\ -2(-2) + 4y &= 8 \\ 4 + 4y &= 8 \\ 4y &= 4 \\ y &= 1 \end{aligned}$$
$$\begin{aligned} x = 0:\ -2(0) + 4y &= 8 \\ 0 + 4y &= 8 \\ 4y &= 8 \\ y &= 2 \end{aligned}$$
$$\begin{aligned} x = 2:\ -2(2) + 4y &= 8 \\ -4 + 4y &= 8 \\ 4y &= 12 \\ y &= 3 \end{aligned}$$
The ordered pairs are $(-2, 1)$, $(0, 2)$, and $(2, 3)$.

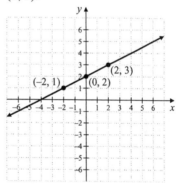

3. $y = 2x$
$$\begin{aligned} x = -2:\ y &= 2(-2) = -4 \\ x = 0:\ y &= 2(0) = 0 \\ x = 3:\ y &= 2(3) = 6 \end{aligned}$$
The ordered pairs are $(-2, -4)$, $(0, 0)$, and $(3, 6)$.

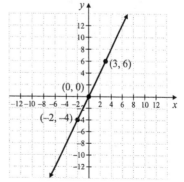

4. $y = -\dfrac{1}{2}x + 4$

$x = -6$: $y = -\dfrac{1}{2}(-6) + 4 = 3 + 4 = 7$

$x = 0$: $-\dfrac{1}{2}(0) + 4 = 0 + 4 = 4$

$x = 4$: $y = -\dfrac{1}{2}(4) + 4 = -2 + 4 = 2$

The ordered pairs are $(-6, 7)$, $(0, 4)$, and $(4, 2)$.

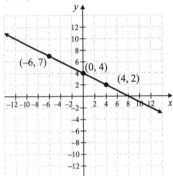

5. The equation $x = 3$ can be written in standard form as $x + 0y = 3$. No matter what value replaces y, x is always 3. It is a vertical line. Plot points $(3, 2)$, $(3, 0)$, and $(3, -4)$, for example.

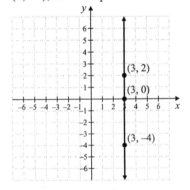

6. $x = 2004 - 1995 = 9$

Find 9 on the x-axis. Move vertically upward to the line and then horizontally to the left. In 2004, we predict that there will be 465 thousand medical assistants.

Calculator Explorations

1. $y = -3x + 7$

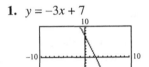

2. $y = -x + 5$

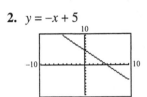

3. $y = 2.5x - 7.9$

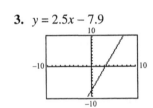

4. $y = -1.3x + 5.2$

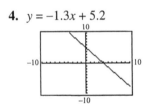

5. $y = -\dfrac{3}{10}x + \dfrac{32}{5}$

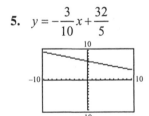

6. $y = \dfrac{2}{9}x - \dfrac{22}{3}$

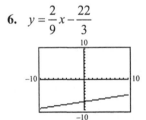

Exercise Set 6.2

1. $x - y = 6$
   $y = 0$: $x - 0 = 6$
            $x = 6$
   $x = 4$: $4 - y = 6$
          $-y = 2$
           $y = -2$
   $y = -1$: $x - (-1) = 6$
           $x + 1 = 6$
             $x = 5$
The ordered pairs are $(6, 0)$, $(4, -2)$, and $(5, -1)$.

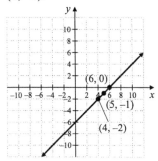

3. $y = -4x$
   $x = 1$: $y = -4(1) = -4$
   $x = 0$: $y = 4(0) = 0$
   $x = -1$: $y = -4(-1) = 4$
The ordered pairs are $(1, -4)$, $(0, 0)$, and $(-1, 4)$.

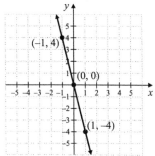

5. $y = \dfrac{1}{3}x$

   $x = 0$: $y = \dfrac{1}{3}(0) = 0$

$x = 6$: $y = \dfrac{1}{3}(6) = 2$

$x = -3$: $y = \dfrac{1}{3}(-3) = -1$

The ordered pairs are $(0, 0)$, $(6, 2)$, and $(-3, -1)$.

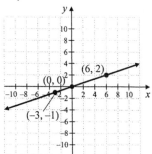

7. $y = -4x + 3$
   $x = 0$: $y = -4(0) + 3 = 0 + 3 = 3$
   $x = 1$: $y = -4(1) + 3 = -4 + 3 = -1$
   $x = 2$: $y = -4(2) + 3 = -8 + 3 = -5$
The ordered pairs are $(0, 3)$, $(1, -1)$, and $(2, -5)$.

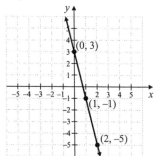

9. $x + y = 1$

11. $x - y = -2$

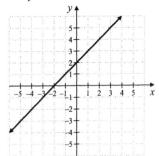

13. $x - 2y = 6$

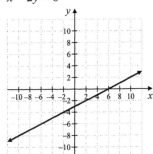

15. $y = 6x + 3$

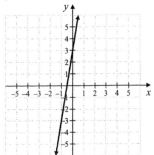

17. $x = -4$

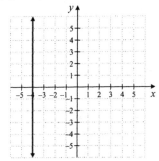

19. $y = 3$

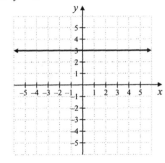

21. $y = x$

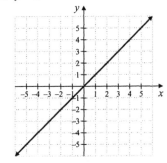

23. $y = 5x$

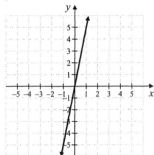

25. $x + 3y = 9$

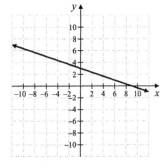

27. $y = \dfrac{1}{2}x - 1$

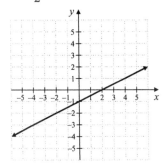

29. $3x - 2y = 12$

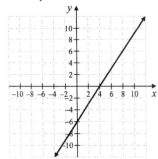

31. $y = -3.5x + 4$

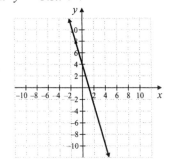

33. a.

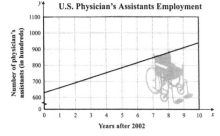

U.S. Physician's Assistants Employment

b. $x = 6$: $y = 31(6) + 630 = 186 + 630 = 816$

Yes, the point (6, 816) lies on the line. answers may vary

35. a.

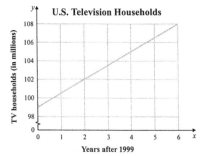

U.S. Television Households

b. $x = 5$

$y = 1.5x + 99$

$\quad = 1.5(5) + 99$

$\quad = 7.5 + 99$

$\quad = 106.5$

The ordered pair is (5, 106.5).

c. Year $= x + 1999 = 5 + 1999 = 2004$

In 2004, there were 106.5 million households in the United States with at least one television.

37. The fourth vertex is the bottom-right corner of the rectangle. The x-coordinate must line the point up with the top-right corner, and the y-coordinate must line the point up with the bottom-left corner. The coordinates are $(4, -1)$.

39. $x - y = -3$

$x = 0$: $0 - y = -3$

$\qquad\qquad y = 3$

$y = 0$: $x - 0 = -3$

$\qquad\qquad x = -3$

x	y
0	3
−3	0

41. $y = 2x$
$x = 0$: $y = 2(0) = 0$
$y = 0$: $0 = 2x$
$0 = x$

x	y
0	0
0	0

43. $y = 5x$
$y = 5x + 4$

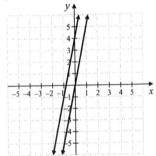

45. $y = -2x$
$y = -2x - 3$

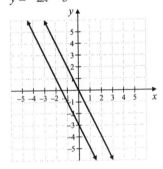

47. $y = x^2$

$x = 0$: $y = 0^2 = 0$

$x = 1$: $y = 1^2 = 1$

$x = -1$: $y = (-1)^2 = 1$

$x = 2$: $y = 2^2 = 4$

$x = -2$: $y = (-2)^2 = 4$

x	y
0	0
1	1
−1	1
2	4
−2	4

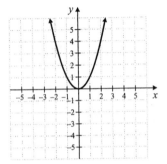

49. The perimeter is the distance around.
$x + 5 + y + 5 = 22$
$x + y + 10 = 22$
$x + y = 12$
$x = 3$: $3 + y = 12$
$y = 9$
If x is 3 centimeters, then y is 9 centimeters.

51. Yes; answers may vary

Section 6.3 Practice Problems

1. x-intercept: $(2, 0)$
y-intercept: $(0, -4)$

2. x-intercepts: $(-4, 0)$, $(2, 0)$
y-intercept: $(0, 2)$

3. x-intercept and y-intercept: $(0, 0)$

4. $2x - y = 4$

$y = 0$: $2x - 0 = 4$

$\qquad 2x = 4$

$\qquad x = 2$

x-intercept: $(2, 0)$

$x = 0$: $2(0) - y = 4$

$\qquad 0 - y = 4$

$\qquad -y = 4$

$\qquad y = -4$

y-intercept: $(0, -4)$

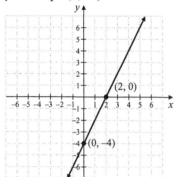

5. $y = 3x$

$y = 0$: $0 = 3x$

$\qquad 0 = x$

x-intercept: $(0, 0)$

$x = 0$: $y = 3(0)$

$\qquad y = 0$

y-intercept: $(0, 0)$

Let $x = 1$ to find a second point.

$x = 1$: $y = 3$

$(1, 3)$ is another point on the line.

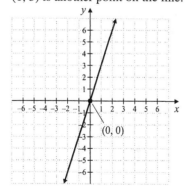

6. $x = -3$

The graph of $x = -3$ is a vertical line with x-intercept $(-3, 0)$.

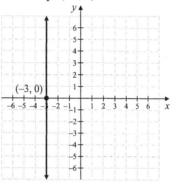

7. $y = 4$

The graph of $y = 4$ is a horizontal line with y-intercept $(0, 4)$.

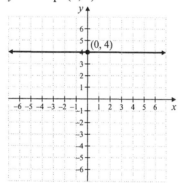

Calculator Explorations

1. $x = 3.78y$

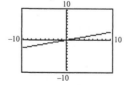

2. $-2.61y = x$

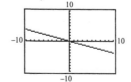

3. $3x + 7y = 21$

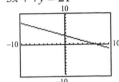

4. $-4x + 6y = 12$

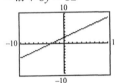

5. $-2.2x + 6.8y = 15.5$

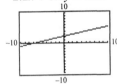

6. $5.9x - 0.8y = -10.4$

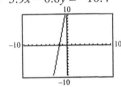

Mental Math

1. False

2. False

3. True

4. True

5. False

6. True

Exercise Set 6.3

1. x-intercept: $(-1, 0)$
    y-intercept: $(0, 1)$

3. x-intercepts: $(-2, 0)$, $(2, 0)$
    y-intercept: $(0, -2)$

5. x-intercepts: $(-2, 0)$, $(1, 0)$, $(3, 0)$
    y-intercept: $(0, 3)$

7. x-intercepts: $(-1, 0)$, $(1, 0)$
    y-intercepts: $(0, 1)$, $(0, -2)$

9. $x - y = 3$
    $y = 0$: $x - 0 = 0$
           $x = 3$
    x-intercept: $(3, 0)$
    $x = 0$: $0 - y = 3$
           $-y = 3$
            $y = -3$
    y-intercept: $(0, -3)$

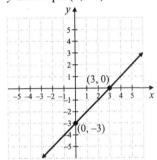

11. $x = 5y$
    $y = 0$: $x = 5(0) = 0$
    x-intercept: $(0, 0)$
    $x = 0$: $0 = 5y$
           $0 = y$
    y-intercept: $(0, 0)$
    Let $y = 1$ to find a second point.
    $y = 1$: $x = 5$
    $(5, 1)$ is another point on the line.

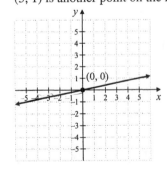

13. $-x + 2y = 6$

$y = 0$: $-x + 2(0) = 6$

$\qquad\qquad -x = 6$

$\qquad\qquad\quad x = -6$

x-intercept: $(-6, 0)$

$x = 0$: $-0 + 2y = 6$

$\qquad\qquad 2y = 6$

$\qquad\qquad\; y = 3$

y-intercept: $(0, 3)$

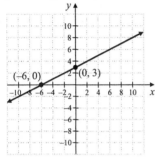

15. $2x - 4y = 8$

$y = 0$: $2x - 4(0) = 8$

$\qquad\qquad 2x - 0 = 8$

$\qquad\qquad\quad 2x = 8$

$\qquad\qquad\quad\; x = 4$

x-intercept: $(4, 0)$

$x = 0$: $2(0) - 4y = 8$

$\qquad\qquad 0 - 4y = 8$

$\qquad\qquad\;\; -4y = 8$

$\qquad\qquad\qquad y = -2$

y-intercept: $(0, -2)$

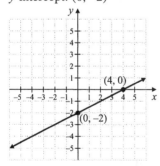

17. $2x - y = 0$

$y = 0$: $2x - 0 = 0$

$\qquad\qquad 2x = 0$

$\qquad\qquad\; x = 0$

x-intercept: $(0, 0)$

$(0, 0)$ is also the y-intercept. Let $x = 1$ to find a second point.

$x = 1$: $2(1) - y = 0$

$\qquad\qquad 2 - y = 0$

$\qquad\qquad\;\; -y = -2$

$\qquad\qquad\qquad y = 2$

$(1, 2)$ is another point on the line.

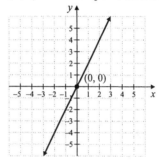

19. $y = 3x + 6$

$y = 0$: $0 = 3x + 6$

$\qquad\quad -6 = 3x$

$\qquad\quad -2 = x$

x-intercept: $(-2, 0)$

$x = 0$: $y = 3(0) + 6$

$\qquad\qquad y = 0 + 6$

$\qquad\qquad y = 6$

y-intercept: $(0, 6)$

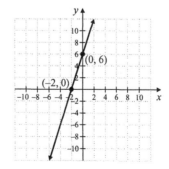

21. The graph of $x = -1$ is a vertical line with x-intercept $(-1, 0)$.

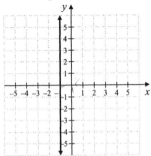

23. The graph of $y = 0$ is a horizontal line with y-intercept $(0, 0)$.

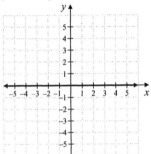

25. $y + 7 = 0$
$$y = -7$$

The graph of $y = -7$ is a horizontal line with y-intercept $(0, -7)$.

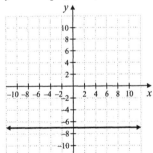

27. $x + 3 = 0$
$$x = -3$$
The graph of $x = -3$ is a vertical line with x-intercept $(0, -3)$.

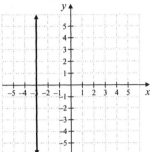

29. $x = y$
$y = 0$: $x = 0$
x-intercept: $(0, 0)$
$(0, 0)$ is also the y-intercept. Let $x = 3$ to find a second point.
$x = 3$: $3 = y$
$(3, 3)$ is another point on the line.

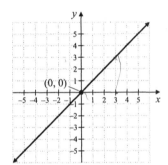

31. $x + 8y = 8$
$y = 0$: $x + 8(0) = 8$
$$x + 0 = 8$$
$$x = 8$$
x-intercept: $(8, 0)$
$x = 0$: $0 + 8y = 8$
$$8y = 8$$
$$y = 1$$
y-intercept: $(0, 1)$

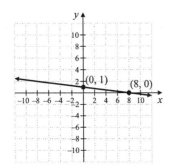

33. $5 = 6x - y$

$y = 0$: $5 = 6x - 0$

$\quad\quad\quad 5 = 6x$

$\quad\quad\quad \dfrac{5}{6} = x$

x-intercept: $\left(\dfrac{5}{6}, 0\right)$

$x = 0$: $\quad 5 = 6(0) - y$

$\quad\quad\quad 5 = -y$

$\quad\quad\quad -5 = y$

y-intercept: $(0, -5)$

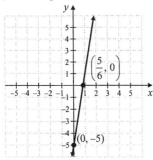

35. $-x + 10y = 11$

$y = 0$: $-x + 10(0) = 11$

$\quad\quad\quad -x + 0 = 11$

$\quad\quad\quad\quad\quad -x = 11$

$\quad\quad\quad\quad\quad\quad x = -11$

x-intercept: $(-11, 0)$

$x = 0$: $-0 + 10y = 11$

$\quad\quad\quad\quad 10y = 11$

$\quad\quad\quad\quad\quad y = \dfrac{11}{10}$

y-intercept: $\left(0, \dfrac{11}{10}\right)$

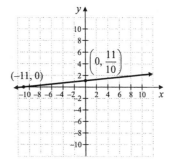

37. $x = -4\dfrac{1}{2}$

This is a vertical line with x-intercept

$\left(-4\dfrac{1}{2}, 0\right)$.

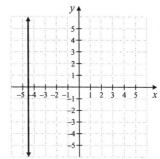

39. $y = 3\dfrac{1}{4}$

This is a horizontal line with y-intercept

$\left(0, 3\dfrac{1}{4}\right)$.

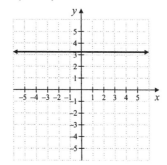

41. $y = -\dfrac{2}{3}x + 1$

$y = 0:$ $0 = -\dfrac{2}{3}x + 1$

$\dfrac{2}{3}x = 1$

$x = \dfrac{3}{2}$

x-intercept: $\left(\dfrac{3}{2}, 0\right)$

$x = 0:$ $y = -\dfrac{2}{3}(0) + 1$

$y = 1$

y-intercept: $(0, 1)$

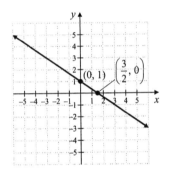

43. $4x - 6y + 2 = 0$

$y = 0:$ $4x - 6(0) + 2 = 0$

$4x - 0 + 2 = 0$

$4x = -2$

$x = -\dfrac{1}{2}$

x-intercept: $\left(-\dfrac{1}{2}, 0\right)$

$x = 0:$ $4(0) - 6y + 2 = 0$

$0 - 6y + 2 = 0$

$-6y = -2$

$y = \dfrac{1}{3}$

y-intercept: $\left(0, \dfrac{1}{3}\right)$

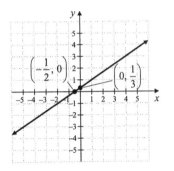

45. $\dfrac{-6 - 3}{2 - 8} = \dfrac{-9}{-6} = \dfrac{3}{2}$

47. $\dfrac{-8 - (-2)}{-3 - (-2)} = \dfrac{-8 + 2}{-3 + 2} = \dfrac{-6}{-1} = 6$

49. $\dfrac{0 - 6}{5 - 0} = \dfrac{-6}{5} = -\dfrac{6}{5}$

51. The graph of $y = 3$ is a horizontal line with y-intercept $(0, 3)$. This is graph c.

53. The graph of $x = 3$ is a vertical line with x-intercept $(3, 0)$. This is graph a.

55. The line can be on the x-axis or the y-axis so it can have infinitely many x- and y-intercepts.

57. A circle can have no x- and y-intercepts. That is, it does not have to intersect the axes.

59. answers may vary

61. a. $3x + 6y = 1200$

$x = 0:$ $3(0) + 6y = 1200$

$6y = 1200$

$y = 200$

The ordered pair $(0, 200)$ corresponds to manufacturing 0 chairs and 200 desks.

b. $3x + 6y = 1200$
$y = 0$: $3x + 6(0) = 1200$
$\qquad\quad 3x = 1200$
$\qquad\qquad x = 400$
The ordered pair (400, 0) corresponds to manufacturing 400 chairs and 0 desks.

c. Manufacturing 50 desks corresponds to $y = 50$.
$3x + 6y = 1200$
$y = 50$: $3x + 6(50) = 1200$
$\qquad\quad 3x + 300 = 1200$
$\qquad\qquad\quad 3x = 900$
$\qquad\qquad\quad\ x = 300$
When 50 desks are manufactured, 300 chairs can be manufactured.

63.

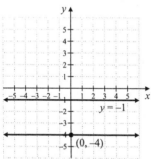

The equation of the line is $y = -4$.

65. a. $y = 29.2x + 919$
$x = 0$: $y = 29.2(0) + 919$
$\qquad\quad y = 919$
The y-intercept is (0, 919).

b. The y-intercept of (0, 919) means that there were 919 stores in 1999 (0 years after 1999).

Section 6.4 Practice Problems

1. $m = \dfrac{y_2 - y_1}{x_2 - x_1} = \dfrac{-1 - 3}{4 - (-2)} = \dfrac{-4}{6} = -\dfrac{2}{3}$

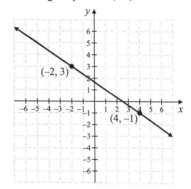

2. $m = \dfrac{y_2 - y_1}{x_2 - x_1} = \dfrac{5 - 1}{3 - (-2)} = \dfrac{4}{5}$

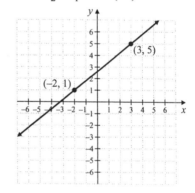

3. $5x + 4y = 10$
$\qquad 4y = -5x + 10$
$\qquad\ y = -\dfrac{5}{4}x + \dfrac{10}{4}$

The slope is $-\dfrac{5}{4}$.

4. $-y = -2x + 7$
$\dfrac{-y}{-1} = \dfrac{-2x}{-1} + \dfrac{7}{-1}$
$\quad y = 2x - 7$

The slope is 2.

5. $y = 3$ is a horizontal line. Horizontal lines have a slope of 0.

6. $x = -2$ is a vertical line. Vertical lines have undefined slopes.

7. a. $x + y = 5$
$$y = -x + 5$$
$$2x + y = 5$$
$$y = -2x + 5$$

slope $= -1$ slope $= -2$
The slopes are not the same, so the lines are not parallel. The product, $(-1)(-2) = 2$, is not -1, so the lines are not perpendicular.

b. $5y = 2x - 3$
$$y = \frac{2}{5}x - \frac{3}{5}$$
$$5x + 2y = 1$$
$$2y = -5x + 1$$
$$y = -\frac{5}{2}x + \frac{1}{2}$$

slope $= \dfrac{2}{5}$ slope $= -\dfrac{5}{2}$

The slopes are not the same, so the lines are not parallel. The product,
$$\left(\frac{2}{5}\right)\left(-\frac{5}{2}\right) = -1, \text{ is } -1, \text{ so the lines are}$$
perpendicular.

c. $y = 2x + 1$ $4x - 2y = 8$
$$-2y = -4x + 8$$
$$y = \frac{-4x}{-2} + \frac{8}{-2}$$
$$y = 2x - 4$$

slope $= 2$ slope $= 2$
The slopes are the same, so the lines are parallel.

8. grade $= \dfrac{\text{rise}}{\text{run}} = \dfrac{3}{20} = 0.15 = 15\%$
The grade is 15%.

9. $m = \dfrac{240 - 120}{1990 - 1980} = \dfrac{120}{10} = 12$
Each year the sales of food and drink from restaurants increases by \$12 billion.

Calculator Explorations

1.

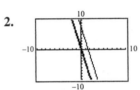

The lines are parallel since they all have a slope of 3.8. The graph of $y = 3.8x - 3$ is the graph of $y = 3.8x$ moved 3 units down with a y-intercept of -3. The graph of $y = 3.8x + 9$ is the graph of $y = 3.8x$ moved 9 units up with a y-intercept of 9.

2.

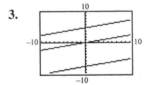

The lines are parallel since they all have a slope of -4.9. The graph of $y = -4.9x + 1$ is the graph of $y = -4.9x$ moved 1 unit up with a y-intercept of 1. The graph of $y = -4.9x + 8$ is the graph of $y = -4.9x$ moved 8 units up with a y-intercept of 8.

3.

The lines are parallel since they all have a slope of $\dfrac{1}{4}$. The graph of $y = \dfrac{1}{4}x + 5$ is the graph of $y = \dfrac{1}{4}x$ moved 5 units up with a y-intercept of 5. The graph of $y = \dfrac{1}{4}x - 8$ is the graph of $y = \dfrac{1}{4}x$ moved 8 units down with a y-intercept of -8.

4.

The lines are parallel since they all have a slope of $-\dfrac{3}{4}$. The graph of $y = -\dfrac{3}{4}x - 5$ is the graph of $y = -\dfrac{3}{4}x$ moved 5 units down with a y-intercept of -5. The graph of $y = -\dfrac{3}{4}x + 6$ is the graph of $y = -\dfrac{3}{4}x$ moved 6 units up with a y-intercept of 6.

Mental Math

1. The slope is negative.

2. The slope is positive.

3. The slope is undefined.

4. The slope is 0.

5. The slope is positive, so it is upward sloping.

6. The slope is negative, so it is downward sloping.

7. The slope is 0, so it is a horizontal line.

8. The slope is undefined, so it is a vertical line.

Exercise Set 6.4

1. $m = \dfrac{y_2 - y_1}{x_2 - x_1} = \dfrac{-2 - 5}{6 - (-1)} = \dfrac{-7}{7} = -1$

3. $m = \dfrac{y_2 - y_1}{x_2 - x_1} = \dfrac{3 - 4}{5 - 1} = \dfrac{-1}{4} = -\dfrac{1}{4}$

5. $m = \dfrac{y_2 - y_1}{x_2 - x_1} = \dfrac{1 - 1}{-2 - 5} = \dfrac{0}{-7} = 0$

7. $m = \dfrac{y_2 - y_1}{x_2 - x_1} = \dfrac{0 - 4}{5 - 5} = \dfrac{-4}{0}$
   The slope is undefined.

9. $(x_1, y_1) = (-1, 2), (x_2, y_2) = (2, -2)$
   $m = \dfrac{y_2 - y_1}{x_2 - x_1} = \dfrac{-2 - 2}{2 - (-1)} = \dfrac{-4}{3} = -\dfrac{4}{3}$

11. $(x_1, y_1) = (1, -2), (x_2, y_2) = (3, 3)$
    $m = \dfrac{y_2 - y_1}{x_2 - x_1} = \dfrac{3 - (-2)}{3 - 1} = \dfrac{3 + 2}{3 - 1} = \dfrac{5}{2}$

13. Line 1 has a positive slope and line 2 has a negative slope, so line 1 has the greater slope.

15. Line 2 is steeper, so it has the greater slope.

17. $y = 5x - 2$
    The slope is $m = 5$.

19. $y = -0.3x + 2.5$
    The slope is $m = -0.3$.

21. $2x + y = 7$
    $y = -2x + 7$
    The slope is $m = -2$.

23. The line is vertical, so it has an undefined slope.

25. $2x - 3y = 10$
    $-3y = -2x + 10$
    $y = \dfrac{2}{3}x - \dfrac{10}{3}$
    The slope is $m = \dfrac{2}{3}$.

27. $x = 1$ is a vertical line, so its slope is undefined.

29. $x = 2y$

$\dfrac{1}{2}x = y$

$y = \dfrac{1}{2}x$

The slope is $m = \dfrac{1}{2}$.

31. $y = -3$ is a horizontal line, so its slope is 0.

33. $-3x - 4y = 6$

$-4y = 3x + 6$

$y = -\dfrac{3}{4}x - \dfrac{3}{2}$

The slope is $m = -\dfrac{3}{4}$.

35. $20x - 5y = 1.2$

$-5y = -20x + 1.2$

$y = \dfrac{-20x}{-5} + \dfrac{1.2}{-5}$

$y = 4x - 0.24$

The slope is $m = 4$.

37. $y = \dfrac{2}{9}x + 3$

$y = -\dfrac{2}{9}x$

$\dfrac{2}{9} \neq -\dfrac{2}{9}$, so the lines are not parallel.

$\left(\dfrac{2}{9}\right)\left(-\dfrac{2}{9}\right) = -\dfrac{4}{81} \neq -1$, so the lines are not

perpendicular.
The lines are neither parallel nor perpendicular.

39. $x - 3y = -6$

$-3y = -x - 6$

$y = \dfrac{1}{3}x + 2$

$y = 3x - 9$

$\dfrac{1}{3} \neq 3$, so the lines are not parallel.

$\left(\dfrac{1}{3}\right)(3) = 1 \neq -1$, so the lines are not

perpendicular. The lines are neither parallel nor perpendicular.

41. $6x = 5y + 1$

$-5y = -6x + 1$

$y = \dfrac{6}{5}x - \dfrac{1}{5}$

$-12x + 10y = 1$

$10y = 12x + 1$

$y = \dfrac{6}{5}x + \dfrac{1}{10}$

Both lines have slope $\dfrac{6}{5}$ and the y-intercepts

are different, so they are parallel.

43. $6 + 4x = 3y$

$2 + \dfrac{4}{3}x = y$ or $y = \dfrac{4}{3}x + 2$

$3x + 4y = 8$

$4y = -3x + 8$

$y = -\dfrac{3}{4}x + 2$

$\left(\dfrac{4}{3}\right)\left(-\dfrac{3}{4}\right) = -1$, so the lines are

perpendicular.

45. $m = \dfrac{y_2 - y_1}{x_2 - x_1} = \dfrac{0 - (-3)}{0 - (-3)} = \dfrac{3}{3} = 1$

  a. The slope of a parallel line is 1.

  b. The slope of a perpendicular line is $-\dfrac{1}{1} = -1$.

47. $m = \dfrac{y_2 - y_1}{x_2 - x_1} = \dfrac{5 - (-4)}{3 - (-8)} = \dfrac{5 + 4}{3 + 8} = \dfrac{9}{11}$

  a. The slope of a parallel line is $\dfrac{9}{11}$.

b. The slope of a perpendicular line is

$$-\frac{1}{\frac{9}{11}} = -\frac{11}{9}.$$

49. $\text{slope} = \dfrac{\text{rise}}{\text{run}} = \dfrac{6 \text{ feet}}{10 \text{ feet}} = \dfrac{3}{5}$

The pitch of the roof is $\dfrac{3}{5}$.

51. $\text{slope} = \dfrac{\text{rise}}{\text{run}} = \dfrac{2}{16} = 0.125 = 12.5\%$

The grade of the road is 12.5%

53. $\text{grade} = \dfrac{\text{rise}}{\text{run}} = \dfrac{2580 \text{ meters}}{6450 \text{ meters}} = 0.40 = 40\%$

The grade of the track is 40%.

55. $\text{grade} = \dfrac{\text{rise}}{\text{run}} = \dfrac{10 \text{ m}}{12.66 \text{ m}} \approx 0.79 = 79\%$

The grade is 79%.

57. $m = \dfrac{y_2 - y_1}{x_2 - x_1} = \dfrac{86 - 74}{2006 - 2002} = \dfrac{12}{4} = 3 = \dfrac{3}{1}$

Every 1 year, there are/should be 3 million more U.S. households with personal computers.

59. $m = \dfrac{y_2 - y_1}{x_2 - x_1}$

$= \dfrac{8400 - 2100}{20,000 - 5000}$

$= \dfrac{6300}{15,000}$

$= 0.42$

It costs $0.42 per 1 mile to own and operate a compact car.

61. $y - (-6) = 2(x - 4)$

$\quad y + 6 = 2x - 8$

$\qquad y = 2x - 14$

63. $y - 1 = -6(x - (-2))$

$\quad y - 1 = -6(x + 2)$

$\quad y - 1 = -6x - 12$

$\qquad y = -6x - 11$

65. $(x_1, y_1) = (0, 0)$, $(x_2, y_2) = (1, 1)$

$m = \dfrac{y_2 - y_1}{x_2 - x_1} = \dfrac{1 - 0}{1 - 0} = \dfrac{1}{1} = 1$

The slope is $m = 1$; d.

67. The line is vertical, so its slope is undefined; b.

69. $(x_1, y_1) = (2, 0)$, $(x_2, y_2) = (4, -1)$

$m = \dfrac{y_2 - y_1}{x_2 - x_1} = \dfrac{-1 - 0}{4 - 2} = \dfrac{-1}{2} = -\dfrac{1}{2}$

The slope is $m = -\dfrac{1}{2}$; e.

71. $m = \dfrac{y_2 - y_1}{x_2 - x_1} = \dfrac{0 - 1}{0 - 2} = \dfrac{-1}{-2} = \dfrac{1}{2}$

$\dfrac{-1 - 1}{-2 - 2} = \dfrac{-2}{-4} = \dfrac{1}{2}$; $\dfrac{-2 - 1}{-4 - 2} = \dfrac{-3}{-6} = \dfrac{1}{2}$

$\dfrac{-1 - 0}{-2 - 0} = \dfrac{-1}{-2} = \dfrac{1}{2}$; $\dfrac{-2 - 0}{-4 - 0} = \dfrac{-2}{-4} = \dfrac{1}{2}$

$\dfrac{-2 - (-1)}{-4 - (-2)} = \dfrac{-2 + 1}{-4 + 2} = \dfrac{-1}{-2} = \dfrac{1}{2}$

73. answers may vary

75. From the graph, for year 2001 the average miles per gallon was 28.5.

77. The lowest points on the graph correspond to 1994 and 2000. The average fuel economy for those years was 28.1 miles per gallon.

79. The line from 2000 to 2001 is the steepest and therefore has the greatest slope.

81. pitch $= \dfrac{\text{rise}}{\text{run}}$

$$\dfrac{2}{5} = \dfrac{4}{\frac{x}{2}}$$

$$2 \cdot \dfrac{x}{2} = 5 \cdot 4$$

$$x = 20$$

83. a. (1993, 10,359) and (2003, 15,139)

b. $m = \dfrac{y_2 - y_1}{x_2 - x_1}$

$= \dfrac{15,139 - 10,359}{2003 - 1993}$

$= \dfrac{4780}{10}$

$= 478$

The slope is 478.

c. For the years 1993 through 2003, the number of kidney transplants increased at a rate of 478 per year.

85. Slope through (1, 3) and (2, 1):

$$m = \dfrac{1-3}{2-1} = \dfrac{-2}{1} = -2$$

Slope through (−4, 0) and (−3, −2):

$$m = \dfrac{-2-0}{-3-(-4)} = \dfrac{-2}{-3+4} = \dfrac{-2}{1} = -2$$

Slope through (1, 3) and (−4, 0):

$$m = \dfrac{0-3}{-4-1} = \dfrac{-3}{-5} = \dfrac{3}{5}$$

Slope through (2, 1) and (−3, −2):

$$m = \dfrac{-2-1}{-3-2} = \dfrac{-3}{-5} = \dfrac{3}{5}$$

Opposite sides are parallel and their slopes are equal.

87. $m = \dfrac{y_2 - y_1}{x_2 - x_1} = \dfrac{4.5-1.2}{-2.2-(-3.8)} = \dfrac{3.3}{1.6} = 2.0625$

89. $m = \dfrac{y_2 - y_1}{x_2 - x_1} = \dfrac{-2.9-(-10.1)}{9.8-14.3} = \dfrac{7.2}{-4.5} = -1.6$

91.

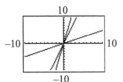

As the slope becomes larger, the line becomes steeper.

Section 6.5 Practice Problems

1. $y = mx + b$

$$y = \dfrac{3}{5}x + (-2)$$

$$y = \dfrac{3}{5}x - 2$$

2. From $y = \dfrac{2}{3}x - 4$, the y-intercept is (0, −4).

The slope is $\dfrac{2}{3}$, so another point on the graph is (0 + 3, −4 + 2) or (3, −2).

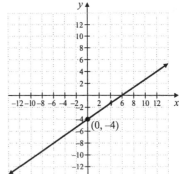

3. $3x + y = 2$

$$y = -3x + 2$$

The slope is $-3 = \dfrac{-3}{1}$ and the y-intercept is (0, 2). Another point on the graph is (0 + 1, 2 − 3) or (1, −1).

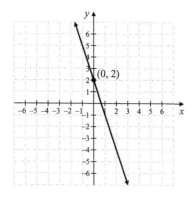

4. $y - y_1 = m(x - x_1)$
$y - (-4) = -3(x - 2)$
$y + 4 = -3(x - 2)$
$y + 4 = -3x + 6$
$y = -3x + 2$

5. $m = \dfrac{-2 - 3}{5 - 1} = \dfrac{-5}{4}$

$y - y_1 = m(x - x_1)$

$y - 3 = \dfrac{-5}{4}(x - 1)$

$4(y - 3) = 4\left[\dfrac{-5}{4}(x - 1)\right]$

$4(y - 3) = -5(x - 1)$
$4y - 12 = -5x + 5$
$5x + 4y = 17$

6. a. (10, 200), (9, 250)

$m = \dfrac{y_2 - y_1}{x_2 - x_1} = \dfrac{250 - 200}{9 - 10} = \dfrac{50}{-1} = -50$

$y - y_1 = m(x - x_1)$
$y - 200 = -50(x - 10)$
$y - 200 = -50x + 500$
$y = -50x + 700$

b. Let $x = 7.50$.
$y = -50x + 700$
$y = -50(7.50) + 700$
$y = -375 + 700$
$y = 325$
The predicted weekly sales is 325.

Calculator Explorations

1.
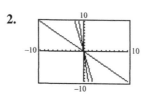

2.
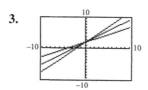

3.
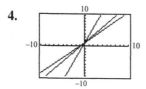

4.

Mental Math

1. $y = mx + b$
$y = 2x - 1$
$m = 2$; y-intercept $(0, -1)$

2. $y = mx + b$
$y = -7x + 3$
$m = -7$; y-intercept $(0, 3)$

3. $y = mx + b$
$y = x + \dfrac{1}{3}$

$m = 1$; y-intercept $\left(0, \dfrac{1}{3}\right)$

4. $y = mx + b$

$$y = -x - \frac{2}{9}$$

$m = -1$; y-intercept $\left(0, -\frac{2}{9}\right)$

5. $y = mx + b$

$$y = \frac{5}{7}x - 4$$

$m = \frac{5}{7}$; y-intercept $(0, -4)$

6. $y = mx + b$

$$y = -\frac{1}{4}x + \frac{3}{5}$$

$m = -\frac{1}{4}$; y-intercept $\left(0, \frac{3}{5}\right)$

7. $y - y_1 = m(x - x_1)$

$$y - 8 = 3(x - 4)$$

$m = 3$; answers may vary, example $(4, 8)$.

8. $y - y_1 = m(x - x_1)$

$$y - 1 = 5(x - 2)$$

$m = 5$; answers may vary, example $(2, 1)$.

9. $y - y_1 = m(x - x_1)$

$$y + 3 = -2(x - 10)$$

$m = -2$; answers may vary, example $(10, -3)$

10. $y - y_1 = m(x - x_1)$

$$y + 6 = -7(x - 2)$$

$m = -7$; answers may vary, example $(2, -6)$

11. $y - y_1 = m(x - x_1)$

$$y = \frac{2}{5}(x + 1)$$

$m = \frac{2}{5}$; answers may vary, example $(-1, 0)$

12. $y - y_1 = m(x - x_1)$

$$y = \frac{3}{7}(x + 4)$$

$m = \frac{3}{7}$; answers may vary; example $(-4, 0)$

Exercise Set 6.5

1. $y = mx + b$

$y = 5x + 3$

3. $y = mx + b$

$$y = -4x + \left(-\frac{1}{6}\right)$$

$$y = -4x - \frac{1}{6}$$

5. $y = mx + b$

$$y = \frac{2}{3}x + 0$$

$$y = \frac{2}{3}x$$

7. $y = mx + b$

$y = 0x + (-8)$

$y = -8$

9. $y = mx + b$

$$y = -\frac{1}{5}x + \frac{1}{9}$$

11. From $y = 2x + 1$, the y-intercept is $(0, 1)$.

The slope is 2 or $\frac{2}{1}$, so another point on the

graph is $(0 + 1, 1 + 2)$ or $(1, 3)$.

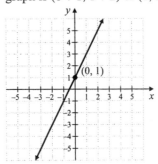

13. From $y = \dfrac{2}{3}x + 5$, the y-intercept is $(0, 5)$.

The slope is $\dfrac{2}{3}$, so another point on the graph is $(0 + 3, 5 + 2)$ or $(3, 7)$.

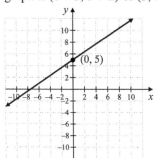

15. From $y = -5x$, the y-intercept is $(0, 0)$. The slope is -5 or $\dfrac{-5}{1}$, so another point on the graph is $(0 + 1, 0 + (-5))$ or $(1, -5)$.

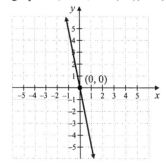

17. $4x + y = 6$
$$y = -4x + 6$$

The slope is $-4 = \dfrac{-4}{1}$ and the y-intercept is $(0, 6)$. Another point on the graph is $(0 + 1, 6 - 4)$ or $(1, 2)$.

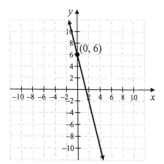

19. $4x - 7y = -14$
$$-7y = -4x - 14$$
$$y = \dfrac{4}{7}x + 2$$

The slope is $\dfrac{4}{7}$ and the y-intercept is $(0, 2)$.

Another point on the graph is $(0 + 7, 2 + 4)$ or $(7, 6)$.

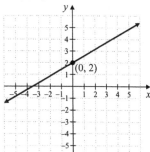

21. $x = \dfrac{5}{4}y$
$$\dfrac{4}{5}x = y$$
$$y = \dfrac{4}{5}x + 0$$

The slope is $\dfrac{4}{5}$ and the y-intercept is $(0, 0)$.

Another point on the graph is $(0 + 5, 0 + 4)$ or $(5, 4)$.

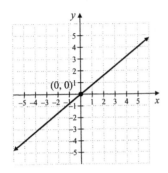

23.
$$y - y_1 = m(x - x_1)$$
$$y - 2 = 6(x - 2)$$
$$y - 2 = 6x - 12$$
$$y = 6x - 10$$
$$-6x + y = -10$$

25.
$$y - y_1 = m(x - x_1)$$
$$y - (-5) = -8[x - (-1)]$$
$$y + 5 = -8(x + 1)$$
$$y + 5 = -8x - 8$$
$$y = -8x - 13$$
$$8x + y = -13$$

27.
$$y - y_1 = m(x - x_1)$$
$$y - (-6) = \frac{3}{2}(x - 5)$$
$$y + 6 = \frac{3}{2}(x - 5)$$
$$2(y + 6) = 2\left[\frac{3}{2}(x - 5)\right]$$
$$2y + 12 = 3x - 15$$
$$2y = 3x - 27$$
$$-3x + 2y = -27 \text{ or } 3x - 2y = 27$$

29.
$$y - y_1 = m(x - x_1)$$
$$y - 0 = -\frac{1}{2}[x - (-3)]$$
$$y = -\frac{1}{2}(x + 3)$$
$$2y = -1(x + 3)$$
$$2y = -x - 3$$
$$x + 2y = -3$$

31. $m = \dfrac{y_2 - y_1}{x_2 - x_1} = \dfrac{6 - 2}{5 - 3} = \dfrac{4}{2} = 2$
$$y - y_1 = m(x - x_1)$$
$$y - 2 = 2(x - 3)$$
$$y - 2 = 2x - 6$$
$$y + 4 = 2x$$
$$4 = 2x - y$$
$$2x - y = 4$$

33. $m = \dfrac{y_2 - y_1}{x_2 - x_1} = \dfrac{-5 - 3}{-2 - (-1)} = \dfrac{-8}{-1} = 8$
$$y - y_1 = m(x - x_1)$$
$$y - 3 = 8[x - (-1)]$$
$$y - 3 = 8(x + 1)$$
$$y - 3 = 8x + 8$$
$$y - 11 = 8x$$
$$-11 = 8x - y$$
$$8x - y = -11$$

35. $m = \dfrac{y_2 - y_1}{x_2 - x_1} = \dfrac{-1 - 3}{-1 - 2} = \dfrac{-4}{-3} = \dfrac{4}{3}$
$$y - y_1 = m(x - x_1)$$
$$y - 3 = \frac{4}{3}(x - 2)$$
$$3(y - 3) = 4(x - 2)$$
$$3y - 9 = 4x - 8$$
$$3y - 1 = 4x$$
$$-1 = 4x - 3y$$
$$4x - 3y = -1$$

37. $m = \dfrac{y_2 - y_1}{x_2 - x_1}$
$$= \dfrac{\frac{1}{13} - 0}{-\frac{1}{8} - 0}$$
$$= \dfrac{\frac{1}{13}}{-\frac{1}{8}}$$
$$= \dfrac{1}{13}\left(-\dfrac{8}{1}\right)$$
$$= -\dfrac{8}{13}$$

$$y - y_1 = m(x - x_1)$$
$$y - 0 = -\frac{8}{13}(x - 0)$$
$$y = -\frac{8}{13}x$$
$$13y = -8x$$
$$8x + 13y = 0$$

39. $y = mx + b$
$$y = -\frac{1}{2}x + \frac{5}{3}$$

41. $m = \dfrac{y_2 - y_1}{x_2 - x_1} = \dfrac{10 - 7}{7 - 10} = \dfrac{3}{-3} = -1$
$$y - y_1 = m(x - x_1)$$
$$y - 7 = -1(x - 10)$$
$$y - 7 = -x + 10$$
$$y = -x + 17$$

43. A line with undefined slope is a vertical line. This one has an x-intercept of $\left(-\dfrac{3}{4}, 0\right)$. The equation is $x = -\dfrac{3}{4}$.

45. $y - y_1 = m(x - x_1)$
$$y - 9 = 1[x - (-7)]$$
$$y - 9 = x + 7$$
$$y = x + 16$$

47. $y = mx + b$
$$y = -5x + 7$$

49. A line parallel to the x-axis is a horizontal line.
$$y = 7$$

51. $m = \dfrac{y_2 - y_1}{x_2 - x_1} = \dfrac{3 - 0}{2 - 0} = \dfrac{3}{2}$
$$y - y_1 = m(x - x_1)$$
$$y - 0 = \frac{3}{2}(x - 0)$$
$$y = \frac{3}{2}x$$

53. A line perpendicular to the y-axis is a horizontal line.
$$y = -3$$

55. $\quad y - y_1 = m(x - x_1)$
$$y - (-2) = -\frac{4}{7}[x - (-1)]$$
$$y + 2 = -\frac{4}{7}(x + 1)$$
$$y + 2 = -\frac{4}{7}x - \frac{4}{7}$$
$$y = -\frac{4}{7}x - \frac{4}{7} - \frac{14}{7}$$
$$y = -\frac{4}{7}x - \frac{18}{7}$$

57. a. The ordered pairs are $(1, 32)$ and $(3, 96)$.
$$m = \frac{s_2 - s_1}{t_2 - t_1} = \frac{96 - 32}{3 - 1} = \frac{64}{2} = 32$$
$$s - s_1 = m(t - t_1)$$
$$s - 32 = 32(t - 1)$$
$$s - 32 = 32t - 32$$
$$s = 32t$$

b. $t = 4$: $s = 32(4)$
$$s = 128$$
The speed of the rock 4 seconds after it was dropped is 128 feet per second.

59. a. The ordered pairs are $(2, 54{,}000)$ and $(0, 22{,}000)$.
$$m = \frac{y_2 - y_1}{x_2 - x_1} = \frac{22{,}000 - 54{,}000}{0 - 2}$$
$$= \frac{-32{,}000}{-2}$$
$$= 16{,}000$$
$$y - y_1 = m(x - x_1)$$
$$y - 54{,}000 = 16{,}000(x - 2)$$
$$y - 54{,}000 = 16{,}000x - 32{,}000$$
$$y = 16{,}000x + 22{,}000$$

b. The year 2006 is 5 years past 2001.

$x = 2006 - 2001 = 5$

$y = 16,000(5) + 22,000$

$y = 80,000 + 22,000$

$y = 102,000$

102,000 vehicles are predicted for 2006.

61. a. The ordered pairs are (4, 5700) and (0, 7032).

$m = \dfrac{y_2 - y_1}{x_2 - x_1}$

$= \dfrac{7032 - 5700}{0 - 4}$

$= \dfrac{1332}{-4}$

$= -333$

$y - y_1 = m(x - x_1)$

$y - 7032 = -333(x - 0)$

$y - 7032 = -333x$

$\qquad y = -333x + 7032$

b. The year 2007 is 8 years past 1999, so it corresponds to $x = 8$.

$x = 8:\; y = -333(8) + 7032$

$\qquad y = -2664 + 7032$

$\qquad y = 4368$

4368 cinema sites are predicted for 2007.

63. a. (0, 1509) and (6, 1456)

b. $m = \dfrac{y_2 - y_1}{x_2 - x_1}$

$= \dfrac{1456 - 1509}{6 - 0}$

$= \dfrac{-53}{6}$

≈ -8.8

$y - y_1 = m(x - x_1)$

$y - 1509 = -8.8(x - 0)$

$y - 1509 = -8.8x$

$\qquad y = -8.8x + 1509$

c. 1999 is 2 years after 1997.

$x = 2:\; y = -8.8(2) + 1509$

$\qquad y = -17.6 + 1509$

$\qquad y \approx 1491$

1491 daily newspapers is estimated for 1999.

65. a. The ordered pairs are (3, 10,000) and (5, 8000).

$m = \dfrac{S_2 - S_1}{p_2 - p_1}$

$= \dfrac{8000 - 10,000}{5 - 3}$

$= \dfrac{-2000}{2}$

$= -1000$

$S - S_1 = m(p - p_1)$

$S - 10,000 = -1000(p - 3)$

$S - 10,000 = -1000p + 3000$

$\qquad S = -1000p + 13,000$

b. $p = 3.50:\; S = -1000(3.50) + 13,000$

$\qquad S = -3500 + 13,000$

$\qquad S = 9500$

9500 Fun Noodles will be sold when the price is $3.50 each.

67. $x = 2$:

$x^2 - 3x + 1 = 2^2 - 3(2) + 1 = 4 - 6 + 1 = -1$

69. $x = -1$: $x^2 - 3x + 1 = (-1)^2 - 3(-1) + 1$

$= 1 + 3 + 1$

$= 5$

71. The graph of $y = 2x + 1$ has slope $m = 2$ and y-intercept (0, 1). This is graph b.

73. The graph of $y = -3x - 2$ has slope $m = -3$ and y-intercept (0, −2). This is graph d.

75. The slope of the line $y = 3x - 1$ is $m = 3$.

$$y - y_1 = m(x - x_1)$$
$$y - 2 = 3[x - (-1)]$$
$$y - 2 = 3(x + 1)$$
$$y - 2 = 3x + 3$$
$$-5 = 3x - y$$
$$3x - y = -5$$

77. a. A line parallel to the line $y = 3x - 1$ will have slope 3.

$$y - y_1 = m(x - x_1)$$
$$y - 2 = 3[x - (-1)]$$
$$y - 2 = 3(x + 1)$$
$$y - 2 = 3x + 3$$
$$y = 3x + 5$$
$$-5 = 3x - y$$
$$3x - y = -5$$

b. A line perpendicular to the line

$y = 3x - 1$ will have slope $-\dfrac{1}{3}$.

$$y - y_1 = m(x - x_1)$$
$$y - 2 = -\frac{1}{3}[x - (-1)]$$
$$y - 2 = -\frac{1}{3}(x + 1)$$
$$3y - 6 = -1(x + 1)$$
$$3y - 6 = -x - 1$$
$$3y = -x + 5$$
$$x + 3y = 5$$

Integrated Review

1. Select two points on the line, such as $(0, 0)$ and $(1, 2)$.

$$m = \frac{y_2 - y_1}{x_2 - x_1} = \frac{2 - 0}{1 - 0} = \frac{2}{1} = 2$$

2. Horizontal lines have slopes of $m = 0$.

3. Select two points on the line, such as $(0, 1)$ and $(-3, 3)$.

$$m = \frac{y_2 - y_1}{x_2 - x_1} = \frac{3 - 1}{-3 - 0} = \frac{2}{-3} = -\frac{2}{3}$$

4. Vertical lines have undefined slopes.

5. $y = -2x$

$y = 0$: $0 = -2x$
$$0 = x$$

The x-intercept is $(0, 0)$.

$x = 0$: $y = -2(0) = 0$

The y-intercept is $(0, 0)$.

Find another point, for example let $x = 1$.

$$y = -2(1) = -2$$

Another point on the line is $(1, -2)$.

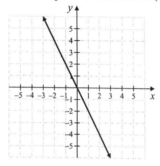

6. $x + y = 3$

$y = 0$: $x + 0 = 3$
$$x = 3$$

The x-intercept is $(3, 0)$.

$x = 0$: $0 + y = 3$
$$y = 3$$

The y-intercept is $(0, 3)$.

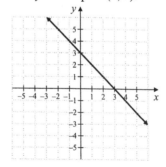

7. The graph of $x = -1$ is a vertical line with an x-intercept of $(-1, 0)$.

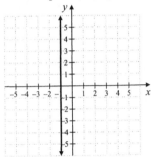

8. The graph of $y = 4$ is a horizontal line with a y-intercept of $(0, 4)$.

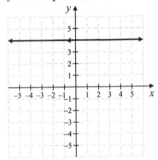

9. $x - 2y = 6$

$y = 0$: $x - 2(0) = 6$

$\quad\quad\quad x - 0 = 6$

$\quad\quad\quad\quad x = 6$

The x-intercept is $(6, 0)$.

$x = 0$: $0 - 2y = 6$

$\quad\quad\quad -2y = 6$

$\quad\quad\quad\quad y = -3$

The y-intercept is $(0, -3)$.

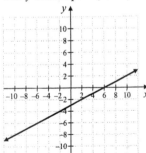

10. $y = 3x + 2$

$y = 0$: $\quad 0 = 3x + 2$

$\quad\quad\quad\quad -2 = 3x$

$$-\frac{2}{3} = x$$

The x-intercept is $\left(-\dfrac{2}{3}, 0\right)$.

$x = 0$: $y = 3(0) + 2$

$\quad\quad\quad\quad y = 0 + 2$

$\quad\quad\quad\quad y = 2$

The y-intercept is $(0, 2)$.

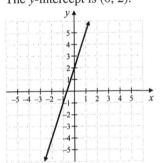

11. $y = -\dfrac{3}{4}x + 3$

$y = 0$: $\quad 0 = -\dfrac{3}{4}x + 3$

$\quad\quad\quad\quad -3 = -\dfrac{3}{4}x$

$\quad\quad\quad\quad 4 = x$

The x-intercept is $(4, 0)$.

$x = 0$: $y = -\dfrac{3}{4}(0) + 3$

$\quad\quad\quad\quad y = 0 + 3$

$\quad\quad\quad\quad y = 3$

The y-intercept is $(0, 3)$.

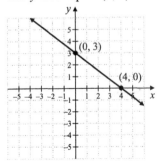

12. $5x - 2y = 8$

$y = 0$: $5x - 2(0) = 8$

$\qquad 5x - 0 = 8$

$\qquad\quad 5x = 8$

$\qquad\qquad x = \dfrac{8}{5}$

The x-intercept is $\left(\dfrac{8}{5}, 0\right)$.

$x = 0$: $5(0) - 2y = 8$

$\qquad 0 - 2y = 8$

$\qquad\quad -2y = 8$

$\qquad\qquad y = -4$

The y-intercept is $(0, -4)$.

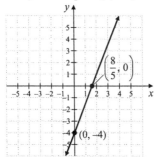

13. $y = mx + b$

$y = 3x - 1$

The slope is $m = 3$.

14. $y = mx + b$

$y = -6x + 2$

The slope is $m = -6$.

15. $y = mx + b$

$7x + 2y = 11$

$\quad 2y = -7x + 11$

$\qquad y = -\dfrac{7}{2}x + \dfrac{11}{2}$

The slope is $m = -\dfrac{7}{2}$.

16. $y = mx + b$

$2x - y = 0$

$\quad -y = -2x$

$\qquad y = 2x$

$\qquad y = 2x + 0$

The slope is $m = 2$.

17. The graph of $x = 2$ is a vertical line. Vertical lines have undefined slopes.

18. The graph of $y = -4$ is a horizontal line. Horizontal lines have slopes of $m = 0$.

19. $y = mx + b$

$y = 2x + \left(-\dfrac{1}{3}\right)$

$y = 2x - \dfrac{1}{3}$

20. $y - y_1 = m(x - x_1)$

$y - 3 = -4[x - (-1)]$

$y - 3 = -4(x + 1)$

$y - 3 = -4x - 4$

$\quad\; y = -4x - 1$

21. $m = \dfrac{y_2 - y_1}{x_2 - x_1} = \dfrac{-3 - 0}{-1 - 2} = \dfrac{-3}{-3} = 1$

$y - y_1 = m(x - x_1)$

$y - 0 = 1(x - 2)$

$\quad\; y = x - 2$

$-x + y = -2$ or $x - y = 2$

22. $6x - y = 7 \qquad\qquad 2x + 3y = 4$

$\quad -y = -6x + 7 \qquad\quad 3y = -2x + 4$

$\qquad y = 6x - 7 \qquad\qquad y = -\dfrac{2}{3}x + \dfrac{4}{3}$

$m = 6 \qquad\qquad\qquad m = -\dfrac{2}{3}$

Since $6 \neq -\dfrac{2}{3}$, the lines are not parallel.

Since $6\left(-\dfrac{2}{3}\right) = -4 \neq -1$, the lines are not perpendicular. The lines are neither parallel nor perpendicular.

23. $3x - 6y = 4$ $y = -2x$
  $-6y = -3x + 4$ $m = -2$

  $$y = \frac{1}{2}x - \frac{2}{3}$$

  $$m = \frac{1}{2}$$

  Since $\left(\dfrac{1}{2}\right)(-2) = -1$, the lines are

  perpendicular.

24. a. The ordered pairs are (1998, 1639) and (2002, 2135).

b. $m = \dfrac{y_2 - y_1}{x_2 - x_1}$

  $$= \frac{2135 - 1639}{2002 - 1998}$$

  $$= \frac{496}{4}$$

  $$= 124$$

c. For the years 1998 through 2002, the amount of yogurt produced increased at a rate of 124 million pounds per year.

Section 6.6 Practice Problems

1. The domain is the set of x-coordinates: $\{-3, 4, 7\}$.
  The range is the set of y-coordinates: $\{0, 1, 5, 6\}$.

2. a. Each x-value is only assigned to one y-value, so the relation is a function.

b. The x-value 1 is paired with two y-values, 4 and -3, so this set of ordered pairs is not a function.

3. a. This is the graph of the relation $\{(-3, -2), (-1, -1), (0, 0), (1, 1)\}$. Each x-coordinate has exactly one y-value, so this is the graph of a function.

b. This is the graph of the relation $\{(-1, -1), (-1, 2), (1, 0), (3, 1)\}$. The x-value -1 is paired with two y-values, -1 and 2, so this is not the graph of a function.

4. a. No vertical line will intersect the graph more than once, so the graph is the graph of a function.

b. No vertical line will intersect the graph more than once, so the graph is the graph of a function.

c. Vertical lines can be drawn that intersect the graph in two points, so the graph is not the graph of a function.

d. A vertical line can be drawn that intersects this line at every point, so the graph is not the graph of a function.

5. a. According to the graph, the time of the sunrise on March 1st is 6:30 A.M.

b. According to the graph, the sun rises at 6 A.M. in the middle of March and the middle of September.

6. $f(x) = x^2 + 1$

a. $f(1) = 1^2 + 1 = 1 + 1 = 2$
  Ordered pair: (1, 2)

b. $f(-3) = (-3)^2 + 1 = 9 + 1 = 10$
  $(-3, 10)$

c. $f(0) = 0^2 + 1 = 0 + 1 = 1$
  $(0, 1)$

Exercise Set 6.6

1. The domain is the set of x-coordinates: $\{-7, 0, 2, 10\}$.
  The range is the set of y-coordinates: $\{-7, 0, 4, 10\}$.

3. The domain is the set of x-coordinates: $\{0, 1, 5\}$
   The range is the set of y-coordinates: $\{-2\}$

5. Each x-value is only assigned to one y-value, so the relation is a function.

7. The x-value -1 is paired with more than one y-value, 0, 6, and 8, so the relation is not a function.

9. The vertical line $x = 1$ will intersect the graph in two points, so the graph is not the graph of a function.

11. No vertical line will intersect the graph more than once, so the graph is the graph of a function.

13. No vertical line will intersect the graph more than once, so the graph is the graph of a function.

15. Vertical lines can be drawn that intersect the graph in two points, so the graph is not the graph of a function.

17. On June 1, the graph shows sunset to be at approximately 9:30 P.M.

19. At 3 P.M., the graph shows this happens on January 1 and December 1.

21. The graph passes the vertical line test, so it is the graph of a function.

23. For 1997, the graph shows the minimum wage is $4.75 per hour.

25. According to the graph, the minimum wage increased to over $5.75 in 2004.

27. Yes; answers may vary. One example: The graph passes the vertical line test, so it is the graph of a function.

29. $f(x) = 2x - 5$
    $f(-2) = 2(-2) - 5 = -4 - 5 = -9$
    $f(0) = 2(0) - 5 = 0 - 5 = -5$
    $f(3) = 2(3) - 5 = 6 - 5 = 1$

31. $f(x) = x^2 + 2$
    $f(-2) = (-2)^2 + 2 = 4 + 2 = 6$
    $f(0) = 0^2 + 2 = 0 + 2 = 2$
    $f(3) = 3^2 + 2 = 9 + 2 = 11$

33. $f(x) = 3x$
    $f(-2) = 3(-2) = -6$
    $f(0) = 3(0) = 0$
    $f(3) = 3(3) = 9$

35. $f(x) = |x|$
    $f(-2) = |-2| = 2$
    $f(0) = |0| = 0$
    $f(3) = |3| = 3$

37. $h(x) = -5x$
    $h(-1) = -5(-1) = 5$
    $h(0) = -5(0) = 0$
    $h(4) = -5(4) = -20$

39. $h(x) = 2x^2 + 3$
    $h(-1) = 2(-1)^2 + 3 = 2(1) + 3 = 2 + 3 = 5$
    $h(0) = 2(0)^2 + 3 = 2(0) + 3 = 0 + 3 = 3$
    $h(4) = 2(4)^2 + 3 = 2(16) + 3 = 32 + 3 = 35$

41. The ordered-pair solution corresponding to $f(3) = 6$ is $(3, 6)$.

43. When $x = 0$, $y = -1$, so the ordered-pair solution is $(0, -1)$.

45. When $x = 0$, $y = -1$, so $f(0) = -1$.

47. When $y = 0$, $x = -1$ and $x = 5$.

49. $2x + 5 < 7$
    $2x < 2$
    $x < 1$

51. $-x + 6 \leq 9$
    $-x \leq 3$
    $x \geq -3$

53. $\dfrac{3}{x} + \dfrac{3}{2x} + \dfrac{5}{x} = \dfrac{3 \cdot 2}{x \cdot 2} + \dfrac{3}{2x} + \dfrac{5 \cdot 2}{x \cdot 2}$

$\qquad\qquad\quad = \dfrac{6}{2x} + \dfrac{3}{2x} + \dfrac{10}{2x}$

$\qquad\qquad\quad = \dfrac{6 + 3 + 10}{2x}$

$\qquad\qquad\quad = \dfrac{19}{2x}$

The perimeter is $\dfrac{19}{2x}$ meters.

55. A function f evaluated at -5 as 12 is written as $f(-5) = 12$.

57. answers may vary

59. $y = x + 7$ written in function notation is $f(x) = x + 7$.

61. $f(x) = \dfrac{136}{25}x$

    a. $f(35) = \dfrac{136}{25}(35) = \dfrac{4760}{25} = \dfrac{952}{5} = 190.4$

    The proper dosage for a 35-pound dog is 190.4 milligrams.

    b. $f(70) = \dfrac{136}{25}(70)$

$\qquad\qquad\; = \dfrac{9520}{25}$

$\qquad\qquad\; = \dfrac{1904}{5}$

$\qquad\qquad\; = 380.8$

    The proper dosage for a 70-pound dog is 380.8 milligrams.

Section 6.7 Practice Problems

1. $x - 4y > 8$

    a. $(-3, 2)$: $-3 - 4(2) > 8$

$\qquad\qquad\quad -3 - 8 > 8$

$\qquad\qquad\qquad\;\; -11 > 8$ False

    $(-3, 2)$ is not a solution of the inequality.

    b. $(9, 0)$: $9 - 4(0) > 8$

$\qquad\qquad\quad 9 - 0 > 8$

$\qquad\qquad\qquad\;\; 9 > 8$ True

    $(9, 0)$ is a solution of the inequality.

2. Graph the boundary line, $x - y = 3$, with a dashed line.

Test $(0, 0)$: $x - y > 3$

$\qquad\qquad\quad 0 - 0 > 3$

$\qquad\qquad\qquad\;\; 0 > 3$ False

Shade the half-plane not containing $(0, 0)$.

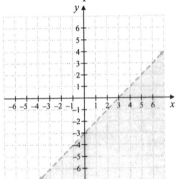

3. Graph the boundary line, $x - 4y = 4$, with a solid line.

Test $(0, 0)$: $\quad x - 4y \le 4$

$\qquad\qquad\quad 0 - 4(0) \le 4$

$\qquad\qquad\qquad\;\; 0 - 0 \le 4$

$\qquad\qquad\qquad\qquad 0 \le 4$ True

Shade the half-plane containing $(0, 0)$.

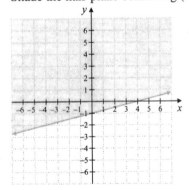

4. Graph the boundary line, $y = 3x$, with a dashed line.

Test $(1, 1)$: $\quad y < 3x$

$\qquad\qquad 1 < 3(1)$

$\qquad\qquad 1 < 3 \quad$ True

Shade the half-plane containing $(1, 1)$.

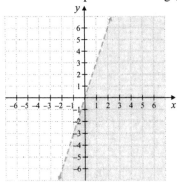

5. Graph the boundary line, $3x + 2y = 12$, with a solid line.

Test $(0, 0)$: $\quad 3x + 2y \geq 12$

$\qquad\qquad 3(0) + 2(0) \geq 12$

$\qquad\qquad 0 + 0 \geq 12$

$\qquad\qquad 0 \geq 12 \quad$ False

Shade the half-plane not containing $(0, 0)$.

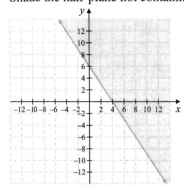

6. Graph the boundary line, $x = 2$, with a dashed line.

Test $(0, 0)$: $x < 2$

$\qquad\qquad 0 < 2 \quad$ True

Shade the half-plane containing $(0, 0)$.

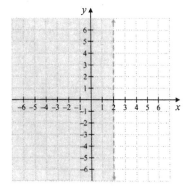

7. Graph the boundary line, $y = \dfrac{1}{4}x + 3$, with a solid line.

Test $(0, 0)$: $\quad y \geq \dfrac{1}{4}x + 3$

$\qquad\qquad 0 \geq \dfrac{1}{4}(0) + 3$

$\qquad\qquad 0 \geq 0 + 3$

$\qquad\qquad 0 \geq 3 \quad$ False

Shade the half-plane not containing $(0, 0)$.

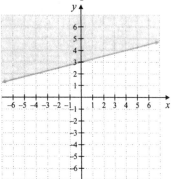

Mental Math

1. $y \geq x + 4$

Since $\geq$ is used, the boundary line is solid, included.

2. $x - y > -7$

Since $>$ is used, the boundary line is dashed, not included.

3. $y \geq x$

Since $\geq$ is used, the boundary line is solid, included.

4. $x > 0$

Since $>$ is used, the boundary line is dashed, not included.

5. $x + y > -5$

$0 + 0 > -5$

$\qquad 0 > -5$ True

Yes

6. $\qquad 2x + 3y < 10$

$2(0) + 3(0) < 10$

$\qquad 0 + 0 < 10$

$\qquad\qquad 0 < 10$ True

Yes

7. $x - y \leq -1$

$0 - 0 \leq -1$

$\qquad 0 \leq -1$ False

No

8. $\qquad \dfrac{2}{3}x + \dfrac{5}{6}y > 4$

$\dfrac{2}{3}(0) + \dfrac{5}{6}(0) > 4$

$\qquad 0 + 0 > 4$

$\qquad\qquad 0 > 4$ False

No

Exercise Set 6.7

1. $x - y > 3$

$(0, 3):\ 0 - 3 > 3$

$\qquad\qquad -3 > 3$ False

$(0, 3)$ is not a solution of the inequality.

$(2, -1):\ 2 - (-1) > 3$

$\qquad\qquad 2 + 1 > 3$

$\qquad\qquad\quad 3 > 3$ False

$(2, -1)$ is not a solution of the inequality.

3. $3x - 5y \leq -4$

$(2, 3):\ 3(2) - 5(3) \leq -4$

$\qquad\qquad 6 - 15 \leq -4$

$\qquad\qquad\quad -9 \leq -4$ True

$(2, 3)$ is a solution of the inequality.

$(-1, -1):\ 3(-1) - 5(-1) \leq -4$

$\qquad\qquad\quad -3 + 5 \leq -4$

$\qquad\qquad\qquad 2 \leq -4$ False

$(-1, -1)$ is not a solution of the inequality.

5. $x < -y$

$(0, 2):\ 0 < -2$ False

$(0, 2)$ is not a solution of the inequality.

$(-5, 1):\ -5 < -1$ True

$(-5, 1)$ is a solution of the inequality.

7. Graph the boundary line, $x + y = 1$, with a solid line.

Test $(0, 0)$: $x + y \leq 1$

$\qquad\qquad 0 + 0 \leq 1$

$\qquad\qquad\quad 0 \leq 1$ True

Shade the half-plane containing $(0, 0)$.

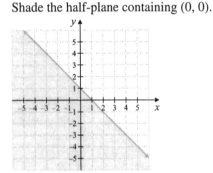

9. Graph the boundary line, $2x - y = -4$, with a dashed line.

Test $(0, 0)$: $2x - y > -4$

$\qquad\qquad 2(0) - 0 > -4$

$\qquad\qquad\quad 0 - 0 > -4$

$\qquad\qquad\qquad 0 > -4$ True

Shade the half-plane containing $(0, 0)$.

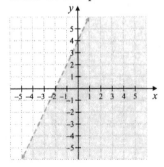

11. Graph the boundary line, $y = 2x$, with a dashed line.

Test $(1, 1)$: $y > 2x$

$$1 > 2(1)$$

$$1 > 2 \quad \text{False}$$

Shade the half-plane not containing $(1, 1)$.

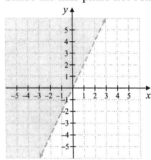

13. Graph the boundary line, $x = -3y$, with a solid line.

Test $(1, 1)$: $x \le -3y$

$$1 \le -3(1)$$

$$1 \le -3 \quad \text{False}$$

Shade the half-plane not containing $(1, 1)$.

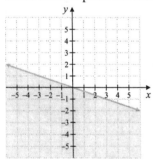

15. Graph the boundary line, $y = x + 5$, with a solid line.

Test $(0, 0)$: $y \ge x + 5$

$$0 \ge 0 + 5$$

$$0 \ge 5 \quad \text{False}$$

Shade the half-plane not containing $(0, 0)$.

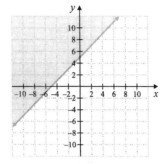

17. Graph the boundary line, $y = 4$, with a dashed line.

Test $(0, 0)$: $y < 4$

$$0 < 4 \quad \text{True}$$

Shade the half-plane containing $(0, 0)$.

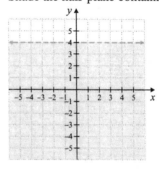

19. Graph the boundary line, $x = -3$, with a solid line.

Test $(0, 0)$: $x \ge -3$

$$0 \ge -3 \quad \text{True}$$

Shade the half-plane containing $(0, 0)$.

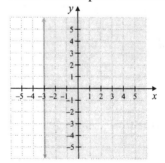

21. Graph the boundary line, $5x + 2y = 10$, with a solid line.

Test $(0, 0)$: $\qquad 5x + 2y \leq 10$

$$5(0) + 2(0) \leq 10$$
$$0 + 0 \leq 10$$
$$0 \leq 10 \quad \text{True}$$

Shade the half-plane containing $(0, 0)$.

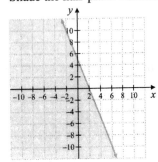

23. Graph the boundary line, $x = y$, with a dashed line.

Test $(1, 4)$: $\quad x > y$

$$1 > 4 \quad \text{False}$$

Shade the half-plane not containing $(1, 4)$.

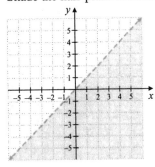

25. Graph the boundary line, $x - y = 6$, with a solid line.

Test $(0, 0)$: $x - y \leq 6$

$$0 - 0 \leq 6$$
$$0 \leq 6 \quad \text{True}$$

Shade the half-plane containing $(0, 0)$.

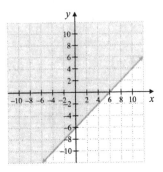

27. Graph the boundary line, $x = 0$, with a solid line.

Test $(1, 1)$: $x \geq 0$

$$1 \geq 0 \quad \text{True}$$

Shade the half-plane containing $(1, 1)$.

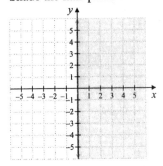

29. Shade the boundary line, $2x + 7y = 5$, with a dashed line.

Test $(0, 0)$: $\qquad 2x + 7y > 5$

$$2(0) + 7(0) > 5$$
$$0 + 0 > 5$$
$$0 > 5 \quad \text{False}$$

Shade the half-plane not containing $(0, 0)$.

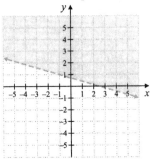

31. Graph the boundary line, $y = \dfrac{1}{2}x - 4$, with a solid line.

Test $(0, 0)$: $y \geq \dfrac{1}{2}x - 4$

$$0 \geq \dfrac{1}{2}(0) - 4$$

$$0 \geq 0 - 4$$

$$0 \geq -4 \quad \text{True}$$

Shade the half-plane containing $(0, 0)$.

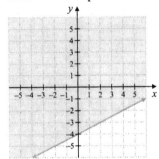

33. The point of intersection appears to be $(-2, 1)$.

35. The point of intersection appears to be $(-3, -1)$.

37. The graph is the half-plane with the dashed boundary line $x = 2$. The choice is a.

39. The graph is the half-plane with the dashed boundary line $y = 2$. The choice is b.

41. answers may vary

43. Test $(1, 1)$: $\quad 3x + 4y < 8$

$$3(1) + 4(1) < 8$$

$$3 + 4 < 8$$

$$7 < 8 \quad \text{True}$$

$(1, 1)$ is included in the graph of $3x + 4y < 8$.

45. Test $(1, 1)$: $y \geq -\dfrac{1}{2}x$

$$1 \geq -\dfrac{1}{2}(1)$$

$$1 \geq -\dfrac{1}{2} \quad \text{True}$$

$(1, 1)$ is included in the graph of $y \geq -\dfrac{1}{2}x$.

47. a. The sum of the number of days, x, times $30, and the number of miles, y, times $0.15, must be at most $500.

$$30x + 0.15y \leq 500$$

b.

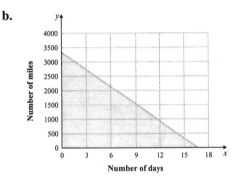

c. answers may vary

Section 6.8 Practice Problems

1. Use $(4, 8)$.

$$y = kx$$

$$8 = k \cdot 4$$

$$\dfrac{8}{4} = \dfrac{k \cdot 4}{4}$$

$$2 = k$$

Since $k = 2$, the equation is $y = 2x$.

2. Let $y = 15$ and $x = 45$.

$$y = kx$$

$$15 = k(45)$$

$$\dfrac{15}{45} = \dfrac{k(45)}{45}$$

$$\dfrac{1}{3} = k$$

The equation is $y = \dfrac{1}{3}x$.

Let $x = 3$.

$y = \dfrac{1}{3}x$

$y = \dfrac{1}{3} \cdot 3$

$y = 1$

Thus, when x is 3, y is 1.

3. Use $(-1, -2)$ and $(0, 0)$.

slope $= \dfrac{0 - (-2)}{0 - (-1)} = \dfrac{2}{1} = 2$

Thus, $k = 2$ and the variation equation is $y = 2x$.

4. Use $(4, 5)$.

$y = \dfrac{k}{x}$

$5 = \dfrac{k}{4}$

$4 \cdot 5 = 4 \cdot \dfrac{k}{4}$

$20 = k$

Since $k = 20$, the equation is $y = \dfrac{20}{x}$.

5. Let $y = 4$ and $x = 0.8$.

$y = \dfrac{k}{x}$

$4 = \dfrac{k}{0.8}$

$0.8(4) = 0.8\left(\dfrac{k}{0.8}\right)$

$3.2 = k$

The equation is $y = \dfrac{3.2}{x}$.

Let $x = 20$.

$y = \dfrac{3.2}{20}$

$y = 0.16$

Thus, when x is 20, y is 0.16.

6. $A = kr^2$

$49\pi = k(7)^2$

$49\pi = 49k$

$\pi = k$

The formula for the area of a circle is $A = \pi r^2$.

Let $r = 4$.

$A = \pi r^2$

$A = \pi \cdot 4^2$

$A = 16\pi$

The area is 16π square feet.

7. $d = kt^2$

$144 = k(3)^2$

$144 = 9k$

$16 = k$

The equation is $d = 16t^2$.

Let $t = 5$.

$d = 16t^2$

$d = 16 \cdot 5^2$

$d = 16 \cdot 25$

$d = 400$

The object will fall 400 feet in 5 seconds.

Mental Math

1. $y = 5x$ represents direct variation.

2. $y = \dfrac{5}{x}$ represents inverse variation.

3. $y = \dfrac{7}{x^2}$ represents inverse variation.

4. $y = 6.5x^4$ represents direct variation.

5. $y = \dfrac{11}{x}$ represents inverse variation.

6. $y = 18x$ represents direct variation.

7. $y = 12x^2$ represents direct variation.

8. $y = \dfrac{20}{x^3}$ represents inverse variation.

Exercise Set 6.8

1. $y = kx$
$3 = k(6)$
$\dfrac{1}{2} = k$
$y = \dfrac{1}{2}x$

3. $y = kx$
$12 = k(2)$
$6 = k$
$y = 6x$

5. $m = \dfrac{y_2 - y_1}{x_2 - x_1} = \dfrac{3 - 0}{1 - 0} = \dfrac{3}{1} = 3$
$y = 3x$

7. $m = \dfrac{y_2 - y_1}{x_2 - x_1} = \dfrac{2 - 0}{3 - 0} = \dfrac{2}{3}$
$y = \dfrac{2}{3}x$

9. $y = \dfrac{k}{x}$
$7 = \dfrac{k}{1}$
$7 = k$
$y = \dfrac{7}{x}$

11. $y = \dfrac{k}{x}$
$0.05 = \dfrac{k}{10}$
$0.5 = k$
$y = \dfrac{0.5}{x}$

13. y varies directly as x is written as $y = kx$.

15. h varies inversely as t is written as $h = \dfrac{k}{t}$.

17. z varies directly as x^2 is written as $z = kx^2$.

19. y varies inversely as z^3 is written as $y = \dfrac{k}{z^3}$.

21. x varies inversely as $\sqrt{y}$ is written as $x = \dfrac{k}{\sqrt{y}}$.

23. $y = kx$
$y = 20$ when $x = 5$: $20 = k(5)$
$\qquad\qquad\qquad\qquad\quad 4 = k$
$y = 4x$
$x = 10$: $y = 4(10) = 40$
$y = 40$ when $x = 10$.

25. $y = \dfrac{k}{x}$
$y = 5$ when $x = 60$: $\quad 5 = \dfrac{k}{60}$
$\qquad\qquad\qquad\qquad\quad 300 = k$
$y = \dfrac{300}{x}$
$x = 100$: $y = \dfrac{300}{100} = 3$
$y = 3$ when $x = 100$.

27. $z = kx^2$
$z = 96$ when $x = 4$: $96 = k(4)^2$
$\qquad\qquad\qquad\qquad\quad 96 = 16k$
$\qquad\qquad\qquad\qquad\quad\; 6 = k$
$z = 6x^2$
$x = 3$: $z = 6(3)^2 = 6(9) = 54$
$z = 54$ when $x = 3$.

29. $a = \dfrac{k}{b^3}$

$a = \dfrac{3}{2}$ when $b = 2$: $\dfrac{3}{2} = \dfrac{k}{2^3}$

$\dfrac{3}{2} = \dfrac{k}{8}$

$12 = k$

$a = \dfrac{12}{b^3}$

$b = 3$: $a = \dfrac{12}{3^3} = \dfrac{12}{27} = \dfrac{4}{9}$

$a = \dfrac{4}{9}$ when $b = 3$.

31. Let p be the paycheck amount when h hours are worked.

$p = kh$

$p = 112.50$ when $h = 18$: $112.50 = k(18)$

$6.25 = k$

$p = 6.25h$

$h = 10$: $p = 6.25(10) = 62.50$

The pay is \$62.50 for 10 hours.

33. Let c be the cost per headphone when h headphones are manufactured.

$c = \dfrac{k}{h}$

$c = 9$ when $h = 5000$: $9 = \dfrac{k}{5000}$

$45,000 = k$

$c = \dfrac{45,000}{h}$

$h = 7500$: $c = \dfrac{45,000}{7500} = 6$

The cost to manufacture 7500 headphones is \$6 per headphone.

35. Let d be the distance when a weight of w is attached.

$d = kw$

$d = 4$ when $w = 60$: $4 = k(60)$

$\dfrac{1}{15} = k$

$d = \dfrac{1}{15}w$

$w = 80$: $d = \dfrac{1}{15}(80) = 5\dfrac{1}{3}$

The spring stretches $5\dfrac{1}{3}$ inches when 80 pounds is attached to the spring.

37. Let w be the weight of an object when it is d miles from the center of the Earth.

$w = \dfrac{k}{d^2}$

$w = 180$ when $d = 4000$:

$180 = \dfrac{k}{4000^2}$

$180 = \dfrac{k}{16,000,000}$

$2,880,000,000 = k$

$w = \dfrac{2,880,000,000}{d^2}$

$w = \dfrac{2,880,000,000}{4010^2}$

$d = 4010$: $= \dfrac{2,880,000,000}{16,080,100}$

≈ 179.1

The man will weigh about 179.1 pounds when he is 10 miles above the surface of the Earth.

39. $d = kt^2$

$d = 64$ when $t = 2$: $64 = k(2)^2$

$64 = 4k$

$16 = k$

$d = 16t^2$

$t = 10$: $d = 16(10)^2 = 16(100) = 1600$

He will fall 1600 feet in 10 seconds.

41. $\begin{array}{r} -3x + 4y = 7 \\ 3x - 2y = 9 \\ \hline 2y = 16 \end{array}$

43. $5x - 0.4y = 0.7$
      $\underline{-9x + 0.4y = -0.2}$
      $\overline{-4x \qquad\ = 0.5}$

45. If y varies directly as x, then $y = kx$. If x is tripled, to become $3x$, then $y = k(3x) = 3(kx)$, and y is multiplied by 3.

47. If p varies directly with the square root of l, then $p = k\sqrt{l}$. If l is quadrupled, to become $4l$, then $k\sqrt{4l} = 2\left(k\sqrt{l}\right)$, and p is doubled.

Chapter 6 Vocabulary Check

1. An ordered pair is a <u>solution</u> of an equation in two variables if replacing the variables by the coordinates of the ordered pair results in a true statement.

2. The vertical number line in the rectangular coordinate system is called the <u>y-axis</u>.

3. A <u>linear</u> equation can be written in the form $Ax + By = C$.

4. A(n) <u>x-intercept</u> is a point of the graph where the graph crosses the x-axis.

5. The form $Ax + By = C$ is called <u>standard</u> form.

6. A(n) <u>y-intercept</u> is a point of the graph where the graph crosses the y-axis.

7. A set of ordered pairs that assigns to each x-value exactly one y-value is called a <u>function</u>.

8. The equation $y = 7x - 5$ is written in <u>slope-intercept</u> form.

9. The set of all x-coordinates of a relation is called the <u>domain</u> of the relation.

10. The set of all y-coordinates of a relation is called the <u>range</u> of the relation.

11. The set of ordered pairs is called a <u>relation</u>.

12. The equation $y + 1 = 7(x - 2)$ is written in <u>point-slope</u> form.

13. To find an x-intercept of a graph, let <u>y</u> = 0.

14. The horizontal number line in the rectangular coordinate system is called the <u>x-axis</u>.

15. To find a y-intercept of a graph, let <u>x</u> = 0.

16. The <u>slope</u> of a line measures the steepness or tilt of a line.

17. The equation $y = kx$ is an example of <u>direct</u> variation.

18. The equation $y = \dfrac{k}{x}$ is an example of <u>inverse</u> variation.

Chapter 6 Review

1–6.

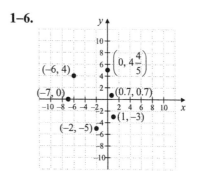

7. $-2 + y = 6x$
In $(7,\ \ )$, the x-coordinate is 7.
$-2 + y = 6(7)$
$-2 + y = 42$
     $y = 44$

The ordered-pair solution is $(7, 44)$.

8. $y = 3x + 5$

In (, −8), the y-coordinate is −8.

$-8 = 3x + 5$

$-13 = 3x$

$-\dfrac{13}{3} = x$

The ordered-pair solution is $\left(-\dfrac{13}{3}, -8\right)$.

9. $9 = -3x + 4y$

$y = 0$: $9 = -3x + 4(0)$

$9 = -3x + 0$

$9 = -3x$

$-3 = x$

$y = 3$: $9 = -3x + 4(3)$

$9 = -3x + 12$

$-3 = -3x$

$1 = x$

$x = 9$: $9 = -3(9) + 4y$

$9 = -27 + 4y$

$36 = 4y$

$9 = y$

x	y
−3	0
1	3
9	9

10. $y = 5$ for each value of x.

x	y
7	5
−7	5
0	5

11. $x = 2y$

$y = 0$: $x = 2(0)$

$x = 0$

$y = 5$: $x = 2(5)$

$x = 10$

$y = -5$: $x = 2(-5)$

$x = -10$

x	y
0	0
10	5
−10	−5

12. a. $y = 5x + 2000$

$x = 1$: $y = 5(1) + 2000 = 5 + 2000 = 2005$

$x = 100$: $y = 5(100) + 2000$

$= 500 + 2000$

$= 2500$

$x = 1000$: $y = 5(1000) + 2000$

$= 5000 + 2000$

$= 7000$

x	1	100	1000
y	2005	2500	7000

b. Let $y = 6430$ and solve for x.

$6430 = 5x + 2000$

$4430 = 5x$

$886 = x$

886 compact disc holders can be produced for \$6430.

13. $x - y = 1$

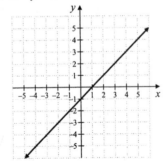

14. $x + y = 6$

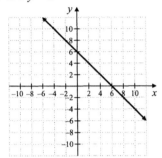

15. $x - 3y = 12$

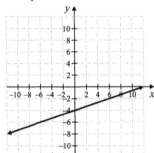

16. $5x - y = -8$

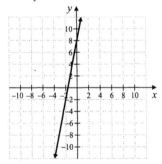

17. $x = 3y$

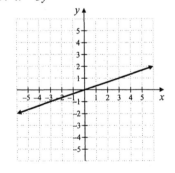

18. $y = -2x$

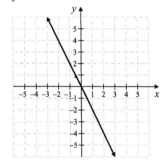

19. The x-intercept is $(4, 0)$.
The y-intercept is $(0, -2)$.

20. The x-intercepts are $(-2, 0)$ and $(2, 0)$.
The y-intercepts are $(0, 2)$ and $(0, -2)$.

21. $y = -3$ is a horizontal line with y-intercept $(0, -3)$.

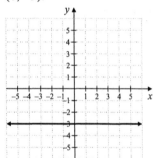

22. $x = 5$ is a vertical line with x-intercept $(5, 0)$.

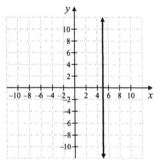

23. $x - 3y = 12$
$$y = 0: \quad x - 3(0) = 12$$
$$x - 0 = 12$$
$$x = 12$$
x-intercept: $(12, 0)$

$x = 0$: $0 - 3y = 12$
$$-3y = 12$$
$$y = -4$$

y-intercept: $(0, -4)$

24. $-4x + y = 8$
$y = 0$: $-4x + 0 = 8$
$$-4x = 8$$
$$x = -2$$

x-intercept: $(-2, 0)$
$x = 0$: $-4(0) + y = 8$
$$0 + y = 8$$
$$y = 8$$

y-intercept: $(0, 8)$

25. $(x_1, y_1) = (-1, 2),\ (x_2, y_2) = (3, -1)$

$$m = \frac{y_2 - y_1}{x_2 - x_1} = \frac{-1 - 2}{3 - (-1)} = \frac{-3}{3 + 1} = -\frac{3}{4}$$

26. $(x_1, y_1) = (-2, -2), (x_2, y_2) = (3, -1)$

$$m = \frac{y_2 - y_1}{x_2 - x_1} = \frac{-1 - (-2)}{3 - (-2)} = \frac{-1 + 2}{3 + 2} = \frac{1}{5}$$

27. When $m = 0$, the line is horizontal. The choice is d.

28. The slope is $m = -1$. The choice is b.

29. When the slope is undefined, the line is vertical. The choice is c.

30. The slope is $m = 4$. The choice is a.

31. $m = \dfrac{y_2 - y_1}{x_2 - x_1} = \dfrac{8 - 5}{6 - 2} = \dfrac{3}{4}$

32. $m = \dfrac{y_2 - y_1}{x_2 - x_1} = \dfrac{2 - 7}{1 - 4} = \dfrac{-5}{-3} = \dfrac{5}{3}$

33. $m = \dfrac{y_2 - y_1}{x_2 - x_1} = \dfrac{-9 - 3}{-2 - 1} = \dfrac{-12}{-3} = 4$

34. $m = \dfrac{y_2 - y_1}{x_2 - x_1} = \dfrac{-6 - 1}{3 - (-4)} = \dfrac{-6 - 1}{3 + 4} = \dfrac{-7}{7} = -1$

35. $y = mx + b$
$y = 3x + 7$
The slope is $m = 3$.

36. $y = mx + b$
$x - 2y = 4$
$$-2y = -x + 4$$
$$y = \frac{1}{2}x - 2$$

The slope is $m = \dfrac{1}{2}$.

37. $y = mx + b$
$y = -2$
$y = 0x - 2$
The slope is $m = 0$.

38. $x = 0$ is a vertical line. The slope is undefined.

39. $x - y = -6$ $\qquad$ $x + y = 3$
$\quad -y = -x - 6$ $\qquad$ $y = -x + 3$
$\quad\ \ y = x + 6$

$m = 1$ $\qquad\qquad$ $m = -1$
Since $(1)(-1) = -1$, the lines are perpendicular.

40. $3x + y = 7$ $\qquad$ $-3x - y = 10$
$\quad y = -3x + 7$ $\qquad\ -y = 3x + 10$
$\qquad\qquad\qquad\qquad\quad y = -3x - 10$

$m = -3$ $\qquad\qquad$ $m = -3$
Since the slopes are equal, the lines are parallel.

41. $y = 4x + \dfrac{1}{2}$ $\qquad$ $4x + 2y = 1$
$\qquad\qquad\qquad\qquad 2y = -4x + 1$
$$y = -2x + \frac{1}{2}$$

$m = 4$ $\qquad\qquad$ $m = -2$
Since $4 \neq -2$ and $(4)(-2) = -8 \neq -1$, the lines are neither parallel nor perpendicular.

42. $m = \dfrac{y_2 - y_1}{x_2 - x_1} = \dfrac{42.92 - 39.2}{1998 - 1995} = \dfrac{3.72}{3} = 1.24$

Every 1 year, 1.24 million more persons have a bachelor's degree or higher.

43. $m = \dfrac{y_2 - y_1}{x_2 - x_1} = \dfrac{859 - 805}{1997 - 1995} = \dfrac{54}{2} = 27$

Every 1 year, 27 million more people go on vacation.

44. $y = mx + b$
$3x + y = 7$
$\quad y = -3x + 7$
$m = -3$; y-intercept $(0, 7)$

45. $y = mx + b$
$x - 6y = -1$
$\quad -6y = -x - 1$
$\qquad y = \dfrac{1}{6}x + \dfrac{1}{6}$
$m = \dfrac{1}{6}$; y-intercept $\left(0, \dfrac{1}{6}\right)$

46. $y = mx + b$
$y = -5x + \dfrac{1}{2}$

47. $y = mx + b$
$y = \dfrac{2}{3}x + 6$

48. $y = mx + b$
$y = 2x + 1$
$m = 2$, y-intercept $(0, 1)$
The choice is d.

49. $y = mx + b$
$y = -4x$
$y = -4x + 0$
$m = -4$; y-intercept $(0, 0)$
The choice is c.

50. $y = mx + b$
$y = 2x$
$y = 2x + 0$
$m = 2$; y-intercept $(0, 0)$
The choice is a.

51. $y = mx + b$
$y = 2x - 1$
$m = 2$; y-intercept $(0, -1)$
The choice is b.

52. $y - y_1 = m(x - x_1)$
$y - 0 = 4(x - 2)$
$\quad y = 4x - 8$
$-4x + y = -8$

53. $y - y_1 = m(x - x_1)$
$y - (-5) = -3(x - 0)$
$\quad y + 5 = -3x$
$\qquad y = -3x - 5$
$3x + y = -5$

54. $y - y_1 = m(x - x_1)$
$y - 4 = \dfrac{3}{5}(x - 1)$
$5(y - 4) = 5 \cdot \dfrac{3}{5}(x - 1)$
$5y - 20 = 3(x - 1)$
$5y - 20 = 3x - 3$
$\quad 5y = 3x + 17$
$-3x + 5y = 17$

55. $y - y_1 = m(x - x_1)$
$y - 3 = -\dfrac{1}{3}[x - (-3)]$
$3y - 9 = -1[x - (-3)]$
$3y - 9 = -(x + 3)$
$3y - 9 = -x - 3$
$\quad 3y = -x + 6$
$x + 3y = 6$

56. $m = \dfrac{y_2 - y_1}{x_2 - x_1} = \dfrac{-7 - 7}{2 - 1} = \dfrac{-14}{1} = -14$

$y - y_1 = m(x - x_1)$
$y - 7 = -14(x - 1)$
$y - 7 = -14x + 14$
$\quad y = -14x + 21$

57. $m = \dfrac{y_2 - y_1}{x_2 - x_1}$

$= \dfrac{6 - 5}{-4 - (-2)}$

$= \dfrac{6 - 5}{-4 + 2}$

$= \dfrac{1}{-2}$

$= -\dfrac{1}{2}$

$y - y_1 = m(x - x_1)$

$y - 5 = -\dfrac{1}{2}[x - (-2)]$

$y - 5 = -\dfrac{1}{2}(x + 2)$

$2(y - 5) = 2\left(-\dfrac{1}{2}\right)(x + 2)$

$2y - 10 = -1(x + 2)$

$2y - 10 = -x - 2$

$2y = -x + 8$

$y = -\dfrac{1}{2}x + 4$

58. The x-value 7 is paired with two y-values, 1 and 5, so the relation is not a function.

59. Each x-value is only assigned to one y-value, so the relation is a function.

60. No vertical line will intersect the graph more than once, so the graph is the graph of a function.

61. No vertical line will intersect the graph more than once, so the graph is the graph of a function.

62. The vertical line $x = 3$ will intersect the graph at more than one point, so the graph is not the graph of a function.

63. No vertical line will intersect the graph more than once, so the graph is the graph of a function.

64. a. $f(x) = -2x + 6$
$f(0) = -2(0) + 6 = 0 + 6 = 6$

b. $f(x) = -2x + 6$
$f(-2) = -2(-2) + 6 = 4 + 6 = 10$

c. $f(x) = -2x + 6$
$f\left(\dfrac{1}{2}\right) = -2\left(\dfrac{1}{2}\right) + 6 = -1 + 6 = 5$

65. Graph the boundary line, $x + 6y = 6$, with a dashed line.
Test $(0, 0)$: $x + 6y < 6$
$0 + 6(0) < 6$
$0 + 0 < 6$
$0 < 6$ True
Shade the half-plane containing $(0, 0)$.

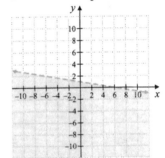

66. Graph the boundary line, $x + y = -2$, with a dashed line.
Test $(0, 0)$: $x + y > -2$
$0 + 0 > -2$
$0 > -2$ True
Shade the half-plane containing $(0, 0)$.

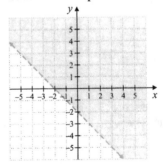

67. Graph the boundary line, $y = -7$, with a solid line.

Test $(0, 0)$: $y \geq -7$

$\qquad 0 \geq -7$ True

Shade the half-plane containing $(0, 0)$.

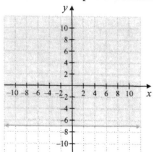

68. Graph the boundary line, $y = -4$, as a solid line.

Test $(0, 0)$: $y \leq -4$

$\qquad 0 \leq -4$ False

Shade the half-plane not containing $(0, 0)$.

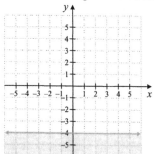

69. Graph the boundary line, $-x = y$, as a solid line.

Test $(1, 1)$: $-x \leq y$

$\qquad -1 \leq 1$ True

Shade the half-plane containing $(1, 1)$.

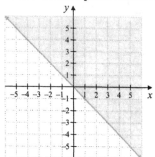

70. Graph the boundary line, $x = -y$, as a solid line.

Test $(1, 1)$: $x \geq -y$

$\qquad 1 \geq -1$ True

Shade the half-plane containing $(1, 1)$.

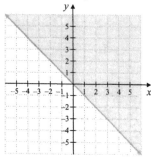

71. $y = kx$

$y = 40$ when $x = 4$: $40 = k(4)$

$\qquad\qquad\qquad\qquad\quad 10 = k$

$y = 10x$

$x = 11$: $y = 10(11) = 110$

$y = 110$ when $x = 11$.

72. $y = \dfrac{k}{x}$

$y = 4$ when $x = 6$: $4 = \dfrac{k}{6}$

$\qquad\qquad\qquad\qquad 24 = k$

$y = \dfrac{24}{x}$

$x = 48$: $y = \dfrac{24}{48} = \dfrac{1}{2}$

$y = \dfrac{1}{2}$ when $x = 48$.

73. $y = \dfrac{k}{x^3}$

$y = 12.5$ when $x = 2$: $12.5 = \dfrac{k}{2^3}$

$\qquad\qquad\qquad\qquad 12.5 = \dfrac{k}{8}$

$\qquad\qquad\qquad\qquad 100 = k$

$y = \dfrac{100}{x^3}$

$x = 3$: $y = \dfrac{100}{3^3} = \dfrac{100}{27}$

$y = \dfrac{100}{27}$ when $x = 3$.

74. $y = kx^2$

$y = 175$ when $x = 5$: $175 = k(5)^2$

$\qquad\qquad\qquad\qquad 175 = 25k$

$\qquad\qquad\qquad\qquad\quad 7 = k$

$y = 7x^2$

$x = 10$: $y = 7(10)^2 = 7(100) = 700$

$y = 700$ when $x = 10$.

75. Let c be the cost for manufacturing m milliliters.

$c = \dfrac{k}{m}$

$c = 6600$ when $m = 3000$:

$\qquad 6600 = \dfrac{k}{3000}$

$19,800,000 = k$

$c = \dfrac{19,800,000}{m}$

Let $m = 5000$: $c = \dfrac{19,800,000}{5000} = 3960$

It costs $3960 to manufacture 5000 milliliters.

76. Let d be the distance when a weight of w is attached.

$d = kw$

$d = 8$ when $w = 150$: $8 = k(150)$

$\qquad\qquad\qquad\qquad\quad \dfrac{4}{75} = k$

$d = \dfrac{4}{75}w$

Let $w = 90$: $d = \dfrac{4}{75} \cdot 90 = 4\dfrac{4}{5}$

The spring stretches $4\dfrac{4}{5}$ inches when 90 pounds is attached.

77. $2x - 5y = 9$

$\quad y = 1$: $2x - 5(1) = 9$

$\qquad\qquad\qquad 2x - 5 = 9$

$\qquad\qquad\qquad\quad 2x = 14$

$\qquad\qquad\qquad\qquad x = 7$

$x = 2$: $2(2) - 5y = 9$

$\qquad\qquad\quad 4 - 5y = 9$

$\qquad\qquad\qquad -5y = 5$

$\qquad\qquad\qquad\quad y = -1$

$y = -3$: $2x - 5(-3) = 9$

$\qquad\qquad\quad 2x + 15 = 9$

$\qquad\qquad\qquad 2x = -6$

$\qquad\qquad\qquad\quad x = -3$

x	y
7	1
2	-1
-3	-3

78. $x = -3y$

$x = 0$: $0 = -3y$

$\qquad\qquad 0 = y$

$y = 1$: $x = -3(1)$

$\qquad\qquad x = -3$

$x = 6$: $6 = -3y$

$\qquad\qquad -2 = y$

x	y
0	0
-3	1
6	-2

79. $2x - 3y = 6$

$y = 0$: $2x - 3(0) = 6$

$\qquad\qquad 2x - 0 = 6$

$\qquad\qquad\qquad 2x = 6$

$\qquad\qquad\qquad\quad x = 3$

x-intercept: (3, 0)

$x = 0$: $2(0) - 3y = 6$

$$0 - 3y = 6$$
$$-3y = 6$$
$$y = -2$$

y-intercept: (0, −2)

80. $-5x + y = 10$

$y = 0$: $-5x + 0 = 10$

$$-5x = 10$$
$$x = -2$$

x-intercept: (−2, 0)

$x = 0$: $-5(0) + y = 10$

$$0 + y = 10$$
$$y = 10$$

y-intercept: (0, 10)

81. $x - 5y = 10$

$y = 0$: $x - 5(0) = 10$

$$x - 0 = 10$$
$$x = 10$$

x-intercept: (10, 0)

$x = 0$: $0 - 5y = 10$

$$-5y = 10$$
$$y = -2$$

y-intercept: (0, −2)

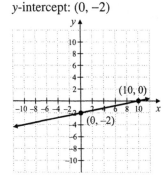

82. $x + y = 4$

$y = 0$: $x + 0 = 4$

$$x = 4$$

x-intercept: (4, 0)

$x = 0$: $0 + y = 4$

$$y = 4$$

y-intercept: (0, 4)

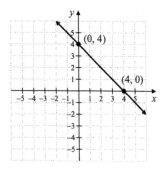

83. $y = -4x$

$y = 0$: $0 = -4x$

$$0 = x$$

x-intercept: (0, 0)

The y-intercept is also (0, 0). Find another point. Let $x = 1$.

$y = -4(1) = -4$

Another point is (1, −4).

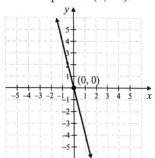

84. $2x + 3y = -6$

$y = 0$: $2x + 3(0) = -6$

$$2x + 0 = -6$$
$$2x = -6$$
$$x = -3$$

x-intercept: (−3, 0)

$x = 0$: $2(0) + 3y = -6$

$$0 + 3y = -6$$
$$y = -2$$

y-intercept: (0, −2)

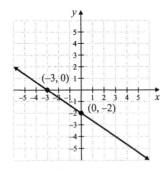

85. $x = 3$ is a vertical line with x-intercept $(3, 0)$.

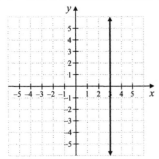

86. $y = -2$ is a horizontal line with y-intercept $(0, -2)$.

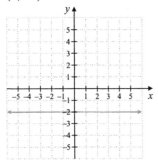

87. $m = \dfrac{y_2 - y_1}{x_2 - x_1} = \dfrac{2 - (-5)}{-4 - 3} = \dfrac{2 + 5}{-4 - 3} = \dfrac{7}{-7} = -1$

88. $m = \dfrac{y_2 - y_1}{x_2 - x_1} = \dfrac{-8 - 3}{-6 - 1} = \dfrac{-11}{-7} = \dfrac{11}{7}$

89. $(x_1, y_1) = (0, -4), (x_2, y_2) = (2, 0)$

$m = \dfrac{y_2 - y_1}{x_2 - x_1} = \dfrac{0 - (-4)}{2 - 0} = \dfrac{4}{2} = 2$

90. $(x_1, y_1) = (0, 2), (x_2, y_2) = (6, 0)$

$m = \dfrac{y_2 - y_1}{x_2 - x_1} = \dfrac{0 - 2}{6 - 0} = \dfrac{-2}{6} = -\dfrac{1}{3}$

91. $y = mx + b$

$-2x + 3y = -15$

$3y = 2x - 15$

$y = \dfrac{2}{3}x - 5$

$m = \dfrac{2}{3};$ y-intercept: $(0, -5)$

92. $y = mx + b$

$6x + y - 2 = 0$

$6x + y = 2$

$y = -6x + 2$

$m = -6;$ y-intercept: $(0, 2)$

93. $y - y_1 = m(x - x_1)$

$y - (-7) = -5(x - 3)$

$y + 7 = -5x + 15$

$y = -5x + 8$

$5x + y = 8$

94. $y - y_1 = m(x - x_1)$

$y - 6 = 3(x - 0)$

$y - 6 = 3x$

$-6 = 3x - y$

$3x - y = -6$

95. $m = \dfrac{y_2 - y_1}{x_2 - x_1} = \dfrac{5 - 9}{-2 - (-3)} = \dfrac{5 - 9}{-2 + 3} = \dfrac{-4}{1} = -4$

$y - y_1 = m(x - x_1)$

$y - 9 = -4[x - (-3)]$

$y - 9 = -4(x + 3)$

$y - 9 = -4x - 12$

$y = -4x - 3$

$4x + y = -3$

96. $m = \dfrac{y_2 - y_1}{x_2 - x_1} = \dfrac{-9 - 1}{5 - 3} = \dfrac{-10}{2} = -5$

$y - y_1 = m(x - x_1)$
$y - 1 = -5(x - 3)$
$y - 1 = -5x + 15$
$y = -5x + 16$
$5x + y = 16$

Chapter 6 Test

1. $12y - 7x = 5$
$x = 1:\ 12y - 7(1) = 5$
$\qquad\quad 12y - 7 = 5$
$\qquad\quad 12y = 12$
$\qquad\qquad y = 1$
The ordered pair is $(1, 1)$.

2. $y = 17$ for each value of x. The ordered pair is $(-4, 17)$.

3. $(x_1,\ y_1) = (-1, -1),\ (x_2,\ y_2) = (4, 1)$

$m = \dfrac{y_2 - y_1}{x_2 - x_1} = \dfrac{1 - (-1)}{4 - (-1)} = \dfrac{1 + 1}{4 + 1} = \dfrac{2}{5}$

4. The slope of a horizontal line is $m = 0$.

5. $m = \dfrac{y_2 - y_1}{x_2 - x_1} = \dfrac{2 - (-5)}{-1 - 6} = \dfrac{7}{-7} = -1$

6. $m = \dfrac{y_2 - y_1}{x_2 - x_1} = \dfrac{-1 - (-8)}{-1 - 0} = \dfrac{-1 + 8}{-1} = \dfrac{7}{-1} = -7$

7. $y = mx + b$
$-3x + y = 5$
$\qquad y = 3x + 5$
$m = 3$

8. $x = 6$ is a vertical line. The slope of a vertical line is undefined.

9. $2x + y = 8$
$y = 0:\ 2x + 0 = 8$
$\qquad\qquad 2x = 8$
$\qquad\qquad\ x = 4$
x-intercept: $(4, 0)$
$x = 0:\ 2(0) + y = 8$
$\qquad\qquad\qquad y = 8$
y-intercept: $(0, 8)$

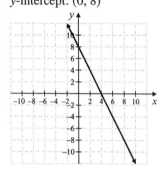

10. $-x + 4y = 5$
$y = 0:\ -x + 4(0) = 5$
$\qquad\qquad\ -x + 0 = 5$
$\qquad\qquad\qquad -x = 5$
$\qquad\qquad\qquad\ x = -5$
x-intercept: $(-5, 0)$
$x = 0:\ -0 + 4y = 5$
$\qquad\qquad\qquad 4y = 5$
$\qquad\qquad\qquad\ y = \dfrac{5}{4}$
y-intercept: $\left(0, \dfrac{5}{4}\right)$

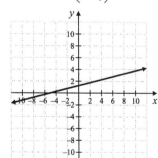

11. Graph the boundary line, $x - y = -2$, with a solid line.

Test $(0, 0)$: $x - y \geq -2$

$$0 - 0 \geq -2$$

$$0 \geq -2 \quad \text{True}$$

Shade the half-plane containing $(0, 0)$.

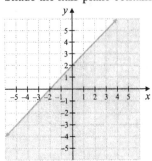

12. Graph the boundary line, $y = -4x$, with a solid line.

Test $(1, 1)$: $y \geq -4x$

$$1 \geq -4(1)$$

$$1 \geq -4 \quad \text{True}$$

Shade the half-plane containing $(1, 1)$.

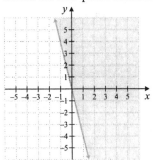

13. $5x - 7y = 10$

$y = 0$: $5x - 7(0) = 10$

$$5x - 0 = 10$$

$$5x = 10$$

$$x = 2$$

x-intercept: $(2, 0)$

$x = 0$: $5(0) - 7y = 10$

$$0 - 7y = 10$$

$$-7y = 10$$

$$y = -\frac{10}{7}$$

y-intercept: $\left(0, -\dfrac{10}{7}\right)$

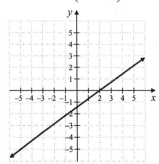

14. Graph the boundary line, $2x - 3y = -6$, with a dashed line.

Test $(0, 0)$: $2x - 3y > -6$

$$2(0) - 3(0) > -6$$

$$0 - 0 > -6$$

$$0 > -6 \quad \text{True}$$

Shade the half-plane containing $(0, 0)$.

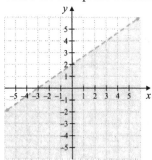

15. Graph the boundary line, $6x + y = -1$, with a dashed line.

Test $(0, 0)$: $6x + y > -1$

$$6(0) + 0 > -1$$

$$0 + 0 > -1$$

$$0 > -1 \quad \text{True}$$

Shade the half-plane containing $(0, 0)$.

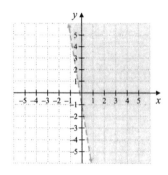

16. The graph of $y = -1$ is a horizontal line with a y-intercept of $(0, -1)$.

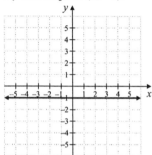

17. $y = 2x - 6$

The slope is $m = 2$ and the y-intercept is $(0, -6)$.

$-4x = 2y$

$-2x = y$

$y = -2x + 0$

The slope is $m = -2$ and the y-intercept is $(0, 0)$. Since the slopes are different, the lines are not parallel. Since $2(-2) = -4 \neq -1$, the lines are not perpendicular. The lines are neither parallel nor perpendicular.

18.
$$y - y_1 = m(x - x_1)$$
$$y - 2 = -\frac{1}{4}(x - 2)$$
$$4(y - 2) = 4\left(-\frac{1}{4}\right)(x - 2)$$
$$4y - 8 = -1(x - 2)$$
$$4y - 8 = -x + 2$$
$$4y = -x + 10$$
$$x + 4y = 10$$

19. $m = \dfrac{y_2 - y_1}{x_2 - x_1} = \dfrac{-7 - 0}{6 - 0} = \dfrac{-7}{6} = -\dfrac{7}{6}$

$$y - y_1 = m(x - x_1)$$
$$y - 0 = -\frac{7}{6}(x - 0)$$
$$y = -\frac{7}{6}x$$
$$6y = 6\left(-\frac{7}{6}x\right)$$
$$6y = -7x$$
$$7x + 6y = 0$$

20. $m = \dfrac{y_2 - y_1}{x_2 - x_1} = \dfrac{3 - (-5)}{1 - 2} = \dfrac{3 + 5}{-1} = -8$

$$y - y_1 = m(x - x_1)$$
$$y - (-5) = -8(x - 2)$$
$$y + 5 = -8x + 16$$
$$y = -8x + 11$$
$$8x + y = 11$$

21. $m = \dfrac{1}{8}; \ b = 12$

$$y = \frac{1}{8}x + 12$$
$$8y = x + 8(12)$$
$$8y = x + 96$$
$$x - 8y = -96$$

22. Each x-value is only assigned to one y-value, so the relation is a function.

23. The x-value -3 is assigned to two y-values, -3 and 2, so the relation is not a function. Note that the x-value 0 is also assigned to two y-values, 5 and 0.

24. No vertical line will intersect the graph more than once, so the graph is the graph of a function.

25. No vertical line will intersect the graph more than once, so the graph is the graph of a function.

26. $f(x) = 2x - 4$

    a. $f(-2) = 2(-2) - 4 = -4 - 4 = -8$

    b. $f(0.2) = 2(0.2) - 4 = 0.4 - 4 = -3.6$

    c. $f(0) = 2(0) - 4 = 0 - 4 = -4$

27. $f(x) = x^3 - x$

    a. $f(-1) = (-1)^3 - (-1) = -1 + 1 = 0$

    b. $f(0) = 0^3 - 0 = 0 - 0 = 0$

    c. $f(4) = 4^3 - 4 = 64 - 4 = 60$

28. $2x + 2(2y) = 42$
$$2(x + 2y) = 2(21)$$
$$x + 2y = 21$$
Let $y = 8$: $x + 2(8) = 21$
$$x + 16 = 21$$
$$x = 5$$
When $y = 8$ meters, $x = 5$ meters.

29. a. The ordered pairs are (2000, 69.3), (2001, 70.0), (2002, 69.9), (2003, 70.1), (2004, 70.3), (2005, 70.5).

    b.

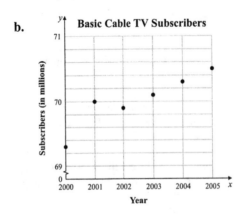

30. $m = \dfrac{y_2 - y_1}{x_2 - x_1} = \dfrac{1570 - 1420}{2003 - 2000} = \dfrac{150}{3} = 50$

Every 1 year, 50 million more movie tickets are sold.

31. $y = kx$
$y = 10$ when $x = 15$: $10 = k(15)$
$$\frac{2}{3} = k$$
$$y = \frac{2}{3}x$$
Let $x = 42$: $y = \dfrac{2}{3}(42) = 28$
When $x = 42$, $y = 28$.

32. $y = \dfrac{k}{x^2}$

$y = 8$ when $x = 5$: $8 = \dfrac{k}{5^2}$
$$8 = \frac{k}{25}$$
$$200 = k$$
$$y = \frac{200}{x^2}$$
Let $x = 15$: $y = \dfrac{200}{15^2} = \dfrac{200}{225} = \dfrac{8}{9}$
When $x = 15$, $y = \dfrac{8}{9}$.

Cumulative Review Chapters 1–6

1. $6 \div 3 + 5^2 = 6 \div 3 + 25 = 2 + 25 = 27$

2. $\dfrac{10}{3} + \dfrac{5}{21} = \dfrac{10}{3} \cdot \dfrac{7}{7} + \dfrac{5}{21} = \dfrac{70}{21} + \dfrac{5}{21} = \dfrac{75}{21} = \dfrac{25}{7}$

3. $1 + 2[5(2 \cdot 3 + 1) - 10] = 1 + 2[5(6 + 1) - 10]$
$$= 1 + 2[5(7) - 10]$$
$$= 1 + 2(35 - 10)$$
$$= 1 + 2(25)$$
$$= 1 + 50$$
$$= 51$$

4. $16 - 3 \cdot 3 + 2^4 = 16 - 3 \cdot 3 + 16$
$$= 16 - 9 + 16$$
$$= 7 + 16$$
$$= 23$$

5. $29,035 - (-1349) = 29,035 + 1349 = 30,384$
The difference in elevation is 30,384 feet.

6. $1.7x - 11 - 0.9x - 25$
$= (1.7x - 0.9x) + (-11 - 25)$
$= 0.8x + (-36)$
$= 0.8x - 36$

7. Twice a number plus 6 is written as $2x + 6$.

8. The product of -15 and the sum of a number and $\frac{2}{3}$ is written as $-15\left(x + \frac{2}{3}\right)$.

9. The difference of a number and 4, divided by 7 is written as $(x - 4) \div 7$.

10. The quotient of -9 and twice a number is written as $\frac{-9}{2x}$.

11. Five plus the sum of a number and 1 is written as $5 + (x + 1) = 5 + x + 1 = 6 + x$.

12. A number subtracted from -86 is written as $-86 - x$.

13. $\frac{5}{2}x = 15$
$\frac{2}{5} \cdot \frac{5}{2}x = \frac{2}{5} \cdot 15$
$x = 6$

14. $\frac{x}{4} - 1 = -7$
$\frac{x}{4} - 1 + 1 = -7 + 1$
$\frac{x}{4} = -6$
$4 \cdot \frac{x}{4} = 4(-6)$
$x = -24$

15. $2x < -4$
$\frac{2x}{2} < \frac{-4}{2}$
$x < -2$
$\{x | x < -2\}$

16. $5(x + 4) \geq 4(2x + 3)$
$5x + 20 \geq 8x + 12$
$5x + 20 - 20 \geq 8x + 12 - 20$
$5x \geq 8x - 8$
$5x - 8x \geq 8x - 8 - 8x$
$-3x \geq -8$
$\frac{-3x}{-3} \leq \frac{-8}{-3}$
$x \leq \frac{8}{3}$

$\left\{x \middle| x \leq \frac{8}{3}\right\}$

17. a. The degree of the trinomial $-2t^2 + 3t + 6$ is 2, the greatest degree of any of its terms.

b. The degree of the binomial $15x - 10$ or $15x^1 - 10$ is 1.

c. The degree of the polynomial $7x + 3x^3 + 2x^2 - 1$ is 3. It is not a monomial, binomial, nor a trinomial, so the answer is none of these.

18. $x + 2y = 6$
$x + 2y - x = 6 - x$
$2y = 6 - x$
$\frac{2y}{2} = \frac{6 - x}{2}$
$y = \frac{6 - x}{2}$

19. $(-2x^2 + 5x - 1) + (-2x^2 + x + 3)$
$= -2x^2 + 5x - 1 - 2x^2 + x + 3$
$= (-2x^2 - 2x^2) + (5x + x) + (-1 + 3)$
$= -4x^2 + 6x + 2$

20. $(-2x^2 + 5x - 1) - (-2x^2 + x + 3)$
$= -2x^2 + 5x - 1 + 2x^2 - x - 3$
$= (-2x^2 + 2x^2) + (5x - x) + (-1 - 3)$
$= 0x^2 + 4x + (-4)$
$= 4x - 4$

21. $(3y + 1)^2 = (3y + 1)(3y + 1)$
$= (3y)(3y) + (3y)(1) + 1(3y) + 1(1)$
$= 9y^2 + 3y + 3y + 1$
$= 9y^2 + 6y + 1$

22. $(x - 12)^2 = x^2 - 2(x)(12) + (12)^2$
$= x^2 - 24x + 144$

23. $-9a^5 + 18a^2 - 3a$
$= 3a(-3a^4) + 3a(6a) + 3a(-1)$
$= 3a(-3a^4 + 6a - 1)$

24. $4x^2 - 36 = 4(x^2 - 9)$
$= 4(x^2 - 3^2)$
$= 4(x + 3)(x - 3)$

25. $x^2 + 4x - 12$
Look for two numbers whose product is -12 and whose sum is 4.
$x^2 + 4x - 12 = (x - 2)(x + 6)$

26. $3x^2 - 20xy - 7y^2 = (3x + y)(x - 7y)$

27. Factors of $8x^2$: $8x^2 = 8x \cdot x$, $8x^2 = 4x \cdot 2x$
Factors of 5: $5 = -1 \cdot -5$
$8x^2 - 22x + 5 = (4x - 1)(2x - 5)$

28. Factors of $18x^2$:
$18x^2 = 18x \cdot x$, $18x^2 = 9x \cdot 2x$,
$18x^2 = 6x \cdot 3x$
Factors of -2: $-2 = -1 \cdot 2$, $-2 = 1 \cdot -2$
$18x^2 + 35x - 2 = (18x - 1)(x + 2)$

29. $x^2 - 9x - 22 = 0$
$(x - 11)(x + 2) = 0$
$x - 11 = 0$ or $x + 2 = 0$
   $x = 11$ $x = -2$
The solutions are 11 and -2.

30. $x^2 = x$
$x^2 - x = 0$
$x(x - 1) = 0$
$x = 0$ or $x - 1 = 0$
        $x = 1$
The solutions are 0 and 1.

31. $\dfrac{2x^2 - 11x + 5}{5x - 25} \div \dfrac{4x - 2}{10}$
$= \dfrac{2x^2 - 11x + 5}{5x - 25} \cdot \dfrac{10}{4x - 2}$
$= \dfrac{(2x - 1)(x - 5) \cdot 2 \cdot 5}{5(x - 5) \cdot 2(2x - 1)}$
$= \dfrac{1}{1}$ or 1

32. $\dfrac{2x^2 - 50}{4x^4 - 20x^3} = \dfrac{2(x^2 - 25)}{4x^3(x - 5)}$
$= \dfrac{2(x - 5)(x + 5)}{2 \cdot 2x^3(x - 5)}$
$= \dfrac{x + 5}{2x^3}$

33. $\dfrac{4b}{9a} = \dfrac{4b}{9a} \cdot 1 = \dfrac{4b}{9a} \cdot \dfrac{3ab}{3ab} = \dfrac{4b(3ab)}{9a(3ab)} = \dfrac{12ab^2}{27a^2b}$

34. $\dfrac{1}{2x} = \dfrac{1}{2x} \cdot 1 = \dfrac{1}{2x} \cdot \dfrac{7x^2}{7x^2} = \dfrac{1(7x^2)}{2x(7x^2)} = \dfrac{7x^2}{14x^3}$

35.
$$1 + \frac{m}{m+1} = \frac{1}{1} + \frac{m}{m+1}$$
$$= \frac{1(m+1)}{1(m+1)} + \frac{m}{m+1}$$
$$= \frac{m+1+m}{m+1}$$
$$= \frac{2m+1}{m+1}$$

36.
$$\frac{2x+1}{x-6} - \frac{x-4}{x-6} = \frac{(2x+1)-(x-4)}{x-6}$$
$$= \frac{2x+1-x+4}{x-6}$$
$$= \frac{x+5}{x-6}$$

37.
$$3 - \frac{6}{x} = x + 8$$
$$x\left(3 - \frac{6}{x}\right) = x(x+8)$$
$$3x - 6 = x^2 + 8x$$
$$0 = x^2 + 5x + 6$$
$$0 = (x+3)(x+2)$$
$$x + 3 = 0 \quad \text{or} \quad x + 2 = 0$$
$$x = -3 \qquad\qquad x = -2$$
The solutions are -3 and -2.

38.
$$3x^2 + 5x = 2$$
$$3x^2 + 5x - 2 = 0$$
$$(3x-1)(x+2) = 0$$
$$3x - 1 = 0 \quad \text{or} \quad x + 2 = 0$$
$$x = \frac{1}{3} \qquad\qquad x = -2$$

The solutions are -2 and $\frac{1}{3}$.

39.
$$\frac{\frac{x+1}{y}}{\frac{x}{y}+2} = \frac{y\left(\frac{x+1}{y}\right)}{y\left(\frac{x}{y}+2\right)} = \frac{y\left(\frac{x+1}{y}\right)}{y\left(\frac{x}{y}\right)+y(2)} = \frac{x+1}{x+2y}$$

40.
$$\frac{\frac{x}{2}-\frac{y}{6}}{\frac{x}{12}-\frac{y}{3}} = \frac{12\left(\frac{x}{2}-\frac{y}{6}\right)}{12\left(\frac{x}{12}-\frac{y}{3}\right)}$$
$$= \frac{12\left(\frac{x}{2}\right)-12\left(\frac{y}{6}\right)}{12\left(\frac{x}{12}\right)-12\left(\frac{y}{3}\right)}$$
$$= \frac{6x-2y}{x-4y} \text{ or } \frac{2(3x-y)}{x-4y}$$

41. $3x + y = 12$

   a. In $(0, \ )$, the x-coordinate is 0.
$$x = 0: \ 3(0) + y = 12$$
$$0 + y = 12$$
$$y = 12$$
The ordered-pair solution is $(0, 12)$.

   b. In $(\ , 6)$, the y-coordinate is 6.
$$y = 6: \ 3x + 6 = 12$$
$$3x = 6$$
$$x = 2$$
The ordered-pair solution is $(2, 6)$.

   c. In $(-1, \ )$, the x-coordinate is -1.
$$x = -1: \ 3(-1) + y = 12$$
$$-3 + y = 12$$
$$y = 15$$
The ordered-pair solution is $(-1, 15)$.

42. $y = -5x$
$$x = 0: y = -5(0) = 0$$
$$x = -1: y = -5(-1) = 5$$
$$y = -10: \ -10 = -5x$$
$$2 = x$$

x	y
0	0
-1	5
2	-10

43. $2x + y = 5$

$y = 0:\ 2x + 0 = 5$

$\qquad\qquad 2x = 5$

$\qquad\qquad x = \dfrac{5}{2}$

The x-intercept is $\left(\dfrac{5}{2}, 0\right)$.

$x = 0:\ 2(0) + y = 5$

$\qquad\qquad 0 + y = 5$

$\qquad\qquad\quad y = 5$

The y-intercept is $(0, 5)$.

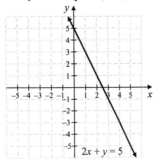

44. $(x_1,\ y_1) = (0, 5),\ (x_2,\ y_2) = (-5, 4)$

$m = \dfrac{y_2 - y_1}{x_2 - x_1} = \dfrac{4 - 5}{-5 - 0} = \dfrac{-1}{-5} = \dfrac{1}{5}$

45. $y = mx + b$

$-2x + 3y = 11$

$\qquad 3y = 2x + 11$

$\qquad\ y = \dfrac{2}{3}x + \dfrac{11}{3}$

The slope is $m = \dfrac{2}{3}$.

46. The graph of $x = -10$ is a vertical line. Slopes of vertical lines are undefined.

47. $y - y_1 = m(x - x_1)$

$y - 5 = -2[x - (-1)]$

$y - 5 = -2(x + 1)$

$y - 5 = -2x - 2$

$\qquad y = -2x + 3$

48. $y = mx + b$

$2x - 5y = 10$

$\quad -5y = -2x + 10$

$\qquad\ y = \dfrac{2}{5}x - 2$

The slope is $m = \dfrac{2}{5}$ and the y-intercept is $(0, -2)$.

49. $g(x) = x^2 - 3$

    a. $g(2) = 2^2 - 3 = 4 - 3 = 1$

        The ordered pair is $(2, 1)$.

    b. $g(-2) = (-2)^2 - 3 = 4 - 3 = 1$

        The ordered pair is $(-2, 1)$.

    c. $g(0) = 0^2 - 3 = 0 - 3 = -3$

        The ordered pair is $(0, -3)$.

50. $(x_1,\ y_1) = (2, 3),\ (x_2,\ y_2) = (0, 0)$

$m = \dfrac{y_2 - y_1}{x_2 - x_1} = \dfrac{0 - 3}{0 - 2} = \dfrac{-3}{-2} = \dfrac{3}{2}$

$y - y_1 = m(x - x_1)$

$y - 3 = \dfrac{3}{2}(x - 2)$

$2(y - 3) = 2\left(\dfrac{3}{2}\right)(x - 2)$

$2(y - 3) = 3(x - 2)$

$2y - 6 = 3x - 6$

$\qquad 2y = 3x$

$\qquad\ 0 = 3x - 2y$

$3x - 2y = 0$

Chapter 7

Section 7.1 Practice Problems

1.
$$5x - 2y = -3 \qquad y = 3x$$
$$5(3) - 2(9) = -3 \qquad 9 = 3(3)$$
$$15 - 18 = -3 \qquad 9 = 9$$
$$-3 = -3$$

(3, 9) is a solution of the system.

2.
$$2x - y = 8 \qquad x + 3y = 4$$
$$2(3) - (-2) = 8 \qquad 3 + 3(-2) = 4$$
$$6 + 2 = 8 \qquad 3 + -6 = 4$$
$$8 = 8 \qquad -3 \neq 4$$

(3, −2) is not a solution of the system.

3.

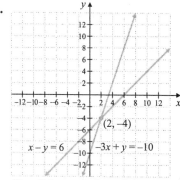

(2, −4) is a solution of the system.

4.

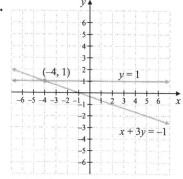

(−4, 1) is a solution of the system.

5.

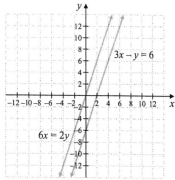

There are no solution because the lines are parallel.

6.

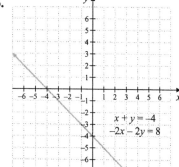

There are an infinite number of solutions because it is the same line.

Calculator Explorations

1. $y = -2.68x + 1.21$
$y = 5.22x - 1.68$

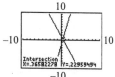

(0.37, 0.23) is the approximate point of intersection.

2. $y = 4.25x + 3.89$
  $y = -1.88x + 3.21$

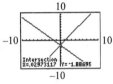

$(-0.11, 3.42)$ is the approximate point of intersection.

3. $4.3x - 2.9y = 5.6 \rightarrow y = -(5.6 - 4.3x)/2.9$
  $8.1x + 7.6y = -14.1$
  $\rightarrow y = (-14.1 - 8.1x)/7.6$

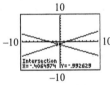

$(0.03, -1.89)$ is the approximate point of intersection.

4. $-3.6x - 8.6y = 10$
  $\rightarrow y = -(10 + 3.6x)/8.6$
  $-4.5x + 9.6y = -7.7$
  $\rightarrow y = (-7.7 + 4.5x)/9.6$

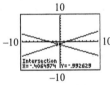

$(-0.41, -0.99)$ is the approximate point of intersection.

Mental Math

1. The lines intersect at $(-1, 3)$; therefore there is one solution.

2. Since the lines are parallel and do not intersect, there is no solutions.

3. Since the lines are the same, there are an infinite number of solutions.

4. The lines intersect at $(3, 4)$; therefore there is one solution.

5. Since the lines are parallel and do not intersect, there is no solution.

6. Since the lines are the same, there are an infinite number of solutions.

7. The lines intersect at $(3, 2)$; therefore there is one solution.

8. The lines intersect at $(0, -3)$; therefore there is one solution.

Exercise Set 7.1

1. a. First equation:
  $x + y = 8$
  $2 + 4 \overset{?}{=} 8$
  $6 = 8$ False
  $(2, 4)$ is not a solution of the first equation, so it is not a solution of the system.

 b. First equation:
  $x + y = 8$
  $5 + 3 \overset{?}{=} 8$
  $8 = 8$ True
  Second equation:
  $3x + 2y = 21$
  $3(5) + 2(3) \overset{?}{=} 21$
  $15 + 6 \overset{?}{=} 21$
  $21 = 21$ True
  Since $(5, 3)$ is a solution of both equations, it is a solution of the system.

3. a. First equation:
  $3x - y = 5$
  $3(3) - 4 \overset{?}{=} 5$
  $9 - 4 \overset{?}{=} 5$
  $5 = 5$ True
  Second equation:
  $x + 2y = 11$
  $3 + 2(4) \overset{?}{=} 11$
  $3 + 8 \overset{?}{=} 11$
  $11 = 11$ True
  Since $(3, 4)$ is a solution of both equations, it is a solution of the system.

b. First equation:

$$3x - y = 5$$
$$3(0) - (-5) \stackrel{?}{=} 5$$
$$0 + 5 \stackrel{?}{=} 5$$
$$5 = 5 \quad \text{True}$$

Second equation:

$$x + 2y = 11$$
$$0 + 2(-5) \stackrel{?}{=} 11$$
$$0 + (-10) \stackrel{?}{=} 11$$
$$-10 = 11 \quad \text{False}$$

$(0, -5)$ is not a solution of the second equation, so it is not a solution of the system.

5. a. First equation:

$$2y = 4x + 6$$
$$2(-3) \stackrel{?}{=} 4(-3) + 6$$
$$-6 \stackrel{?}{=} -12 + 6$$
$$-6 = -6 \quad \text{True}$$

Second equation:

$$2x - y = -3$$
$$2(-3) - (-3) \stackrel{?}{=} -3$$
$$-6 + 3 \stackrel{?}{=} -3$$
$$-3 = -3 \quad \text{True}$$

$(-3, -3)$ is a solution of both equations, it is a solution of the system.

b. First equation:

$$2y = 4x + 6$$
$$2(3) \stackrel{?}{=} 4(0) + 6$$
$$6 \stackrel{?}{=} 0 + 6$$
$$6 = 6 \quad \text{True}$$

Second equation:

$$2x - y = -3$$
$$2(0) - 3 \stackrel{?}{=} -3$$
$$0 - 3 \stackrel{?}{=} -3$$
$$-3 = -3 \quad \text{True}$$

Since $(0, 3)$ is a solution of both equations, it is a solution of the system.

7. a. First equation:

$$-2 = x - 7y$$
$$-2 \stackrel{?}{=} -2 - 7(0)$$
$$-2 \stackrel{?}{=} -2 - 0$$
$$-2 = -2 \quad \text{True}$$

Second equation:

$$6x - y = 13$$
$$6(-2) - 0 \stackrel{?}{=} 13$$
$$-12 - 0 \stackrel{?}{=} 13$$
$$-12 = 13 \quad \text{False}$$

$(-2, 0)$ is not a solution of the second equation, so it is not a solution of the system.

b. First equation:

$$-2 = x - 7y$$
$$-2 \stackrel{?}{=} \frac{1}{2} - 7\left(\frac{5}{14}\right)$$
$$-2 \stackrel{?}{=} \frac{1}{2} - \frac{5}{2}$$
$$-2 = -\frac{4}{2}$$
$$-2 = -2 \quad \text{True}$$

Second equation:

$$6x - y = 13$$
$$6\left(\frac{1}{2}\right) - \frac{5}{14} \stackrel{?}{=} 13$$
$$\frac{42}{14} - \frac{5}{14} \stackrel{?}{=} 13$$
$$\frac{37}{14} = 13 \quad \text{False}$$

$\left(\frac{1}{2}, \frac{5}{14}\right)$ is not a solution of the second equation, so it is not a solution of the system.

9.

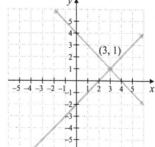

First equation:

$$x + y = 4$$
$$3 + 1 \stackrel{?}{=} 4$$
$$4 = 4 \quad \text{True}$$

Second equation:
$x - y = 2$
$3 - 1 \overset{?}{=} 2$
$\quad 2 = 2$ True
The solution of the system is (3, 1).

11.

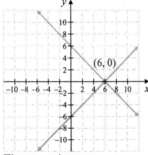

First equation:
$x + y = 6$
$6 + 0 \overset{?}{=} 6$
$\quad 6 = 6$ True
Second equation:
$-x + y = -6$
$-6 + 0 = -6$
$\quad -6 = -6$ True
The solution of the system is (6, 0).

13.

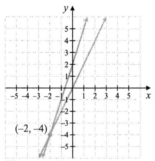

First equation:
$y = 2x$
$-4 \overset{?}{=} 2(-2)$
$-4 = -4$ True
Second equation:
$3x - y = -2$
$3(-2) - (-4) \overset{?}{=} -2$
$\quad -6 + 4 \overset{?}{=} -2$
$\quad\quad -2 = -2$ True
The solution of the system is (−2, −4).

15.

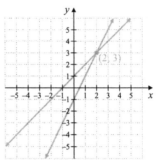

First equation:
$y = x + 1$
$3 \overset{?}{=} 2 + 1$
$3 = 3$ True
Second equation:
$y = 2x - 1$
$3 \overset{?}{=} 2(2) - 1$
$3 \overset{?}{=} 4 - 1$
$3 = 3$ True
The solution of the system is (2, 3).

17.

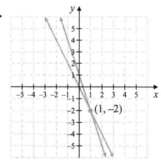

First equation:
$\quad 2x + y = 0$
$2(1) + (-2) \overset{?}{=} 0$
$\quad\quad 2 - 2 \overset{?}{=} 0$
$\quad\quad\quad 0 = 0$ True
Second equation:
$\quad 3x + y = 1$
$3(1) + (-2) \overset{?}{=} 1$
$\quad\quad 3 - 2 \overset{?}{=} 1$
$\quad\quad\quad 1 = 1$ True
The solution of the system is (1, −2).

19.

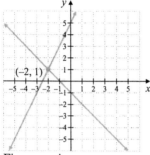

First equation:
$y = -x - 1$
$1 \stackrel{?}{=} -(-2) - 1$
$1 \stackrel{?}{=} 2 - 1$
$1 = 1$ True
Second equation:
$y = 2x + 5$
$1 \stackrel{?}{=} 2(-2) + 5$
$1 \stackrel{?}{=} -4 + 5$
$1 = 1$ True
The solution of the system is $(-2, 1)$.

21.

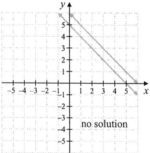

First equation:
$x + y = 5$
$\quad y = -x + 5$

Second equation:
$x + y = 6$
$\quad y = -x + 6$

The lines have the same slope, but different y-intercepts, so they are parallel. The system has no solution.

23.

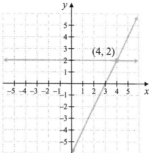

First equation:
$\quad 2x - y = 6$
$\quad 2(4) - 2 \stackrel{?}{=} 6$
$\quad\quad 8 - 2 \stackrel{?}{=} 6$
$\quad\quad\quad 6 = 6$ True
Second equation:
$y = 2$
$2 = 2$ True
The solution of the system is $(4, 2)$.

25.

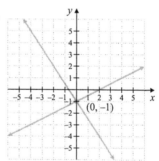

First equation:
$\quad x - 2y = 2$
$\quad 0 - 2(-1) \stackrel{?}{=} 2$
$\quad\quad 0 + 2 \stackrel{?}{=} 2$
$\quad\quad\quad 2 = 2$ True
Second equation:
$\quad 3x + 2y = -2$
$\quad 3(0) + 2(-1) \stackrel{?}{=} -2$
$\quad\quad 0 - 2 \stackrel{?}{=} -2$
$\quad\quad\quad -2 = -2$ True
The solution of the system is $(0, -1)$.

27.

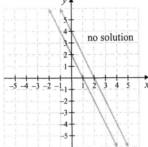

First equation:
$2x + y = 4$
$\qquad y = -2x + 4$

Second equation:
$6x = -3y + 6$
$3y = -6x + 6$
$\quad y = -2x + 2$

The lines have the same slope, but different y-intercepts, so they are parallel. The system has no solution.

29.

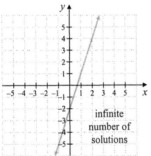

First equation:
$y - 3x = -2$
$\qquad y = 3x - 2$

Second equation:
$6x - 2y = 4$
$\quad -2y = -6x + 4$
$\qquad y = 3x - 2$

The graphs of the equations are the same line, so the system has an infinite number of solutions.

31.

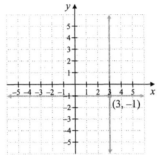

First equation:
$x = 3$
$3 = 3$ True
Second equation:
$y = -1$
$-1 = -1$ True
The solution of the system is $(3, -1)$.

33.

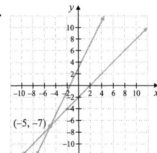

First equation:
$\quad y = x - 2$
$-7 \overset{?}{=} -5 - 2$
$-7 = -7$ True
Second equation:
$\quad y = 2x + 3$
$-7 \overset{?}{=} 2(-5) + 3$
$-7 \overset{?}{=} -10 + 3$
$-7 = -7$ True
The solution of the system is $(-5, -7)$.

35.

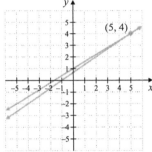

First equation:
$$2x - 3y = -2$$
$$2(5) - 3(4) \overset{?}{=} -2$$
$$10 - 12 \overset{?}{=} -2$$
$$-2 = -2 \quad \text{True}$$
Second equation:
$$-3x + 5y = 5$$
$$-3(5) + 5(4) \overset{?}{=} 5$$
$$-15 + 20 \overset{?}{=} 5$$
$$5 = 5 \quad \text{True}$$
The solution of the system is (5, 4).

37.

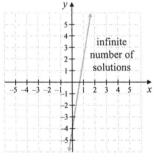

First equation:
$$6x - y = 4$$
$$-y = -6x + 4$$
$$y = 6x - 4$$
Second equation:
$$\frac{1}{2}y = -2 + 3x$$
$$y = -4 + 6x$$
$$y = 6x - 4$$
The graphs of the equations are the same line, so the system has an infinite number of solutions.

39. $\quad 5(x - 3) + 3x = 1$
$$5x - 15 + 3x = 1$$
$$8x = 16$$
$$x = 2$$

41. $\quad 4\left(\dfrac{y+1}{2}\right) + 3y = 0$
$$2(y + 1) + 3y = 0$$
$$2y + 2 + 3y = 0$$
$$5y + 2 = 0$$
$$5y = -2$$
$$y = -\frac{2}{5}$$

43. $\quad 8a - 2(3a - 1) = 6$
$$8a - 6a + 2 = 6$$
$$2a = 4$$
$$a = 2$$

45. Answers may vary. Possible answer:

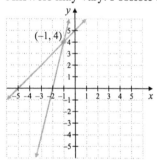

47. Answers may vary. Possible answer:

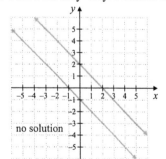

49. The lines cross at points corresponding to the years 1984 and 1988. The number of pounds of imported fishery products was equal to the number of pounds of domestic catch in 1984 and 1988.

51. In 2003 the number of Target stores approximately equal to the number of Wal-Mart discount stores.

53. answers may vary

55. answers may vary

57. answers may vary

Section 7.2 Practice Problems

1. $\begin{cases} 2x+3y=13 \\ x=y+4 \end{cases}$

Substitute $y + 4$ for x in the first equation and solve for y.
$$2(y+4)+3y=13$$
$$2y+8+3y=13$$
$$5y=5$$
$$y=1$$
Solve for x.
$$x=y+4=1+4=5$$
The solution of the system is (5, 1).

2. $\begin{cases} 4x-y=2 \\ y=5x \end{cases}$

Substitute $5x$ for y in the first equation and solve for x.
$$4x-5x=2$$
$$-x=2$$
$$x=-2$$
Solve for y.
$$y=5x=5(-2)=-10$$
The solution of the system is (−2, −10).

3. $\begin{cases} 3x+y=5 \\ 3x-2y=-7 \end{cases}$

Solve the first equation for y.
$$3x+y=5$$
$$y=-3x+5$$
Substitute $-3x + 5$ for y in the second equation.
$$3x-2y=-7$$
$$3x-2(-3x+5)=-7$$
$$3x+6x-10=-7$$
$$9x=3$$
$$x=\frac{3}{9}=\frac{1}{3}$$
Solve for y.
$$y=-3x+5=-3\left(\frac{1}{3}\right)+5=-1+5=4$$
The solution of the system is $\left(\frac{1}{3}, 4\right)$.

4. $\begin{cases} 5x-2y=6 \\ -3x+y=-3 \end{cases}$

Solve the second equation for y.
$$-3x+y=-3$$
$$y=3x-3$$
Substitute $3x - 3$ for y in the first equation.
$$5x-2y=6$$
$$5x-2(3x-3)=6$$
$$5x-6x+6=6$$
$$-x=0$$
$$x=0$$
Solve for y.
$$y=3x-3=3(0)-3=0-3=-3$$
The solution of the system is (0, −3).

5. $\begin{cases} -x+3y=6 \\ y=\frac{1}{3}x+2 \end{cases}$

Substitute $\frac{1}{3}x+2$ for y in the first equation.

$$-x + 3y = 6$$
$$-x + 3\left(\frac{1}{3}x + 2\right) = 6$$
$$-x + x + 6 = 6$$
$$0x = 0$$
$$0 = 0$$

The statement $0 = 0$ indicates that this system has an infinite number of solutions. It is a graph of the same line.

6. $\begin{cases} 2x - 3y = 6 \\ -4x + 6y = -12 \end{cases}$

$$2x - 3y = 6$$
$$-3y = -2x + 6$$
$$y = \frac{2}{3}x - 2$$

Substitute $\frac{2}{3}x - 2$ for y in the second

equation.
$$-4x + 6y = -12$$
$$-4x + 6\left(\frac{2}{3}x - 2\right) = -12$$
$$-4x + 4x - 12 = -12$$
$$0x = 0$$
$$0 = 0$$

The statement $0 = 0$ indicates that this system has an infinite number of solutions. It is a graph of the same line.

Exercise Set 7.2

1. $\begin{cases} x + y = 3 \\ x = 2y \end{cases}$

Substitute $2y$ for x in the first equation and solve for y.
$$x + y = 3$$
$$2y + y = 3$$
$$3y = 3$$
$$y = 1$$
Now solve for x.
$$x = 2y = 2(1) = 2$$
The solution of the system is $(2, 1)$.

3. $\begin{cases} x + y = 6 \\ y = -3x \end{cases}$

Substitute $-3x$ for y in the first equation and solve for x.
$$x + y = 6$$
$$x + (-3x) = 6$$
$$-2x = 6$$
$$x = -3$$
Now solve for y.
$$y = -3x = -3(-3) = 9$$
The solution of the system is $(-3, 9)$.

5. $\begin{cases} y = 3x + 1 \\ 4y - 8x = 12 \end{cases}$

Substitute $3x + 1$ for y in the second equation and solve for x.
$$4y - 8x = 12$$
$$4(3x + 1) - 8x = 12$$
$$12x + 4 - 8x = 12$$
$$4x + 4 = 12$$
$$4x = 8$$
$$x = 2$$
Now solve for y.
$$y = 3x + 1 = 3(2) + 1 = 6 + 1 = 7$$
The solution of the system is $(2, 7)$.

7. $\begin{cases} y = 2x + 9 \\ y = 7x + 10 \end{cases}$

Substitute $2x + 9$ for y in the second equation and solve for x.
$$y = 7x + 10$$
$$2x + 9 = 7x + 10$$
$$-5x = 1$$
$$y = -\frac{1}{5}$$
Now solve for y.
$$y = 2x + 9 = 2\left(-\frac{1}{5}\right) + 9 = \frac{-2}{5} + \frac{45}{5} = \frac{43}{5}$$

The solution of the system is $\left(-\frac{1}{5}, \frac{43}{5}\right)$.

9. $\begin{cases} 3x-4y=10 \\ y=x-3 \end{cases}$

Substitute $x-3$ for y in the first equation and solve for x.

$3x-4y=10$
$3x-4(x-3)=10$
$3x-4x+12=10$
$-x+12=10$
$-x=-2$
$x=2$

Now solve for y.
$y=x-3=2-3=-1$
The solution of the system is $(2,-1)$.

11. $\begin{cases} x+2y=6 \\ 2x+3y=8 \end{cases}$

Solve the first equation for x.
$x+2y=6$
$x=-2y+6$

Substitute $-2y+6$ for x in the second equation and solve for y.
$2x+3y=8$
$2(-2y+6)+3y=8$
$-4y+12+3y=8$
$-y=-4$
$y=4$

Now solve for x.
$x=-2y+6=-2(4)+6=-8+6=-2$
The solution of the system is $(-2,4)$.

13. $\begin{cases} 3x+2y=16 \\ x=3y-2 \end{cases}$

Substitute $3y-2$ for x in the first equation and solve for y.
$3x+2y=16$
$3(3y-2)+2y=16$
$9y-6+2y=16$
$11y-6=16$
$11y=22$
$y=2$

Now solve for x.
$x=3y-2=3(2)-2=6-2=4$
The solution of the system is $(4,2)$.

15. $\begin{cases} 2x-5y=1 \\ 3x+y=-7 \end{cases}$

Solve the second equation for y.
$3x+y=-7$
$y=-3x-7$

Substitute $-3x-7$ for y in the first equation and solve for y.
$2x-5y=1$
$2x-5(-3x-7)=1$
$2x+15x+35=1$
$17x=-34$
$x=-2$

Now solve for y.
$y=-3x-7=-3(-2)-7=6-7=-1$
The solution of the system is $(-2,-1)$.

17. $\begin{cases} 4x+2y=5 \\ -2x=y+4 \end{cases}$

Solve the second equation for x.
$-2x=y+4$
$x=-\dfrac{1}{2}y-2$

Substitute $-\dfrac{1}{2}y-2$ for x in the first equation and solve for y.
$4x+2y=5$
$4\left(-\dfrac{1}{2}y-2\right)+2y=5$
$-2y-8+2y=5$
$-8=5$ False

Since the statement $-8=5$ is false, the system has no solution.

19. $\begin{cases} 4x+y=11 \\ 2x+5y=1 \end{cases}$

Solve the first equation for y.
$4x+y=11$
$y=-4x+11$

Substitute $-4x+11$ for y in the second equation and solve for y.

$$2x + 5y = 1$$
$$2x + 5(-4x + 11) = 1$$
$$2x + (-20x) + 55 = 1$$
$$-18x = -54$$
$$x = 3$$

Now solve for y.
$$y = -4x + 11 = -4(3) + 11 = -12 + 11 = -1$$
The solution of the system is $(3, -1)$.

21. $\begin{cases} x + 2y + 5 = -4 + 5y - x \\ 2x + x = y + 4 \end{cases}$

Simplify each equation.
$$\begin{cases} 2x + 9 = 3y \\ 3x = y + 4 \end{cases}$$

Solve the second simplified equation for y.
$$3x = y + 4$$
$$3x - 4 = y$$

Substitute $3x - 4$ for y in the first simplified equation and solve for x.
$$2x + 9 = 3y$$
$$2x + 9 = 3(3x - 4)$$
$$2x + 9 = 9x - 12$$
$$2x + 21 = 9x$$
$$21 = 7x$$
$$3 = x$$

Now solve for y.
$$y = 3x - 4 = 3(3) - 4 = 9 - 4 = 5$$
The solution of the system is $(3, 5)$.

23. $\begin{cases} 6x - 3y = 5 \\ x + 2y = 0 \end{cases}$

Solve the second equation for x.
$$x + 2y = 0$$
$$x = -2y$$

Substitute $-2y$ for x in the first equation and solve for y.
$$6x - 3y = 5$$
$$6(-2y) - 3y = 5$$
$$-12y - 3y = 5$$
$$-15y = 5$$
$$y = -\frac{5}{15} = -\frac{1}{3}$$

Now solve for x.

$$x = -2y = -2\left(-\frac{1}{3}\right) = \frac{2}{3}$$

The solution of the system is $\left(\frac{2}{3}, -\frac{1}{3}\right)$.

25. $\begin{cases} 3x - y = 1 \\ 2x - 3y = 10 \end{cases}$

Solve the first equation for y.
$$3x - y = 1$$
$$-y = -3x + 1$$
$$y = 3x - 1$$

Substitute $3x - 1$ for y in the second equation and solve for x.
$$2x - 3y = 10$$
$$2x - 3(3x - 1) = 10$$
$$2x - 9x + 3 = 10$$
$$-7x + 3 = 10$$
$$-7x = 7$$
$$x = -1$$

Now solve for y.
$$y = 3x - 1 = 3(-1) - 1 = -3 - 1 = -4$$
The solution of the system is $(-1, -4)$.

27. $\begin{cases} -x + 2y = 10 \\ -2x + 3y = 18 \end{cases}$

Solve the first equation for x.
$$-x + 2y = 10$$
$$2y - 10 = x$$

Substitute $2y - 10$ for x in the second equation and solve for y.
$$-2x + 3y = 18$$
$$-2(2y - 10) + 3y = 18$$
$$-4y + 20 + 3y = 18$$
$$-y + 20 = 18$$
$$2 = y$$

Now solve for x.
$$x = 2y - 10 = 2(2) - 10 = 4 - 10 = -6$$
The solution of the system is $(-6, 2)$.

29. $\begin{cases} 5x + 10y = 20 \\ 2x + 6y = 10 \end{cases}$

Solve the first equation for x. (Note that the second equation could also be easily solved for x.)

$5x + 10y = 20$

$\quad 5x = -10y + 20$

$\quad\quad x = -2y + 4$

Substitute $-2y + 4$ for x in the second equation and solve for y.

$2x + 6y = 10$

$2(-2y + 4) + 6y = 10$

$-4y + 8 + 6y = 10$

$2y + 8 = 10$

$2y = 2$

$y = 1$

Now solve for x.

$x = -2y + 4 = -2(1) + 4 = -2 + 4 = 2$

The solution of the system is $(2, 1)$.

31. $\begin{cases} 3x + 6y = 9 \\ 4x + 8y = 16 \end{cases}$

Solve the first equation for x.

$3x + 6y = 9$

$3x = -6y + 9$

$x = -2y + 3$

Substitute $-2y + 3$ for x in the second equation and solve for y.

$4x + 8y = 16$

$4(-2y + 3) + 8y = 16$

$-8y + 12 + 8y = 16$

$0 = 4 \quad$ False

Since the statement $0 = 4$ is false, the system has no solution.

33. $\begin{cases} \dfrac{1}{3}x - y = 2 \\ x - 3y = 6 \end{cases}$

Solve the second equation for x.

$x - 3y = 6$

$x = 3y + 6$

Substitute $3y + 6$ for x in the first equation and solve for y.

$\dfrac{1}{3}x - y = 2$

$\dfrac{1}{3}(3y + 6) - y = 2$

$y + 2 - y = 2$

$2 = 2$

Since $2 = 2$ is a true statement, the two equations in the original system are equivalent. The system has an infinite number of solutions.

35. $\begin{cases} x = \dfrac{3}{4}y - 1 \\ 8x - 5y = -6 \end{cases}$

Substitute $\dfrac{3}{4}y - 1$ for x in the second equation and solve for y.

$8x - 5y = -6$

$8\left(\dfrac{3}{4}y - 1\right) - 5y = -6$

$6y - 8 - 5y = -6$

$y = 2$

Now solve for x.

$x = \dfrac{3}{4}y - 1 = \dfrac{3}{4}(2) - 1 = \dfrac{3}{2} - 1 = \dfrac{1}{2}$

The solution of the system is $\left(\dfrac{1}{2}, 2\right)$.

37. $\quad 3x + 2y = 6$

$-2(3x + 2y) = -2(6)$

$-6x - 4y = -12$

39. $\quad -4x + y = 3$

$3(-4x + y) = 3(3)$

$-12x + 3y = 9$

41. $\quad 3n + 6m$

$\underline{+\ 2n - 6m}$

$\quad 5n$

43. $\quad -5a - 7b$

$\underline{\quad 5a - 8b}$

$\quad -15b$

45. $\begin{cases} -5y + 6y = 3x + 2(x - 5) - 3x + 5 \\ 4(x + y) - x + y = -12 \end{cases}$

Simplify each equation.

$\begin{cases} y = 2x - 5 \\ 3x + 5y = -12 \end{cases}$

Substitute $2x - 5$ for y in the second

simplified equation and solve for x.

$$3x + 5y = -12$$
$$3x + 5(2x - 5) = -12$$
$$3x + 10x - 25 = -12$$
$$13x - 25 = -12$$
$$13x = 13$$
$$x = 1$$

Now solve for y.

$$y = 2x - 5 = 2(1) - 5 = 2 - 5 = -3$$

The solution of the system is $(1, -3)$.

47. answers may vary

49. no; answers may vary

51. c; answers may vary

53. Using a graphing calculator, the solution of the system is $(-2.6, 1.3)$.

55. Using a graphing calculator, the solution of the system is $(3.28, 2.1)$.

57. a.
$$\begin{cases} y = -0.50x + 21.92 \\ y = 0.71x + 11.03 \end{cases}$$

Substitute $-0.50x + 21.92$ for y in the second equation and solve for x.

$$y = 0.71x + 11.03$$
$$-0.50x + 21.92 = 0.71x + 11.03$$
$$-0.50x + 10.89 = 0.71x$$
$$10.89 = 1.21x$$
$$9 = x$$

Now solve for y.

$$y = -0.50x + 21.92$$
$$= -0.50(9) + 21.92$$
$$= -4.50 + 21.92$$
$$= 17.42$$

Rounded to the nearest whole numbers, the solution of the system is $(9, 17)$.

b. In $1973 + 9 = 1982$, the percent of households that used fuel oil and electricity was the same, 17%.

c.

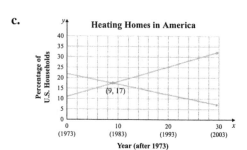

answers may vary

Section 7.3 Practice Problems

1.
$$\begin{cases} x + y = 13 \\ x - y = 5 \end{cases}$$

Add the equations to eliminate y; then solve for x.

$$\begin{array}{r} x + y = 13 \\ x - y = 5 \\ \hline 2x \quad\;\; = 18 \\ x \quad\;\; = 9 \end{array}$$

Now solve for y.

$$x + y = 13$$
$$9 + y = 13$$
$$y = 4$$

The solution of the system is $(9, 4)$.

2.
$$\begin{cases} 2x - y = -6 \\ -x + 4y = 17 \end{cases}$$

Multiply the first equation by 4.

$$\begin{cases} 4(2x - y) = 4(-6) \\ -x + 4y = 17 \end{cases} \Rightarrow \begin{cases} 8x - 4y = -24 \\ -x + 4y = 17 \end{cases}$$

Add the equations to eliminate y; then solve for x.

$$\begin{array}{r} 8x - 4y = -24 \\ -x + 4y = 17 \\ \hline 7x \quad\quad = -7 \\ x = -1 \end{array}$$

Now solve for y.

$$2x - y = -6$$
$$2(-1) - y = -6$$
$$-2 - y = -6$$
$$-y = -4$$
$$y = 4$$

The solution of the system is $(-1, 4)$.

3. $\begin{cases} x - 3y = -2 \\ -3x + 9y = 5 \end{cases}$

Multiply the first equation by 3.

$\begin{cases} 3(x - 3y) = 3(-2) \\ -3x + 9y = 5 \end{cases} \Rightarrow \begin{cases} 3x - 9y = -6 \\ -3x + 9y = 5 \end{cases}$

Add the equations to eliminate y; then solve for x.

$\begin{aligned} 3x - 9y &= -6 \\ \underline{-3x + 9y} &= \underline{5} \\ 0 &= -1 \end{aligned}$

Since the statement $0 = -1$ is false, there is no solution to the system.

4. $\begin{cases} 2x + 5y = 1 \\ -4x - 10y = -2 \end{cases}$

Multiply the first equation by 2.

$\begin{cases} 2(2x + 5y) = 2(1) \\ -4x - 10y = -2 \end{cases} \Rightarrow \begin{cases} 4x + 10y = 2 \\ -4x - 10y = -2 \end{cases}$

Add the equations to eliminate y; then solve for x.

$\begin{aligned} 4x + 10y &= 2 \\ \underline{-4x - 10y} &= \underline{-2} \\ 0 &= 0 \end{aligned}$

Since the statement $0 = 0$ is true, there are an infinite number of solutions.

5. $\begin{cases} 4x + 5y = 14 \\ 3x - 2y = -1 \end{cases}$

Multiply the first equation by 2 and multiply the second equation by 5.

$\begin{cases} 2(4x + 5y) = 2(14) \\ 5(3x - 2y) = 5(-1) \end{cases} \Rightarrow \begin{cases} 8x + 10y = 28 \\ 15x - 10y = -5 \end{cases}$

Add the equations to eliminate y; then solve for x.

$\begin{aligned} 8x + 10y &= 28 \\ \underline{15x - 10y} &= \underline{-5} \\ 23x &= 23 \\ x &= 1 \end{aligned}$

Now solve for y.

$\begin{aligned} 4x + 5y &= 14 \\ 4(1) + 5y &= 14 \\ 5y &= 10 \\ y &= 2 \end{aligned}$

The solution of the system is $(1, 2)$.

6. $\begin{cases} -\dfrac{x}{3} + y = \dfrac{4}{3} \\ \dfrac{x}{2} - \dfrac{5}{2}y = -\dfrac{1}{2} \end{cases}$

Multiply the first equation by 3 and the second equation by 2.

$\begin{cases} 3\left(-\dfrac{x}{3} + y\right) = 3\left(\dfrac{4}{3}\right) \\ 2\left(\dfrac{x}{2} - \dfrac{5}{2}y\right) = 2\left(-\dfrac{1}{2}\right) \end{cases} \Rightarrow \begin{cases} -x + 3y = 4 \\ x - 5y = -1 \end{cases}$

Add the equations to eliminate x; then solve for y.

$\begin{aligned} -x + 3y &= 4 \\ \underline{x - 5y} &= \underline{-1} \\ -2y &= 3 \\ y &= -\dfrac{3}{2} \end{aligned}$

Now solve for y through elimination. Multiply the first equation by 15 and the second equation by 6.

$\begin{cases} 15\left(-\dfrac{x}{3} + y\right) = 15\left(\dfrac{4}{3}\right) \\ 6\left(\dfrac{x}{2} - \dfrac{5}{2}y\right) = 6\left(-\dfrac{1}{2}\right) \end{cases} \Rightarrow \begin{cases} -5x + 15y = 20 \\ 3x - 15y = -3 \end{cases}$

Add the equations to eliminate y; then solve for x.

$\begin{aligned} -5x + 15y &= 20 \\ \underline{3x - 15y} &= \underline{-3} \\ -2x &= 17 \\ x &= -\dfrac{17}{2} \end{aligned}$

The solution to the system is $\left(-\dfrac{17}{2}, -\dfrac{3}{2}\right)$.

Exercise Set 7.3

1. $\begin{cases} 3x + y = 5 \\ 6x - y = 4 \end{cases}$

Add the equations to eliminate y; then solve for x.

$\begin{aligned} 3x + y &= 5 \\ \underline{6x - y} &= \underline{4} \\ 9x &= 9 \\ x &= 1 \end{aligned}$

Now solve for y.
$$3x + y = 5$$
$$3(1) + y = 5$$
$$3 + y = 5$$
$$y = 2$$
The solution of the system is $(1, 2)$.

3. $\begin{cases} x - 2y = 8 \\ -x + 5y = -17 \end{cases}$

Add the equations to eliminate x; then solve for y.
$$x - 2y = 8$$
$$\underline{-x + 5y = -17}$$
$$3y = -9$$
$$y = -3$$
Now solve for x.
$$x - 2y = 8$$
$$x - 2(-3) = 8$$
$$x + 6 = 8$$
$$x = 2$$
The solution of the system is $(2, -3)$.

5. $\begin{cases} 3x + y = -11 \\ 6x - 2y = -2 \end{cases}$

Multiply the first equation by 2.
$$\begin{cases} 2(3x + y) = 2(-11) \\ 6x - 2y = -2 \end{cases} \rightarrow \begin{cases} 6x + 2y = -22 \\ 6x - 2y = -2 \end{cases}$$
Add the equations to eliminate y; then solve for x.
$$6x + 2y = -22$$
$$\underline{6x - 2y = -2}$$
$$12x \quad = -24$$
$$x = -2$$
Now solve for y.
$$3x + y = -11$$
$$3(-2) + y = -11$$
$$-6 + y = -11$$
$$y = -5$$
The solution of the system is $(-2, -5)$.

7. $\begin{cases} 3x + 2y = 11 \\ 5x - 2y = 29 \end{cases}$

Add the equations to eliminate y; then solve for x.

$$3x + 2y = 11$$
$$\underline{5x - 2y = 29}$$
$$8x = 40$$
$$x = 5$$
Now solve for y.
$$3x + 2y = 11$$
$$3(5) + 2y = 11$$
$$15 + 2y = 11$$
$$2y = -4$$
$$y = -2$$
The solution of the system is $(5, -2)$.

9. $\begin{cases} x + 5y = 18 \\ 3x + 2y = -11 \end{cases}$

Multiply the first equation by -3.
$$\begin{cases} -3(x + 5y) = -3(18) \\ 3x + 2y = -11 \end{cases} \rightarrow \begin{cases} -3x - 15y = -54 \\ 3x + 2y = -11 \end{cases}$$
Add the equations to eliminate x; then solve for y.
$$-3x - 15y = -54$$
$$\underline{3x + 2y = -11}$$
$$-13y = -65$$
$$y = 5$$
Now solve for x.
$$x + 5y = 18$$
$$x + 5(5) = 18$$
$$x + 25 = 18$$
$$x = -7$$
The solution of the system is $(-7, 5)$.

11. $\begin{cases} x + y = 6 \\ x - y = 6 \end{cases}$

Add the equations to eliminate y; then solve for x.
$$x + y = 6$$
$$\underline{x - y = 6}$$
$$2x \quad = 12$$
$$x = 6$$
Now solve for y.
$$x + y = 6$$
$$6 + y = 6$$
$$y = 0$$
The solution of the system is $(6, 0)$.

13. $\begin{cases} 2x+3y=0 \\ 4x+6y=3 \end{cases}$

Multiply the first equation by -2.

$\begin{cases} -2(2x+3y)=-2(0) \\ 4x+6y=3 \end{cases} \rightarrow \begin{cases} -4x-6y=0 \\ 4x+6y=3 \end{cases}$

Add the equations to eliminate x.

$\begin{aligned} -4x-6y&=0 \\ 4x+6y&=3 \\ \hline 0&=3 \end{aligned}$

Since the statement $0=3$ is false, the system has no solution.

15. $\begin{cases} -x+5y=-1 \\ 3x-15y=3 \end{cases}$

Multiply the first equation by 3.

$\begin{cases} 3(-x+5y)=3(-1) \\ 3x-15y=3 \end{cases} \rightarrow \begin{cases} -3x+15y=-3 \\ 3x-15y=3 \end{cases}$

Add the equations to eliminate x.

$\begin{aligned} -3x+15y&=-3 \\ 3x-15y&=3 \\ \hline 0&=0 \end{aligned}$

Since the statement $0=0$ is true, the system has an infinite number of solutions.

17. $\begin{cases} 3x-2y=7 \\ 5x+4y=8 \end{cases}$

Multiply the first equation by 2.

$\begin{cases} 2(3x-2y)=2(7) \\ 5x+4y=8 \end{cases} \rightarrow \begin{cases} 6x-4y=14 \\ 5x+4y=8 \end{cases}$

Add the equations to eliminate y; then solve for x.

$\begin{aligned} 6x-4y&=14 \\ 5x+4y&=8 \\ \hline 11x&=22 \\ x&=2 \end{aligned}$

Now solve for y.

$\begin{aligned} 3x-2y&=7 \\ 3(2)-2y&=7 \\ 6-2y&=7 \\ -2y&=1 \\ y&=-\frac{1}{2} \end{aligned}$

The solution of the system is $\left(2,\ -\dfrac{1}{2}\right)$.

19. $\begin{cases} 8x=-11y-16 \\ 2x+3y=-4 \end{cases} \rightarrow \begin{cases} 8x+11y=-16 \\ 2x+3y=-4 \end{cases}$

Multiply the second equation by -4.

$\begin{cases} 8x+11y=-16 \\ -4(2x+3y)=-4(-4) \end{cases} \rightarrow \begin{cases} 8x+11y=-16 \\ -8x-12y=16 \end{cases}$

Add the equations to eliminate x; then solve for y.

$\begin{aligned} 8x+11y&=-16 \\ -8x-12y&=16 \\ \hline -y&=0 \\ y&=0 \end{aligned}$

Now solve for x.

$\begin{aligned} 8x+11y&=-16 \\ 8x+11(0)&=-16 \\ 8x&=-16 \\ x&=-2 \end{aligned}$

The solution of the system is $(-2,\ 0)$.

21. $\begin{cases} 4x-3y=7 \\ 7x+5y=2 \end{cases}$

Multiply the first equation by 5 and the second equation by 3.

$\begin{cases} 5(4x-3y)=5(7) \\ 3(7x+5y)=3(2) \end{cases} \rightarrow \begin{cases} 20x-15y=35 \\ 21x+15y=6 \end{cases}$

Add the equations to eliminate y; then solve for x.

$\begin{aligned} 20x-15y&=35 \\ 21x+15y&=6 \\ \hline 41x&=41 \\ x&=1 \end{aligned}$

Now solve for y.

$\begin{aligned} 4x-3y&=7 \\ 4(1)-3y&=7 \\ 4-3y&=7 \\ -3y&=3 \\ y&=-1 \end{aligned}$

The solution of the system is $(1,\ -1)$.

23. $\begin{cases} 4x-6y=8 \\ 6x-9y=12 \end{cases}$

Multiply the first equation by -3 and the second equation by 2.

$\begin{cases} -3(4x-6y)=-3(8) \\ 2(6x-9y)=2(12) \end{cases} \rightarrow \begin{cases} -12x+18y=-24 \\ 12x-18y=24 \end{cases}$

Add the equations to eliminate x.

$$-12x + 18y = -24$$
$$\underline{12x - 18y = 24}$$
$$0 = 0$$

Since the statement $0 = 0$ is true, the system has an infinite number of solutions.

25. $\begin{cases} 2x - 5y = 4 \\ 3x - 2y = 4 \end{cases}$

Multiply the first equation by -3 and the second equation by 2.

$$\begin{cases} -3(2x - 5y) = -3(4) \\ 2(3x - 2y) = 2(4) \end{cases} \rightarrow \begin{cases} -6x + 15y = -12 \\ 6x - 4y = 8 \end{cases}$$

Add the equations to eliminate x; then solve for y.

$$-6x + 15y = -12$$
$$\underline{6x - 4y = 8}$$
$$11y = -4$$
$$y = -\frac{4}{11}$$

Multiply the first original equation by -2 and the second original equation by 5.

$$\begin{cases} -2(2x - 5y) = -2(4) \\ 5(3x - 2y) = 5(4) \end{cases} \rightarrow$$

$$\begin{cases} -4x + 10y = -8 \\ 15x - 10y = 20 \end{cases}$$

Add the equations to eliminate y; then solve for x.

$$-4x + 10y = -8$$
$$\underline{15x - 10y = 20}$$
$$11x \qquad = 12$$
$$x = \frac{12}{11}$$

The solution of the system is $\left(\frac{12}{11}, -\frac{4}{11} \right)$.

27. $\begin{cases} \dfrac{x}{3} + \dfrac{y}{6} = 1 \\ \dfrac{x}{2} - \dfrac{y}{4} = 0 \end{cases}$

Multiply the first equation by 6 and the second equation by 4.

$$\begin{cases} 6\left(\dfrac{x}{3} + \dfrac{y}{6} \right) = 6(1) \\ 4\left(\dfrac{x}{2} - \dfrac{y}{4} \right) = 4(0) \end{cases} \rightarrow \begin{cases} 2x + y = 6 \\ 2x - y = 0 \end{cases}$$

Add the equations to eliminate y; then solve for x.

$$2x + y = 6$$
$$\underline{2x - y = 0}$$
$$4x = 6$$
$$x = \frac{6}{4} = \frac{3}{2}$$

Multiply the original first equation by 6 and the second original equation by -4.

$$\begin{cases} 6\left(\dfrac{x}{3} + \dfrac{y}{6} \right) = 6(1) \\ -4\left(\dfrac{x}{2} - \dfrac{y}{4} \right) = -4(0) \end{cases} \rightarrow \begin{cases} 2x + y = 6 \\ -2x + y = 0 \end{cases}$$

Add the equations to eliminate x; then solve for y.

$$2x + y = 6$$
$$\underline{-2x + y = 0}$$
$$2y = 6$$
$$y = 3$$

The solution of the system is $\left(\dfrac{3}{2}, 3 \right)$.

29. $\begin{cases} \dfrac{10}{3}x + 4y = -4 \\ 5x + 6y = -6 \end{cases}$

Multiply the first equation by 3 and the second equation by -2.

$$\begin{cases} 3\left(\dfrac{10}{3}x + 4y \right) = 3(-4) \\ -2(5x + 6y) = -2(-6) \end{cases}$$

$$\rightarrow \begin{cases} 10x + 12y = -12 \\ -10x - 12y = 12 \end{cases}$$

Add the equations to eliminate x.

$$10x + 12y = -12$$
$$\underline{-10x - 12y = 12}$$
$$0 = 0$$

Since the statement $0 = 0$ is true, the system has an infinite number of solutions.

31. $\begin{cases} x - \dfrac{y}{3} = -1 \\ -\dfrac{x}{2} + \dfrac{y}{8} = \dfrac{1}{4} \end{cases}$

Multiply the first equation by 3 and the second equation by 8.

$\begin{cases} 3\left(x - \dfrac{y}{3}\right) = 3(-1) \\ 8\left(-\dfrac{x}{2} + \dfrac{y}{8}\right) = 8\left(\dfrac{1}{4}\right) \end{cases} \rightarrow \begin{cases} 3x - y = -3 \\ -4x + y = 2 \end{cases}$

Add the equations to eliminate y; then solve for x.

$\begin{aligned} 3x - y &= -3 \\ \underline{-4x + y} &= \underline{2} \\ -x &= -1 \\ x &= 1 \end{aligned}$

Now solve for y.

$x - \dfrac{y}{3} = -1$

$1 - \dfrac{y}{3} = -1$

$-\dfrac{y}{3} = -2$

$y = 6$

The solution of the system is $(1, 6)$.

33. $\begin{cases} -4(x + 2) = 3y \\ 2x - 2y = 3 \end{cases}$

Rewrite the first equation.

$-4(x + 2) = 3y$

$-4x - 8 = 3y$

$-4x = 3y + 8$

$-4x - 3y = 8$

$\begin{cases} -4x - 3y = 8 \\ 2x - 2y = 3 \end{cases}$

Multiply the second equation by 2.

$\begin{cases} -4x - 3y = 8 \\ 2(2x - 2y) = 2(3) \end{cases} \rightarrow \begin{cases} -4x - 3y = 8 \\ 4x - 4y = 6 \end{cases}$

Add the equations to eliminate x; then solve for y.

$\begin{aligned} -4x - 3y &= 8 \\ \underline{4x - 4y} &= \underline{6} \\ -7y &= 14 \\ y &= -2 \end{aligned}$

Now solve for x.

$2x - 2y = 3$

$2x - 2(-2) = 3$

$2x + 4 = 3$

$2x = -1$

$x = -\dfrac{1}{2}$

The solution of the system is $\left(-\dfrac{1}{2}, -2\right)$.

35. $\begin{cases} \dfrac{x}{3} - y = 2 \\ -\dfrac{x}{2} + \dfrac{3y}{2} = -3 \end{cases}$

Multiply the first equation by 3 and the second by 2.

$\begin{cases} 3\left(\dfrac{x}{3} - y\right) = 3(2) \\ 2\left(-\dfrac{x}{2} + \dfrac{3y}{2}\right) = 2(-3) \end{cases} \rightarrow \begin{cases} x - 3y = 6 \\ -x + 3y = -6 \end{cases}$

Add the equations to eliminate y.

$\begin{aligned} x - 3y &= 6 \\ \underline{-x + 3y} &= \underline{-6} \\ 0 &= 0 \end{aligned}$

Since the statement $0 = 0$ is true, the system has an infinite number of solutions.

37. $\begin{cases} \dfrac{3}{5}x - y = -\dfrac{4}{5} \\ 3x + \dfrac{y}{2} = -\dfrac{9}{5} \end{cases}$

Multiply the first equation by 5 and the second equation by 10 to eliminate fractions.

$\begin{cases} 5\left(\dfrac{3}{5}x - y\right) = 5\left(-\dfrac{4}{5}\right) \\ 10\left(3x + \dfrac{y}{2}\right) = 10\left(-\dfrac{9}{5}\right) \end{cases} \rightarrow$

$\begin{cases} 3x - 5y = -4 \\ 30x + 5y = -18 \end{cases}$

Add the equations to eliminate y; then solve for x.

$$3x - 5y = -4$$
$$\underline{30x + 5y = -18}$$
$$33x \qquad = -22$$
$$x = -\frac{22}{33} = -\frac{2}{3}$$

Now use the equation $3x - 5y = -4$ to solve for y.

$$3x - 5y = -4$$
$$3\left(-\frac{2}{3}\right) - 5y = -4$$
$$-2 - 5y = -4$$
$$-5y = -2$$
$$y = \frac{2}{5}$$

The solution of the system is $\left(-\frac{2}{3}, \frac{2}{5}\right)$.

39. $\begin{cases} 3.5x + 2.5y = 17 \\ -1.5x - 7.5y = -33 \end{cases}$

Multiply the first equation by 30 and the second equation by 10.

$$\begin{cases} 30(3.5x + 2.5y) = 30(17) \\ 10(-1.5x - 7.5y) = 10(-33) \end{cases}$$
$$\rightarrow \begin{cases} 105x + 75y = 510 \\ -15x - 75y = -330 \end{cases}$$

Add the equations to eliminate y; then solve for x.

$$105x + 75y = 510$$
$$\underline{-15x - 75y = -330}$$
$$90x \qquad = 180$$
$$x = 2$$

Now solve for y.

$$3.5x + 2.5y = 17$$
$$3.5(2) + 2.5y = 17$$
$$7 + 2.5y = 17$$
$$2.5y = 10$$
$$y = 4$$

The solution of the system is $(2, 4)$.

41. $\begin{cases} 0.02x + 0.04y = 0.09 \\ -0.1x + 0.3y = 0.8 \end{cases}$

Multiply the first equation by 100 and the second equation by 10 to eliminate decimals.

$$\begin{cases} 100(0.02x + 0.04y) = 100(0.09) \\ 10(-0.1x + 0.3y) = 10(0.8) \end{cases}$$
$$\rightarrow \begin{cases} 2x + 4y = 9 \\ -x + 3y = 8 \end{cases}$$

Multiply the second equation by 2.

$$\begin{cases} 2x + 4y = 9 \\ 2(-x + 3y) = 2(8) \end{cases} \rightarrow \begin{cases} 2x + 4y = 9 \\ -2x + 6y = 16 \end{cases}$$

Add the equations to eliminate x; then solve for y.

$$2x + 4y = 9$$
$$\underline{-2x + 6y = 16}$$
$$10y = 25$$
$$y = 2.5$$

Use the equation $-x + 3y = 8$ to solve for x.

$$-x + 3y = 8$$
$$-x + 3(2.5) = 8$$
$$-x + 7.5 = 8$$
$$-x = 0.5$$
$$x = -0.5$$

The solution of the system is $(-0.5, 2.5)$.

43. Twice a number, added to 6, is 3 less than the number is written as $2x + 6 = x - 3$.

45. Three times a number, subtracted from 20, is 2 is written as $20 - 3x = 2$.

47. The product of 4 and the sum of a number and 6 is twice the number is written as $4(x + 6) = 2x$.

49. To eliminate the variable y, multiply the second equation by 2.

$$3x - y = -12$$
$$2(3x - y) = 2(-12)$$
$$6x - 2y = -24$$

51. b; answers may vary

53. answers may vary

55. a. When $b = 15$, the system has an infinite number of solutions.

   b. When b is any real number except 15, the system has no solutions.

57. $\begin{cases} 2x+3y=14 \\ 3x-4y=-69.1 \end{cases}$

Multiply the first equation by -3 and the second equation by 2.

$\begin{cases} -3(2x+3y)=-3(14) \\ 2(3x-4y)=2(-69.1) \end{cases}$

$\rightarrow \begin{cases} -6x-9y=-42 \\ 6x-8y=-138.2 \end{cases}$

Add the equations to eliminate x; then solve for y.

$-6x-9y=-42$
$\underline{6x-8y=-138.2}$
$-17y=-180.2$
$y=10.6$

Now solve for x.

$2x+3y=14$
$2x+3(10.6)=14$
$2x+31.8=14$
$2x=-17.8$
$x=-8.9$

The solution of the system is $(-8.9, 10.6)$.

59. a. $\begin{cases} 21.4x-y=-365 \\ 17.9x-y=-394 \end{cases}$

Multiply the second equation by -1.

$\begin{cases} 21.4x-y=-365 \\ -1(17.9x-y)=-1(-394) \end{cases}$

$\rightarrow \begin{cases} 21.4x-y=-365 \\ -17.9x+y=394 \end{cases}$

Add the equations to eliminate y; then solve for x.

$21.4x-y=-365$
$\underline{-17.9x+y=394}$
$3.5x\quad=29$
$x\approx 8.3 \text{ or } 8$

Now solve for y.

$21.4x-y=-365$
$21.4(8)-y=-365$
$171.2-y=-365$
$y=536.2 \text{ or } 536$

The solution of the system is $(8, 536)$.

b. In 2010 (2002 + 8), the number of medical assistant jobs equals the number of computer software engineer jobs.

c. In 2010, there are 536 thousand medical assistant and computer software engineer jobs.

Integrated Review

1. $\begin{cases} 2x-3y=-11 \\ y=4x-3 \end{cases}$

Substitute $4x-3$ for y in the first equation and solve for x.

$2x-3y=-11$
$2x-3(4x-3)=-11$
$2x-12x+9=-11$
$-10x=-20$
$x=2$

Now solve for y.

$y=4x-3=4(2)-3=8-3=5$

The solution to the system is $(2, 5)$.

2. $\begin{cases} 4x-5y=6 \\ y=3x-10 \end{cases}$

Substitute $3x-10$ for y in the first equation and solve for x.

$4x-5y=6$
$4x-5(3x-10)=6$
$4x-15x+50=6$
$-11x=-44$
$x=4$

Now solve for y.

$y=3x-10=3(4)-10=12-10=2$

The solution of the system is $(4, 2)$.

3. $\begin{cases} x+y=3 \\ x-y=7 \end{cases}$

Add the equations to eliminate y; then solve for x.

$x+y=3$
$\underline{x-y=7}$
$2x\quad=10$
$x=5$

Now solve for y.

$$x + y = 3$$
$$5 + y = 3$$
$$y = -2$$

The solution of the system is $(5, -2)$.

4. $\begin{cases} x - y = 20 \\ x + y = -8 \end{cases}$

Add the equations to eliminate y; then solve for x.

$$x - y = 20$$
$$\underline{x + y = -8}$$
$$2x = 12$$
$$x = 6$$

Now solve for y.

$$x + y = -8$$
$$6 + y = -8$$
$$y = -14$$

The solution of the system is $(6, -14)$.

5. $\begin{cases} x + 2y = 1 \\ 3x + 4y = -1 \end{cases}$

Solve the first equation for x.

$$x + 2y = 1$$
$$x = -2y + 1$$

Substitute $-2y + 1$ for x in the second equation and solve for y.

$$3x + 4y = -1$$
$$3(-2y + 1) + 4y = -1$$
$$-6y + 3 + 4y = -1$$
$$-2y = -4$$
$$y = 2$$

Now solve for x.

$$x = -2y + 1 = -2(2) + 1 = -4 + 1 = -3$$

The solution of the system is $(-3, 2)$.

6. $\begin{cases} x + 3y = 5 \\ 5x + 6y = -2 \end{cases}$

Solve the first equation for x.

$$x + 3y = 5$$
$$x = -3y + 5$$

Substitute $-3y + 5$ for x in the second equation and solve for y.

$$5x + 6y = -2$$
$$5(-3y + 5) + 6y = -2$$
$$-15y + 25 + 6y = -2$$
$$-9y = -27$$
$$y = 3$$

Now solve for x.

$$x = -3y + 5 = -3(3) + 5 = -9 + 5 = -4$$

The solution of the system is $(-4, 3)$.

7. $y = x + 3$
$3x = 2y - 6$

Substitute $x + 3$ for y in the second equation and solve for x.

$$3x = 2y - 6$$
$$3x = 2(x + 3) - 6$$
$$3x = 2x + 6 - 6$$
$$x = 0$$

Now solve for y.

$$y = x + 3 = 0 + 3 = 3$$

The solution of the system is $(0, 3)$.

8. $\begin{cases} y = -2x \\ 2x - 3y = -16 \end{cases}$

Substitute $-2x$ for y in the second equation and solve for x.

$$2x - 3y = -16$$
$$2x - 3(-2x) = -16$$
$$2x + 6x = -16$$
$$8x = -16$$
$$x = -2$$

Now solve for y.

$$y = -2x = -2(-2) = 4$$

The solution of the system is $(-2, 4)$.

9. $\begin{cases} y = 2x - 3 \\ y = 5x - 18 \end{cases}$

Substitute $2x - 3$ for y in the second equation and solve for x.

$$y = 5x - 18$$
$$2x - 3 = 5x - 18$$
$$15 = 3x$$
$$5 = x$$

Now solve for y.

$$y = 2x - 3 = 2(5) - 3 = 10 - 3 = 7$$

The solution of the system is $(5, 7)$.

10. $\begin{cases} y = 6x - 5 \\ y = 4x - 11 \end{cases}$

Substitute $6x - 5$ for y in the second equation and solve for x.

$y = 4x - 11$
$6x - 5 = 4x - 11$
$2x = -6$
$x = -3$

Now solve for y.

$y = 6x - 5 = 6(-3) - 5 = -18 - 5 = -23$

The solution of the system is $(-3, -23)$.

11. $\begin{cases} x + \dfrac{1}{6}y = \dfrac{1}{2} \\ 3x + 2y = 3 \end{cases}$

Multiply the first equation by 6 to eliminate fractions.

$$6\left(x + \frac{1}{6}y\right) = 6\left(\frac{1}{2}\right)$$
$$6x + y = 3$$

Now solve for y.

$y = -6x + 3$

Substitute $-6x + 3$ for y in the second equation and solve for x.

$3x + 2y = 3$
$3x + 2(-6x + 3) = 3$
$3x - 12x + 6 = 3$
$-9x = -3$
$$x = \frac{3}{9} = \frac{1}{3}$$

Now solve for y.

$$y = -6x + 3 = -6\left(\frac{1}{3}\right) + 3 = -2 + 3 = 1$$

The solution of the system is $\left(\dfrac{1}{3}, 1\right)$.

12. $x + \dfrac{1}{3}y = \dfrac{5}{12}$
$8x + 3y = 4$

Multiply the first equation by 12 to eliminate fractions.

$$12\left(x + \frac{1}{3}y\right) = 12\left(\frac{5}{12}\right)$$
$$12x + 4y = 5$$

Multiply the revised first equation by -3 and the second original equation by 4.

$\begin{cases} -3(12x + 4y) = -3(5) \\ 4(8x + 3y) = 4(4) \end{cases}$

$\rightarrow \begin{cases} -36x - 12y = -15 \\ 32x + 12y = 16 \end{cases}$

Add the equations to eliminate y; then solve for x.

$\begin{array}{r} -36x - 12y = -15 \\ 32x + 12y = 16 \\ \hline -4x \qquad\quad = 1 \end{array}$

$$x = -\frac{1}{4}$$

Now solve for y.

$8x + 3y = 4$
$$8\left(-\frac{1}{4}\right) + 3y = 4$$
$-2 + 3y = 4$
$3y = 6$
$y = 2$

The solution of the system is $\left(-\dfrac{1}{4}, 2\right)$.

13. $\begin{cases} x - 5y = 1 \\ -2x + 10y = 3 \end{cases}$

Solve the first equation for x.

$x - 5y = 1$
$x = 5y + 1$

Substitute $5y + 1$ for x in the second equation and then solve for y.

$-2x + 10y = 3$
$-2(5y + 1) + 10y = 3$
$-10y - 2 + 10y = 3$
$-2 = 3$

Since the statement is false, the system has no solution.

14. $\begin{cases} -x + 2y = 3 \\ 3x - 6y = -9 \end{cases}$

Solve the first equation for x.

$-x + 2y = 3$
$x = 2y - 3$

Substitute $2y - 3$ for x in the second equation and solve for y.

$$3x - 6y = -9$$
$$3(2y - 3) - 6y = -9$$
$$6y - 9 - 6y = -9$$
$$0 = 0$$

Since the statement $0 = 0$ is true, the system has an infinite number of solutions.

15. $\begin{cases} 0.2x - 0.3y = -0.95 \\ 0.4x + 0.1y = 0.55 \end{cases}$

Multiply both equations by 100 to eliminate decimals.

$$\begin{cases} 100(0.2x - 0.3y) = 100(-0.95) \\ 100(0.4x + 0.1y) = 100(0.55) \end{cases}$$

$$\rightarrow \begin{cases} 20x - 30y = -95 \\ 40x + 10y = 55 \end{cases}$$

Multiply the second revised equation by 3.

$$\begin{cases} 20x - 30y = -95 \\ 3(40x + 10y) = 3(55) \end{cases} \rightarrow \begin{cases} 20x - 30y = -95 \\ 120x + 30y = 165 \end{cases}$$

Add the equations to eliminate y; then solve for x.

$$20x - 30y = -95$$
$$\underline{120x + 30y = 165}$$
$$140x \quad\quad = 70$$
$$x = 0.5$$

Now solve for y.

$$0.4x + 0.1y = 0.55$$
$$0.4(0.5) + 0.1y = 0.55$$
$$0.2 + 0.1y = 0.55$$
$$0.1y = 0.35$$
$$y = 3.5$$

The solution of the system is (0.5, 3.5).

16. $\begin{cases} 0.08x - 0.04y = -0.11 \\ 0.02x - 0.06y = -0.09 \end{cases}$

Multiply both equations by 100 to eliminate decimals.

$$\begin{cases} 100(0.08x - 0.04y) = 100(-0.11) \\ 100(0.02x - 0.06y = 100(-0.09) \end{cases}$$

$$\rightarrow \begin{cases} 8x - 4y = -11 \\ 2x - 6y = -9 \end{cases}$$

Multiply the second revised equation by -4.

$$\begin{cases} 8x - 4y = -11 \\ -4(2x - 6y) = -4(-9) \end{cases} \rightarrow \begin{cases} 8x - 4y = -11 \\ -8x + 24y = 36 \end{cases}$$

Add the equations to eliminate x; then solve for y.

$$8x - 4y = -11$$
$$\underline{-8x + 24y = 36}$$
$$20y = 25$$
$$y = 1.25$$

Now solve for x.

$$0.08x - 0.04y = -0.11$$
$$0.08x - 0.04(1.25) = -0.11$$
$$0.08x - 0.05 = -0.11$$
$$0.08x = -0.06$$
$$x = -0.75$$

The solution of the system is $(-0.75, 1.25)$.

17. $\begin{cases} x = 3y - 7 \\ 2x - 6y = -14 \end{cases}$

Substitute $3y - 7$ for x in the second equation and solve for y.

$$2x - 6y = -14$$
$$2(3y - 7) - 6y = -14$$
$$6y - 14 - 6y = -14$$
$$0 = 0$$

Since the statement $0 = 0$ is true, the system has an infinite number of solutions.

18. $\begin{cases} y = \dfrac{x}{2} - 3 \\ 2x - 4y = 0 \end{cases}$

Substitute $\dfrac{x}{2} - 3$ for y in the second equation and solve for x.

$$2x - 4y = 0$$
$$2x - 4\left(\frac{x}{2} - 3\right) = 0$$
$$2x - 2x + 12 = 0$$
$$12 = 0$$

Since the statement $12 = 0$ is false, the system has no solution.

19. $\begin{cases} 2x+5y=-1 \\ 3x-4y=33 \end{cases}$

Multiply the first equation by 3 and the second equation by -2.

$\begin{cases} 3(2x+5y)=3(-1) \\ -2(3x-4y)=-2(33) \end{cases} \rightarrow \begin{cases} 6x+15y=-3 \\ -6x+8y=-66 \end{cases}$

Add the equations to eliminate x; then solve for y.

$\begin{array}{r} 6x+15y=-3 \\ -6x+8y=-66 \\ \hline 23y=-69 \\ y=-3 \end{array}$

Now solve for x.

$2x+5y=-1$
$2x+5(-3)=-1$
$2x-15=-1$
$2x=14$
$x=7$

The solution of the system is $(7, -3)$.

20. $\begin{cases} 7x-3y=2 \\ 6x+5y=-21 \end{cases}$

Multiply the first equation by 5 and the second equation by 3.

$\begin{cases} 5(7x-3y)=5(2) \\ 3(6x+5y)=3(-21) \end{cases} \rightarrow \begin{cases} 35x-15y=10 \\ 18x+15y=-63 \end{cases}$

Add the equations to eliminate y; then solve for x.

$\begin{array}{r} 35x-15y=10 \\ 18x+15y=-63 \\ \hline 53x=-53 \\ x=-1 \end{array}$

Now solve for y.

$7x-3y=2$
$7(-1)-3y=2$
$-7-3y=2$
$-3y=9$
$y=-3$

The solution of the system is $(-1, -3)$.

21. answers may vary

22. answers may vary

Section 7.4 Practice Problems

1. $\begin{cases} x+y=50 \\ x-y=22 \end{cases}$

Add the equations to eliminate y; then solve for x.

$\begin{array}{r} x+y=50 \\ x-y=22 \\ \hline 2x=72 \\ x=36 \end{array}$

Now solve for y.

$x+y=50$
$36+y=50$
$y=14$

The two numbers are 36 and 14.

2. $\begin{cases} A+C=587 \\ 7A+5C=3379 \end{cases}$

Multiply the first equation by -5.

$\begin{cases} -5(A+C)=-5(587) \\ 7A+5C=3379 \end{cases}$

$\rightarrow \begin{cases} -5A-5C=-2935 \\ 7A+5C=3379 \end{cases}$

Add the equations to eliminate C; then solve for A.

$\begin{array}{r} -5A-5C=-2935 \\ 7A+5C=3379 \\ \hline 2A=444 \\ A=222 \end{array}$

Now solve for C.

$A+C=587$
$222+C=587$
$C=365$

There were 222 adults and 365 children.

3.

	r	$\cdot$ t $=$	d
Faster car	x	3	$3x$
Slower car	y	3	$3y$

$3x+3y=440$
$x=y+10$

Substitute $y+1$ for x in the first equation and solve for y.

$$3x + 3y = 440$$
$$3(y+10) + 3y = 440$$
$$3y + 30 + 3y = 440$$
$$6y = 410$$
$$y = 68\frac{1}{3}$$

Now solve for x.

$$x = y + 10 = 68\frac{1}{3} + 10 = 78\frac{1}{3}$$

One car's speed is $68\frac{1}{3}$ mph and the other

car's speed is $78\frac{1}{3}$ mph.

4. Let x be the liters of 20% solution.
   Let y be the liters of 70% solution.
$$\begin{cases} x + y = 50 \\ 0.2x + 0.7y = 0.6(50) \end{cases}$$

Multiply the first equation by -2 and the second equation by 10.
$$\begin{cases} -2(x+y) = -2(50) \\ 10(0.2x + 0.7y) = 10(30) \end{cases}$$
$$\rightarrow \begin{cases} -2x - 2y = -100 \\ 2x + 7y = 300 \end{cases}$$

Add the equations to eliminate x; then solve for y.
$$\begin{array}{r} -2x - 2y = -100 \\ 2x + 7y = 300 \\ \hline 5y = 200 \\ y = 40 \end{array}$$

Now solve for x.
$$x + y = 50$$
$$x + 40 = 50$$
$$x = 10$$

10 liters of the 20% alcohol solution and 40 liters of the 70% alcohol solution make 50 liters of the 60% alcohol solution.

Exercise Set 7.4

1. In choice b, the length is not 3 feet longer than the width. In choice a, the perimeter is $2(8 + 5) = 2(13) = 26$ feet, not 30 feet. Choice c gives the solution, since $9 = 6 + 3$ and $2(9 + 6) = 2(15) = 30$.

3. In choice a, the total cost is
   $2(3) + 3(4) = 6 + 12 = \$18$, not \$17.
   In choice c, the total cost is
   $2(2) + 3(5) = 4 + 15 = \$19$, not \$17.
   Choice b gives the solution, since
   $2(4) + 3(3) = 8 + 9 = \$17$ and
   $5(4) + 4(3) = 20 + 12 = \$32$.

5. In choice b, the total number of coins is
   $20 + 44 = 64$, not 100.
   In choice c, the total value of the coins is
   $60(0.10) + 40(0.25) = 6.00 + 10.00$
   $\qquad\qquad\qquad\qquad = \$16.00$, not \$13.00.
   Choice a gives the solution, since
   $80 + 20 = 100$ and
   $80(0.10) + 20(0.25) = 8.00 + 5.00 = \$13.00$.

7. Let x be the first number and y the second.
$$\begin{cases} x + y = 15 \\ x - y = 7 \end{cases}$$

9. Let x be the amount in the larger account and y be the amount in the smaller account.
$$\begin{cases} x + y = 6500 \\ x = y + 800 \end{cases}$$

11. Let x be the first number and y be the second.
$$x + y = 83$$
$$x - y = 17$$

Add the equations to eliminate y then solve for x.
$$\begin{array}{r} x + y = 83 \\ x - y = 17 \\ \hline 2x = 100 \\ x = 50 \end{array}$$

Now solve for y.
$$x + y = 83$$
$$50 + y = 83$$
$$y = 33$$

The numbers are 50 and 33.

13. Let x be the first number and y the second.
$$\begin{cases} x + 2y = 8 \\ 2x + y = 25 \end{cases}$$
Solve the first equation for x.
$$x + 2y = 8$$
$$x = 8 - 2y$$
Substitute $8 - 2y$ for x in the second equation and solve for y.
$$2x + y = 25$$
$$2(8 - 2y) + y = 25$$
$$16 - 4y + y = 25$$
$$16 - 3y = 25$$
$$-3y = 9$$
$$y = -3$$
Now solve for x.
$$x = 8 - 2y = 8 - 2(-3) = 8 + 6 = 14$$
The numbers are 14 and -3.

15. Let x be the points scored by Leslie and y be the points scored by Jackson.
$$\begin{cases} y = x + 36 \\ x + y = 1232 \end{cases}$$
Substitute $x + 36$ for y in the second equation and solve for x.
$$x + y = 1232$$
$$x + x + 36 = 1232$$
$$2x = 1196$$
$$x = 598$$
Now solve for y.
$$x + y = 1232$$
$$598 + y = 1232$$
$$y = 634$$
Over the course of the season Leslie scored 598 points and Jackson scored 634 points.

17. Let a be the price of an adult's ticket and c be the price of a child's ticket.
$$\begin{cases} 3a + 4c = 159 \\ 2a + 3c = 112 \end{cases}$$
Multiply the first equation by -2 and the second equation by 3.

$$\begin{cases} -2(3a + 4c) = -2(159) \\ 3(2a + 3c) = 3(112) \end{cases}$$
$$\rightarrow \begin{cases} -6a - 8c = -318 \\ 6a + 9c = 336 \end{cases}$$
Add the equations to eliminate a and solve for c.
$$\begin{array}{r} -6a - 8c = -318 \\ 6a + 9c = 336 \\ \hline c = 18 \end{array}$$
Now solve for a.
$$2a + 3c = 112$$
$$2a + 3(18) = 112$$
$$2a + 54 = 112$$
$$2a = 58$$
$$a = 29$$
The price of an adult's ticket is $29 and the price of a child's ticket is $18.

19. Let x be quarters and y be nickels.
$$\begin{cases} x + y = 80 \\ 0.25x + 0.05y = 14.60 \end{cases}$$
Solve the first equation in terms of y.
$$x + y = 80$$
$$y = -x + 80$$
Substitute $-x + 80$ for y in the second equation and solve for x.
$$0.25x + 0.05y = 14.60$$
$$0.25x + 0.05(-x + 80) = 14.60$$
$$0.25x - 0.05x + 4 = 14.60$$
$$0.20x = 10.60$$
$$x = 53$$
Now solve for y.
$$y = -x + 80 = -53 + 80 = 27$$
There are 53 quarters and 27 nickels.

21. Let x be the value of one McDonald's share and let y be the value of one Ohio Art Company share.
$$\begin{cases} 35x + 69y = 1551 \\ x = y + 25 \end{cases}$$
Substitute $y + 25$ for x in the first equation and solve for y.

$$35x + 69y = 1551$$
$$35(y + 25) + 69y = 1551$$
$$35y + 875 + 69y = 1551$$
$$875 + 104y = 1551$$
$$104y = 676$$
$$y = 6.5$$

Now solve for x.
$$x = y + 25 = 6.5 + 25 = 31.5$$
On that day, the closing price of the McDonald's stock was $31.5 per share and the closing price of The Ohio Art Company stock was $6.50 per share.

23. Let x be the daily fee and y be the mileage charge.
$$\begin{cases} 4x + 450y = 240.50 \\ 3x + 200y = 146.00 \end{cases}$$

Multiply the first equation by 3 and the second by −4.
$$\begin{cases} 3(4x + 450y) = 3(240.50) \\ -4(3x + 200y) = -4(146.00) \end{cases}$$

$$\rightarrow \begin{cases} 12x + 1350y = 721.5 \\ -12x - 800y = -584 \end{cases}$$

Add the equations to eliminate x and solve for y.
$$\begin{array}{r} 12x + 1350y = 721.5 \\ -12x - 800y = -584 \\ \hline 550y = 137.5 \\ y = 0.25 \end{array}$$

Now solve for x.
$$3x + 200y = 146$$
$$3x + 200(0.25) = 146$$
$$3x + 50 = 146$$
$$3x = 96$$
$$x = 32$$
There is a $32 daily fee and a $0.25 per mile mileage charge.

25. $\begin{cases} 18 = 2(x + y) \\ 18 = \dfrac{9}{2}(x - y) \end{cases}$

Multiply the first equation by $\dfrac{1}{2}$ and the

second equation by $\dfrac{2}{9}$.

$$\begin{cases} \dfrac{1}{2}(18) = \dfrac{1}{2}[2(x+y)] \\ \dfrac{2}{9}(18) = \dfrac{2}{9}\left[\dfrac{9}{2}(x-y)\right] \end{cases} \rightarrow \begin{cases} 9 = x + y \\ 4 = x - y \end{cases}$$

Add the equations to eliminate y; then solve for x.
$$\begin{array}{r} 9 = x + y \\ 4 = x - y \\ \hline 13 = 2x \\ 6.5 = x \end{array}$$

Now solve for y.
$$9 = x + y$$
$$9 = 6.5 + y$$
$$2.5 = y$$
The rate that Pratap can row in still water is 6.5 miles per hour and the rate of the current is 2.5 miles per hour.

27. Let x = rate of flight in still wind. Then y = rate of wind.
$$\begin{cases} 780 = \dfrac{3}{2}(x + y) \\ 780 = 2(x - y) \end{cases}$$

Multiply the first equation by $\dfrac{2}{3}$ and the

second equation by $\dfrac{1}{2}$.

$$\begin{cases} \dfrac{2}{3}(780) = \dfrac{2}{3}\left[\dfrac{3}{2}(x+y)\right] \\ \dfrac{1}{2}(780) = \dfrac{1}{2}[2(x-y)] \end{cases} \rightarrow \begin{cases} 520 = x + y \\ 390 = x - y \end{cases}$$

Add the equations to eliminate y; then solve for x.
$$\begin{array}{r} 520 = x + y \\ 390 = x - y \\ \hline 910 = 2x \\ 455 = x \end{array}$$

Now solve for y.
$$780 = 2(x - y)$$
$$780 = 2(455) - 2y$$
$$780 = 910 - 2y$$
$$-130 = -2y$$
$$65 = y$$
The speed of the plane in still air is 455 mph and the speed of the wind is 65 mph.

29. Let x be the number of hours that Jim spent on his bicycle and y be the number of hours he spent walking. Then the distance he rode was $40x$ miles and the distance he walked was $4y$ miles.

$$\begin{cases} x+y=6 \\ 40x+4y=186 \end{cases}$$

Solve the first equation for y.

$x+y=6$

$\quad y=6-x$

Substitute $6-x$ for y in the second equation and solve for x.

$40x+4y=186$

$40x+4(6-x)=186$

$40x+24-4x=186$

$\quad 36x+24=186$

$\quad\quad 36x=162$

$\quad\quad\quad x=4.5$

Jim spent 4.5 hours on his bike.

31. Let x be ounces of 4% solution, and let y be ounces of 12% solution.

$$\begin{cases} x+y=12 \\ 0.04x+0.12y=0.09(12) \end{cases}$$

Solve the first equation in terms of x.

$x+y=12$

$\quad x=-y+12$

Substitute $-y+12$ for x in the second equation and solve for y.

$\quad\quad 0.04x+0.12y=1.08$

$0.04(-y+12)+0.12y=1.08$

$-0.04y+0.48+0.12y=1.08$

$\quad\quad\quad\quad 0.08y=0.60$

$\quad\quad\quad\quad\quad y=7.5$

Now solve for x.

$x=-y+12=-7.5+12=4.5$

Darren will need 4.5 ounces of the 4% solution and 7.5 ounces of the 12% solution to make 12 ounces of the 9% solution.

33. Let x be the number of pounds of high-quality coffee, and let y be the number of pounds of the cheaper coffee.

$$\begin{cases} x+y=200 \\ 4.95x+2.65y=200(3.95) \end{cases}$$

Solve the first equation for x.

$x+y=200$

$\quad x=200-y$

Substitute $200-y$ for x in the second equation and solve for y.

$\quad\quad 4.95x+2.65y=200(3.95)$

$4.95(200-y)+2.65y=790$

$990-4.95y+2.65y=790$

$\quad\quad 990-2.30y=790$

$\quad\quad\quad -2.30y=-200$

$\quad\quad\quad\quad\quad y=\dfrac{200}{2.30}$

$\quad\quad\quad\quad\quad y\approx 87$

$x=200-y\approx 200-87=113$

Wayne should blend 113 pounds of the coffee that sells for $4.95 per pound with 87 pounds of the cheaper coffee.

35. Let x be one angle, and let y be the other angle.

$x+y=90$

$y=2x$

Substitute $2x$ for y in the first equation and solve for x.

$\quad x+y=90$

$x+2x=90$

$\quad\quad 3x=90$

$\quad\quad\quad x=30$

Now solve for y.

$y=2x=2(30)=60$

One angle is 30° and the other is 60°.

37. Let x be the measure of one angle and y be the measure of the other.

$$\begin{cases} x+y=90 \\ x=10+3y \end{cases}$$

Substitute $10+3y$ for x in the first equation and solve for y.

$$x + y = 90$$
$$10 + 3y + y = 90$$
$$10 + 4y = 90$$
$$4y = 80$$
$$y = 20$$
$$x = 10 + 3y = 10 + 3(20) = 10 + 60 = 70$$
The angles measure $20°$ and $70°$.

39. Let x be the number of pieces sold at the original price, and let y be the number of pieces sold at the discounted price.
$$\begin{cases} x + y = 90 \\ 9.5x + 7.5y = 721 \end{cases}$$

Solve the first equation in terms of x.
$$x + y = 90$$
$$x = -y + 90$$

Substitute $-y + 90$ for x in the second equation and solve for y.
$$9.5x + 7.5y = 721$$
$$9.5(-y + 90) + 7.5y = 721$$
$$-9.5y + 855 + 7.5y = 721$$
$$-2.0y = -134$$
$$y = 67$$

Now solve for x.
$$x = -y + 90 = -67 + 90 = 23$$
They sold 23 pieces at \$9.50 each and 67 pieces at \$7.50 each.

41. Let x be the rate of the faster group and y be the rate of the slower group.
$$\begin{cases} 240x + 240y = 1200 \\ y = x - \dfrac{1}{2} \end{cases}$$

Substitute $x - \dfrac{1}{2}$ for y in the fist equation and solve for x.
$$240x + 240y = 1200$$
$$240x + 240\left(x - \frac{1}{2} \right) = 1200$$
$$240x + 240x - 120 = 1200$$
$$480x - 120 = 1200$$
$$480x = 1320$$
$$x = 2.75$$

$$y = x - \frac{1}{2} = x - 0.5 = 2.75 - 0.5 = 2.25$$

The hiking rates are $2.75 = 2\dfrac{3}{4}$ miles per hour and $2.25 = 2\dfrac{1}{4}$ miles per hour.

43. Let x be the number of gallons of 30% solution and y be the number of gallons of 60% solution.
$$\begin{cases} x + y = 150 \\ 0.3x + 0.6y = 0.5(150) \end{cases}$$

Solve the first equation in terms of x.
$$x = 150 - y$$
Substitute $150 - y$ for x in the second equation and solve for y.
$$0.3x + 0.6y = 0.5(150)$$
$$0.3x + 0.6y = 75$$
$$0.3(150 - y) + 0.6y = 75$$
$$45 - 0.3y + 0.6y = 75$$
$$0.3y = 30$$
$$y = 100$$

Now solve for x.
$$x = 150 - y = 150 - 100 = 50$$
Combining 50 gallons of the 30% solution and 100 gallons of the 60% solution is necessary to create 150 gallons of a 50% solution.

45. Let x be the length and y the width.
$$\begin{cases} 2(x + y) = 144 \\ x = y + 12 \end{cases}$$

Substitute $y + 12$ for x in the first equation and solve for y.
$$2(x + y) = 144$$
$$2(y + 12 + y) = 144$$
$$2(2y + 12) = 144$$
$$4y + 24 = 144$$
$$4y = 120$$
$$y = 30$$
$$x = y + 12 = 30 + 12 = 42$$
The length is 42 inches and the width is 30 inches.

47. $4^2 = 4 \cdot 4 = 16$

49. $(6x)^2 = (6x)(6x) = 36x^2$

51. $(10y^3)^2 = (10y^3)(10y^3) = 100y^6$

53. The price of the result must be between $0.49 and $0.65, so choice a is the only possibility.

55. $y + 2x = 33$
$\quad\quad y = 2x - 3$
Substitute $2x - 3$ for y in the first equation and solve for y.
$$y + 2x = 33$$
$$2x - 3 + 2x = 33$$
$$4x = 36$$
$$x = 9$$
Now solve for y.
$y = 2x - 3 = 2(9) - 3 = 18 - 3 = 15$
The width is 9 feet and the length is 15 feet.

Chapter 7 Vocabulary Check

1. In a system of linear equations in two variables, if the graphs of the equations are the same, the equations are <u>dependent</u> equations.

2. Two or more linear equations are called a <u>system of linear equations</u>.

3. A system of equations that has at least one solution is called a(n) <u>consistent</u> system.

4. A <u>solution</u> of a system of two equations in two variables is an ordered pair of numbers that is a solution of both equations in the system.

5. Two algebraic methods for solving systems of equations are <u>addition</u> and <u>substitution</u>.

6. A system of equations that has no solution is called a(n) <u>inconsistent</u> system.

7. In a system of linear equations in two variables, if the graphs of the equations are different, the equations are <u>independent</u> equations.

Chapter 7 Review

1. a. $\begin{cases} 2x - 3y = 12 \\ 3x + 4y = 1 \end{cases}$

First equation:
$$2x - 3y = 12$$
$$2(12) - 3(4) \stackrel{?}{=} 12$$
$$24 - 12 \stackrel{?}{=} 12$$
$$12 = 12 \quad \text{True}$$
Second equation:
$$3x + 4y = 1$$
$$3(12) - 4(4) \stackrel{?}{=} 1$$
$$36 - 16 \stackrel{?}{=} 1$$
$$20 = 1 \quad \text{False}$$
$(12, 4)$ is not a solution of the system.

b. First equation:
$$2x - 3y = 12$$
$$2(3) - 3(-2) \stackrel{?}{=} 12$$
$$6 + 6 \stackrel{?}{=} 12$$
$$12 = 12 \quad \text{True}$$
Second equation:
$$3x + 4y = 1$$
$$3(3) + 4(-2) \stackrel{?}{=} 1$$
$$9 - 8 \stackrel{?}{=} 1$$
$$1 = 1 \quad \text{True}$$
$(3, -2)$ is a solution of the system.

2. a. $\begin{cases} 4x + y = 0 \\ -8x - 5y = 9 \end{cases}$

First equation:
$$4x + y = 0$$
$$4\left(\frac{3}{4}\right) + (-3) \stackrel{?}{=} 0$$
$$3 - 3 \stackrel{?}{=} 0$$
$$0 = 0 \quad \text{True}$$

Second equation:
$$-8x - 5y = 9$$
$$-8\left(\frac{3}{4}\right) - 5(-3) \stackrel{?}{=} 9$$
$$-6 + 15 \stackrel{?}{=} 9$$
$$9 = 9 \quad \text{True}$$
$\left(\frac{3}{4}, -3\right)$ is a solution of the system.

b. First equation:
$$4x + y = 0$$
$$4(-2) + 8 \stackrel{?}{=} 0$$
$$-8 + 8 \stackrel{?}{=} 0$$
$$0 = 0 \quad \text{True}$$
Second equation:
$$-8x - 5y = 9$$
$$-8(-2) - 5(8) \stackrel{?}{=} 9$$
$$16 - 40 \stackrel{?}{=} 9$$
$$-24 = 9 \quad \text{False}$$
$(-2, 8)$ is not a solution of the system.

3. a. $\begin{cases} 5x - 6y = 18 \\ 2y - x = -4 \end{cases}$

First equation:
$$5x - 6y = 18$$
$$5(-6) - 6(-8) \stackrel{?}{=} 18$$
$$-30 + 48 \stackrel{?}{=} 18$$
$$18 = 18 \quad \text{True}$$
Second equation:
$$2y - x = -4$$
$$2(-8) - (-6) \stackrel{?}{=} -4$$
$$-16 + 6 \stackrel{?}{=} -4$$
$$-10 = -4 \quad \text{False}$$
$(-6, -8)$ is not a solution of the system.

b. First equation:
$$5x - 6y = 18$$
$$5(3) - 6\left(\frac{5}{2}\right) \stackrel{?}{=} 18$$
$$15 - 15 \stackrel{?}{=} 18$$
$$0 = 18 \quad \text{False}$$
$\left(3, \frac{5}{2}\right)$ is not a solution of the system.

4. a. $\begin{cases} 2x + 3y = 1 \\ 3y - x = 4 \end{cases}$

First equation:
$$2x + 3y = 1$$
$$2(2) + 3(2) \stackrel{?}{=} 1$$
$$4 + 6 \stackrel{?}{=} 1$$
$$10 = 1 \quad \text{False}$$
$(2, 2)$ is not a solution of the system.

b. First equation:
$$2x + 3y = 1$$
$$2(-1) + 3(1) \stackrel{?}{=} 1$$
$$-2 + 3 \stackrel{?}{=} 1$$
$$1 = 1 \quad \text{True}$$
Second equation:
$$3y - x = 4$$
$$3(1) - (-1) \stackrel{?}{=} 4$$
$$3 + 1 \stackrel{?}{=} 4$$
$$4 = 4 \quad \text{True}$$
$(-1, 1)$ is a solution of the system.

5. $\begin{cases} x + y = 5 \\ x - y = 1 \end{cases}$

The solution of the system is $(3, 2)$.

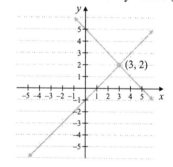

6. $\begin{cases} x + y = 3 \\ x - y = -1 \end{cases}$

The solution of the system is $(1, 2)$.

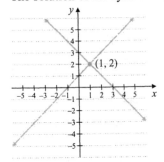

7. $\begin{cases} x = 5 \\ y = -1 \end{cases}$

The solution of the system is $(5, -1)$.

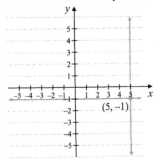

8. $\begin{cases} x = -3 \\ y = 2 \end{cases}$

The solution of the system is $(-3, 2)$.

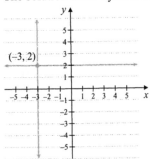

9. $\begin{cases} 2x + y = 5 \\ x = -3y \end{cases}$

The solution of the system is $(3, -1)$.

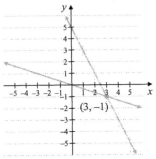

10. $\begin{cases} 3x + y = -2 \\ y = -5x \end{cases}$

The solution of the system is $(1, -5)$.

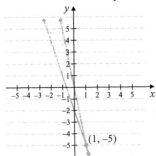

11. $\begin{cases} y = 3x \\ -6x + 2y = 6 \end{cases}$

There are no solutions to the system because the lines are parallel.

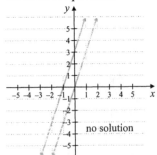

12. $\begin{cases} x - 2y = 2 \\ -2x + 4y = -4 \end{cases}$

Since they are the same lines, there are an infinite number of solutions.

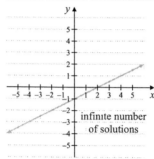

13. $\begin{cases} y = 2x + 6 \\ 3x - 2y = -11 \end{cases}$

Substitute $2x + 6$ for y in the second equation and solve for x.
$$3x - 2y = -11$$
$$3x - 2(2x + 6) = -11$$
$$3x - 4x - 12 = -11$$
$$-x = 1$$
$$x = -1$$
Now solve for y.
$$y = 2x + 6 = 2(-1) + 6 = -2 + 6 = 4$$
The solution of the system is $(-1, 4)$.

14. $\begin{cases} y = 3x - 7 \\ 2x - 3y = 7 \end{cases}$

Substitute $3x - 7$ for y in the second equation and solve for x.
$$2x - 3y = 7$$
$$2x - 3(3x - 7) = 7$$
$$2x - 9x + 21 = 7$$
$$-7x = -14$$
$$x = 2$$
Now solve for y.
$$y = 3x - 7 = 3(2) - 7 = 6 - 7 = -1$$
The solution of the system is $(2, -1)$.

15. $\begin{cases} x + 3y = -3 \\ 2x + y = 4 \end{cases}$

Solve the second equation in terms of y.

$$2x + y = 4$$
$$y = -2x + 4$$
Substitute $-2x + 4$ for y in the first equation and solve for x.
$$x + 3y = -3$$
$$x + 3(-2x + 4) = -3$$
$$x - 6x + 12 = -3$$
$$-5x = -15$$
$$x = 3$$
Now solve for y.
$$y = -2x + 4 = -2(3) + 4 = -6 + 4 = -2$$
The solution of the system is $(3, -2)$.

16. $\begin{cases} 3x + y = 11 \\ x + 2y = 12 \end{cases}$

Solve the first equation in terms of y.
$$3x + y = 11$$
$$y = -3x + 11$$
Substitute $-3x + 11$ for y in the second equation and solve for x.
$$x + 2y = 12$$
$$x + 2(-3x + 11) = 12$$
$$x - 6x + 22 = 12$$
$$-5x = -10$$
$$x = 2$$
Now solve for y.
$$y = -3x + 11 = -3(2) + 11 = -6 + 11 = 5$$
The solution of the system is $(2, 5)$.

17. $\begin{cases} 4y = 2x + 6 \\ x - 2y = -3 \end{cases}$

Solve the second equation in terms of x.
$$x - 2y = -3$$
$$x = 2y - 3$$
Substitute $2y - 3$ for x in the first equation and solve for y.
$$4y = 2x + 6$$
$$4y = 2(2y - 3) + 6$$
$$4y = 4y - 6 + 6$$
$$0 = 0$$
Since the statement $0 = 0$ is true, there are an infinite number of solutions for the system.

18. $\begin{cases} 9x = 6y + 3 \\ 6x - 4y = 2 \end{cases}$

Solve the first equation in terms of x.

$9x = 6y + 3$

$x = \dfrac{6}{9}y + \dfrac{3}{9}$

$x = \dfrac{2}{3}y + \dfrac{1}{3}$

Substitute $\dfrac{2}{3}y + \dfrac{1}{3}$ for x in the second

equation and solve for y.

$6x - 4y = 2$

$6\left(\dfrac{2}{3}y + \dfrac{1}{3}\right) - 4y = 2$

$\qquad 4y + 2 - 4y = 2$

$\qquad\qquad\qquad 0 = 0$

Since the statement $0 = 0$ is true, there are an infinite number of solutions for the system.

19. $\begin{cases} x + y = 6 \\ y = -x - 4 \end{cases}$

Substitute $-x - 4$ for y in the first equation and solve for x.

$x + y = 6$

$x + (-x - 4) = 6$

$\qquad\qquad 0 = 10 \quad$ False

Since the statement $0 = 10$ is false, there is no solution for the system.

20. $\begin{cases} -3x + y = 6 \\ y = 3x + 2 \end{cases}$

Substitute $3x + 2$ for y in the first equation and solve for x.

$-3x + y = 6$

$-3x + 3x + 2 = 6$

$\qquad\qquad 2 = 6 \quad$ False

Since the statement $2 = 6$ is false, there is no solution for the system.

21. $\begin{cases} 2x + 3y = -6 \\ x - 3y = -12 \end{cases}$

Add the equations to eliminate y and solve for x.

$\begin{aligned} 2x + 3y &= -6 \\ \underline{x - 3y} &= \underline{-12} \\ 3x\phantom{{}+3y} &= -18 \\ x &= -6 \end{aligned}$

Now solve for y.

$x - 3y = -12$

$-6 - 3y = -12$

$\quad -3y = -6$

$\qquad y = 2$

The solution for the system is $(-6, 2)$.

22. $\begin{cases} 4x + y = 15 \\ -4x + 3y = -19 \end{cases}$

Add the equations to eliminate x and solve for y.

$\begin{aligned} 4x + y &= 15 \\ \underline{-4x + 3y} &= \underline{-19} \\ 4y &= -4 \\ y &= -1 \end{aligned}$

Now solve for x.

$4x + y = 15$

$4x - 1 = 15$

$\quad 4x = 16$

$\qquad x = 4$

The solution for the system is $(4, -1)$.

23. $\begin{cases} 2x - 3y = -15 \\ x + 4y = 31 \end{cases}$

Multiply the second equation by -2.

$\begin{cases} 2x - 3y = -15 \\ -2(x + 4y) = -2(31) \end{cases} \rightarrow \begin{cases} 2x - 3y = -15 \\ -2x - 8y = -62 \end{cases}$

Add the equations to eliminate x and solve for y.

$\begin{aligned} 2x - 3y &= -15 \\ \underline{-2x - 8y} &= \underline{-62} \\ -11y &= -77 \\ y &= 7 \end{aligned}$

Now solve for x.

$x + 4y = 31$

$x + 4(7) = 31$

$x + 28 = 31$

$\qquad x = 3$

The solution of the system is $(3, 7)$.

24. $\begin{cases} x - 5y = -22 \\ 4x + 3y = 4 \end{cases}$

Multiply the first equation by -4.

$\begin{cases} -4(x-5y) = -4(-22) \\ 4x + 3y = 4 \end{cases} \rightarrow \begin{cases} -4x + 20y = 88 \\ 4x + 3y = 4 \end{cases}$

Add the equations to eliminate x and solve for y.

$\begin{array}{r} -4x + 20y = 88 \\ 4x + 3y = 4 \\ \hline 23y = 92 \\ y = 4 \end{array}$

Now solve for x.

$x - 5y = -22$
$x - 5(4) = -22$
$x - 20 = -22$
$x = -2$

The solution of the system is $(-2, 4)$.

25. $\begin{cases} 2x - 6y = -1 \\ -x + 3y = \dfrac{1}{2} \end{cases}$

Multiply the second equation by 2.

$\begin{cases} 2x - 6y = -1 \\ 2(-x + 3y) = 2\left(\dfrac{1}{2}\right) \end{cases} \rightarrow \begin{cases} 2x - 6y = -1 \\ -2x + 6y = 1 \end{cases}$

Add the equations to eliminate x.

$\begin{array}{r} 2x - 6y = -1 \\ -2x + 6y = 1 \\ \hline 0 = 0 \end{array}$

Since the statement $0 = 0$ is true, there are an infinite number of solutions for the system.

26. $\begin{cases} 0.6x - 0.3y = -1.5 \\ 0.04x - 0.02y = -0.1 \end{cases}$

Multiply the first equation by 10 and the second by 100 to eliminate decimals.

$\begin{cases} 10(0.6x - 0.3y) = 10(-1.5) \\ 100(0.04x - 0.02y) = 100(-0.1) \end{cases}$

$\rightarrow \begin{cases} 6x - 3y = -15 \\ 4x - 2y = -10 \end{cases}$

Multiply the first revised equation by 2 and the second revised equation by -3.

$\begin{cases} 2(6x - 3y) = 2(-15) \\ -3(4x - 2y) = -3(-10) \end{cases}$

$\rightarrow \begin{cases} 12x - 6y = -30 \\ -12x + 6y = 30 \end{cases}$

Add the equations to eliminate y.

$\begin{array}{r} 12x - 6y = -30 \\ -12x + 6y = 30 \\ \hline 0 = 0 \end{array}$

Since the statement $0 = 0$ is true, there are an infinite number of solutions for the system.

27. $\begin{cases} \dfrac{3}{4}x + \dfrac{2}{3}y = 2 \\ x + \dfrac{y}{3} = 6 \end{cases}$

Multiply the first equation by 12 and the second by 3 to eliminate fractions.

$\begin{cases} 12\left(\dfrac{3}{4}x + \dfrac{2}{3}y\right) = 12(2) \\ 3\left(x + \dfrac{y}{3}\right) = 3(6) \end{cases} \rightarrow \begin{cases} 9x + 8y = 24 \\ 3x + y = 18 \end{cases}$

Multiply the second revised equation by -3.

$\begin{cases} 9x + 8y = 24 \\ -3(3x + y) = -3(18) \end{cases} \rightarrow \begin{cases} 9x + 8y = 24 \\ -9x - 3y = -54 \end{cases}$

Add the equations to eliminate x and solve for y.

$\begin{array}{r} 9x + 8y = 24 \\ -9x - 3y = -54 \\ \hline 5y = -30 \\ y = -6 \end{array}$

Now solve for x.

$x + \dfrac{y}{3} = 6$
$x + \left(\dfrac{-6}{3}\right) = 6$
$x + (-2) = 6$
$x = 8$

The solution of the system is $(8, -6)$.

28. $\begin{cases} 10x + 2y = 0 \\ 3x + 5y = 33 \end{cases}$

Multiply the first equation by 5 and the second by -2.

$\begin{cases} 5(10x+2y)=5(0) \\ -2(3x+5y)=-2(33) \end{cases} \rightarrow \begin{cases} 50x+10y=0 \\ -6x-10y=-66 \end{cases}$

Add the equations to eliminate y and solve for x.

$50x+10y=0$
$\underline{-6x-10y=-66}$
$44x\qquad=-66$

$$x=\frac{-66}{44}=-\frac{3}{2}$$

Now solve for y.
$$10x+2y=0$$
$$10\left(-\frac{3}{2}\right)+2y=0$$
$$-15+2y=0$$
$$2y=15$$
$$y=\frac{15}{2}$$

The solution of the system is $\left(-\frac{3}{2},\frac{15}{2}\right)$.

29. Let x be the smaller number and y be the larger.

$$\begin{cases} x+y=16 \\ 3y-x=72 \end{cases}$$

Solve the first equation in terms of y.
$$x+y=16$$
$$y=-x+16$$

Substitute $-x+16$ for y in the second equation and solve for x.
$$3y-x=72$$
$$3(-x+16)-x=72$$
$$-3x+48-x=72$$
$$-4x=24$$
$$x=-6$$

Now solve for y.
$$y=-x+16=-(-6)+16=6+16=22$$
The two numbers are -6 and 22.

30. Let x be the number of orchestra seats and y be the number of balcony seats.

$$\begin{cases} x+y=360 \\ 45x+35y=15,150 \end{cases}$$

Solve the first equation in terms of x.
$$x=-y+360$$

Substitute $-y+360$ for x in the second equation and solve for y.
$$45x+35y=15,150$$
$$45(-y+360)+35y=15,150$$
$$-45y+16,200+35y=15,150$$
$$-10y=-1050$$
$$y=105$$

Now solve for x.
$$x+y=360$$
$$x+105=360$$
$$x=255$$
There are 255 orchestra seats and 105 balcony seats.

31. Let x be the rate of the current and y be the speed in still water.
$$19(x-y)=340$$
$$14(x+y)=340$$

Multiply the first equation by $\frac{1}{19}$ and the second by $\frac{1}{14}$.

$$\begin{cases} \frac{1}{19}[19(x-y)]=\frac{1}{19}(340) \\ \frac{1}{14}[14(x+y)]=\frac{1}{14}(340) \end{cases} \rightarrow \begin{cases} x-y=17.9 \\ x+y=24.3 \end{cases}$$

Add the equations to eliminate y and solve for x.
$$x-y=17.9$$
$$\underline{x+y=24.3}$$
$$2x\qquad=42.2$$
$$x=21.1$$
Now solve for x.
$$x+y=24.3$$
$$21.1+y=24.3$$
$$y=3.2$$
The speed in still water is 21.1 mph and the current of the river is 3.2 mph.

32. Let x = number of cc of 6% acid solution and y = number of cc of 14% acid solution.
$$x+y=50$$
$$0.06x+0.14y=0.12(50)$$
$$\rightarrow 0.06x+0.14y=6$$
Solve the first equation in terms of x.

$$x + y = 50$$
$$x = -y + 50$$

Substitute $-y + 50$ for x in the second equation and solve for y.

$$0.06x + 0.14y = 6$$
$$0.06(-y + 50) + 0.14y = 6$$
$$-0.06y + 3 + 0.14y = 6$$
$$0.08y = 3$$
$$y = 37.5$$

Now solve for x.

$$x = -y + 50 = -37.5 + 50 = 12.5$$

12.5 cc of the 6% solution and 37.5 cc of the 14% solution are needed to make 50 cc of the 12% solution.

33. Let x be the cost of an egg and y be the cost of a strip of bacon.

$$\begin{cases} 3x + 4y = 3.80 \\ 2x + 3y = 2.75 \end{cases}$$

Multiply the first equation by 2 and the second by -3.

$$\begin{cases} 2(3x + 4y) = 2(3.80) \\ -3(2x + 3y) = -3(2.75) \end{cases}$$

$$\rightarrow \begin{cases} 6x + 8y = 7.60 \\ -6x - 9y = -8.25 \end{cases}$$

Add the equations to eliminate x and solve for y.

$$\begin{aligned} 6x + 8y &= 7.60 \\ \underline{-6x - 9y} &= \underline{-8.25} \\ -y &= -0.65 \\ y &= 0.65 \end{aligned}$$

Now solve for x.

$$2x + 3y = 2.75$$
$$2x + 3(0.65) = 2.75$$
$$2x + 1.95 = 2.75$$
$$2x = 0.80$$
$$x = 0.40$$

Each egg costs $0.40 and each strip of bacon costs $0.65.

34. Let x be time spent jogging, and let y be the time spent walking.

$$\begin{cases} x + y = 3 \\ 7.5x + 4y = 15 \end{cases}$$

Solve the first equation in terms of x.

$$x + y = 3$$
$$x = -y + 3$$

Substitute $-y + 3$ for x in the second equation and solve for y.

$$7.5x + 4y = 15$$
$$7.5(-y + 3) + 4y = 15$$
$$-7.5y + 22.5 + 4y = 15$$
$$-3.5y = -7.5$$
$$y = 2.14$$

Now solve for x.

$$x + y = 3$$
$$x + 2.14 = 3$$
$$x = 0.86$$

He spent 0.86 hour jogging and 2.14 hours walking.

35. $\begin{cases} x - 2y = 1 \\ 2x + 3y = -12 \end{cases}$

The solution to the system is $(-3, -2)$.

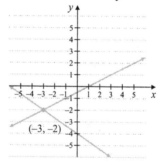

36. $\begin{cases} 3x - y = -4 \\ 6x - 2y = -8 \end{cases}$

The system has an infinite number of solutions because it is the same line.

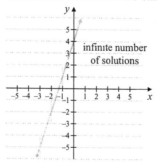

37. $\begin{cases} x + 4y = 11 \\ 5x - 9y = -3 \end{cases}$

Solve the first equation in terms of *x*.

$x + 4y = 11$

$\quad x = -4y + 11$

Substitute −4*y* + 11 for *x* in the second equation and solve for *y*.

$5x - 9y = -3$

$5(-4y + 11) - 9y = -3$

$-20y + 55 - 9y = -3$

$\quad\quad\quad -29y = -58$

$\quad\quad\quad\quad\quad y = 2$

Now solve for *x*.

$x = -4y + 11 = -4(2) + 11 = -8 + 11 = 3$

The solution of the system is (3, 2).

38. $\begin{cases} x + 9y = 16 \\ 3x - 8y = 13 \end{cases}$

Solve the first equation in terms of *x*.

$x + 9y = 16$

$\quad x = -9y + 16$

Substitute −9*y* + 16 for *x* in the second equation and solve for *y*.

$3x - 8y = 13$

$3(-9y + 16) - 8y = 13$

$-27y + 48 - 8y = 13$

$\quad\quad\quad -35y = -35$

$\quad\quad\quad\quad\quad y = 1$

Now solve for *x*.

$x = -9y + 16 = -9(1) + 16 = -9 + 16 = 7$

The solution of the system is (7, 1).

39. $y = -2x$

$4x + 7y = -15$

Substitute −2*x* for *y* in the second equation and solve for *x*.

$4x + 7y = -15$

$4x + 7(-2x) = -15$

$4x - 14x = -15$

$\quad -10x = -15$

$\quad\quad\quad x = \dfrac{-15}{-10} = \dfrac{3}{2}$

Now solve for *y*.

$y = -2x = -2\left(\dfrac{3}{2}\right) = -3$

The solution of the system is $\left(\dfrac{3}{2}, -3\right)$.

40. $\begin{cases} 3y = 2x + 15 \\ -2x + 3y = 21 \end{cases} \rightarrow \begin{cases} 2x - 3y = -15 \\ -2x + 3y = 21 \end{cases}$

Add the equations to eliminate *x*.

$\quad 2x - 3y = -15$

$\underline{-2x + 3y = 21}$

$\quad\quad\quad 0 = 6 \quad$ False

Since the statement 0 = 6 in false, there is no solution for the system.

41. $\begin{cases} 3x - y = 4 \\ 4y = 12x - 16 \end{cases}$

Solve the first equation in terms of *y*.

$3x - y = 4$

$3x - 4 = y$

Substitute 3*x* − 4 for *y* in the second equation and solve for *x*.

$\quad 4y = 12x - 16$

$4(3x - 4) = 12x - 16$

$12x - 16 = 12x - 16$

$\quad\quad\quad 0 = 0$

Since the statement 0 = 0 is true, there are an infinite number of solutions for the system.

42. $\begin{cases} x + y = 19 \\ x - y = -3 \end{cases}$

Add the equations to eliminate *y* and solve for *x*.

$\quad x + y = 19$

$\underline{\quad x - y = -3}$

$\quad 2x \quad\;\; = 16$

$\quad\quad\;\; x = 8$

Now solve for *y*.

$x + y = 19$

$8 + y = 19$

$\quad\;\; y = 11$

The solution of the system is (8, 11).

43. $\begin{cases} x - 3y = -11 \\ 4x + 5y = -10 \end{cases}$

Solve the first equation in terms of x.

$x - 3y = -11$

$\qquad x = 3y - 11$

Substitute $3y - 11$ for x in the second equation and solve for y.

$\qquad\qquad 4x + 5y = -10$

$4(3y - 11) + 5y = -10$

$12y - 44 + 5y = -10$

$\qquad\qquad\quad 17y = 34$

$\qquad\qquad\qquad y = 2$

Now solve for x.

$x = 3y - 11 = 3(2) - 11 = 6 - 11 = -5$

The solution of the system is $(-5, 2)$.

44. $\begin{cases} -x - 15y = 44 \\ 2x + 3y = 20 \end{cases}$

Solve the first equation in terms of x.

$-x - 15y = 44$

$-15y - 44 = x$

Substitute $-15y - 44$ for x in the second equation and solve for y.

$\qquad\qquad\quad 2x + 3y = 20$

$2(-15y - 44) + 3y = 20$

$-30y - 88 + 3y = 20$

$\qquad\qquad -27y = 108$

$\qquad\qquad\qquad y = -4$

Now solve for x.

$x = -15y - 44$

$\quad = -15(-4) - 44$

$\quad = 60 - 44$

$\quad = 16$

The solution of the system is $(16, -4)$.

45. $\begin{cases} 2x + y = 3 \\ 6x + 3y = 9 \end{cases}$

Solve the first equation in terms of y.

$2x + y = 3$

$\qquad y = -2x + 3$

Substitute $-2x + 3$ for y in the second equation and solve for x.

$6x + 3y = 9$

$6x + 3(-2x + 3) = 9$

$6x - 6x + 9 = 9$

$\qquad\qquad 0 = 0$

Since the statement $0 = 0$ is true, there are an infinite number of solutions for the system.

46. $\begin{cases} -3x + y = 5 \\ -3x + y = -2 \end{cases}$

Solve the first equation in terms of y.

$-3x + y = 5$

$\qquad y = 3x + 5$

Substitute $3x + 5$ for y in the second equation and solve for x.

$\qquad\qquad -3x + y = -2$

$-3x + 3x + 5 = -2$

$\qquad\qquad\quad 5 = -2 \quad$ False

Since the statement $5 = -2$ is false, there is no solution for the system.

47. Let x be the smaller number and y be the larger number.

$\begin{cases} x + y = 12 \\ 3x + y = 20 \end{cases}$

Solve the first equation in terms of y.

$x + y = 12$

$\qquad y = -x + 12$

Substitute $-x + 12$ for y in the second equation and solve for x.

$\qquad\qquad 3x + y = 20$

$3x + (-x + 12) = 20$

$\qquad\quad 2x + 12 = 20$

$\qquad\qquad\quad 2x = 8$

$\qquad\qquad\qquad x = 4$

Now solve for y.

$y = -x + 12 = -4 + 12 = 8$

The two numbers are 4 and 8.

48. Let x be the smaller number and y be the larger number.

$x - y = -18$

$2x - y = -23$

Solve the first equation in terms of x.

$x - y = -18$

$x = y - 18$

Substitute $y - 18$ for x in the second equation and solve for y.

$2x - y = -23$

$2(y - 18) - y = -23$

$2y - 36 - y = -23$

$y = 13$

Now solve for x.

$x = y - 18 = 13 - 18 = -5$

The two numbers are -5 and 13.

49. Let x be nickels and y be dimes.

$x + y = 65$

$0.05x + 0.1y = 5.30$

Solve the first equation in terms of x.

$x + y = 65$

$x = -y + 65$

Substitute $-y + 65$ for x in the second equation and solve for y.

$0.05x + 0.1y = 5.30$

$0.05(-y + 65) + 0.1y = 5.30$

$-0.05y + 3.25 + 0.1y = 5.30$

$0.05y = 2.05$

$y = 41$

Now solve for x.

$x = -y + 65 = -41 + 65 = 24.$

There are 24 nickels and 41 dimes.

50. Let x be the number of 13¢ stamps and y be the number of 22¢ stamps.

$\begin{cases} x + y = 26 \\ 0.13x + 0.22y = 4.19 \end{cases}$

Solve the first equation in terms of x.

$x + y = 26$

$x = -y + 26$

Substitute $-y + 26$ for x in the second equation and solve for y.

$0.13x + 0.22y = 4.19$

$0.13(-y + 26) + 0.22y = 4.19$

$-0.13y + 3.38 + 0.22y = 4.19$

$0.09y = 0.81$

$y = 9$

Now solve for x.

$x = -y + 26 = -9 + 26 = 17$

They purchased 17 13¢ stamps and 9 22¢ stamps.

Chapter 7 Test

1. False; a system of two linear equations can have no solutions, exactly one solution, or infinitely many solutions.

2. False; a solution has to be a solution to both equations to be a solution of the system.

3. True; when the resulting statement is false the system has no solutions.

4. False; when $3x = 0 \to x = 0$; the system does have a solution.

5. First equation:

$2x - 3y = 5$

$2(1) - 3(-1) \overset{?}{=} 5$

$2 + 3 \overset{?}{=} 5$

$5 = 5$　True

Second equation:

$6x + y = 1$

$6(1) + (-1) \overset{?}{=} 1$

$6 - 1 \overset{?}{=} 1$

$5 = 1$　False

Since the statement $5 = 1$ is false, $(1, -1)$ is not a solution of the system.

6. $\begin{cases} 4x - 3y = 24 \\ 4x + 5y = -8 \end{cases}$

First equation:

$4x - 3y = 24$

$4(3) - 3(-4) \overset{?}{=} 24$

$12 + 12 \overset{?}{=} 24$

$24 = 24$　True

Second equation:

$4x + 5y = -8$

$4(3) + 5(-4) \overset{?}{=} -8$

$12 - 20 \overset{?}{=} -8$

$-8 = -8$　True

$(3, -4)$ is a solution of the system.

7. $\begin{cases} x - y = 2 \\ 3x - y = -2 \end{cases}$

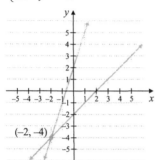

$(-2, -4)$

8. $\begin{cases} y = -3x \\ 3x + y = 6 \end{cases}$

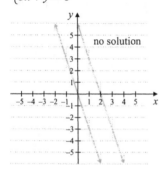

no solution

9. $\begin{cases} 3x - 2y = -14 \\ y = x + 5 \end{cases}$

Substitute $x + 5$ for y in the first equation and solve for x.

$$3x - 2y = -14$$
$$3x - 2(x + 5) = -14$$
$$3x - 2x - 10 = -14$$
$$x - 10 = -14$$
$$x = -4$$

$y = x + 5 = -4 + 5 = 1$

The solution of the system is $(-4, 1)$.

10. $\begin{cases} \dfrac{1}{2}x + 2y = -\dfrac{15}{4} \\ 4x = -y \end{cases}$

Multiply the second equation by -1.

$$(-1)4x = (-1)(-y)$$
$$-4x = y$$

Substitute $-4x$ for y in the first equation.

$$\frac{1}{2}x + 2y = -\frac{15}{4}$$
$$\frac{1}{2}x + 2(-4x) = -\frac{15}{4}$$
$$\frac{1}{2}x - 8x = -\frac{15}{4}$$
$$4\left(\frac{1}{2}x - 8x\right) = 4\left(-\frac{15}{4}\right)$$
$$2x - 32x = -15$$
$$-30x = -15$$
$$x = \frac{15}{30} = \frac{1}{2}$$

Now solve for y.

$$y = -4x = -4\left(\frac{1}{2}\right) = -2$$

The solution of the system is $\left(\dfrac{1}{2}, -2\right)$.

11. $\begin{cases} x + y = 28 \\ x - y = 12 \end{cases}$

Add the equations to eliminate y.

$$\begin{array}{r} x + y = 28 \\ x - y = 12 \\ \hline 2x = 40 \\ x = 20 \end{array}$$

Now solve for y.

$$x + y = 28$$
$$20 + y = 28$$
$$y = 8$$

The solution of the system is $(20, 8)$.

12. $\begin{cases} 4x - 6y = 7 \\ -2x + 3y = 0 \end{cases}$

Multiply the second equation by 2.

$$\begin{cases} 4x - 6y = 7 \\ 2(-2x + 3y) = 2(0) \end{cases} \rightarrow \begin{cases} 4x - 6y = 7 \\ -4x + 6y = 0 \end{cases}$$

Add the equations to eliminate x.

$$\begin{array}{r} 4x - 6y = 7 \\ -4x + 6y = 0 \\ \hline 0 = 7 \end{array}$$

Since the statement $0 = 7$ is false, the system has no solution.

13. $\begin{cases} 3x + y = 7 \\ 4x + 3y = 1 \end{cases}$

Solve the first equation for y.

$3x + y = 7$

$y = 7 - 3x$

Substitute $7 - 3x$ for y in the second equation and solve for x.

$4x + 3y = 1$

$4x + 3(7 - 3x) = 1$

$4x + 21 - 9x = 1$

$21 - 5x = 1$

$-5x = -20$

$x = 4$

$y = 7 - 3x = 7 - 3(4) = 7 - 12 = -5$

The solution of the system is $(4, -5)$.

14. $\begin{cases} 3(2x + y) = 4x + 20 \\ x - 2y = 3 \end{cases}$

Simplify the first equation.

$3(2x + y) = 4x + 20$

$6x + 3y = 4x + 20$

$2x + 3y = 20$

$\begin{cases} 2x + 3y = 20 \\ x - 2y = 3 \end{cases}$

Multiply the second equation by -2.

$\begin{cases} 2x + 3y = 20 \\ -2(x - 2y) = -2(3) \end{cases} \rightarrow \begin{cases} 2x + 3y = 20 \\ -2x + 4y = -6 \end{cases}$

Add the equations to eliminate x; then solve for y.

$2x + 3y = 20$

$\underline{-2x + 4y = -6}$

$7y = 14$

$y = 2$

Now solve for x.

$x - 2y = 3$

$x - 2(2) = 3$

$x - 4 = 3$

$x = 7$

The solution of the system is $(7, 2)$.

15. $\begin{cases} \dfrac{x-3}{2} = \dfrac{2-y}{4} \\ \dfrac{7-2x}{3} = \dfrac{y}{2} \end{cases}$

Multiply the first equation by 4 and the second equation by 6 to eliminate fractions and simplify.

$\begin{cases} 4\left(\dfrac{x-3}{2}\right) = 4\left(\dfrac{2-y}{4}\right) \\ 6\left(\dfrac{7-2x}{3}\right) = 6\left(\dfrac{y}{2}\right) \end{cases} \rightarrow \begin{cases} 2x - 6 = 2 - y \\ 14 - 4x = 3y \end{cases}$

$\rightarrow \begin{cases} 2x + y = 8 \\ 4x + 3y = 14 \end{cases}$

Multiply the first revised equation by -3.

$\begin{cases} -3(2x + y) = -3(8) \\ 4x + 3y = 14 \end{cases} \rightarrow \begin{cases} -6x - 3y = -24 \\ 4x + 3y = 14 \end{cases}$

Add the equations to eliminate y.

$-6x - 3y = -24$

$\underline{4x + 3y = 14}$

$-2x = -10$

$x = 5$

Now solve for y.

$2x + y = 8$

$2(5) + y = 8$

$10 + y = 8$

$y = -2$

The solution of the system is $(5, -2)$.

16. $\begin{cases} 8x - 4y = 12 \\ y = 2x - 3 \end{cases}$

Substitute $2x - 3$ for y in the first equation and solve for x.

$8x - 4y = 12$

$8x - 4(2x - 3) = 12$

$8x - 8x + 12 = 12$

$0 = 0$

Since the statement $0 = 0$ is true, the system has an infinite number of solutions.

17. $\begin{cases} 0.01x - 0.06y = -0.23 \\ 0.2x + 0.4y = 0.2 \end{cases}$

Multiply the first equation by 100 and the second equation by 10 to eliminate decimals.

$$\begin{cases} 100(0.01x - 0.06y) = 100(-0.23) \\ 10(0.2x + 0.4y) = 10(0.2) \end{cases}$$

$$\rightarrow \begin{cases} x - 6y = -23 \\ 2x + 4y = 2 \end{cases}$$

Multiply the first equation by -2.

$$\begin{cases} -2(x - 6y) = -2(-23) \\ 2x + 4y = 2 \end{cases} \rightarrow \begin{cases} -2x + 12y = 46 \\ 2x + 4y = 2 \end{cases}$$

Add the equations to eliminate x; then solve for y.

$$\begin{array}{r} -2x + 12y = 46 \\ 2x + 4y = 2 \\ \hline 16y = 48 \\ y = 3 \end{array}$$

Now solve for x.

$$x - 6y = -23$$
$$x - 6(3) = -23$$
$$x - 18 = -23$$
$$x = -5$$

The solution of the system is $(-5, 3)$.

18. $$\begin{cases} x - \dfrac{2}{3}y = 3 \\ -2x + 3y = 10 \end{cases}$$

Solve the first equation in terms of x.

$$x - \frac{2}{3}y = 3$$
$$x = \frac{2}{3}y + 3$$

Substitute $\dfrac{2}{3}y + 3$ for x in the second

equation and solve for y.
$$-2x + 3y = 10$$
$$-2\left(\frac{2}{3}y + 3\right) + 3y = 10$$
$$-\frac{4}{3}y - 6 + 3y = 10$$
$$-4y - 18 + 9y = 30$$
$$5y = 48$$
$$y = \frac{48}{5}$$

Now solve for x.

$$x = \frac{2}{3}y + 3 = \frac{2}{3}\left(\frac{48}{5}\right) + 3 = \frac{32}{5} + \frac{15}{5} = \frac{47}{5}$$

The solution of the system is $\left(\dfrac{47}{5}, \dfrac{48}{5}\right)$.

19. Let x be the first number and the y be the second.

$$\begin{cases} x + y = 124 \\ x - y = 32 \end{cases}$$

Add the equations to eliminate y then solve for x.

$$\begin{array}{r} x + y = 124 \\ x - y = 32 \\ \hline 2x = 156 \\ x = 78 \end{array}$$

Now solve for y.
$$x + y = 124$$
$$78 + y = 124$$
$$y = 46$$

The numbers are 78 and 46.

20. Let x = number of cc of 12% solution and y = number of cc of 16% solution.

$$\begin{cases} x + 80 = y \\ 0.12x + 0.22(80) = 0.16y \end{cases}$$

Substitute $x + 80$ for y in the second equation; then solve for x.
$$0.12x + 17.6 = 0.16y$$
$$0.12x + 17.6 = 0.16(x + 80)$$
$$0.12x + 17.6 = 0.16x + 12.8$$
$$4.8 = 0.04x$$
$$120 = x$$

120 cc of the 12% saline is needed to make the 16% solution.

21. Let t be the number of farms in Texas and let m be the number of farms in Missouri.

$$\begin{cases} t + m = 336 \\ t = m + 116 \end{cases}$$

Substitute $m + 116$ for t in the first equation and solve for m.

$$t + m = 336$$
$$m + 116 + m = 336$$
$$2m + 116 = 336$$
$$2m = 220$$
$$m = 110$$
$$t = m + 116 = 110 + 116 = 226$$
There are 226 thousand farms in Texas and 110 thousand in Missouri.

22. Let x be speed of one hiker and y be the speed of the other hiker.
$$\begin{cases} 4x + 4y = 36 \\ y = 2x \end{cases}$$
Substitute $2x$ for y in the first equation; then solve for x.
$$4x + 4y = 36$$
$$4x + 4(2x) = 36$$
$$4x + 8x = 36$$
$$12x = 36$$
$$x = 3$$
Now solve for y.
$$y = 2x = 2(3) = 6$$
One hiker hikes at 3 mph and the other hikes at 6 mph.

23. In 1999 the purchases of country music equals the purchases of rap/hip-hop music.

24. In the period of 1991–1999, the purchases of country music were more than those of rap/hip-hop music.

Cumulative Review Chapters 1–7

1. a. $-14 - 8 + 10 - (-6) = -6$

b. $1.6 - (-10.3) + (-5.6) = 6.3$

2. a. $5^2 = 25$

b. $2^5 = 32$

3. reciprocal of $22 = \dfrac{1}{22}$

4. opposite of $22 = -22$

5. reciprocal of $\dfrac{3}{16} = \dfrac{16}{3}$

6. opposite of $\dfrac{3}{16} = -\dfrac{3}{16}$

7. reciprocal of $-10 = -\dfrac{1}{10}$

8. opposite of $-10 = 10$

9. reciprocal of $-\dfrac{9}{13} = -\dfrac{13}{9}$

10. opposite of $-\dfrac{9}{13} = \dfrac{9}{13}$

11. reciprocal of $1.7 = \dfrac{1}{1.7}$

12. opposite of $1.7 = -1.7$

13. a. $x + y = 8$
$$x + 3 = 8$$
$$x = 5$$

b. $x + y = 8$
$$y = 8 - x$$

14. $5(x-1) = 6x$
$$5x - 5 = 6x$$
$$-5 = x$$

15. $-2(x-5) + 10 = -3(x+2) + x$
$$-2x + 10 + 10 \stackrel{?}{=} -3x - 6 + x$$
$$-2x + 20 \stackrel{?}{=} -2x - 6$$
$$0x \stackrel{?}{=} -26$$
$$0 = -26 \quad \text{False}$$
Since the statement $0 = -26$ is false, there is no solution.

16. $5(y-5) = 5y+10$

$5y-25 \overset{?}{=} 5y+10$

$0y \overset{?}{=} 35$

$0 = 35$ False

Since the statement $0 = 35$ is false, there is no solution.

17. $\dfrac{x}{2} - 1 = \dfrac{2}{3}x - 3$

$6\left(\dfrac{x}{2} - 1\right) = 6\left(\dfrac{2}{3}x - 3\right)$

$3x - 6 = 4x - 18$

$12 = x$

18. $7(x-2) - 6(x+1) = 20$

$7x - 14 - 6x - 6 = 20$

$x - 20 = 20$

$x = 40$

19. $-5x + 7 < 2(x-3)$

$-5x + 7 < 2x - 6$

$13 < 7x$

$\dfrac{13}{7} < x$

$\left\{ x \middle| x > \dfrac{13}{7} \right\}$

20. $P = a + b + c$ for b.

$b = P - a - c$

21. $\left(\dfrac{m}{n}\right)^7 = \dfrac{m^7}{n^7}$ where $n \neq 0$

22. $\dfrac{a^7 b^{10}}{ab^{15}} = \dfrac{a^6}{b^5}$ where $b \neq 0$

23. $\left(\dfrac{2x^4}{3y^5}\right)^4 = \dfrac{2^4 x^{4\cdot4}}{3^4 y^{5\cdot4}} = \dfrac{16x^{16}}{81y^{20}}$ where $y \neq 0$

24. $(7a^2 b^{-3})^2 = 7^2 a^{2\cdot2} b^{-3\cdot2}$

$= 49a^4 b^{-6}$

$= \dfrac{49a^4}{b^6}$

where $b \neq 0$.

25. $(2x^3 + 8x^2 - 6x) - (2x^3 - x^2 + 1)$

$= 2x^3 + 8x^2 - 6x - 2x^3 + x^2 - 1$

$= 9x^2 - 6x - 1$

26. $\left(5x^2 + 6x + \dfrac{1}{2}\right) + \left(x^2 - \dfrac{4}{3}x - \dfrac{10}{21}\right)$

$= 6x^2 + \dfrac{14}{3}x + \dfrac{1}{42}$

27. $3x - 1 \overline{\smash{\big)}\, 6x^2 + 10x - 5}$

$\begin{array}{r} 2x+4 \\ \underline{6x^2 - 2x} \\ 12x - 5 \\ \underline{12x - 4} \\ -1 \end{array}$

$\dfrac{6x^2 + 10x - 1}{3x - 1} = 2x + 4 - \dfrac{1}{3x - 1}$

28. $9x^2 = 3 \cdot 3 \cdot x \cdot x$

$6x^3 = 2 \cdot 3 \cdot x \cdot x \cdot x$

$21x^5 = 3 \cdot 7 \cdot x \cdot x \cdot x \cdot x \cdot x$

The common factors are 3 and $x \cdot x = x^2$, so the GCF is $3x^2$.

29. $x(2x - 7) = 4$

$2x^2 - 7x = 4$

$2x^2 - 7x - 4 = 0$

$(2x + 1)(x - 4) = 0$

$2x + 1 = 0$ or $x - 4 = 0$

$2x = -1$ $x = 4$

$x = -\dfrac{1}{2}$

The solutions are $x = -\dfrac{1}{2}$ or $x = 4$.

30. $x(2x-7)=0$

$x=0$ or $2x-7=0$

$$2x=7$$

$$x=\frac{7}{2}$$

The solutions are $x=0$ or $x=\dfrac{7}{2}$.

31. $x=$ one leg

$x+2=$ other leg

$x+4=$ hypotenuse

$$x^2+(x+2)^2=(x+4)^2$$

$$x^2+x^2+4x+4=x^2+8x+16$$

$$x^2-4x-12=0$$

$$(x-6)(x+2)=0$$

$x-6=0$ or $x+2=0$

$\quad x=6 \qquad\qquad x=-2$

Discard $x=-2$ because length cannot be negative.

one leg: $x=6$

other leg: $x+2=6+2=8$

hypotenuse: $x+4=6+4=10$

The lengths of the sides of the right triangle are 6, 8, and 10 units.

32. $A=bh$

Let $x=$ base; then $h=3x+5$.

$$A=x(3x+5)=182$$

$$3x^2+5x-182=0$$

$$(3x+26)(x-7)=0$$

$3x+26=0$ or $x-7=0$

$\quad 3x=-26 \qquad\qquad x=7$

$$x=-\frac{26}{3}$$

Discard $x=-\dfrac{26}{3}$ because length cannot be negative.

base $=7$ ft

height $=3x+5=3(7)+5=21+5=26$ ft

33. $\dfrac{2y}{2y-7}-\dfrac{7}{2y-7}=\dfrac{2y-7}{2y-7}=1$

34. $\dfrac{2}{x-6}+\dfrac{3}{x+1}=\dfrac{2(x+1)}{(x-6)(x+1)}+\dfrac{3(x-6)}{(x+1)(x-6)}$

$$=\frac{2x+2+3x-18}{(x-6)(x+1)}$$

$$=\frac{5x-16}{(x-6)(x+1)}$$

35. $\dfrac{\frac{x}{y}+\frac{3}{2x}}{\frac{x}{2}+y}=\dfrac{(2xy)\left(\frac{x}{y}+\frac{3}{2x}\right)}{(2xy)\left(\frac{x}{2}+y\right)}=\dfrac{2x^2+3y}{x^2y+2xy^2}$

36. $m=\dfrac{y_2-y_1}{x_2-x_1}=\dfrac{3-(-8)}{-1-2}=-\dfrac{11}{3}$

37. $y=mx+b$

If $y=-1$, then $m=0$.

38. Pick any 2 points on the line $x=2$.

Let $(x_1,\,y_1)=(2,\,0)$ and $(x_2,\,y_2)=(2,\,5)$.

$$m=\frac{y_2-y_1}{x_2-x_1}=\frac{5-0}{2-2}=\frac{5}{0}=\text{ undefined}$$

39. $m=\dfrac{y_2-y_1}{x_2-x_1}=\dfrac{4-5}{-3-2}=\dfrac{-1}{-5}=\dfrac{1}{5}$

$$y-y_1=m(x-x_1)$$

$$y-5=\frac{1}{5}(x-2)$$

$$y-5=\frac{1}{5}x-\frac{2}{5}$$

$$5(y-5)=5\left(\frac{1}{5}x-\frac{2}{5}\right)$$

$$5y-25=x-2$$

$$-x+5y=23$$

40. $y-y_1=m(x-x_1)$

$$y-3=-5(x-(-2))$$

$$y-3=-5x-10$$

$$y=-5x-7$$

41. $\{(0,2),(3,3),(-1,0),(3,-2)\}$

Domain: $\{-1,0,3\}$

Range: $\{-2,0,2,3\}$

42. $f(x) = 5x^2 - 6$

$f(0) = 5(0)^2 - 6 = -6$

$f(-2) = 5(-2)^2 - 6 = 5(4) - 6 = 20 - 6 = 14$

43. $\begin{cases} 2x - 3y = 6 \\ x = 2y \end{cases}$

First equation:

$2x - 3y = 6$

$2(12) - 3(6) \stackrel{?}{=} 6$

$24 - 18 \stackrel{?}{=} 6$

$ 6 = 6$ True

Second equation:

$x = 2y$

$12 \stackrel{?}{=} 2(6)$

$12 = 12$ True

$(12, 6)$ is a solution of the system.

44. a. $\begin{cases} 2x - y = 6 \\ 3x + 2y = -5 \end{cases}$

First equation:

$2(1) - (-4) \stackrel{?}{=} 6$

$2 + 4 \stackrel{?}{=} 6$

$ 6 = 6$ True

Second equation:

$3x + 2y = -5$

$3(1) + 2(-4) \stackrel{?}{=} -5$

$3 + (-8) \stackrel{?}{=} -5$

$ -5 = -5$ True

$(1, -4)$ is a solution of the system.

b. First equation:

$2x - y = 6$

$2(0) - (6) \stackrel{?}{=} 6$

$ -6 = 6$ False

$(0, 6)$ is not a solution of the system.

c. First equation:

$2x - y = 6$

$2(3) - 0 = 6$

$ 6 = 6$ True

Second equation:

$3x + 2y = -5$

$3(3) + 2(0) \stackrel{?}{=} -5$

$9 + 0 \stackrel{?}{=} -5$

$9 = -5$ False

$(3, 0)$ is not a solution of the system.

45. $\begin{cases} x + 2y = 7 \\ 2x + 2y = 13 \end{cases}$

Multiply the first equation by -1.

$\begin{cases} -1(x + 2y) = -1(7) \\ 2x + 2y = 13 \end{cases} \rightarrow \begin{cases} -x - 2y = -7 \\ 2x + 2y = 13 \end{cases}$

Add the equations to eliminate y; then solve for x.

$-x - 2y = -7$

$\underline{2x + 2y = 13}$

$x = 6$

Now solve for y.

$x + 2y = 7$

$6 + 2y = 7$

$ 2y = 1$

$ y = \dfrac{1}{2}$

The solution of the system is $\left(6, \dfrac{1}{2}\right)$.

46. $\begin{cases} 3x - 4y = 10 \\ y = 2x \end{cases}$

Substitute $2x$ for y in the first equation and solve for x.

$3x - 4y = 10$

$3x - 4(2x) = 10$

$3x - 8x = 10$

$ -5x = 10$

$ x = -2$

Now solve for y.

$y = 2x = 2(-2) = -4$

The solution of the system is $(-2, -4)$.

47. $\begin{cases} -x - \dfrac{y}{2} = \dfrac{5}{2} \\ \dfrac{x}{6} - \dfrac{y}{2} = 0 \end{cases}$

Multiply the first equation by -6 and the

second equation by 6.

$$\begin{cases} -6\left(-x-\dfrac{y}{2}\right) = -6\left(\dfrac{5}{2}\right) \\ 6\left(\dfrac{x}{6}-\dfrac{y}{2}\right) = 6(0) \end{cases} \rightarrow \begin{cases} 6x+3y=-15 \\ x-3y=0 \end{cases}$$

Add the equations to eliminate y; then solve for x.

$$6x+3y=-15$$
$$\underline{x-3y=0}$$
$$7x \qquad =-15$$
$$x=-\frac{15}{7}$$

Now solve for y.

$$-x-\frac{y}{2}=\frac{5}{2}$$

$$-\left(-\frac{15}{7}\right)-\frac{y}{2}=\frac{5}{2}$$

$$14\left(\frac{15}{7}-\frac{y}{2}\right)=14\left(\frac{5}{2}\right)$$

$$30-7y=35$$

$$-7y=5$$

$$y=-\frac{5}{7}$$

The solution of the system is $\left(-\dfrac{15}{7}, -\dfrac{5}{7}\right)$.

48. $\begin{cases} x=5y-3 \\ x=8y+4 \end{cases}$

Substitute $5y-3$ for x in the second equation and solve for y.

$$x=8y+4$$
$$5y-3=8y+4$$
$$-7=3y$$
$$-\frac{7}{3}=y$$

Now solve for x.

$$x=5y-3=5\left(-\frac{7}{3}\right)-3=-\frac{35}{3}-\frac{9}{3}=-\frac{44}{3}$$

The solution of the system is $\left(-\dfrac{44}{3}, -\dfrac{7}{3}\right)$.

49. Let x be the first number and y be the second.

$$\begin{cases} x+y=37 \\ x-y=21 \end{cases}$$

Add the equations to eliminate y; then solve for x.

$$x+y=37$$
$$\underline{x-y=21}$$
$$2x \qquad =58$$
$$x=29$$

Now solve for y.

$$x+y=37$$
$$29+y=37$$
$$y=8$$

The two numbers are 29 and 8.

50. **a.** It is not a function because each x-coordinate (except 0) has more than one y-coordinate.

   b. It is a function because each x-coordinate has exactly one y-coordinate.

   c. It is not a function because all but two x-coordinates have more than one y-coordinate.

Chapter 8

Section 8.1 Practice Problems

1. $\sqrt{100} = 10$, because $10^2 = 100$ and 10 is positive.

2. $-\sqrt{36} = -6$. The negative sign in front of the radical indicates the negative square root of 36.

3. $\sqrt{\dfrac{25}{81}} = \dfrac{5}{9}$, because $\left(\dfrac{5}{9}\right)^2 = \dfrac{25}{81}$ and $\dfrac{5}{9}$ is positive.

4. $\sqrt{1} = 1$, because $1^2 = 1$ and 1 is positive.

5. $\sqrt{0.81} = 0.9$ because $(0.9)^2 = 0.81$ and 0.9 is positive.

6. $\sqrt[3]{27} = 3$ because $3^3 = 27$.

7. $\sqrt[3]{-8} = -2$ because $(-2)^3 = -8$.

8. $\sqrt[3]{\dfrac{1}{64}} = \dfrac{1}{4}$ because $\left(\dfrac{1}{4}\right)^3 = \dfrac{1}{64}$.

9. $\sqrt[4]{-16}$ is not a real number since the index 4 is even and the radicand -16 is negative.

10. $\sqrt[5]{-1} = -1$ because $(-1)^5 = -1$.

11. $\sqrt[4]{81} = 3$ because $3^4 = 81$ and 3 is positive.

12. $\sqrt[6]{-1}$ is not a real number because the index 6 is even and the radicand -1 is negative.

13. To three decimal places, $\sqrt{22} \approx 4.690$.

14. $\sqrt{z^8} = z^4$ because $(z^4)^2 = z^8$.

15. $\sqrt{x^{20}} = x^{10}$ because $(x^{10})^2 = x^{20}$.

16. $\sqrt{4x^6} = 2x^3$ because $(2x^3)^2 = 4x^6$.

17. $\sqrt[3]{8y^{12}} = 2y^4$ because $(2y^4)^3 = 8y^{12}$.

18. $\sqrt[3]{-64x^9 y^{24}} = -4x^3 y^8$ because $(-4x^3 y^8)^3 = -64x^9 y^{24}$.

Calculator Explorations

1. $\sqrt{6} \approx 2.449$; since 6 is between perfect squares 4 and 9, $\sqrt{6}$ is between $\sqrt{4} = 2$ and $\sqrt{9} = 3$.

2. $\sqrt{14} \approx 3.742$; since 14 is between perfect squares 9 and 16, $\sqrt{14}$ is between $\sqrt{9} = 3$ and $\sqrt{16} = 4$.

3. $\sqrt{11} \approx 3.317$; since 11 is between perfect squares 9 and 16, $\sqrt{11}$ is between $\sqrt{9} = 3$ and $\sqrt{16} = 4$.

4. $\sqrt{200} \approx 14.142$; since 200 is between perfect squares 196 and 225, $\sqrt{200}$ is between $\sqrt{196} = 14$ and $\sqrt{225} = 15$.

5. $\sqrt{82} \approx 9.055$; since 82 is between perfect squares 81 and 100, $\sqrt{82}$ is between $\sqrt{81} = 9$ and $\sqrt{100} = 10$.

6. $\sqrt{46} \approx 6.782$; since 46 is between perfect squares 36 and 49, $\sqrt{46}$ is between $\sqrt{36} = 6$ and $\sqrt{49} = 7$.

7. $\sqrt[3]{40} \approx 3.420$

8. $\sqrt[3]{71} \approx 4.141$

9. $\sqrt[4]{20} \approx 2.115$

10. $\sqrt[4]{15} \approx 1.968$

11. $\sqrt[5]{18} \approx 1.783$

12. $\sqrt[6]{2} \approx 1.122$

Exercise Set 8.1

1. $\sqrt{16} = 4$, because $4^2 = 16$ and 4 is positive.

3. $\sqrt{\dfrac{1}{25}} = \dfrac{1}{5}$, because $\left(\dfrac{1}{5}\right)^2 = \dfrac{1}{25}$ and $\dfrac{1}{5}$ is positive.

5. $-\sqrt{100} = -10$. The negative sign indicates the negative square root of 100.

7. $\sqrt{-4}$ is not a real number, because there is no real number whose square is -4.

9. $-\sqrt{121} = -11$. The negative sign indicates the negative square root of 121.

11. $\sqrt{\dfrac{9}{25}} = \dfrac{3}{5}$, because $\left(\dfrac{3}{5}\right)^2 = \dfrac{9}{5}$ and $\dfrac{3}{5}$ is positive.

13. $\sqrt{900} = 30$, because $30^2 = 900$ and 30 is positive.

15. $\sqrt{144} = 12$, because $12^2 = 144$ and 12 is positive.

17. $\sqrt{\dfrac{1}{100}} = \dfrac{1}{10}$, because $\left(\dfrac{1}{10}\right)^2 = \dfrac{1}{100}$ and $\dfrac{1}{10}$ is positive.

19. $\sqrt{0.25} = 0.5$, because $0.5^2 = 0.25$ and 0.5 is positive.

21. $\sqrt[3]{125} = 5$, because $5^3 = 125$.

23. $\sqrt[3]{-64} = -4$, because $(-4)^3 = -64$.

25. $-\sqrt[3]{8} = -2$, because $2^3 = 8$.

27. $\sqrt[3]{\dfrac{1}{8}} = \dfrac{1}{2}$, because $\left(\dfrac{1}{2}\right)^3 = \dfrac{1}{8}$.

29. $\sqrt[3]{-125} = -5$, because $(-5)^3 = -125$.

31. $\sqrt[5]{32} = 2$, because $2^5 = 32$.

33. $\sqrt{81} = 9$, because $9^2 = 81$ and 9 is positive.

35. $\sqrt[4]{-16}$ is not a real number since the index 4 is even and the radicand -16 is negative.

37. $\sqrt[3]{-\dfrac{27}{64}} = -\dfrac{3}{4}$, because $\left(-\dfrac{3}{4}\right)^3 = -\dfrac{27}{64}$.

39. $-\sqrt[4]{625} = -5$, because $5^4 = 625$.

41. $\sqrt[6]{1} = 1$, because $1^6 = 1$ and 1 is positive.

43. $\sqrt{7} \approx 2.646$

45. $\sqrt{37} \approx 6.083$

47. $\sqrt{136} \approx 11.662$

49. $\sqrt{2} \approx 1.41$
$90\sqrt{2} \approx 90(1.41) = 126.90$
The distance from home plate to second base is approximately 126.90 feet.

51. $\sqrt{m^2} = m$, because $m^2 = m^2$.

53. $\sqrt{x^4} = x^2$, because $(x^2)^2 = x^4$.

55. $\sqrt{9x^8} = 3x^4$, because $(3x^4)^2 = 9x^8$.

57. $\sqrt{81x^2} = 9x$ because $(9x)^2 = 81x^2$.

59. $\sqrt{a^2b^4} = ab^2$, because $(ab^2)^2 = a^2b^4$.

61. $\sqrt{16a^6b^4} = 4a^3b^2$, because
$(4a^3b^2)^2 = 16a^6b^4$

63. $\sqrt[3]{a^6b^{18}} = a^2b^6$, because $(a^2b^6)^3 = a^6b^{18}$.

65. $\sqrt[3]{-8x^3y^{27}} = -2xy^9$, because
$(-2xy^9)^3 = -8x^3y^{27}$

67. $50 = 25 \cdot 2$

69. $32 = 16 \cdot 2$ or
$32 = 4 \cdot 8$

71. $28 = 4 \cdot 7$

73. $27 = 9 \cdot 3$

75. a. $\sqrt[7]{-1}$ is a real number because the index is odd.

    b. $\sqrt[3]{-125}$ is a real number because the index is odd.

    c. $\sqrt[6]{-128}$ is not a real number because the index is even and the radicand is negative.

    d. $\sqrt[8]{-1}$ is not a real number because the index is even and the radicand is negative.

77. The length of the side is $\sqrt{49}$. Since $7^2 = 49$, $\sqrt{49} = 7$ and the sides of the square have length 7 miles.

79. The length of a side is $\sqrt{9.0601}$ inches. Since $(3.01)^2 = 9.0601$, $\sqrt{9.0601} = 3.01$. The length of a side is 3.01 inches.

81. $\sqrt{\sqrt{81}} = \sqrt{9} = 3$, since $3^2 = 9$ and $9^2 = 81$.

83. $\sqrt{\sqrt{10,000}} = 10$ since $10^2 = 100$ and $100^2 = 10,000$.

85. $T = 2\pi\sqrt{\dfrac{L}{g}} = 2\pi\sqrt{\dfrac{30}{32}} \approx 2(3.14)(0.968) \approx 6.1$

The period of the pendulum is 6.1 seconds.

87. answers may vary. Possible answer: There is no real number whose square is a negative number.

89.

x	$y = \sqrt{x}$
0	$\sqrt{0} = 0$
1	$\sqrt{1} = 1$
3	$\sqrt{3} \approx 1.7$
4	$\sqrt{4} = 2$
9	$\sqrt{9} = 3$

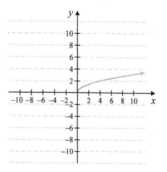

91. The graph of $y = \sqrt{x-2}$ 'starts' at $(2, 0)$.

93. The graph of $y = \sqrt{x+4}$ 'starts' at $(-4, 0)$.

Section 8.2 Practice Problems

1. $\sqrt{40} = \sqrt{4 \cdot 10} = \sqrt{4} \cdot \sqrt{10} = 2\sqrt{10}$

2. $\sqrt{18} = \sqrt{9 \cdot 2} = \sqrt{9} \cdot \sqrt{2} = 3\sqrt{2}$

3. $\sqrt{700} = \sqrt{100 \cdot 7} = \sqrt{100} \cdot \sqrt{7} = 10\sqrt{7}$

4. $\sqrt{15}$

   The radicand 15 contains no perfect square factors other than 1. Thus $\sqrt{15}$ is in simplest form.

5. $\begin{aligned} 7\sqrt{75} &= 7 \cdot \sqrt{25 \cdot 3} \\ &= 7 \cdot \sqrt{25} \cdot \sqrt{3} \\ &= 7 \cdot 5 \cdot \sqrt{3} \\ &= 35\sqrt{3} \end{aligned}$

6. $\sqrt{\dfrac{16}{81}} = \dfrac{\sqrt{16}}{\sqrt{81}} = \dfrac{4}{9}$

7. $\sqrt{\dfrac{2}{25}} = \dfrac{\sqrt{2}}{\sqrt{25}} = \dfrac{\sqrt{2}}{5}$

8. $\sqrt{\dfrac{45}{49}} = \dfrac{\sqrt{45}}{\sqrt{49}} = \dfrac{\sqrt{9} \cdot \sqrt{5}}{7} = \dfrac{3\sqrt{5}}{7}$

9. $\sqrt{x^{11}} = \sqrt{x^{10} \cdot x} = \sqrt{x^{10}} \cdot \sqrt{x} = x^5\sqrt{x}$

10. $\begin{aligned} \sqrt{18x^4} &= \sqrt{9 \cdot 2 \cdot x^4} \\ &= \sqrt{9x^4 \cdot 2} \\ &= \sqrt{9x^4} \cdot \sqrt{2} \\ &= 3x^2\sqrt{2} \end{aligned}$

11. $\sqrt{\dfrac{27}{x^8}} = \dfrac{\sqrt{27}}{\sqrt{x^8}} = \dfrac{\sqrt{9 \cdot 3}}{x^4} = \dfrac{\sqrt{9} \cdot \sqrt{3}}{x^4} = \dfrac{3\sqrt{3}}{x^4}$

12. $\begin{aligned} \sqrt{\dfrac{7y^7}{25}} &= \dfrac{\sqrt{7y^7}}{\sqrt{25}} \\ &= \dfrac{\sqrt{y^6 \cdot 7y}}{5} \\ &= \dfrac{\sqrt{y^6} \cdot \sqrt{7y}}{5} \\ &= \dfrac{y^3\sqrt{7y}}{5} \end{aligned}$

13. $\sqrt[3]{88} = \sqrt[3]{8 \cdot 11} = \sqrt[3]{8} \cdot \sqrt[3]{11} = 2\sqrt[3]{11}$

14. $\sqrt[3]{50}$

    The number 50 contains no perfect cube factors, so $\sqrt[3]{50}$ cannot be simplified further.

15. $\sqrt[3]{\dfrac{10}{27}} = \dfrac{\sqrt[3]{10}}{\sqrt[3]{27}} = \dfrac{\sqrt[3]{10}}{3}$

16. $\sqrt[3]{\dfrac{81}{8}} = \dfrac{\sqrt[3]{81}}{\sqrt[3]{8}} = \dfrac{\sqrt[3]{27 \cdot 3}}{2} = \dfrac{\sqrt[3]{27} \cdot \sqrt[3]{3}}{2} = \dfrac{3\sqrt[3]{3}}{2}$

Mental Math

1. $\sqrt{4 \cdot 9} = 2 \cdot 3 = 6$

2. $\sqrt{9 \cdot 36} = 3 \cdot 6 = 18$

3. $\sqrt{x^2} = x$

4. $\sqrt{y^4} = y^2$

5. $\sqrt{0} = 0$

6. $\sqrt{1} = 1$

7. $\sqrt{25x^4} = 5x^2$

8. $\sqrt{49x^2} = 7x$

Exercise Set 8.2

1. $\sqrt{20} = \sqrt{4 \cdot 5} = \sqrt{4}\sqrt{5} = 2\sqrt{5}$

3. $\sqrt{50} = \sqrt{25 \cdot 2} = \sqrt{25} \cdot \sqrt{2} = 5\sqrt{2}$

5. $\sqrt{33}$ is in simplest form.

7. $\sqrt{98} = \sqrt{49 \cdot 2} = \sqrt{49} \cdot \sqrt{2} = 7\sqrt{2}$

9. $\sqrt{60} = \sqrt{4 \cdot 15} = \sqrt{4} \cdot \sqrt{15} = 2\sqrt{15}$

11. $\sqrt{180} = \sqrt{36 \cdot 5} = \sqrt{36} \cdot \sqrt{5} = 6\sqrt{5}$

13. $\sqrt{52} = \sqrt{4 \cdot 13} = \sqrt{4} \cdot \sqrt{13} = 2\sqrt{13}$

15. $3\sqrt{25} = 3 \cdot 5 = 15$

17. $7\sqrt{63} = 7\sqrt{9 \cdot 7}$
$= 7 \cdot \sqrt{9} \cdot \sqrt{7}$
$= 7 \cdot 3 \cdot \sqrt{7}$
$= 21\sqrt{7}$

19. $-5\sqrt{27} = -5 \cdot \sqrt{9 \cdot 3}$
$= -5 \cdot \sqrt{9} \cdot \sqrt{3}$
$= -5 \cdot 3 \cdot \sqrt{3}$
$= -15\sqrt{3}$

21. $\sqrt{\dfrac{8}{25}} = \dfrac{\sqrt{8}}{\sqrt{25}} = \dfrac{\sqrt{4 \cdot 2}}{5} = \dfrac{\sqrt{4}\sqrt{2}}{5} = \dfrac{2\sqrt{2}}{5}$

23. $\sqrt{\dfrac{27}{121}} = \dfrac{\sqrt{27}}{\sqrt{121}} = \dfrac{\sqrt{9 \cdot 3}}{11} = \dfrac{\sqrt{9} \cdot \sqrt{3}}{11} = \dfrac{3\sqrt{3}}{11}$

25. $\sqrt{\dfrac{9}{4}} = \dfrac{\sqrt{9}}{\sqrt{4}} = \dfrac{3}{2}$

27. $\sqrt{\dfrac{125}{9}} = \dfrac{\sqrt{125}}{\sqrt{9}} = \dfrac{\sqrt{25 \cdot 5}}{3} = \dfrac{\sqrt{25} \cdot \sqrt{5}}{3} = \dfrac{5\sqrt{5}}{3}$

29. $\sqrt{\dfrac{11}{36}} = \dfrac{\sqrt{11}}{\sqrt{36}} = \dfrac{\sqrt{11}}{6}$

31. $-\sqrt{\dfrac{27}{144}} = -\dfrac{\sqrt{27}}{\sqrt{144}}$
$= -\dfrac{\sqrt{9 \cdot 3}}{12}$
$= -\dfrac{\sqrt{9} \cdot \sqrt{3}}{12}$
$= -\dfrac{3\sqrt{3}}{12}$
$= -\dfrac{\sqrt{3}}{4}$

33. $\sqrt{x^7} = \sqrt{x^6 \cdot x} = \sqrt{x^6}\sqrt{x} = x^3\sqrt{x}$

35. $\sqrt{x^{13}} = \sqrt{x^{12} \cdot x} = \sqrt{x^{12}} \cdot \sqrt{x} = x^6\sqrt{x}$

37. $\sqrt{36a^3} = \sqrt{36a^2 \cdot a} = \sqrt{36a^2}\sqrt{a} = 6a\sqrt{a}$

39. $\sqrt{96x^4} = \sqrt{16x^4 \cdot 6} = \sqrt{16x^4} \cdot \sqrt{6} = 4x^2\sqrt{6}$

41. $\sqrt{\dfrac{12}{m^2}} = \dfrac{\sqrt{12}}{\sqrt{m^2}} = \dfrac{\sqrt{4 \cdot 3}}{m} = \dfrac{\sqrt{4}\sqrt{3}}{m} = \dfrac{2\sqrt{3}}{m}$

43. $\sqrt{\dfrac{9x}{y^{10}}} = \dfrac{\sqrt{9x}}{\sqrt{y^{10}}} = \dfrac{\sqrt{9}\sqrt{x}}{y^5} = \dfrac{3\sqrt{x}}{y^5}$

45. $\sqrt{\dfrac{88}{x^{12}}} = \dfrac{\sqrt{88}}{\sqrt{x^{12}}} = \dfrac{\sqrt{4 \cdot 22}}{x^6} = \dfrac{\sqrt{4}\sqrt{22}}{x^6} = \dfrac{2\sqrt{22}}{x^6}$

47. $8\sqrt{4} = 8 \cdot 2 = 16$

49. $\sqrt{\dfrac{36}{121}} = \dfrac{\sqrt{36}}{\sqrt{121}} = \dfrac{6}{11}$

51. $\sqrt{175} = \sqrt{25 \cdot 7} = \sqrt{25} \cdot \sqrt{7} = 5\sqrt{7}$

53. $\sqrt{\dfrac{20}{9}} = \dfrac{\sqrt{20}}{\sqrt{9}} = \dfrac{\sqrt{4 \cdot 5}}{3} = \dfrac{\sqrt{4}\sqrt{5}}{3} = \dfrac{2\sqrt{5}}{3}$

55. $\sqrt{24m^7} = \sqrt{4m^6 \cdot 6m}$
$= \sqrt{4m^6} \cdot \sqrt{6m}$
$= 2m^3 \sqrt{6m}$

57. $\sqrt{\dfrac{23y^3}{4x^6}} = \dfrac{\sqrt{23y^3}}{\sqrt{4x^6}}$
$= \dfrac{\sqrt{y^2 \cdot 23y}}{2x^3}$
$= \dfrac{\sqrt{y^2}\sqrt{23y}}{2x^3}$
$= \dfrac{y\sqrt{23y}}{2x^3}$

59. $\sqrt[3]{24} = \sqrt[3]{8 \cdot 3} = \sqrt[3]{8} \cdot \sqrt[3]{3} = 2\sqrt[3]{3}$

61. $\sqrt[3]{250} = \sqrt[3]{125 \cdot 2} = \sqrt[3]{125} \cdot \sqrt[3]{2} = 5\sqrt[3]{2}$

63. $\sqrt[3]{\dfrac{5}{64}} = \dfrac{\sqrt[3]{5}}{\sqrt[3]{64}} = \dfrac{\sqrt[3]{5}}{4}$

65. $\sqrt[3]{\dfrac{23}{8}} = \dfrac{\sqrt[3]{23}}{\sqrt[3]{8}} = \dfrac{\sqrt[3]{23}}{2}$

67. $\sqrt[3]{\dfrac{15}{64}} = \dfrac{\sqrt[3]{15}}{\sqrt[3]{64}} = \dfrac{\sqrt[3]{15}}{4}$

69. $\sqrt[3]{80} = \sqrt[3]{8 \cdot 10} = \sqrt[3]{8} \cdot \sqrt[3]{10} = 2\sqrt[3]{10}$

71. $6x + 8x = 14x$

73. $(2x+3)(x-5) = 2x^2 - 10x + 3x - 15$
$= 2x^2 - 7x - 15$

75. $9y^2 - 9y^2 = 0$

77. $\sqrt{x^6 y^3} = \sqrt{x^6 y^2 y}$
$= \sqrt{x^6} \cdot \sqrt{y^2} \cdot \sqrt{y}$
$= x^3 y\sqrt{y}$

79. $\sqrt{98x^5 y^4} = \sqrt{49x^4 y^4 \cdot 2x}$
$= \sqrt{49x^4 y^4} \cdot \sqrt{2x}$
$= 7x^2 y^2 \sqrt{2x}$

81. $\sqrt[3]{-8x^6} = \sqrt[3]{-8} \cdot \sqrt[3]{x^6} = -2x^3$

83. $\sqrt[3]{80} = \sqrt[3]{8 \cdot 10} = \sqrt[3]{8} \cdot \sqrt[3]{10} = 2\sqrt[3]{10}$
Each side length is $2\sqrt[3]{10}$ inches.

85. answers may vary; possible answer: let $a = 1$ and $b = 1$ so $\sqrt{a^2 + b^2} = \sqrt{2} \neq a + b = 2$.

87. $\dfrac{\sqrt{6A}}{6} = \dfrac{\sqrt{6 \cdot 120}}{6}$
$= \dfrac{\sqrt{720}}{6}$
$= \dfrac{\sqrt{144 \cdot 5}}{6}$
$= \dfrac{\sqrt{144} \cdot \sqrt{5}}{6}$
$= \dfrac{12\sqrt{5}}{6}$
$= 2\sqrt{5}$
The length of a side is $2\sqrt{5}$ inches.

89. Use $\dfrac{\sqrt{6A}}{6}$ with $A = 30.375$.
$\dfrac{\sqrt{6 \cdot 30.375}}{6} = \dfrac{\sqrt{182.25}}{6} = \dfrac{13.5}{6} = 2.25$

91. $C = 100\sqrt[3]{n} + 700$
$= 100\sqrt[3]{1000} + 700$
$= 100 \cdot 10 + 700$
$= 1000 + 700$
$= 1700$
The cost is $1700.

93. $h = 169$ and $w = 64$.

$$B = \sqrt{\frac{hw}{3600}}$$

$$= \sqrt{\frac{169 \cdot 64}{3600}}$$

$$= \frac{\sqrt{169 \cdot 64}}{\sqrt{3600}}$$

$$= \frac{\sqrt{169}\sqrt{64}}{60}$$

$$= \frac{13 \cdot 8}{60}$$

$$= \frac{104}{60}$$

$$= \frac{26}{15} \approx 1.7$$

The body surface area is about 1.7 square meters.

Section 8.3 Practice Problems

1. $6\sqrt{11} + 9\sqrt{11} = (6+9)\sqrt{11} = 15\sqrt{11}$

2. $\sqrt{7} - 3\sqrt{7} = 1\sqrt{7} - 3\sqrt{7} = (1-3)\sqrt{7} = -2\sqrt{7}$

3. $\sqrt{2} + \sqrt{2} = 1\sqrt{2} + 1\sqrt{2} = (1+1)\sqrt{2} = 2\sqrt{2}$

4. $3\sqrt{3} - 3\sqrt{2}$ cannot be simplified further since the radicands are not the same.

5. $\sqrt{27} + \sqrt{75} = \sqrt{9 \cdot 3} + \sqrt{25 \cdot 3}$
$$= \sqrt{9} \cdot \sqrt{3} + \sqrt{25} \cdot \sqrt{3}$$
$$= 3\sqrt{3} + 5\sqrt{3}$$
$$= 8\sqrt{3}$$

6. $3\sqrt{20} - 7\sqrt{45} = 3\sqrt{4 \cdot 5} - 7\sqrt{9 \cdot 5}$
$$= 3\sqrt{4} \cdot \sqrt{5} - 7\sqrt{9} \cdot \sqrt{5}$$
$$= 3 \cdot 2\sqrt{5} - 7 \cdot 3\sqrt{5}$$
$$= 6\sqrt{5} - 21\sqrt{5}$$
$$= -15\sqrt{5}$$

7. $\sqrt{36} - \sqrt{48} - 4\sqrt{3} - \sqrt{9}$
$$= 6 - \sqrt{16 \cdot 3} - 4\sqrt{3} - 3$$
$$= 6 - \sqrt{16} \cdot \sqrt{3} - 4\sqrt{3} - 3$$
$$= 6 - 4\sqrt{3} - 4\sqrt{3} - 3$$
$$= 3 - 8\sqrt{3}$$

8. $\sqrt{9x^4} - \sqrt{36x^3} + \sqrt{x^3}$
$$= 3x^2 - \sqrt{36x^2 \cdot x} + \sqrt{x^2 \cdot x}$$
$$= 3x^2 - \sqrt{36x^2} \cdot \sqrt{x} + \sqrt{x^2} \cdot \sqrt{x}$$
$$= 3x^2 - 6x\sqrt{x} + x\sqrt{x}$$
$$= 3x^2 - 5x\sqrt{x}$$

9. $10\sqrt[3]{81p^6} - \sqrt[3]{24p^6}$
$$= 10\sqrt[3]{27p^6 \cdot 3} - \sqrt[3]{8p^6 \cdot 3}$$
$$= 10\sqrt[3]{27p^6} \cdot \sqrt[3]{3} - \sqrt[3]{8p^6} \cdot \sqrt[3]{3}$$
$$= 10 \cdot 3p^2\sqrt[3]{3} - 2p^2\sqrt[3]{3}$$
$$= 30p^2\sqrt[3]{3} - 2p^2\sqrt[3]{3}$$
$$= 28p^2\sqrt[3]{3}$$

Mental Math

1. $3\sqrt{2} + 5\sqrt{2} = (3+5)\sqrt{2} = 8\sqrt{2}$

2. $3\sqrt{5} + 7\sqrt{5} = (3+7)\sqrt{5} = 10\sqrt{5}$

3. $5\sqrt{x} + 2\sqrt{x} = (5+2)\sqrt{x} = 7\sqrt{x}$

4. $8\sqrt{x} + 3\sqrt{x} = (8+3)\sqrt{x} = 11\sqrt{x}$

5. $5\sqrt{7} - 2\sqrt{7} = (5-2)\sqrt{7} = 3\sqrt{7}$

6. $8\sqrt{6} - 5\sqrt{6} = (8-5)\sqrt{6} = 3\sqrt{6}$

Exercise Set 8.3

1. $4\sqrt{3} - 8\sqrt{3} = (4-8)\sqrt{3} = -4\sqrt{3}$

3. $3\sqrt{6} + 8\sqrt{6} - 2\sqrt{6} - 5 = (3+8-2)\sqrt{6} - 5$
$$= 9\sqrt{6} - 5$$

5. $6\sqrt{5} - 5\sqrt{5} + \sqrt{2} = (6-5)\sqrt{5} + \sqrt{2}$
$= \sqrt{5} + \sqrt{2}$

7. $2\sqrt{3} + 5\sqrt{3} - \sqrt{2} = (2+5)\sqrt{3} - \sqrt{2}$
$= 7\sqrt{3} - \sqrt{2}$

9. $2\sqrt{2} - 7\sqrt{2} - 6 = (2-7)\sqrt{2} - 6 = -5\sqrt{2} - 6$

11. $\sqrt{12} + \sqrt{27} = \sqrt{4\cdot3} + \sqrt{9\cdot3}$
$= \sqrt{4}\sqrt{3} + \sqrt{9}\sqrt{3}$
$= 2\sqrt{3} + 3\sqrt{3}$
$= (2+3)\sqrt{3}$
$= 5\sqrt{3}$

13. $\sqrt{45} + 3\sqrt{20} = \sqrt{9\cdot5} + 3\sqrt{4\cdot5}$
$= \sqrt{9}\sqrt{5} + 3\sqrt{4}\sqrt{5}$
$= 3\sqrt{5} + 3\cdot2\sqrt{5}$
$= 3\sqrt{5} + 6\sqrt{5}$
$= (3+6)\sqrt{5}$
$= 9\sqrt{5}$

15. $2\sqrt{54} - \sqrt{20} + \sqrt{45} - \sqrt{24}$
$= 2\sqrt{9\cdot6} - \sqrt{4\cdot5} + \sqrt{9\cdot5} - \sqrt{4\cdot6}$
$= 2\sqrt{9}\sqrt{6} - \sqrt{4}\sqrt{5} + \sqrt{9}\sqrt{5} - \sqrt{4}\sqrt{6}$
$= 2\cdot3\sqrt{6} - 2\sqrt{5} + 3\sqrt{5} - 2\sqrt{6}$
$= 6\sqrt{6} - 2\sqrt{5} + 3\sqrt{5} - 2\sqrt{6}$
$= (6-2)\sqrt{6} + (3-2)\sqrt{5}$
$= 4\sqrt{6} + 1\sqrt{5}$
$= 4\sqrt{6} + \sqrt{5}$

17. $4x - 3\sqrt{x^2} + \sqrt{x} = 4x - 3x + \sqrt{x} = x + \sqrt{x}$

19. $\sqrt{25x} + \sqrt{36x} - 11\sqrt{x}$
$= \sqrt{25}\sqrt{x} + \sqrt{36}\sqrt{x} - 11\sqrt{x}$
$= 5\sqrt{x} + 6\sqrt{x} - 11\sqrt{x}$
$= (5+6-11)\sqrt{x}$
$= 0$

21. $\sqrt{\dfrac{5}{9}} + \sqrt{\dfrac{5}{81}} = \dfrac{\sqrt{5}}{\sqrt{9}} + \dfrac{\sqrt{5}}{\sqrt{81}}$
$= \dfrac{\sqrt{5}}{3} + \dfrac{\sqrt{5}}{9}$
$= \dfrac{3\sqrt{5}}{9} + \dfrac{\sqrt{5}}{9}$
$= \left(\dfrac{3}{9} + \dfrac{1}{9}\right)\sqrt{5}$
$= \dfrac{4}{9}\sqrt{5}$
$= \dfrac{4\sqrt{5}}{9}$

23. $\sqrt{\dfrac{3}{4}} - \sqrt{\dfrac{3}{64}} = \dfrac{\sqrt{3}}{\sqrt{4}} - \dfrac{\sqrt{3}}{\sqrt{64}}$
$= \dfrac{\sqrt{3}}{2} - \dfrac{\sqrt{3}}{8}$
$= \dfrac{4\sqrt{3}}{8} - \dfrac{\sqrt{3}}{8}$
$= \left(\dfrac{4}{8} - \dfrac{1}{8}\right)\sqrt{3}$
$= \dfrac{3}{8}\sqrt{3}$
$= \dfrac{3\sqrt{3}}{8}$

25. $12\sqrt{5} - \sqrt{5} - 4\sqrt{5} = (12-1-4)\sqrt{5} = 7\sqrt{5}$

27. $\sqrt{75} + \sqrt{48} = \sqrt{25\cdot3} + \sqrt{16\cdot3}$
$= \sqrt{25}\sqrt{3} + \sqrt{16}\sqrt{3}$
$= 5\sqrt{3} + 4\sqrt{3}$
$= (5+4)\sqrt{3}$
$= 9\sqrt{3}$

29. $\sqrt{5} + \sqrt{15}$ is in simplest form.

31. $3\sqrt{x^3} - x\sqrt{4x} = 3\sqrt{x^2 \cdot x} - x\sqrt{4 \cdot x}$
$$= 3\sqrt{x^2}\sqrt{x} - x\sqrt{4}\sqrt{x}$$
$$= 3x\sqrt{x} - 2x\sqrt{x}$$
$$= (3x - 2x)\sqrt{x}$$
$$= x\sqrt{x}$$

33. $\sqrt{8} + \sqrt{9} + \sqrt{18} + \sqrt{81}$
$$= \sqrt{4 \cdot 2} + 3 + \sqrt{9 \cdot 2} + 9$$
$$= \sqrt{4}\sqrt{2} + 3 + \sqrt{9}\sqrt{2} + 9$$
$$= 2\sqrt{2} + 3 + 3\sqrt{2} + 9$$
$$= (2 + 3)\sqrt{2} + 3 + 9$$
$$= 5\sqrt{2} + 12$$

35. $4 + 8\sqrt{2} - 9 = 8\sqrt{2} - 5$

37. $2\sqrt{45} - 2\sqrt{20} = 2\sqrt{9 \cdot 5} - 2\sqrt{4 \cdot 5}$
$$= 2\sqrt{9}\sqrt{5} - 2\sqrt{4}\sqrt{5}$$
$$= 2 \cdot 3\sqrt{5} - 2 \cdot 2\sqrt{5}$$
$$= 6\sqrt{5} - 4\sqrt{5}$$
$$= (6 - 4)\sqrt{5}$$
$$= 2\sqrt{5}$$

39. $\sqrt{35} - \sqrt{140} = \sqrt{35} - \sqrt{4 \cdot 35}$
$$= \sqrt{35} - \sqrt{4}\sqrt{35}$$
$$= \sqrt{35} - 2\sqrt{35}$$
$$= (1 - 2)\sqrt{35}$$
$$= -1\sqrt{35}$$
$$= -\sqrt{35}$$

41. $6 - 2\sqrt{3} - \sqrt{3} = 6 + (-2 - 1)\sqrt{3} = 6 - 3\sqrt{3}$

43. $3\sqrt{9x} + 2\sqrt{x} = 3\sqrt{9}\sqrt{x} + 2\sqrt{x}$
$$= 3 \cdot 3\sqrt{x} + 2\sqrt{x}$$
$$= 9\sqrt{x} + 2\sqrt{x}$$
$$= (9 + 2)\sqrt{x}$$
$$= 11\sqrt{x}$$

45. $\sqrt{9x^2} + \sqrt{81x^2} - 11\sqrt{x} = 3x + 9x - 11\sqrt{x}$
$$= 12x - 11\sqrt{x}$$

47. $\sqrt{3x^3} + 3x\sqrt{x} = \sqrt{x^2 \cdot 3x} + 3x\sqrt{x}$
$$= \sqrt{x^2}\sqrt{3x} + 3x\sqrt{x}$$
$$= x\sqrt{3x} + 3x\sqrt{x}$$

49. $\sqrt{32x^2} + \sqrt{32x^2} + \sqrt{4x^2}$
$$= \sqrt{16x^2 \cdot 2} + \sqrt{16x^2 \cdot 2} + 2x$$
$$= \sqrt{16x^2}\sqrt{2} + \sqrt{16x^2}\sqrt{2} + 2x$$
$$= 4x\sqrt{2} + 4x\sqrt{2} + 2x$$
$$= (4x + 4x)\sqrt{2} + 2x$$
$$= 8x\sqrt{2} + 2x$$

51. $\sqrt{40x} + \sqrt{40x^4} - 2\sqrt{10x} - \sqrt{5x^4}$
$$= \sqrt{4 \cdot 10x} + \sqrt{4x^4 \cdot 10} - 2\sqrt{10x} - \sqrt{x^4 \cdot 5}$$
$$= \sqrt{4}\sqrt{10x} + \sqrt{4x^4}\sqrt{10} - 2\sqrt{10x} - \sqrt{x^4}\sqrt{5}$$
$$= 2\sqrt{10x} + 2x^2\sqrt{10} - 2\sqrt{10x} - x^2\sqrt{5}$$
$$= (2 - 2)\sqrt{10x} + 2x^2\sqrt{10} - x^2\sqrt{5}$$
$$= 0\sqrt{10x} + 2x^2\sqrt{10} - x^2\sqrt{5}$$
$$= 2x^2\sqrt{10} - x^2\sqrt{5}$$

53. $2\sqrt[3]{9} + 5\sqrt[3]{9} - \sqrt[3]{25} = (2 + 5)\sqrt[3]{9} - \sqrt[3]{25}$
$$= 7\sqrt[3]{9} - \sqrt[3]{25}$$

55. $2\sqrt[3]{2} - 7\sqrt[3]{2} - 6 = (2 - 7)\sqrt[3]{2} - 6 = -5\sqrt[3]{2} - 6$

57. $\sqrt[3]{81} + \sqrt[3]{24} = \sqrt[3]{27 \cdot 3} + \sqrt[3]{8 \cdot 3}$
$$= \sqrt[3]{27}\sqrt[3]{3} + \sqrt[3]{8}\sqrt[3]{3}$$
$$= 3\sqrt[3]{3} + 2\sqrt[3]{3}$$
$$= (3 + 2)\sqrt[3]{3}$$
$$= 5\sqrt[3]{3}$$

59. $2\sqrt[3]{8x^3} + 2\sqrt[3]{16x^3} = 2 \cdot 2x + 2\sqrt[3]{8x^3 \cdot 2}$
$$= 4x + 2\sqrt[3]{8x^3}\sqrt[3]{2}$$
$$= 4x + 2 \cdot 2x\sqrt[3]{2}$$
$$= 4x + 4x\sqrt[3]{2}$$

61. $\sqrt{40x} + x\sqrt[3]{40} - 2\sqrt{10x} - x\sqrt[3]{5}$

$= \sqrt{4 \cdot 10x} + x\sqrt[3]{8 \cdot 5} - 2\sqrt{10x} - x\sqrt[3]{5}$

$= \sqrt{4}\sqrt{10x} + x\sqrt[3]{8}\sqrt[3]{5} - 2\sqrt{10x} - x\sqrt[3]{5}$

$= 2\sqrt{10x} + 2x\sqrt[3]{5} - 2\sqrt{10x} - x\sqrt[3]{5}$

$= (2-2)\sqrt{10x} + (2x-x)\sqrt[3]{5}$

$= x\sqrt[3]{5}$

63. $12\sqrt[3]{y^7} - y^2\sqrt[3]{8y} = 12\sqrt[3]{y^6 \cdot y} - y^2\sqrt[3]{8}\sqrt[3]{y}$

$= 12\sqrt[3]{y^6}\sqrt[3]{y} - 2y^2\sqrt[3]{y}$

$= 12y^2\sqrt[3]{y} - 2y^2\sqrt[3]{y}$

$= (12y^2 - 2y^2)\sqrt[3]{y}$

$= 10y^2\sqrt[3]{y}$

65. $(x+6)^2 = (x)^2 + 2(x)(6) + (6)^2$

$= x^2 + 12x + 36$

67. $(2x-1)^2 = (2x)^2 - 2(2x)(1) + (1)^2$

$= 4x^2 - 4x + 1$

69. answers may vary

71. $P = 2l + 2w$

$= 2 \cdot 3\sqrt{5} + 2 \cdot \sqrt{5}$

$= 6\sqrt{5} + 2\sqrt{5}$

$= (6+2)\sqrt{5}$

$= 8\sqrt{5}$

The perimeter is $8\sqrt{5}$ inches.

73. Two triangular end pieces and two rectangular side panels are needed. Each side panel has area $8 \cdot 3 = 24$ square feet.

$2 \cdot 24 + 2 \cdot \dfrac{3\sqrt{27}}{4} = 48 + \dfrac{3\sqrt{9 \cdot 3}}{2}$

$= 48 + \dfrac{3\sqrt{9}\sqrt{3}}{2}$

$= 48 + \dfrac{3 \cdot 3\sqrt{3}}{2}$

$= 48 + \dfrac{9\sqrt{3}}{2}$

The total area of wood needed is

$\left(48 + \dfrac{9\sqrt{3}}{2}\right)$ square feet.

75. The expression can be simplified.

$4\sqrt{2} + 3\sqrt{2} = (4+3)\sqrt{2} = 7\sqrt{2}$

77. The expression $6 + 7\sqrt{6}$ cannot be simplified.

79. The expression can be simplified.

$\sqrt{7} + \sqrt{7} + \sqrt{7} = (1+1+1)\sqrt{7} = 3\sqrt{7}$

81. $\sqrt{\dfrac{x^3}{16}} - x\sqrt{\dfrac{9x}{25}} + \dfrac{\sqrt{81x^3}}{2}$

$= \dfrac{\sqrt{x^3}}{\sqrt{16}} - x\dfrac{\sqrt{9x}}{\sqrt{25}} + \dfrac{\sqrt{81x^3}}{2}$

$= \dfrac{\sqrt{x^2 \cdot x}}{4} - x\dfrac{\sqrt{9 \cdot x}}{5} + \dfrac{\sqrt{81x^2 \cdot x}}{2}$

$= \dfrac{\sqrt{x^2}\sqrt{x}}{4} - x\dfrac{\sqrt{9}\sqrt{x}}{5} + \dfrac{\sqrt{81x^2}\sqrt{x}}{2}$

$= \dfrac{x\sqrt{x}}{4} - \dfrac{3x\sqrt{x}}{5} + \dfrac{9x\sqrt{x}}{2}$

$= \left(\dfrac{1}{4} - \dfrac{3}{5} + \dfrac{9}{2}\right)x\sqrt{x}$

$= \left(\dfrac{5}{20} - \dfrac{12}{20} + \dfrac{90}{20}\right)x\sqrt{x}$

$= \dfrac{83}{20}x\sqrt{x}$

$= \dfrac{83x\sqrt{x}}{20}$

Section 8.4 Practice Problems

1. $\sqrt{5} \cdot \sqrt{2} = \sqrt{5 \cdot 2} = \sqrt{10}$

2. $\sqrt{7} \cdot \sqrt{7} = \sqrt{7 \cdot 7} = \sqrt{49} = 7$

3. $\sqrt{6} \cdot \sqrt{3} = \sqrt{18} = \sqrt{9 \cdot 2} = \sqrt{9} \cdot \sqrt{2} = 3\sqrt{2}$

4. $\sqrt{10x} \cdot \sqrt{2x} = \sqrt{10x \cdot 2x}$

$\quad\quad\quad\quad\quad = \sqrt{20x^2}$

$\quad\quad\quad\quad\quad = \sqrt{4x^2 \cdot 5}$

$\quad\quad\quad\quad\quad = \sqrt{4x^2} \cdot \sqrt{5}$

$\quad\quad\quad\quad\quad = 2x\sqrt{5}$

5. a. $\sqrt{7}\left(\sqrt{7} - \sqrt{3}\right) = \sqrt{7} \cdot \sqrt{7} - \sqrt{7} \cdot \sqrt{3}$

$\quad\quad\quad\quad\quad\quad\quad = 7 - \sqrt{21}$

b. $\sqrt{5x}\left(\sqrt{x} - 3\sqrt{5}\right)$

$\quad = \sqrt{5x} \cdot \sqrt{x} - \sqrt{5x} \cdot 3\sqrt{5}$

$\quad = \sqrt{5x \cdot x} - 3\sqrt{5x \cdot 5}$

$\quad = \sqrt{5 \cdot x^2} - 3\sqrt{25 \cdot x}$

$\quad = \sqrt{5} \cdot \sqrt{x^2} - 3 \cdot \sqrt{25} \cdot \sqrt{x}$

$\quad = x\sqrt{5} - 3 \cdot 5 \cdot \sqrt{x}$

$\quad = x\sqrt{5} - 15\sqrt{x}$

c. $\left(\sqrt{x} + \sqrt{5}\right)\left(\sqrt{x} - \sqrt{3}\right)$

$\quad = \sqrt{x} \cdot \sqrt{x} - \sqrt{x} \cdot \sqrt{3} + \sqrt{5} \cdot \sqrt{x} - \sqrt{5} \cdot \sqrt{3}$

$\quad = x - \sqrt{3x} + \sqrt{5x} - \sqrt{15}$

6. a. $\left(\sqrt{3} + 6\right)\left(\sqrt{3} - 6\right) = \left(\sqrt{3}\right)^2 - 6^2$

$\quad\quad\quad\quad\quad\quad\quad\quad = 3 - 36$

$\quad\quad\quad\quad\quad\quad\quad\quad = -33$

b. $\left(\sqrt{5x} + 4\right)^2$

$\quad = \left(\sqrt{5x}\right)^2 + 2\left(\sqrt{5x}\right)(4) + (4)^2$

$\quad = 5x + 8\sqrt{5x} + 16$

7. $\dfrac{\sqrt{15}}{\sqrt{3}} = \sqrt{\dfrac{15}{3}} = \sqrt{5}$

8. $\dfrac{\sqrt{90}}{\sqrt{2}} = \sqrt{\dfrac{90}{2}} = \sqrt{45} = \sqrt{9 \cdot 5} = \sqrt{9} \cdot \sqrt{5} = 3\sqrt{5}$

9. $\dfrac{\sqrt{75x^3}}{\sqrt{5x}} = \sqrt{\dfrac{75x^3}{5x}}$

$\quad\quad\quad\quad = \sqrt{15x^2}$

$\quad\quad\quad\quad = \sqrt{15}\sqrt{x^2}$

$\quad\quad\quad\quad = x\sqrt{15}$

10. $\dfrac{5}{\sqrt{3}} = \dfrac{5}{\sqrt{3}} \cdot \dfrac{\sqrt{3}}{\sqrt{3}} = \dfrac{5 \cdot \sqrt{3}}{\sqrt{3} \cdot \sqrt{3}} = \dfrac{5\sqrt{3}}{3}$

11. $\dfrac{\sqrt{7}}{\sqrt{20}} = \dfrac{\sqrt{7}}{\sqrt{4 \cdot 5}}$

$\quad\quad\quad = \dfrac{\sqrt{7}}{2\sqrt{5}} \cdot \dfrac{\sqrt{5}}{\sqrt{5}}$

$\quad\quad\quad = \dfrac{\sqrt{7} \cdot \sqrt{5}}{2\sqrt{5} \cdot \sqrt{5}}$

$\quad\quad\quad = \dfrac{\sqrt{35}}{2 \cdot 5}$

$\quad\quad\quad = \dfrac{\sqrt{35}}{10}$

12. $\sqrt{\dfrac{2}{45x}} = \dfrac{\sqrt{2}}{\sqrt{45x}}$

$\quad\quad\quad = \dfrac{\sqrt{2}}{\sqrt{9} \cdot \sqrt{5x}}$

$\quad\quad\quad = \dfrac{\sqrt{2}}{3\sqrt{5x}} \cdot \dfrac{\sqrt{5x}}{\sqrt{5x}}$

$\quad\quad\quad = \dfrac{\sqrt{2} \cdot \sqrt{5x}}{3\sqrt{5x} \cdot \sqrt{5x}}$

$\quad\quad\quad = \dfrac{\sqrt{10x}}{3 \cdot 5x}$

$\quad\quad\quad = \dfrac{\sqrt{10x}}{15x}$

13. $\dfrac{3}{2+\sqrt{7}} = \dfrac{3\left(2-\sqrt{7}\right)}{\left(2+\sqrt{7}\right)\left(2-\sqrt{7}\right)}$

$\phantom{\dfrac{3}{2+\sqrt{7}}} = \dfrac{3\left(2-\sqrt{7}\right)}{2^{2}-\left(\sqrt{7}\right)^{2}}$

$\phantom{\dfrac{3}{2+\sqrt{7}}} = \dfrac{3\left(2-\sqrt{7}\right)}{4-7}$

$\phantom{\dfrac{3}{2+\sqrt{7}}} = \dfrac{3\left(2-\sqrt{7}\right)}{-3}$

$\phantom{\dfrac{3}{2+\sqrt{7}}} = -\dfrac{3\left(2-\sqrt{7}\right)}{3}$

$\phantom{\dfrac{3}{2+\sqrt{7}}} = -1\left(2-\sqrt{7}\right)$

$\phantom{\dfrac{3}{2+\sqrt{7}}} = -2+\sqrt{7}$

14. $\dfrac{\sqrt{2}+5}{\sqrt{2}-1} = \dfrac{\left(\sqrt{2}+5\right)\left(\sqrt{2}+1\right)}{\left(\sqrt{2}-1\right)\left(\sqrt{2}+1\right)}$

$\phantom{\dfrac{\sqrt{2}+5}{\sqrt{2}-1}} = \dfrac{2+\sqrt{2}+5\sqrt{2}+5}{2-1}$

$\phantom{\dfrac{\sqrt{2}+5}{\sqrt{2}-1}} = \dfrac{7+6\sqrt{2}}{1}$

$\phantom{\dfrac{\sqrt{2}+5}{\sqrt{2}-1}} = 7+6\sqrt{2}$

15. $\dfrac{7}{2-\sqrt{x}} = \dfrac{7\left(2+\sqrt{x}\right)}{\left(2-\sqrt{x}\right)\left(2+\sqrt{x}\right)} = \dfrac{7\left(2+\sqrt{x}\right)}{4-x}$

Mental Math

1. $\sqrt{2}\cdot\sqrt{11} = \sqrt{2\cdot11} = \sqrt{22}$

2. $\sqrt{5}\cdot\sqrt{7} = \sqrt{5\cdot7} = \sqrt{35}$

3. $\sqrt{1}\cdot\sqrt{6} = \sqrt{1\cdot6} = \sqrt{6}$

4. $\sqrt{7}\cdot\sqrt{x} = \sqrt{7\cdot x} = \sqrt{7x}$

5. $\sqrt{10}\cdot\sqrt{y} = \sqrt{10\cdot y} = \sqrt{10y}$

6. $\sqrt{x}\cdot\sqrt{y} = \sqrt{x\cdot y} = \sqrt{xy}$

Exercise Set 8.4

1. $\sqrt{8}\cdot\sqrt{2} = \sqrt{16} = 4$

3. $\sqrt{10}\cdot\sqrt{5} = \sqrt{50} = \sqrt{25\cdot2} = \sqrt{25}\sqrt{2} = 5\sqrt{2}$

5. $\left(\sqrt{6}\right)^{2} = 6$

7. $\sqrt{2x}\cdot\sqrt{2x} = \left(\sqrt{2x}\right)^{2} = 2x$

9. $\left(2\sqrt{5}\right)^{2} = \left(2\sqrt{5}\right)\left(2\sqrt{5}\right)$

$\phantom{\left(2\sqrt{5}\right)^{2}} = 4\left(\sqrt{5}\right)^{2}$

$\phantom{\left(2\sqrt{5}\right)^{2}} = 4\cdot5$

$\phantom{\left(2\sqrt{5}\right)^{2}} = 20$

11. $\left(6\sqrt{x}\right)^{2} = \left(6\sqrt{x}\right)\left(6\sqrt{x}\right) = 36\left(\sqrt{x}\right)^{2} = 36x$

13. $\sqrt{3x^{5}}\cdot\sqrt{6x} = \sqrt{3x^{5}\cdot6x}$

$\phantom{\sqrt{3x^{5}}\cdot\sqrt{6x}} = \sqrt{18x^{6}}$

$\phantom{\sqrt{3x^{5}}\cdot\sqrt{6x}} = \sqrt{9x^{6}\cdot2}$

$\phantom{\sqrt{3x^{5}}\cdot\sqrt{6x}} = \sqrt{9x^{6}}\sqrt{2}$

$\phantom{\sqrt{3x^{5}}\cdot\sqrt{6x}} = 3x^{3}\sqrt{2}$

15. $\sqrt{2xy^{2}}\cdot\sqrt{8xy} = \sqrt{2xy^{2}\cdot8xy}$

$\phantom{\sqrt{2xy^{2}}\cdot\sqrt{8xy}} = \sqrt{16x^{2}y^{3}}$

$\phantom{\sqrt{2xy^{2}}\cdot\sqrt{8xy}} = \sqrt{16x^{2}y^{2}\cdot y}$

$\phantom{\sqrt{2xy^{2}}\cdot\sqrt{8xy}} = \sqrt{16x^{2}y^{2}}\sqrt{y}$

$\phantom{\sqrt{2xy^{2}}\cdot\sqrt{8xy}} = 4xy\sqrt{y}$

17. $\sqrt{6}\left(\sqrt{5}+\sqrt{7}\right) = \sqrt{6}\cdot\sqrt{5}+\sqrt{6}\cdot\sqrt{7}$

$\phantom{\sqrt{6}\left(\sqrt{5}+\sqrt{7}\right)} = \sqrt{30}+\sqrt{42}$

19. $\sqrt{10}\left(\sqrt{2}+\sqrt{5}\right) = \sqrt{10}\cdot\sqrt{2}+\sqrt{10}\cdot\sqrt{5}$
$$= \sqrt{20}+\sqrt{50}$$
$$= \sqrt{4\cdot5}+\sqrt{25\cdot2}$$
$$= \sqrt{4}\sqrt{5}+\sqrt{25}\sqrt{2}$$
$$= 2\sqrt{5}+5\sqrt{2}$$

21. $\sqrt{7y}\left(\sqrt{y}-2\sqrt{7}\right) = \sqrt{7y}\cdot\sqrt{y}-\sqrt{7y}\cdot2\sqrt{7}$
$$= \sqrt{7y\cdot y}-2\sqrt{7y\cdot7}$$
$$= \sqrt{7y^2}-2\sqrt{49y}$$
$$= \sqrt{y^2\cdot7}-2\sqrt{49\cdot y}$$
$$= \sqrt{y^2}\sqrt{7}-2\sqrt{49}\sqrt{y}$$
$$= y\sqrt{7}-2\cdot7\sqrt{y}$$
$$= y\sqrt{7}-14\sqrt{y}$$

23. $\left(\sqrt{3}+6\right)\left(\sqrt{3}-6\right) = \left(\sqrt{3}\right)^2-6^2$
$$= 3-36$$
$$= -33$$

25. $\left(\sqrt{3}+\sqrt{5}\right)\left(\sqrt{2}-\sqrt{5}\right)$
$$= \sqrt{3}\cdot\sqrt{2}-\sqrt{3}\sqrt{5}+\sqrt{5}\cdot\sqrt{2}-\sqrt{5}\cdot\sqrt{5}$$
$$= \sqrt{6}-\sqrt{15}+\sqrt{10}-\sqrt{25}$$
$$= \sqrt{6}-\sqrt{15}+\sqrt{10}-5$$

27. $\left(2\sqrt{11}+1\right)\left(\sqrt{11}-6\right)$
$$= 2\sqrt{11}\cdot\sqrt{11}-2\sqrt{11}\cdot6+1\cdot\sqrt{11}-1\cdot6$$
$$= 2\cdot11-12\sqrt{11}+\sqrt{11}-6$$
$$= 22-11\sqrt{11}-6$$
$$= 16-11\sqrt{11}$$

29. $\left(\sqrt{x}+6\right)\left(\sqrt{x}-6\right) = \left(\sqrt{x}\right)^2-(6)^2 = x-36$

31. $\left(\sqrt{x}-7\right)^2 = \left(\sqrt{x}\right)^2-2\left(\sqrt{x}\right)(7)+(7)^2$
$$= x-14\sqrt{x}+49$$

33. $\left(\sqrt{6y}+1\right)^2 = \left(\sqrt{6y}\right)^2+2\left(\sqrt{6y}\right)(1)+(1)^2$
$$= 6y+2\sqrt{6y}+1$$

35. $\dfrac{\sqrt{32}}{\sqrt{2}} = \sqrt{\dfrac{32}{2}} = \sqrt{16} = 4$

37. $\dfrac{\sqrt{21}}{\sqrt{3}} = \sqrt{\dfrac{21}{3}} = \sqrt{7}$

39. $\dfrac{\sqrt{90}}{\sqrt{5}} = \sqrt{\dfrac{90}{5}} = \sqrt{18} = \sqrt{9\cdot2} = \sqrt{9}\sqrt{2} = 3\sqrt{2}$

41. $\dfrac{\sqrt{75y^5}}{\sqrt{3y}} = \sqrt{\dfrac{75y^5}{3y}} = \sqrt{25y^4} = 5y^2$

43. $\dfrac{\sqrt{150}}{\sqrt{2}} = \sqrt{\dfrac{150}{2}}$
$$= \sqrt{75}$$
$$= \sqrt{25\cdot3}$$
$$= \sqrt{25}\sqrt{3}$$
$$= 5\sqrt{3}$$

45. $\dfrac{\sqrt{72y^5}}{\sqrt{3y^3}} = \sqrt{\dfrac{72y^5}{3y^3}}$
$$= \sqrt{24y^2}$$
$$= \sqrt{4y^2\cdot6}$$
$$= \sqrt{4y^2}\sqrt{6}$$
$$= 2y\sqrt{6}$$

47. $\dfrac{\sqrt{24x^3y^4}}{\sqrt{2xy}} = \sqrt{\dfrac{24x^3y^4}{2xy}}$
$$= \sqrt{12x^2y^3}$$
$$= \sqrt{4x^2y^2\cdot3y}$$
$$= \sqrt{4x^2y^2}\sqrt{3y}$$
$$= 2xy\sqrt{3y}$$

49. $\dfrac{\sqrt{3}}{\sqrt{5}} = \dfrac{\sqrt{3}}{\sqrt{5}} \cdot \dfrac{\sqrt{5}}{\sqrt{5}} = \dfrac{\sqrt{15}}{5}$

51. $\dfrac{7}{\sqrt{2}} = \dfrac{7}{\sqrt{2}} \cdot \dfrac{\sqrt{2}}{\sqrt{2}} = \dfrac{7\sqrt{2}}{2}$

53. $\dfrac{1}{\sqrt{6y}} = \dfrac{1}{\sqrt{6y}} \cdot \dfrac{\sqrt{6y}}{\sqrt{6y}} = \dfrac{\sqrt{6y}}{6y}$

55. $\sqrt{\dfrac{5}{18}} = \dfrac{\sqrt{5}}{\sqrt{18}}$

$= \dfrac{\sqrt{5}}{\sqrt{9 \cdot 2}}$

$= \dfrac{\sqrt{5}}{3\sqrt{2}} \cdot \dfrac{\sqrt{2}}{\sqrt{2}}$

$= \dfrac{\sqrt{5}\sqrt{2}}{3\sqrt{2} \cdot \sqrt{2}}$

$= \dfrac{\sqrt{10}}{3 \cdot 2}$

$= \dfrac{\sqrt{10}}{6}$

57. $\sqrt{\dfrac{3}{x}} = \dfrac{\sqrt{3}}{\sqrt{x}} = \dfrac{\sqrt{3}}{\sqrt{x}} \cdot \dfrac{\sqrt{x}}{\sqrt{x}} = \dfrac{\sqrt{3x}}{x}$

59. $\sqrt{\dfrac{1}{8}} = \dfrac{\sqrt{1}}{\sqrt{8}}$

$= \dfrac{1}{\sqrt{4 \cdot 2}}$

$= \dfrac{1}{2\sqrt{2}}$

$= \dfrac{1}{2\sqrt{2}} \cdot \dfrac{\sqrt{2}}{\sqrt{2}}$

$= \dfrac{\sqrt{2}}{2 \cdot 2}$

$= \dfrac{\sqrt{2}}{4}$

61. $\sqrt{\dfrac{2}{15}} = \dfrac{\sqrt{2}}{\sqrt{15}} = \dfrac{\sqrt{2}}{\sqrt{15}} \cdot \dfrac{\sqrt{15}}{\sqrt{15}} = \dfrac{\sqrt{30}}{15}$

63. $\sqrt{\dfrac{3}{20}} = \dfrac{\sqrt{3}}{\sqrt{20}}$

$= \dfrac{\sqrt{3}}{\sqrt{4 \cdot 5}}$

$= \dfrac{\sqrt{3}}{2\sqrt{5}}$

$= \dfrac{\sqrt{3}}{2\sqrt{5}} \cdot \dfrac{\sqrt{5}}{\sqrt{5}}$

$= \dfrac{\sqrt{15}}{2 \cdot 5}$

$= \dfrac{\sqrt{15}}{10}$

65. $\dfrac{3x}{\sqrt{2x}} = \dfrac{3x}{\sqrt{2x}} \cdot \dfrac{\sqrt{2x}}{\sqrt{2x}} = \dfrac{3x\sqrt{2x}}{2x} = \dfrac{3\sqrt{2x}}{2}$

67. $\dfrac{8y}{\sqrt{5}} = \dfrac{8y}{\sqrt{5}} \cdot \dfrac{\sqrt{5}}{\sqrt{5}} = \dfrac{8y\sqrt{5}}{5}$

69. $\sqrt{\dfrac{y}{12x}} = \dfrac{\sqrt{y}}{\sqrt{12x}}$

$= \dfrac{\sqrt{y}}{\sqrt{4}\sqrt{3x}}$

$= \dfrac{\sqrt{y}}{2\sqrt{3x}}$

$= \dfrac{\sqrt{y}}{2\sqrt{3x}} \cdot \dfrac{\sqrt{3x}}{\sqrt{3x}}$

$= \dfrac{\sqrt{3xy}}{2 \cdot 3x}$

$= \dfrac{\sqrt{3xy}}{6x}$

71.
$$\frac{3}{\sqrt{2}+1} = \frac{3}{\sqrt{2}+1} \cdot \frac{\sqrt{2}-1}{\sqrt{2}-1}$$
$$= \frac{3\left(\sqrt{2}-1\right)}{\left(\sqrt{2}\right)^2 - 1^2}$$
$$= \frac{3\sqrt{2}-3}{2-1}$$
$$= \frac{3\sqrt{2}-3}{1}$$
$$= 3\sqrt{2}-3$$

73.
$$\frac{4}{2-\sqrt{5}} = \frac{4}{2-\sqrt{5}} \cdot \frac{2+\sqrt{5}}{2+\sqrt{5}}$$
$$= \frac{4\left(2+\sqrt{5}\right)}{2^2 - \left(\sqrt{5}\right)^2}$$
$$= \frac{8+4\sqrt{5}}{4-5}$$
$$= \frac{8+4\sqrt{5}}{-1}$$
$$= -8-4\sqrt{5}$$

75.
$$\frac{\sqrt{5}+1}{\sqrt{6}-\sqrt{5}} = \frac{\sqrt{5}+1}{\sqrt{6}-\sqrt{5}} \cdot \frac{\sqrt{6}+\sqrt{5}}{\sqrt{6}+\sqrt{5}}$$
$$= \frac{\left(\sqrt{5}+1\right)\left(\sqrt{6}+\sqrt{5}\right)}{\left(\sqrt{6}\right)^2 - \left(\sqrt{5}\right)^2}$$
$$= \frac{\sqrt{5}\sqrt{6}+\sqrt{5}\sqrt{5}+1 \cdot \sqrt{6}+1 \cdot \sqrt{5}}{6-5}$$
$$= \frac{\sqrt{30}+5+\sqrt{6}+\sqrt{5}}{1}$$
$$= 5+\sqrt{30}+\sqrt{6}+\sqrt{5}$$

77.
$$\frac{\sqrt{3}+1}{\sqrt{2}-1} = \frac{\sqrt{3}+1}{\sqrt{2}-1} \cdot \frac{\sqrt{2}+1}{\sqrt{2}+1}$$
$$= \frac{\left(\sqrt{3}+1\right)\left(\sqrt{2}+1\right)}{\left(\sqrt{2}\right)^2 - 1}$$
$$= \frac{\sqrt{3}\sqrt{2}+\sqrt{3} \cdot 1+1 \cdot \sqrt{2}+1^2}{2-1}$$
$$= \frac{\sqrt{6}+\sqrt{3}+\sqrt{2}+1}{1}$$
$$= \sqrt{6}+\sqrt{3}+\sqrt{2}+1$$

79.
$$\frac{5}{2+\sqrt{x}} = \frac{5}{2+\sqrt{x}} \cdot \frac{2-\sqrt{x}}{2-\sqrt{x}}$$
$$= \frac{5\left(2-\sqrt{x}\right)}{2^2 - \left(\sqrt{x}\right)^2}$$
$$= \frac{10-5\sqrt{x}}{4-x}$$

81.
$$\frac{3}{\sqrt{x}-4} = \frac{3}{\sqrt{x}-4} \cdot \frac{\sqrt{x}+4}{\sqrt{x}+4}$$
$$= \frac{3\left(\sqrt{x}+4\right)}{\left(\sqrt{x}\right)^2 - (4)^2}$$
$$= \frac{3\sqrt{x}+12}{x-16}$$

83. $x+5 = 7^2$
$x = 49-5$
$x = 44$

85. $4z^2 + 6z - 12 = (2z)^2$
$4z^2 + 6z - 12 = 4z^2$
$6z - 12 = 0$
$6z = 12$
$z = 2$

87. $9x^2 + 5x + 4 = (3x + 1)^2$

$$9x^2 + 5x + 4 = 9x^2 + 6x + 1$$
$$5x + 4 = 6x + 1$$
$$4 = x + 1$$
$$3 = x$$

89. Area = (length)(width)

$$13\sqrt{2} \cdot 5\sqrt{6} = 13 \cdot 5 \cdot \sqrt{2} \cdot \sqrt{6}$$
$$= 65\sqrt{12}$$
$$= 65\sqrt{4}\sqrt{3}$$
$$= 65 \cdot 2\sqrt{3}$$
$$= 130\sqrt{3}$$

The area is $130\sqrt{3}$ square meters.

91. $\sqrt{\dfrac{A}{\pi}} = \dfrac{\sqrt{A}}{\sqrt{\pi}} = \dfrac{\sqrt{A}}{\sqrt{\pi}} \cdot \dfrac{\sqrt{\pi}}{\sqrt{\pi}} = \dfrac{\sqrt{A\pi}}{\pi}$

93. $\sqrt{5} \cdot \sqrt{5} = \left(\sqrt{5}\right)^2 = 5$

The statement is true.

95. $\sqrt{3x} \cdot \sqrt{3x} = \left(\sqrt{3x}\right)^2 = 3x$

The statement is false.

97. $\sqrt{11} + \sqrt{2}$ cannot be simplified because the radicands are different. The statement is false.

99. answers may vary

101. answers may vary

103. $\dfrac{\sqrt{3}+1}{\sqrt{2}-1} = \dfrac{\sqrt{3}+1}{\sqrt{2}-1} \cdot \dfrac{\sqrt{3}-1}{\sqrt{3}-1}$

$$= \dfrac{\left(\sqrt{3}\right)^2 - 1^2}{\left(\sqrt{2}-1\right)\left(\sqrt{3}-1\right)}$$

$$= \dfrac{3-1}{\sqrt{2}\sqrt{3} - 1 \cdot \sqrt{2} - 1 \cdot \sqrt{3} + 1 \cdot 1}$$

$$= \dfrac{2}{\sqrt{6} - \sqrt{2} - \sqrt{3} + 1}$$

Integrated Review

1. $\sqrt{36} = 6$, because $6^2 = 36$ and 6 is positive.

2. $\sqrt{48} = \sqrt{16 \cdot 3} = \sqrt{16} \cdot \sqrt{3} = 4\sqrt{3}$

3. $\sqrt{x^4} = x^2$, because $(x^2)^2 = x^4$.

4. $\sqrt{y^7} = \sqrt{y^6 \cdot y} = \sqrt{y^6}\sqrt{y} = y^3\sqrt{y}$

5. $\sqrt{16x^2} = 4x$, because $(4x)^2 = 16x^2$.

6. $\sqrt{18x^{11}} = \sqrt{9x^{10} \cdot 2x}$

$$= \sqrt{9x^{10}}\sqrt{2x}$$
$$= 3x^5\sqrt{2x}$$

7. $\sqrt[3]{8} = 2$, because $2^3 = 8$.

8. $\sqrt[4]{81} = 3$, because $3^4 = 81$.

9. $\sqrt[3]{-27} = -3$, because $(-3)^3 = -27$.

10. $\sqrt{-4}$ is not a real number.

11. $\sqrt{\dfrac{11}{9}} = \dfrac{\sqrt{11}}{\sqrt{9}} = \dfrac{\sqrt{11}}{3}$

12. $\sqrt[3]{\dfrac{7}{64}} = \dfrac{\sqrt[3]{7}}{\sqrt[3]{64}} = \dfrac{\sqrt[3]{7}}{4}$

13. $-\sqrt{16} = -4$. The negative sign indicates the negative square root of 16.

14. $-\sqrt{25} = -5$. The negative sign indicates the negative square root of 25.

15. $\sqrt{\dfrac{9}{49}} = \dfrac{\sqrt{9}}{\sqrt{49}} = \dfrac{3}{7}$

16. $\sqrt{\dfrac{1}{64}} = \dfrac{\sqrt{1}}{\sqrt{64}} = \dfrac{1}{8}$

17. $\sqrt{a^8 a^2} = \sqrt{a^8}\sqrt{b^2} = a^4 b$

18. $\sqrt{x^{10}y^{20}} = \sqrt{x^{10}}\sqrt{y^{20}} = x^5 y^{10}$

19. $\sqrt{25m^6} = \sqrt{25}\sqrt{m^6} = 5m^3$

20. $\sqrt{9n^{16}} = \sqrt{9}\sqrt{n^{16}} = 3n^8$

21. $5\sqrt{7} + \sqrt{7} = (5+1)\sqrt{7} = 6\sqrt{7}$

22. $\sqrt{50} - \sqrt{8} = \sqrt{25\cdot 2} - \sqrt{4\cdot 2}$
$\qquad = \sqrt{25}\sqrt{2} - \sqrt{4}\sqrt{2}$
$\qquad = 5\sqrt{2} - 2\sqrt{2}$
$\qquad = (5-2)\sqrt{2}$
$\qquad = 3\sqrt{2}$

23. $5\sqrt{2} - 5\sqrt{3}$ cannot be simplified.

24. $2\sqrt{x} + \sqrt{25x} - \sqrt{36x} + 3x$
$\qquad = 2\sqrt{x} + \sqrt{25}\sqrt{x} - \sqrt{36}\sqrt{x} + 3x$
$\qquad = 2\sqrt{x} + 5\sqrt{x} - 6\sqrt{x} + 3x$
$\qquad = (2+5-6)\sqrt{x} + 3x$
$\qquad = \sqrt{x} + 3x$

25. $\sqrt{2}\cdot\sqrt{15} = \sqrt{2\cdot 15} = \sqrt{30}$

26. $\sqrt{3}\cdot\sqrt{3} = \sqrt{3\cdot 3} = \sqrt{9} = 3$

27. $\left(2\sqrt{7}\right)^2 = \left(2\sqrt{7}\right)\left(2\sqrt{7}\right)$
$\qquad = 4\left(\sqrt{7}\right)^2$
$\qquad = 4\cdot 7$
$\qquad = 28$

28. $\left(3\sqrt{5}\right)^2 = \left(3\sqrt{5}\right)\left(3\sqrt{5}\right) = 9\left(\sqrt{5}\right)^2 = 9\cdot 5 = 45$

29. $\sqrt{3}\left(\sqrt{11}+1\right) = \sqrt{3}\cdot\sqrt{11} + \sqrt{3}\cdot 1 = \sqrt{33} + \sqrt{3}$

30. $\sqrt{6}\left(\sqrt{3}-2\right) = \sqrt{6}\cdot\sqrt{3} - \sqrt{6}\cdot 2$
$\qquad = \sqrt{18} - 2\sqrt{6}$
$\qquad = \sqrt{9\cdot 2} - 2\sqrt{6}$
$\qquad = 3\sqrt{2} - 2\sqrt{6}$

31. $\sqrt{8y}\sqrt{2y} = \sqrt{8y\cdot 2y} = \sqrt{16y^2} = 4y$

32. $\sqrt{15x^2}\cdot\sqrt{3x^2} = \sqrt{15x^2\cdot 3x^2}$
$\qquad = \sqrt{45x^4}$
$\qquad = \sqrt{9x^4\cdot 5}$
$\qquad = 3x^2\sqrt{5}$

33. $\left(\sqrt{x}-5\right)\left(\sqrt{x}+2\right)$
$\qquad = \sqrt{x}\cdot\sqrt{x} + 2\sqrt{x} - 5\sqrt{x} - 5\cdot 2$
$\qquad = x - 3\sqrt{x} - 10$

34. $\left(3+\sqrt{2}\right)^2 = (3)^2 + 2(3)\left(\sqrt{2}\right) + \left(\sqrt{2}\right)^2$
$\qquad = 9 + 6\sqrt{2} + 2$
$\qquad = 11 + 6\sqrt{2}$

35. $\dfrac{\sqrt{8}}{\sqrt{2}} = \sqrt{\dfrac{8}{2}} = \sqrt{4} = 2$

36. $\dfrac{\sqrt{45}}{\sqrt{15}} = \sqrt{\dfrac{45}{15}} = \sqrt{3}$

37. $\dfrac{\sqrt{24x^5}}{\sqrt{2x}} = \sqrt{\dfrac{24x^5}{2x}}$
$\qquad = \sqrt{12x^4}$
$\qquad = \sqrt{4x^4\cdot 3}$
$\qquad = 2x^2\sqrt{3}$

38. $\dfrac{\sqrt{75a^4b^5}}{\sqrt{5ab}} = \sqrt{\dfrac{75a^4b^5}{5ab}}$

$\qquad\qquad = \sqrt{15a^3b^4}$

$\qquad\qquad = \sqrt{a^2b^4 \cdot 15a}$

$\qquad\qquad = ab^2\sqrt{15a}$

39. $\sqrt{\dfrac{1}{6}} = \dfrac{\sqrt{1}}{\sqrt{6}} = \dfrac{1}{\sqrt{6}} = \dfrac{1}{\sqrt{6}} \cdot \dfrac{\sqrt{6}}{\sqrt{6}} = \dfrac{\sqrt{6}}{6}$

40. $\dfrac{x}{\sqrt{20}} = \dfrac{x}{\sqrt{4 \cdot 5}}$

$\qquad = \dfrac{x}{2\sqrt{5}}$

$\qquad = \dfrac{x}{2\sqrt{5}} \cdot \dfrac{\sqrt{5}}{\sqrt{5}}$

$\qquad = \dfrac{x\sqrt{5}}{2 \cdot 5}$

$\qquad = \dfrac{x\sqrt{5}}{10}$

41. $\dfrac{4}{\sqrt{6}+1} = \dfrac{4}{\sqrt{6}+1} \cdot \dfrac{\sqrt{6}-1}{\sqrt{6}-1}$

$\qquad = \dfrac{4\left(\sqrt{6}-1\right)}{\left(\sqrt{6}\right)^2 - 1^2}$

$\qquad = \dfrac{4\sqrt{6}-4}{6-1}$

$\qquad = \dfrac{4\sqrt{6}-4}{5}$

42. $\dfrac{\sqrt{2}+1}{\sqrt{x}-5} = \dfrac{\sqrt{2}+1}{\sqrt{x}-5} \cdot \dfrac{\sqrt{x}+5}{\sqrt{x}+5}$

$\qquad = \dfrac{\left(\sqrt{2}+1\right)\left(\sqrt{x}+5\right)}{\left(\sqrt{x}\right)^2 - 5^2}$

$\qquad = \dfrac{\sqrt{2}\sqrt{x}+5\sqrt{2}+1\sqrt{x}+1 \cdot 5}{x-25}$

$\qquad = \dfrac{\sqrt{2x}+5\sqrt{2}+\sqrt{x}+5}{x-25}$

Section 8.5 Practice Problems

1. $\sqrt{x-2} = 7$

$\qquad \left(\sqrt{x-2}\right)^2 = 7^2$

$\qquad\qquad x-2 = 49$

$\qquad\qquad\quad x = 51$

2. $\sqrt{6x-1} = \sqrt{x}$

$\qquad \left(\sqrt{6x-1}\right)^2 = \left(\sqrt{x}\right)^2$

$\qquad\qquad 6x-1 = x$

$\qquad\qquad 5x-1 = 0$

$\qquad\qquad\quad 5x = 1$

$\qquad\qquad\quad\ x = \dfrac{1}{5}$

3. $\sqrt{x}+9 = 2$

$\qquad\quad \sqrt{x} = -7$

$\sqrt{x}$ cannot equal -7. Thus, the equation has no solution.

4. $\sqrt{9y^2 + 2y - 10} = 3y$

$\qquad \left(\sqrt{9y^2 + 2y - 10}\right)^2 = (3y)^2$

$\qquad\qquad 9y^2 + 2y - 10 = 9y^2$

$\qquad\qquad\qquad 2y - 10 = 0$

$\qquad\qquad\qquad\qquad 2y = 10$

$\qquad\qquad\qquad\qquad\ y = 5$

5. $\sqrt{x+1} - x = -5$

$\qquad\quad \sqrt{x+1} = x-5$

$\qquad \left(\sqrt{x+1}\right)^2 = (x-5)^2$

$\qquad\qquad x+1 = x^2 - 10x + 25$

$\qquad\qquad\quad 0 = x^2 - 11x + 24$

$\qquad\qquad\quad 0 = (x-8)(x-3)$

$0 = x-8 \quad$ or $\quad 0 = x-3$

$8 = x \qquad\qquad\quad 3 = x$

Replacing x with 3 results in a false statement. 3 is an extraneous solution. The only solution is 8.

6. $\sqrt{x} + 3 = \sqrt{x + 15}$

$$\left(\sqrt{x} + 3\right)^2 = \left(\sqrt{x + 15}\right)^2$$

$$x + 6\sqrt{x} + 9 = x + 15$$

$$6\sqrt{x} = 6$$

$$\sqrt{x} = 1$$

$$x = 1$$

Exercise Set 8.5

1. $\sqrt{x} = 9$

$$\left(\sqrt{x}\right)^2 = 9^2$$

$$x = 81$$

3. $\sqrt{x + 5} = 2$

$$\left(\sqrt{x + 5}\right)^2 = 2^2$$

$$x + 5 = 4$$

$$x = -1$$

5. $\sqrt{x} - 2 = 5$

$$\sqrt{x} = 7$$

$$\left(\sqrt{x}\right)^2 = 7^2$$

$$x = 49$$

7. $3\sqrt{x} + 5 = 2$

$$3\sqrt{x} = -3$$

$$\sqrt{x} = -1$$

$\sqrt{x}$ cannot equal -1. Thus, the equation has no solution.

9. $\sqrt{x} = \sqrt{3x - 8}$

$$\left(\sqrt{x}\right)^2 = \left(\sqrt{3x - 8}\right)^2$$

$$x = 3x - 8$$

$$-2x = -8$$

$$x = 4$$

11. $\sqrt{4x - 3} = \sqrt{x + 3}$

$$\left(\sqrt{4x - 3}\right)^2 = \left(\sqrt{x + 3}\right)^2$$

$$4x - 3 = x + 3$$

$$3x - 3 = 3$$

$$3x = 6$$

$$x = 2$$

13. $\sqrt{9x^2 + 2x - 4} = 3x$

$$\left(\sqrt{9x^2 + 2x - 4}\right)^2 = (3x)^2$$

$$9x^2 + 2x - 4 = 9x^2$$

$$2x - 4 = 0$$

$$2x = 4$$

$$x = 2$$

15. $\sqrt{x} = x - 6$

$$\left(\sqrt{x}\right)^2 = (x - 6)^2$$

$$x = x^2 - 12x + 36$$

$$0 = x^2 - 13x + 36$$

$$0 = (x - 9)(x - 4)$$

$$0 = x - 9 \quad \text{or} \quad 0 = x - 4$$

$$9 = x \qquad\qquad 4 = x$$

$x = 4$ does not check, so the solution is $x = 9$.

17. $\sqrt{x + 7} = x + 5$

$$\left(\sqrt{x + 7}\right)^2 = (x + 5)^2$$

$$x + 7 = x^2 + 10x + 25$$

$$0 = x^2 + 9x + 18$$

$$0 = (x + 6)(x + 3)$$

$$x + 6 = 0 \quad \text{or} \quad x + 3 = 0$$

$$x = -6 \qquad\qquad x = -3$$

$x = -6$ does not check, so the solution is $x = -3$.

19. $\sqrt{3x+7} - x = 3$

$\sqrt{3x+7} = x+3$

$\left(\sqrt{3x+7}\right)^2 = (x+3)^2$

$3x+7 = x^2 + 6x + 9$

$0 = x^2 + 3x + 2$

$0 = (x+2)(x+1)$

$x+2 = 0 \quad \text{or} \quad x+1 = 0$

$x = -2 \qquad\qquad x = -1$

21. $\sqrt{16x^2 + 2x + 2} = 4x$

$\left(\sqrt{16x^2 + 2x + 2}\right)^2 = (4x)^2$

$16x^2 + 2x + 2 = 16x^2$

$2x + 2 = 0$

$2x = -2$

$x = -1$

$x = -1$ does not check, so the equation has no solution.

23. $\sqrt{2x^2 + 6x + 9} = 3$

$\left(\sqrt{2x^2 + 6x + 9}\right)^2 = 3^2$

$2x^2 + 6x + 9 = 9$

$2x^2 + 6x = 0$

$2x(x+3) = 0$

$2x = 0 \quad \text{or} \quad x+3 = 0$

$x = 0 \qquad\qquad x = -3$

25. $\sqrt{x-7} = \sqrt{x} - 1$

$\left(\sqrt{x-7}\right)^2 = \left(\sqrt{x} - 1\right)^2$

$x - 7 = x - 2\sqrt{x} + 1$

$2\sqrt{x} = 8$

$\sqrt{x} = 4$

$\left(\sqrt{x}\right)^2 = 4^2$

$x = 16$

27. $\sqrt{x} + 2 = \sqrt{x+24}$

$\left(\sqrt{x} + 2\right)^2 = \left(\sqrt{x+24}\right)^2$

$x + 4\sqrt{x} + 4 = x + 24$

$4\sqrt{x} = 20$

$\sqrt{x} = 5$

$\left(\sqrt{x}\right)^2 = 5^2$

$x = 25$

29. $\sqrt{x+8} = \sqrt{x} + 2$

$\left(\sqrt{x+8}\right)^2 = \left(\sqrt{x} + 2\right)^2$

$x + 8 = x + 4\sqrt{x} + 4$

$4 = 4\sqrt{x}$

$1 = \sqrt{x}$

$1^2 = \left(\sqrt{x}\right)^2$

$1 = x$

31. $\sqrt{2x+6} = 4$

$\left(\sqrt{2x+6}\right)^2 = 4^2$

$2x + 6 = 16$

$2x = 10$

$x = 5$

33. $\sqrt{x+6} + 1 = 3$

$\sqrt{x+6} = 2$

$\left(\sqrt{x+6}\right)^2 = 2^2$

$x + 6 = 4$

$x = -2$

35. $\sqrt{x+6} + 5 = 3$

$\sqrt{x+6} = -2$

$\sqrt{x+6}$ cannot equal -2. Thus, the equation has no solution.

37. $\sqrt{16x^2 - 3x + 6} = 4x$

$$\left(\sqrt{16x^2 - 3x + 6}\right)^2 = (4x)^2$$
$$16x^2 - 3x + 6 = 16x^2$$
$$-3x + 6 = 0$$
$$-3x = -6$$
$$x = 2$$

39. $-\sqrt{x} = -6$

$$\sqrt{x} = 6$$
$$\left(\sqrt{x}\right)^2 = 6^2$$
$$x = 36$$

41. $\sqrt{x + 9} = \sqrt{x} - 3$

$$\left(\sqrt{x + 9}\right)^2 = \left(\sqrt{x} - 3\right)^2$$
$$x + 9 = x - 6\sqrt{x} + 9$$
$$0 = -6\sqrt{x}$$
$$0 = \sqrt{x}$$
$$0^2 = \left(\sqrt{x}\right)^2$$
$$0 = x$$

$x = 0$ does not check, so the equation has no solution.

43. $\sqrt{2x + 1} + 3 = 5$

$$\sqrt{2x + 1} = 2$$
$$\left(\sqrt{2x + 1}\right)^2 = 2^2$$
$$2x + 1 = 4$$
$$2x = 3$$
$$x = \frac{3}{2}$$

45. $\sqrt{x} + 3 = 7$

$$\sqrt{x} = 4$$
$$\left(\sqrt{x}\right)^2 = 4^2$$
$$x = 16$$

47. $\sqrt{4x} = \sqrt{2x + 6}$

$$\left(\sqrt{4x}\right)^2 = \left(\sqrt{2x + 6}\right)^2$$
$$4x = 2x + 6$$
$$2x = 6$$
$$x = 3$$

49. $\sqrt{2x + 1} = x - 7$

$$\left(\sqrt{2x + 1}\right)^2 = (x - 7)^2$$
$$2x + 1 = x^2 - 14x + 49$$
$$0 = x^2 - 16x + 48$$
$$0 = (x - 4)(x - 12)$$
$$x - 4 = 0 \quad \text{or} \quad x - 12 = 0$$
$$x = 4 \qquad\qquad x = 12$$

$x = 4$ does not check, so the only solution is $x = 12$.

51.
$$x = \sqrt{2x - 2} + 1$$
$$x - 1 = \sqrt{2x - 2}$$
$$(x - 1)^2 = \left(\sqrt{2x - 2}\right)^2$$
$$x^2 - 2x + 1 = 2x - 2$$
$$x^2 - 4x + 3 = 0$$
$$(x - 3)(x - 1) = 0$$
$$x - 3 = 0 \quad \text{or} \quad x - 1 = 0$$
$$x = 3 \qquad\qquad x = 1$$

53. $\sqrt{1 - 8x} - x = 4$

$$\sqrt{1 - 8x} = x + 4$$
$$\left(\sqrt{1 - 8x}\right)^2 = (x + 4)^2$$
$$1 - 8x = x^2 + 8x + 16$$
$$0 = x^2 + 16x + 15$$
$$0 = (x + 15)(x + 1)$$
$$x + 15 = 0 \quad \text{or} \quad x + 1 = 0$$
$$x = -15 \qquad\qquad x = -1$$

$x = -15$ does not check, so the solution is $x = -1$.

55. $3x - 8 = 19$

$$3x = 27$$
$$x = 9$$

57. Let x be the width of the rectangle, then the length is $2x$.
$$2(2x + x) = 24$$
$$2(3x) = 24$$
$$6x = 24$$
$$x = 4$$
$$2x = 2(4) = 8$$
The length of the rectangle is 8 inches.

59.
$$\sqrt{x-3} + 3 = \sqrt{3x+4}$$
$$\left(\sqrt{x-3}+3\right)^2 = \left(\sqrt{3x+4}\right)^2$$
$$x - 3 + 6\sqrt{x-3} + 9 = 3x + 4$$
$$6\sqrt{x-3} = 2x - 2$$
$$\left(6\sqrt{x-3}\right)^2 = (2x-2)^2$$
$$36(x-3) = 4x^2 - 8x + 4$$
$$36x - 108 = 4x^2 - 8x + 4$$
$$0 = 4x^2 - 44x + 112$$
$$0 = 4(x-7)(x-4)$$
$$x - 7 = 0 \quad \text{or} \quad x - 4 = 0$$
$$x = 7 \qquad\qquad x = 4$$

61. answers may vary

63. a. For $V = 20$, $b = \sqrt{\dfrac{20}{2}} = \sqrt{10} \approx 3.2$.

For $V = 200$, $b = \sqrt{\dfrac{200}{2}} = \sqrt{100} = 10$.

For $V = 2000$,
$$b = \sqrt{\dfrac{2000}{2}} = \sqrt{1000} \approx 31.6.$$

V	20	200	2000
b	3.2	10	31.6

b. No, the volume increases by a factor of $\sqrt{10}$.

65.

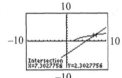

The solution of $\sqrt{x-2} = x - 5$ is $x \approx 7.30$.

67.

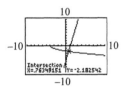

The solution of $-\sqrt{x+4} = 5x - 6$ is $x \approx 0.76$.

The Bigger Picture

1. $\sqrt{56} = \sqrt{4 \cdot 14} = \sqrt{4}\sqrt{14} = 2\sqrt{14}$

2.
$$\sqrt{\dfrac{20x^5}{49}} = \dfrac{\sqrt{20x^5}}{\sqrt{49}}$$
$$= \dfrac{\sqrt{4x^4 \cdot 5x}}{7}$$
$$= \dfrac{\sqrt{4x^4}\sqrt{5x}}{7}$$
$$= \dfrac{2x^2\sqrt{5x}}{7}$$

3. $(-5x^{12}y^{-3})(3x^{-7}y^{14}) = -15x^{12-7}y^{-3+14}$
$$= -15x^5 y^{11}$$

4. $\sqrt{\dfrac{10}{11}} = \dfrac{\sqrt{10}}{\sqrt{11}} = \dfrac{\sqrt{10}}{\sqrt{11}} \cdot \dfrac{\sqrt{11}}{\sqrt{11}} = \dfrac{\sqrt{110}}{11}$

5. $\dfrac{8}{\sqrt{5}-1} = \dfrac{8}{\sqrt{5}-1} \cdot \dfrac{\sqrt{5}+1}{\sqrt{5}+1}$

$= \dfrac{8\left(\sqrt{5}+1\right)}{5-1}$

$= \dfrac{8\left(\sqrt{5}+1\right)}{4}$

$= 2\left(\sqrt{5}+1\right) \text{ or } 2\sqrt{5}+2$

6. $\dfrac{1}{2}(6x^2-4) + \dfrac{1}{3}(6x^2-9) - 14$

$= 3x^2 - 2 + 2x^2 - 3 - 14$

$= 5x^2 - 19$

7. $9x - 7 = 7x - 9$

$2x - 7 = -9$

$2x = -2$

$x = -1$

8. $\dfrac{x}{5} = \dfrac{x-3}{11}$

$55\left(\dfrac{x}{5}\right) = 55\left(\dfrac{x-3}{11}\right)$

$11x = 5(x-3)$

$11x = 5x - 15$

$6x = -15$

$x = -\dfrac{15}{6} = -\dfrac{5}{2}$

9. $-5(2y+1) \le 3y - 2 - 2y + 1$

$-10y - 5 \le y - 1$

$-11y - 5 \le -1$

$-11y \le 4$

$y \ge -\dfrac{4}{11}$

The solution set is $\left\{ y \mid y \ge -\dfrac{4}{11} \right\}$.

10. $x(x+1) = 42$

$x^2 + x = 42$

$x^2 + x - 42 = 0$

$(x+7)(x-6) = 0$

$x + 7 = 0 \quad \text{or} \quad x - 6 = 0$

$x = -7 \qquad\qquad x = 6$

11. $\dfrac{-6}{x-7} + \dfrac{8}{x} = \dfrac{-4}{x-7}$

$x(x-7)\left(\dfrac{-6}{x-7} + \dfrac{8}{x} \right) = x(x-7)\left(\dfrac{-4}{x-7} \right)$

$-6x + 8(x-7) = -4x$

$-6x + 8x - 56 = -4x$

$2x - 56 = -4x$

$-56 = -6x$

$\dfrac{-56}{-6} = x$

$\dfrac{28}{3} = x$

12. $1 + 4(x-2) = x(x-6) - x^2 + 13$

$1 + 4x - 8 = x^2 - 6x - x^2 + 13$

$4x - 7 = -6x + 13$

$10x - 7 = 13$

$10x = 20$

$x = 2$

Section 8.6 Practice Problems

1. $a^2 + b^2 = c^2$

$3^2 + 4^2 = c^2$

$9 + 16 = c^2$

$25 = c^2$

$\sqrt{25} = c$

$5 = c$

The hypotenuse has a length of 5 centimeters.

2. $a^2 + b^2 = c^2$
$5^2 + 3^2 = c^2$
$25 + 9 = c^2$
$34 = c^2$
$\sqrt{34} = c$
$5.83 \approx c$

3. The property owner is using a right triangle with one leg measuring 40 feet and hypotenuse measuring 65 feet to find the unknown distance.
$a^2 + b^2 = c^2$
$40^2 + b^2 = 65^2$
$1600 + b^2 = 4225$
$b^2 = 2625$
$b = \sqrt{2625} = 5\sqrt{105}$

The distance across the pond is $5\sqrt{105}$ feet or approximately 51.2 feet.

4. $v = \sqrt{2gh}$
$\quad = \sqrt{2 \cdot 32 \cdot 20}$
$\quad = \sqrt{1280}$
$\quad = 16\sqrt{5}$

The velocity of the object after falling 20 feet is exactly $16\sqrt{5}$ feet per second or approximately 35.8 feet per second.

Exercise Set 8.6

1. $a^2 + b^2 = c^2$
$2^2 + 3^2 = c^2$
$4 + 9 = c^2$
$13 = c^2$
$\sqrt{13} = \sqrt{c^2}$
$\sqrt{13} = c$
$c = \sqrt{13} \approx 3.61$

3. $a^2 + b^2 = c^2$
$3^2 + b^2 = 6^2$
$9 + b^2 = 36$
$b^2 = 27$
$\sqrt{b^2} = \sqrt{27}$
$b = 3\sqrt{3} \approx 5.20$

5. $a^2 + b^2 = c^2$
$7^2 + 24^2 = c^2$
$49 + 576 = c^2$
$625 = c^2$
$\sqrt{625} = \sqrt{c^2}$
$25 = c$

7. $a^2 + b^2 = c^2$
$\left(\sqrt{3}\right)^2 + b^2 = 5^2$
$3 + b^2 = 25$
$b^2 = 22$
$\sqrt{b^2} = \sqrt{22}$
$b = \sqrt{22} \approx 4.69$

9. $a^2 + b^2 = c^2$
$4^2 + b^2 = 13^2$
$16 + b^2 = 169$
$b^2 = 153$
$\sqrt{b^2} = \sqrt{153}$
$b = 3\sqrt{17}$
$b = 3\sqrt{17} \approx 12.37$

11. $a^2 + b^2 = c^2$
$4^2 + 5^2 = c^2$
$16 + 25 = c^2$
$41 = c^2$
$\sqrt{41} = \sqrt{c^2}$
$c = \sqrt{41} \approx 6.40$

13. $a^2 + b^2 = c^2$

$a^2 + 2^2 = 6^2$

$a^2 + 4 = 36$

$a^2 = 32$

$\sqrt{a^2} = \sqrt{32}$

$a = 4\sqrt{2}$

$a = 4\sqrt{2} \approx 5.66$

15. $a^2 + b^2 = c^2$

$\left(\sqrt{10}\right)^2 + b^2 = 10^2$

$10 + b^2 = 100$

$b^2 = 90$

$\sqrt{b^2} = \sqrt{90}$

$b = 3\sqrt{10} \approx 9.49$

17. The pole, wire, and ground form a right triangle with legs of 5 feet and 20 feet.

$a^2 + b^2 = c^2$

$5^2 + 20^2 = c^2$

$25 + 400 = c^2$

$425 = c^2$

$\sqrt{425} = \sqrt{c^2}$

$\sqrt{425} = c$

$c = \sqrt{425} \approx 20.6$

The length of the wire is 20.6 feet.

19. The diagonal brace is the hypotenuse of a right triangle with legs measuring 6 feet and 10 feet.

$a^2 + b^2 = c^2$

$6^2 + 10^2 = c^2$

$36 + 100 = c^2$

$136 = c^2$

$\sqrt{136} = \sqrt{c^2}$

$c = \sqrt{136} \approx 11.7$

The brace needs to be 11.7 feet.

21. $b = \sqrt{\dfrac{3V}{h}}$

$6 = \sqrt{\dfrac{3V}{2}}$

$6^2 = \left(\sqrt{\dfrac{3V}{2}}\right)^2$

$36 = \dfrac{3V}{2}$

$72 = 3V$

$24 = V$

The volume is 24 cubic feet.

23. $s = \sqrt{30fd}$

$s = \sqrt{30 \cdot 0.35 \cdot 280}$

$s = \sqrt{2940}$

$s \approx 54$

The car was moving at a speed of 54 miles per hour.

25. $v = \sqrt{2.5r}$

$v = \sqrt{2.5(300)}$

$v = \sqrt{750}$

$v \approx 27.4$

The maximum safe speed is 27 miles per hour.

27. $d = 3.5\sqrt{h}$

$d = 3.5\sqrt{285.4}$

$d \approx 59.1$

You can see a distance of 59.1 kilometers.

29. $d = 3.5\sqrt{h}$

$d = 3.5\sqrt{295.7}$

$d \approx 60.2$

You can see a distance of 60.2 kilometers.

31. $\sqrt{9} = 3$ and $-\sqrt{9} = -3$, so -3 and 3 are numbers whose square is 9.

33. $\sqrt{100} = 10$ and $-\sqrt{100} = -10$, so -10 and 10 are numbers whose square is 100.

35. $\sqrt{64} = 8$ and $-\sqrt{64} = -8$, so -8 and 8 are numbers whose square is 64.

37. First find y.

$$a^2 + b^2 = c^2$$
$$3^2 + y^2 = 7^2$$
$$9 + y^2 = 49$$
$$y^2 = 40$$
$$y = \sqrt{40}$$
$$y = \sqrt{4 \cdot 10}$$
$$y = 2\sqrt{10}$$

Let b be the second leg of the right triangle with hypotenuse 5.

$$a^2 + b^2 = c^2$$
$$3^2 + b^2 = 5^2$$
$$9 + b^2 = 25$$
$$b^2 = 16$$
$$\sqrt{b^2} = \sqrt{16}$$
$$b = 4$$

Now find x.

$$x = y - 4$$
$$x = 2\sqrt{10} - 4$$

39. The distance is the length of the hypotenuse of a right triangle. One leg has length $3 \cdot 30 = 90$ miles, and the other leg has length $3 \cdot 60 = 180$ miles.

$$a^2 + b^2 = c^2$$
$$90^2 + 180^2 = c^2$$
$$8100 + 32,400 = c^2$$
$$40,500 = c^2$$
$$\sqrt{40,500} = \sqrt{c^2}$$
$$201 \approx c$$

They are 201 miles apart.

41. answers may vary

Chapter 8 Vocabulary Check

1. The expressions $5\sqrt{x}$ and $7\sqrt{x}$ are examples of <u>like radicals</u>.

2. In the expression $\sqrt[3]{45}$ the number 3 is the <u>index</u>, the number 45 is the <u>radicand</u>, and $\sqrt{}$ is called the <u>radical</u> sign.

3. The <u>conjugate</u> of $a + b$ is $a - b$.

4. The <u>principal square root</u> of 25 is 5.

5. The process eliminating the radical in the denominator of a radical expression is called <u>rationalizing the denominator</u>.

Chapter 8 Review

1. $\sqrt{81} = 9$, because $9^2 = 81$ and 9 is positive.

2. $-\sqrt{49} = -7$. The negative indicates the negative square root of 49.

3. $\sqrt[3]{27} = 3$, because $3^3 = 27$.

4. $\sqrt[4]{81} = 3$, because $3^4 = 81$.

5. $-\sqrt{\dfrac{9}{64}} = -\dfrac{3}{8}$ because $\left(\dfrac{3}{8}\right)^2 = \dfrac{9}{64}$.

6. $\sqrt{\dfrac{36}{81}} = \dfrac{6}{9}$ because $\left(\dfrac{6}{9}\right)^2 = \dfrac{36}{81}$, and $\dfrac{6}{9} = \dfrac{2}{3}$.

7. $\sqrt[4]{16} = 2$ because $2^4 = 16$.

8. $\sqrt[3]{-8} = -2$ because $(-2)^3 = -8$.

9. c; $\sqrt{-4}$ is not a real number because the radicand is negative and the index is even.

10. a, c; $\sqrt{-5}$ and $\sqrt[4]{-5}$ are not real numbers because the radicands are negatives and the indexes are even.

11. $\sqrt{x^{12}} = x^6$, because $(x^6)^2 = x^{12}$.

12. $\sqrt{x^8} = x^4$, because $(x^4)^2 = x^8$.

13. $\sqrt{9y^2} = 3y$, because $(3y)^2 = 9y^2$.

14. $\sqrt{25x^4} = 5x^2$, because $(5x^2)^2 = 25x^4$.

15. $\sqrt{40} = \sqrt{4 \cdot 10} = \sqrt{4}\sqrt{10} = 2\sqrt{10}$

16. $\sqrt{24} = \sqrt{4 \cdot 6} = \sqrt{4}\sqrt{6} = 2\sqrt{6}$

17. $\sqrt{54} = \sqrt{9 \cdot 6} = \sqrt{9}\sqrt{6} = 3\sqrt{6}$

18. $\sqrt{88} = \sqrt{4 \cdot 22} = \sqrt{4}\sqrt{22} = 2\sqrt{22}$

19. $\sqrt{x^5} = \sqrt{x^4 \cdot x} = \sqrt{x^4}\sqrt{x} = x^2\sqrt{x}$

20. $\sqrt{y^7} = \sqrt{y^6 \cdot y} = \sqrt{y^6}\sqrt{y} = y^3\sqrt{y}$

21. $\sqrt{20x^2} = \sqrt{4x^2 \cdot 5} = \sqrt{4x^2}\sqrt{5} = 2x\sqrt{5}$

22. $\sqrt{50y^4} = \sqrt{25y^4 \cdot 2} = \sqrt{25y^4}\sqrt{2} = 5y^2\sqrt{2}$

23. $\sqrt[3]{54} = \sqrt[3]{27 \cdot 2} = \sqrt[3]{27}\sqrt[3]{2} = 3\sqrt[3]{2}$

24. $\sqrt[3]{88} = \sqrt[3]{8 \cdot 11} = \sqrt[3]{8}\sqrt[3]{11} = 2\sqrt[3]{11}$

25. $\sqrt{\dfrac{18}{25}} = \dfrac{\sqrt{18}}{\sqrt{25}} = \dfrac{\sqrt{9 \cdot 2}}{5} = \dfrac{\sqrt{9}\sqrt{2}}{5} = \dfrac{3\sqrt{2}}{5}$

26. $\sqrt{\dfrac{75}{64}} = \dfrac{\sqrt{75}}{\sqrt{64}} = \dfrac{\sqrt{25 \cdot 3}}{8} = \dfrac{\sqrt{25}\sqrt{3}}{8} = \dfrac{5\sqrt{3}}{8}$

27.
$$-\sqrt{\frac{50}{9}} = -\frac{\sqrt{50}}{\sqrt{9}}$$
$$= -\frac{\sqrt{25 \cdot 2}}{3}$$
$$= -\frac{\sqrt{25}\sqrt{2}}{3}$$
$$= -\frac{5\sqrt{2}}{3}$$

28.
$$-\sqrt{\frac{12}{49}} = -\frac{\sqrt{12}}{\sqrt{49}}$$
$$= -\frac{\sqrt{4 \cdot 3}}{7}$$
$$= -\frac{\sqrt{4}\sqrt{3}}{7}$$
$$= -\frac{2\sqrt{3}}{7}$$

29. $\sqrt{\dfrac{11}{x^2}} = \dfrac{\sqrt{11}}{\sqrt{x^2}} = \dfrac{\sqrt{11}}{x}$

30. $\sqrt{\dfrac{7}{y^4}} = \dfrac{\sqrt{7}}{\sqrt{y^4}} = \dfrac{\sqrt{7}}{y^2}$

31.
$$\sqrt{\frac{y^5}{100}} = \frac{\sqrt{y^5}}{\sqrt{100}}$$
$$= \frac{\sqrt{y^4 \cdot y}}{10}$$
$$= \frac{\sqrt{y^4}\sqrt{y}}{10}$$
$$= \frac{y^2\sqrt{y}}{10}$$

32. $\sqrt{\dfrac{x^3}{81}} = \dfrac{\sqrt{x^3}}{\sqrt{81}} = \dfrac{\sqrt{x^2 \cdot x}}{9} = \dfrac{\sqrt{x^2}\sqrt{x}}{9} = \dfrac{x\sqrt{x}}{9}$

33. $5\sqrt{8} - 8\sqrt{2} = (5-8)\sqrt{2} = -3\sqrt{2}$

34. $\sqrt{3} - 6\sqrt{3} = (1-6)\sqrt{3} = -5\sqrt{3}$

35. $6\sqrt{5} + 3\sqrt{6} - 2\sqrt{5} + \sqrt{6}$
$= (6-2)\sqrt{5} + (3+1)\sqrt{6}$
$= 4\sqrt{5} + 4\sqrt{6}$

36. $-\sqrt{7} + 8\sqrt{2} - \sqrt{7} - 6\sqrt{2}$
$= (-1-1)\sqrt{7} + (8-6)\sqrt{2}$
$= -2\sqrt{7} + 2\sqrt{2}$

37. $\sqrt{28} + \sqrt{63} + \sqrt{56}$
$= \sqrt{4 \cdot 7} + \sqrt{9 \cdot 7} + \sqrt{4 \cdot 14}$
$= \sqrt{4}\sqrt{7} + \sqrt{9}\sqrt{7} + \sqrt{4}\sqrt{14}$
$= 2\sqrt{7} + 3\sqrt{7} + 2\sqrt{14}$
$= (2+3)\sqrt{7} + 2\sqrt{14}$
$= 5\sqrt{7} + 2\sqrt{14}$

38. $\sqrt{75} + \sqrt{48} - \sqrt{16} = \sqrt{25 \cdot 3} + \sqrt{16 \cdot 3} - 4$
$= \sqrt{25}\sqrt{3} + \sqrt{16}\sqrt{3} - 4$
$= 5\sqrt{3} + 4\sqrt{3} - 4$
$= (5+4)\sqrt{3} - 4$
$= 9\sqrt{3} - 4$

39. $\sqrt{\dfrac{5}{9}} - \sqrt{\dfrac{5}{36}} = \dfrac{\sqrt{5}}{\sqrt{9}} - \dfrac{\sqrt{5}}{\sqrt{36}}$
$= \dfrac{\sqrt{5}}{3} - \dfrac{\sqrt{5}}{6}$
$= \dfrac{2\sqrt{5}}{6} - \dfrac{\sqrt{5}}{6}$
$= \dfrac{\sqrt{5}}{6}$

40. $\sqrt{\dfrac{11}{25}} + \sqrt{\dfrac{11}{16}} = \dfrac{\sqrt{11}}{\sqrt{25}} + \dfrac{\sqrt{11}}{\sqrt{16}}$
$= \dfrac{\sqrt{11}}{5} + \dfrac{\sqrt{11}}{4}$
$= \dfrac{4\sqrt{11}}{20} + \dfrac{5\sqrt{11}}{20}$
$= \dfrac{9\sqrt{11}}{20}$

41. $\sqrt{45x^2} + 3\sqrt{5x^2} - 7x\sqrt{5} + 10$
$= \sqrt{9x^2 \cdot 5} + 3\sqrt{x^2 \cdot 5} - 7x\sqrt{5} + 10$
$= \sqrt{9x^2}\sqrt{5} + 3\sqrt{x^2}\sqrt{5} - 7x\sqrt{5} + 10$
$= 3x\sqrt{5} + 3x\sqrt{5} - 7x\sqrt{5} + 10$
$= (3x + 3x - 7x)\sqrt{5} + 10$
$= -x\sqrt{5} + 10$ or $10 - x\sqrt{5}$

42. $\sqrt{50x} - 9\sqrt{2x} + \sqrt{72x} - \sqrt{3x}$
$= \sqrt{25 \cdot 2x} - 9\sqrt{2x} + \sqrt{36 \cdot 2x} - \sqrt{3x}$
$= \sqrt{25}\sqrt{2x} - 9\sqrt{2x} + \sqrt{36}\sqrt{2x} - \sqrt{3x}$
$= 5\sqrt{2x} - 9\sqrt{2x} + 6\sqrt{2x} - \sqrt{3x}$
$= (5 - 9 + 6)\sqrt{2x} - \sqrt{3x}$
$= 2\sqrt{2x} - \sqrt{3x}$

43. $\sqrt{3} \cdot \sqrt{6} = \sqrt{3 \cdot 6}$
$= \sqrt{18}$
$= \sqrt{9 \cdot 2}$
$= \sqrt{9} \cdot \sqrt{2}$
$= 3\sqrt{2}$

44. $\sqrt{5} \cdot \sqrt{15} = \sqrt{5 \cdot 15}$
$= \sqrt{75}$
$= \sqrt{25 \cdot 3}$
$= \sqrt{25} \cdot \sqrt{3}$
$= 5\sqrt{3}$

45. $\sqrt{2}\left(\sqrt{5} - \sqrt{7}\right) = \sqrt{2}\sqrt{5} - \sqrt{2}\sqrt{7} = \sqrt{10} - \sqrt{14}$

46. $\sqrt{5}\left(\sqrt{11} + \sqrt{3}\right) = \sqrt{5}\sqrt{11} + \sqrt{5}\sqrt{3}$
$= \sqrt{55} + \sqrt{15}$

47. $\left(\sqrt{3} + 2\right)\left(\sqrt{6} - 5\right)$
$= \sqrt{3}\sqrt{6} - 5\sqrt{3} + 2\sqrt{6} - 2 \cdot 5$
$= \sqrt{18} - 5\sqrt{3} + 2\sqrt{6} - 10$
$= \sqrt{9 \cdot 2} - 5\sqrt{3} + 2\sqrt{6} - 10$
$= 3\sqrt{2} - 5\sqrt{3} + 2\sqrt{6} - 10$

48. $\left(\sqrt{5} + 1\right)\left(\sqrt{5} - 3\right) = \sqrt{5}\sqrt{5} - 3\sqrt{5} + 1\sqrt{5} - 1 \cdot 3$
$= 5 + (-3 + 1)\sqrt{5} - 3$
$= 2 - 2\sqrt{5}$

49. $\left(\sqrt{x} - 2\right)^2 = \left(\sqrt{x}\right)^2 - 2\left(\sqrt{x}\right)(2) + 2^2$
$= x - 4\sqrt{x} + 4$

50. $\left(\sqrt{y}+4\right)^2 = \left(\sqrt{y}\right)^2 + 2\left(\sqrt{y}\right)(4)+4^2$
$$= y+8\sqrt{y}+16$$

51. $\dfrac{\sqrt{27}}{\sqrt{3}} = \sqrt{\dfrac{27}{3}} = \sqrt{9} = 3$

52. $\dfrac{\sqrt{20}}{\sqrt{5}} = \sqrt{\dfrac{20}{5}} = \sqrt{4} = 2$

53. $\dfrac{\sqrt{160}}{\sqrt{8}} = \sqrt{\dfrac{160}{8}}$
$$= \sqrt{20}$$
$$= \sqrt{4 \cdot 5}$$
$$= \sqrt{4}\sqrt{5}$$
$$= 2\sqrt{5}$$

54. $\dfrac{\sqrt{96}}{\sqrt{3}} = \sqrt{\dfrac{96}{3}}$
$$= \sqrt{32}$$
$$= \sqrt{16 \cdot 2}$$
$$= \sqrt{16}\sqrt{2}$$
$$= 4\sqrt{2}$$

55. $\dfrac{\sqrt{30x^6}}{\sqrt{2x^3}} = \sqrt{\dfrac{30x^6}{2x^3}}$
$$= \sqrt{15x^3}$$
$$= \sqrt{x^2 \cdot 15x}$$
$$= \sqrt{x^2}\sqrt{15x}$$
$$= x\sqrt{15x}$$

56. $\dfrac{\sqrt{54x^5y^2}}{\sqrt{3xy^2}} = \sqrt{\dfrac{54x^5y^2}{3xy^2}}$
$$= \sqrt{18x^4}$$
$$= \sqrt{9x^4 \cdot 2}$$
$$= \sqrt{9x^4}\sqrt{2}$$
$$= 3x^2\sqrt{2}$$

57. $\dfrac{\sqrt{2}}{\sqrt{11}} = \dfrac{\sqrt{2}}{\sqrt{11}} \cdot \dfrac{\sqrt{11}}{\sqrt{11}} = \dfrac{\sqrt{2 \cdot 11}}{11} = \dfrac{\sqrt{22}}{11}$

58. $\dfrac{\sqrt{3}}{\sqrt{13}} = \dfrac{\sqrt{3}}{\sqrt{13}} \cdot \dfrac{\sqrt{13}}{\sqrt{13}} = \dfrac{\sqrt{3 \cdot 13}}{13} = \dfrac{\sqrt{39}}{13}$

59. $\sqrt{\dfrac{5}{6}} = \dfrac{\sqrt{5}}{\sqrt{6}} = \dfrac{\sqrt{5}}{\sqrt{6}} \cdot \dfrac{\sqrt{6}}{\sqrt{6}} = \dfrac{\sqrt{5 \cdot 6}}{6} = \dfrac{\sqrt{30}}{6}$

60. $\sqrt{\dfrac{7}{10}} = \dfrac{\sqrt{7}}{\sqrt{10}} = \dfrac{\sqrt{7}}{\sqrt{10}} \cdot \dfrac{\sqrt{10}}{\sqrt{10}} = \dfrac{\sqrt{7 \cdot 10}}{10} = \dfrac{\sqrt{70}}{10}$

61. $\dfrac{1}{\sqrt{5x}} = \dfrac{1}{\sqrt{5x}} \cdot \dfrac{\sqrt{5x}}{\sqrt{5x}} = \dfrac{\sqrt{5x}}{5x}$

62. $\dfrac{5}{\sqrt{3y}} = \dfrac{5}{\sqrt{3y}} \cdot \dfrac{\sqrt{3y}}{\sqrt{3y}} = \dfrac{5\sqrt{3y}}{3y}$

63. $\sqrt{\dfrac{3}{x}} = \dfrac{\sqrt{3}}{\sqrt{x}} = \dfrac{\sqrt{3}}{\sqrt{x}} \cdot \dfrac{\sqrt{x}}{\sqrt{x}} = \dfrac{\sqrt{3x}}{x}$

64. $\sqrt{\dfrac{6}{y}} = \dfrac{\sqrt{6}}{\sqrt{y}} = \dfrac{\sqrt{6}}{\sqrt{y}} \cdot \dfrac{\sqrt{y}}{\sqrt{y}} = \dfrac{\sqrt{6y}}{y}$

65. $\dfrac{3}{\sqrt{5}-2} = \dfrac{3}{\sqrt{5}-2} \cdot \dfrac{\sqrt{5}+2}{\sqrt{5}+2}$
$$= \dfrac{3\left(\sqrt{5}+2\right)}{5-4}$$
$$= \dfrac{3\left(\sqrt{5}+2\right)}{1}$$
$$= 3\left(\sqrt{5}+2\right) \text{ or } 3\sqrt{5}+6$$

66. $\dfrac{8}{\sqrt{10}-3} = \dfrac{8}{\sqrt{10}-3} \cdot \dfrac{\sqrt{10}+3}{\sqrt{10}+3}$

$\qquad = \dfrac{8\left(\sqrt{10}+3\right)}{10-9}$

$\qquad = \dfrac{8\left(\sqrt{10}+3\right)}{1}$

$\qquad = 8\left(\sqrt{10}+3\right)$ or $8\sqrt{10}+24$

67. $\dfrac{\sqrt{2}+1}{\sqrt{3}-1} = \dfrac{\sqrt{2}+1}{\sqrt{3}-1} \cdot \dfrac{\sqrt{3}+1}{\sqrt{3}+1}$

$\qquad = \dfrac{\sqrt{2}\sqrt{3}+1\cdot\sqrt{2}+1\cdot\sqrt{3}+1\cdot 1}{3-1}$

$\qquad = \dfrac{\sqrt{6}+\sqrt{2}+\sqrt{3}+1}{2}$

68. $\dfrac{\sqrt{3}-2}{\sqrt{5}+2} = \dfrac{\sqrt{3}-2}{\sqrt{5}+2} \cdot \dfrac{\sqrt{5}-2}{\sqrt{5}-2}$

$\qquad = \dfrac{\sqrt{3}\sqrt{5}-2\sqrt{3}-2\sqrt{5}+2\cdot 2}{5-4}$

$\qquad = \dfrac{\sqrt{15}-2\sqrt{3}-2\sqrt{5}+4}{1}$

$\qquad = \sqrt{15}-2\sqrt{3}-2\sqrt{5}+4$

69. $\dfrac{10}{\sqrt{x}+5} = \dfrac{10}{\sqrt{x}+5} \cdot \dfrac{\sqrt{x}-5}{\sqrt{x}-5}$

$\qquad = \dfrac{10\left(\sqrt{x}-5\right)}{x-25}$

$\qquad = \dfrac{10\sqrt{x}-50}{x-25}$

70. $\dfrac{8}{\sqrt{x}-1} = \dfrac{8}{\sqrt{x}-1} \cdot \dfrac{\sqrt{x}+1}{\sqrt{x}+1}$

$\qquad = \dfrac{8\left(\sqrt{x}+1\right)}{x-1}$

$\qquad = \dfrac{8\sqrt{x}+8}{x-1}$

71. $\sqrt{2x} = 6$

$\qquad \left(\sqrt{2x}\right)^2 = 6^2$

$\qquad\qquad 2x = 36$

$\qquad\qquad\;\; x = 18$

72. $\sqrt{x+3} = 4$

$\qquad \left(\sqrt{x+3}\right)^2 = 4^2$

$\qquad\qquad x+3 = 16$

$\qquad\qquad\quad\; x = 13$

73. $\sqrt{x}+3 = 8$

$\qquad \sqrt{x} = 5$

$\qquad \left(\sqrt{x}\right)^2 = 5^2$

$\qquad\qquad x = 25$

74. $\sqrt{x}+8 = 3$

$\qquad \sqrt{x} = -5$

$\sqrt{x}$ cannot equal -5. Thus, the equation has no solution.

75. $\sqrt{2x+1} = x-7$

$\qquad \left(\sqrt{2x+1}\right)^2 = (x-7)^2$

$\qquad\qquad 2x+1 = x^2-14x+49$

$\qquad\qquad\quad 0 = x^2-16x+48$

$\qquad\qquad\quad 0 = (x-12)(x-4)$

$\qquad x-12 = 0 \quad$ or $\quad x-4 = 0$

$\qquad\quad x = 12 \qquad\qquad x = 4$

$x = 4$ does not check, so the solution is $x = 12$.

76. $\sqrt{3x+1} = x-1$

$\qquad \left(\sqrt{3x+1}\right)^2 = (x-1)^2$

$\qquad\qquad 3x+1 = x^2-2x+1$

$\qquad\qquad\quad 0 = x^2-5x$

$\qquad\qquad\quad 0 = x(x-5)$

$\qquad x = 0 \quad$ or $\quad x-5 = 0$

$\qquad\qquad\qquad\qquad x = 5$

$x = 0$ does not check, so the solution is $x = 5$.

77. $\sqrt{x} + 3 = \sqrt{x+15}$

$\left(\sqrt{x}+3\right)^2 = \left(\sqrt{x+15}\right)^2$

$x + 6\sqrt{x} + 9 = x + 15$

$6\sqrt{x} = 6$

$\sqrt{x} = 1$

$\left(\sqrt{x}\right)^2 = 1^2$

$x = 1$

78. $\sqrt{x-5} = \sqrt{x} - 1$

$\left(\sqrt{x-5}\right)^2 = \left(\sqrt{x}-1\right)^2$

$x - 5 = x - 2\sqrt{x} + 1$

$-6 = -2\sqrt{x}$

$3 = \sqrt{x}$

$3^2 = \left(\sqrt{x}\right)^2$

$9 = x$

79. $a^2 + b^2 = c^2$

$5^2 + b^2 = 9^2$

$25 + b^2 = 81$

$b^2 = 56$

$\sqrt{b^2} = \sqrt{56}$

$b = 2\sqrt{14} \approx 7.48$

80. $a^2 + b^2 = c^2$

$6^2 + 9^2 = c^2$

$36 + 81 = c^2$

$117 = c^2$

$\sqrt{117} = \sqrt{c^2}$

$c = 3\sqrt{13} \approx 10.82$

81. The distance between Romeo and Juliet is the length of the hypotenuse of a right triangle with legs of length 20 feet and 12 feet.

$a^2 + b^2 = c^2$

$20^2 + 12^2 = c^2$

$400 + 144 = c^2$

$544 = c^2$

$\sqrt{544} = \sqrt{c^2}$

$c = 4\sqrt{34} \approx 23.32$

The distance is exactly $4\sqrt{34}$ feet or approximately 23.32 feet.

82. The diagonal of a rectangle forms right triangles with the sides of the rectangle.

$a^2 + b^2 = c^2$

$5^2 + b^2 = 10^2$

$25 + b^2 = 100$

$b^2 = 75$

$\sqrt{b^2} = \sqrt{75}$

$b = 5\sqrt{3} \approx 8.66$

The length of the rectangle is exactly $5\sqrt{3}$ inches or approximately 8.66 inches.

83. $r = \sqrt{\dfrac{S}{4\pi}}$

$r = \sqrt{\dfrac{72}{4\pi}}$

$r \approx 2.4$

The radius is 2.4 inches.

84. $r = \sqrt{\dfrac{S}{4\pi}}$

$6 = \sqrt{\dfrac{S}{4\pi}}$

$6^2 = \left(\sqrt{\dfrac{S}{4\pi}} \right)^2$

$36 = \dfrac{S}{4\pi}$

$144\pi = S$

The surface area is 144π square inches.

85. $\sqrt{144} = 12$, because $12^2 = 144$ and 12 is positive.

86. $-\sqrt[3]{64} = -4$, because $4^3 = 64$.

87. $\sqrt{16x^{16}} = 4x^8$, because $(4x^8)^2 = 16x^{16}$.

88. $\sqrt{4x^{24}} = 2x^{12}$, because $(2x^{12})^2 = 4x^{24}$.

89. $\sqrt{18x^7} = \sqrt{9x^6 \cdot 2x^1} = \sqrt{9x^6}\sqrt{2x} = 3x^3\sqrt{2x}$

90. $\sqrt{48y^6} = \sqrt{16y^6 \cdot 3} = \sqrt{16y^6}\sqrt{3} = 4y^3\sqrt{3}$

91. $\sqrt{\dfrac{y^4}{81}} = \dfrac{\sqrt{y^4}}{\sqrt{81}} = \dfrac{y^2}{9}$

92. $\sqrt{\dfrac{x^9}{9}} = \dfrac{\sqrt{x^9}}{\sqrt{9}} = \dfrac{\sqrt{x^8 \cdot x}}{3} = \dfrac{\sqrt{x^8}\sqrt{x}}{3} = \dfrac{x^4\sqrt{x}}{3}$

93. $\sqrt{12} + \sqrt{75} = \sqrt{4 \cdot 3} + \sqrt{25 \cdot 3}$
$= \sqrt{4}\sqrt{3} + \sqrt{25}\sqrt{3}$
$= 2\sqrt{3} + 5\sqrt{3}$
$= (2+5)\sqrt{3}$
$= 7\sqrt{3}$

94. $\sqrt{63} + \sqrt{28} - \sqrt{9} = \sqrt{9 \cdot 7} + \sqrt{4 \cdot 7} - 3$
$= \sqrt{9}\sqrt{7} + \sqrt{4}\sqrt{7} - 3$
$= 3\sqrt{7} + 2\sqrt{7} - 3$
$= (3+2)\sqrt{7} - 3$
$= 5\sqrt{7} - 3$

95. $\sqrt{\dfrac{3}{16}} - \sqrt{\dfrac{3}{4}} = \dfrac{\sqrt{3}}{\sqrt{16}} - \dfrac{\sqrt{3}}{\sqrt{4}}$
$= \dfrac{\sqrt{3}}{4} - \dfrac{\sqrt{3}}{2}$
$= \dfrac{\sqrt{3}}{4} - \dfrac{2\sqrt{3}}{4}$
$= -\dfrac{\sqrt{3}}{4}$

96. $\sqrt{45x^3} + x\sqrt{20x} - \sqrt{5x^3}$
$= \sqrt{9x^2 \cdot 5x} + x\sqrt{4 \cdot 5x} - \sqrt{x^2 \cdot 5x}$
$= \sqrt{9x^2}\sqrt{5x} + x\sqrt{4}\sqrt{5x} - \sqrt{x^2}\sqrt{5x}$
$= 3x\sqrt{5x} + 2x\sqrt{5x} - x\sqrt{5x}$
$= (3x + 2x - x)\sqrt{5x}$
$= 4x\sqrt{5x}$

97. $\sqrt{7} \cdot \sqrt{14} = \sqrt{7 \cdot 14}$
$= \sqrt{98}$
$= \sqrt{49 \cdot 2}$
$= \sqrt{49}\sqrt{2}$
$= 7\sqrt{2}$

98. $\sqrt{3}\left(\sqrt{9} - \sqrt{2}\right) = \sqrt{3}\left(3 - \sqrt{2}\right)$
$= 3\sqrt{3} - \sqrt{3}\sqrt{2}$
$= 3\sqrt{3} - \sqrt{6}$

99. $\left(\sqrt{2} + 4\right)\left(\sqrt{5} - 1\right) = \sqrt{2}\sqrt{5} - 1\sqrt{2} + 4\sqrt{5} - 4 \cdot 1$
$= \sqrt{10} - \sqrt{2} + 4\sqrt{5} - 4$

100. $\left(\sqrt{x} + 3\right)^2 = \left(\sqrt{x}\right)^2 + 2\left(\sqrt{x}\right)(3) + (3)^2$
$= x + 6\sqrt{x} + 9$

101. $\dfrac{\sqrt{120}}{\sqrt{5}} = \sqrt{\dfrac{120}{5}} = \sqrt{24} = \sqrt{4 \cdot 6} = 2\sqrt{6}$

102. $\dfrac{\sqrt{60x^9}}{\sqrt{15x^4}} = \sqrt{\dfrac{60x^9}{15x^7}} = \sqrt{4x^2} = 2x$

103. $\sqrt{\dfrac{2}{7}} = \dfrac{\sqrt{2}}{\sqrt{7}} = \dfrac{\sqrt{2}}{\sqrt{7}} \cdot \dfrac{\sqrt{7}}{\sqrt{7}} = \dfrac{\sqrt{14}}{7}$

104. $\dfrac{3}{\sqrt{2x}} = \dfrac{3}{\sqrt{2x}} \cdot \dfrac{\sqrt{2x}}{\sqrt{2x}} = \dfrac{3\sqrt{2x}}{2x}$

105. $\dfrac{3}{\sqrt{x} - 6} = \dfrac{3}{\sqrt{x} - 6} \cdot \dfrac{\sqrt{x} + 6}{\sqrt{x} + 6}$

$= \dfrac{3\left(\sqrt{x} + 6\right)}{x - 36}$

$= \dfrac{3\sqrt{x} + 18}{x - 36}$

106. $\dfrac{\sqrt{7} - 5}{\sqrt{5} + 3} = \dfrac{\sqrt{7} - 5}{\sqrt{5} + 3} \cdot \dfrac{\sqrt{5} - 3}{\sqrt{5} - 3}$

$= \dfrac{\sqrt{7}\sqrt{5} - 3\sqrt{7} - 5\sqrt{5} + 5 \cdot 3}{5 - 9}$

$= \dfrac{\sqrt{35} - 3\sqrt{7} - 5\sqrt{5} + 15}{-4}$

107. $\sqrt{4x} = 2$

$\left(\sqrt{4x}\right)^2 = 2^2$

$4x = 4$

$x = 1$

108. $\sqrt{x - 4} = 3$

$\left(\sqrt{x - 4}\right)^2 = 3^2$

$x - 4 = 9$

$x = 13$

109. $\sqrt{4x + 8} + 6 = x$

$\sqrt{4x + 8} = x - 6$

$\left(\sqrt{4x + 8}\right)^2 = (x - 6)^2$

$4x + 8 = x^2 - 12x + 36$

$0 = x^2 - 16x + 28$

$0 = (x - 14)(x - 2)$

$x - 14 = 0 \quad$ or $\quad x - 2 = 0$

$x = 14 \qquad\qquad x = 2$

$x = 2$ does not check, so the solution is $x = 14$.

110. $\sqrt{x - 8} = \sqrt{x} - 2$

$\left(\sqrt{x - 8}\right)^2 = \left(\sqrt{x} - 2\right)^2$

$x - 8 = x - 4\sqrt{x} + 4$

$-12 = -4\sqrt{x}$

$3 = \sqrt{x}$

$3^2 = \left(\sqrt{x}\right)^2$

$9 = x$

111. $a^2 + b^2 = c^2$

$3^2 + 7^2 = c^2$

$9 + 49 = c^2$

$58 = c^2$

$\sqrt{58} = \sqrt{c^2}$

$c = \sqrt{58} \approx 7.62$

112. The diagonal is the hypotenuse of a right triangle with the sides of the rectangle as its legs.

$a^2 + b^2 = c^2$

$2^2 + b^2 = 6^2$

$4 + b^2 = 36$

$b^2 = 32$

$\sqrt{b^2} = \sqrt{32}$

$b = 4\sqrt{2} \approx 5.66$

The length of the rectangle is exactly $4\sqrt{2}$ inches or approximately 5.66 inches.

Chapter 8 Test

1. $\sqrt{16} = 4$, because $4^2 = 16$ and 4 is positive.

2. $\sqrt[3]{125} = 5$, because $5^3 = 125$.

3. $\sqrt[4]{81} = 3$, because $3^4 = 81$.

4. $\sqrt{\dfrac{9}{16}} = \dfrac{3}{4}$, because $\left(\dfrac{3}{4}\right)^2 = \dfrac{9}{16}$.

5. $\sqrt[4]{-81}$ is not a real number since the index 4 is even and the radicand -81 is negative.

6. $\sqrt{x^{10}} = x^5$, because $(x^5)^2 = x^{10}$.

7. $\sqrt{54} = \sqrt{9 \cdot 6} = \sqrt{9}\sqrt{6} = 3\sqrt{6}$

8. $\sqrt{92} = \sqrt{4 \cdot 23} = \sqrt{4}\sqrt{23} = 2\sqrt{23}$

9. $\sqrt{y^7} = \sqrt{y^6 \cdot y} = \sqrt{y^6}\sqrt{y} = y^3\sqrt{y}$

10. $\sqrt{24x^8} = \sqrt{4x^8 \cdot 6} = \sqrt{4x^8}\sqrt{6} = 2x^4\sqrt{6}$

11. $\sqrt[3]{27} = 3$

12. $\sqrt[3]{16} = \sqrt[3]{8 \cdot 2} = \sqrt[3]{8}\sqrt[3]{2} = 2\sqrt[3]{2}$

13. $\sqrt{\dfrac{5}{16}} = \dfrac{\sqrt{5}}{\sqrt{16}} = \dfrac{\sqrt{5}}{4}$

14. $\sqrt{\dfrac{y^3}{25}} = \dfrac{\sqrt{y^3}}{\sqrt{25}} = \dfrac{\sqrt{y^2 \cdot y}}{5} = \dfrac{\sqrt{y^2}\sqrt{y}}{5} = \dfrac{y\sqrt{y}}{5}$

15. $\sqrt{13} + \sqrt{13} - 4\sqrt{13} = (1 + 1 - 4)\sqrt{13} = -2\sqrt{13}$

16. $\sqrt{18} - \sqrt{75} + 7\sqrt{3} - \sqrt{8}$
$= \sqrt{9 \cdot 2} - \sqrt{25 \cdot 3} + 7\sqrt{3} - \sqrt{4 \cdot 2}$
$= \sqrt{9}\sqrt{2} - \sqrt{25}\sqrt{3} + 7\sqrt{3} - \sqrt{4}\sqrt{2}$
$= 3\sqrt{2} - 5\sqrt{3} + 7\sqrt{3} - 2\sqrt{2}$
$= (3 - 2)\sqrt{2} + (-5 + 7)\sqrt{3}$
$= \sqrt{2} + 2\sqrt{3}$

17. $\sqrt{\dfrac{3}{4}} + \sqrt{\dfrac{3}{25}} = \dfrac{\sqrt{3}}{\sqrt{4}} + \dfrac{\sqrt{3}}{\sqrt{25}}$
$= \dfrac{\sqrt{3}}{2} + \dfrac{\sqrt{3}}{5}$
$= \dfrac{5\sqrt{3}}{10} + \dfrac{2\sqrt{3}}{10}$
$= \dfrac{5\sqrt{3} + 2\sqrt{3}}{10}$
$= \dfrac{(5 + 2)\sqrt{3}}{10}$
$= \dfrac{7\sqrt{3}}{10}$

18. $\sqrt{7} \cdot \sqrt{14} = \sqrt{7 \cdot 14}$
$= \sqrt{98}$
$= \sqrt{49 \cdot 2}$
$= \sqrt{49}\sqrt{2}$
$= 7\sqrt{2}$

19. $\sqrt{2}\left(\sqrt{6} - \sqrt{5}\right) = \sqrt{2}\sqrt{6} - \sqrt{2}\sqrt{5}$
$= \sqrt{12} - \sqrt{10}$
$= \sqrt{4 \cdot 3} - \sqrt{10}$
$= 2\sqrt{3} - \sqrt{10}$

20. $\left(\sqrt{x} + 2\right)\left(\sqrt{x} - 3\right)$
$= \sqrt{x} \cdot \sqrt{x} - 3\sqrt{x} + 2\sqrt{x} - 2 \cdot 3$
$= x - 3\sqrt{x} + 2\sqrt{x} - 6$
$= x - \sqrt{x} - 6$

21. $\dfrac{\sqrt{50}}{\sqrt{10}} = \sqrt{\dfrac{50}{10}} = \sqrt{5}$

22. $\dfrac{\sqrt{40x^4}}{\sqrt{2x}} = \sqrt{\dfrac{40x^4}{2x}}$

$= \sqrt{20x^3}$

$= \sqrt{4x^2 \cdot 5x}$

$= \sqrt{4x^2}\sqrt{5x}$

$= 2x\sqrt{5x}$

23. $\sqrt{\dfrac{2}{3}} = \dfrac{\sqrt{2}}{\sqrt{3}} = \dfrac{\sqrt{2}}{\sqrt{3}} \cdot \dfrac{\sqrt{3}}{\sqrt{3}} = \dfrac{\sqrt{6}}{3}$

24. $\dfrac{8}{\sqrt{5y}} = \dfrac{8}{\sqrt{5y}} \cdot \dfrac{\sqrt{5y}}{\sqrt{5y}} = \dfrac{8\sqrt{5y}}{5y}$

25. $\dfrac{8}{\sqrt{6}+2} = \dfrac{8}{\sqrt{6}+2} \cdot \dfrac{\sqrt{6}-2}{\sqrt{6}-2}$

$= \dfrac{8\left(\sqrt{6}-2\right)}{\left(\sqrt{6}\right)^2 - 2^2}$

$= \dfrac{8\left(\sqrt{6}-2\right)}{6-4}$

$= \dfrac{8\left(\sqrt{6}-2\right)}{2}$

$= 4\left(\sqrt{6}-2\right)$

$= 4\sqrt{6}-8$

26. $\dfrac{1}{3-\sqrt{x}} = \dfrac{1}{3-\sqrt{x}} \cdot \dfrac{3+\sqrt{x}}{3+\sqrt{x}}$

$= \dfrac{1\left(3+\sqrt{x}\right)}{3^2 - \left(\sqrt{x}\right)^2}$

$= \dfrac{3+\sqrt{x}}{9-x}$

27. $\sqrt{x}+8 = 11$

$\sqrt{x} = 3$

$\left(\sqrt{x}\right)^2 = 3^2$

$x = 9$

28. $\sqrt{3x-6} = \sqrt{x+4}$

$\left(\sqrt{3x-6}\right)^2 = \left(\sqrt{x+4}\right)^2$

$3x-6 = x+4$

$2x-6 = 4$

$2x = 10$

$x = 5$

29. $\sqrt{2x-2} = x-5$

$\left(\sqrt{2x-2}\right)^2 = (x-5)^2$

$2x-2 = x^2 - 10x + 25$

$0 = x^2 - 12x + 27$

$0 = (x-3)(x-9)$

$x-3=0$ or $x-9=0$

$x=3 \qquad x=9$

$x = 3$ does not check, so the solution is $x = 9$.

30. $a^2 + b^2 = c^2$

$8^2 + b^2 = 12^2$

$64 + b^2 = 144$

$b^2 = 80$

$\sqrt{b^2} = \sqrt{80}$

$b = 4\sqrt{5}$

The length is $4\sqrt{5}$ inches.

31. $r = \sqrt{\dfrac{A}{\pi}}$

$r = \sqrt{\dfrac{15}{\pi}}$

$r \approx 2.19$

The radius is 2.19 meters.

Cumulative Review Chapters 1–8

1. $-2(-14) = 2 \cdot 14 = 28$

2. $9(-5.2) = -9 \cdot 5.2 = -46.8$

3. $-\dfrac{2}{3} \cdot \dfrac{4}{7} = -\dfrac{2 \cdot 4}{3 \cdot 7} = -\dfrac{8}{21}$

4. $-3\dfrac{3}{8}\cdot 5\dfrac{1}{3} = -\dfrac{27}{8}\cdot\dfrac{16}{3}$

$\qquad = -\dfrac{27\cdot 16}{8\cdot 3}$

$\qquad = -9\cdot 2$

$\qquad = -18$

5. $4(2x-3)+7 = 3x+5$

$\quad 8x-12+7 = 3x+5$

$\qquad 8x-5 = 3x+5$

$\qquad 5x-5 = 5$

$\qquad\quad 5x = 10$

$\qquad\quad x = 2$

6. $6y-11+4+2y = 8+15y-8y$

$\qquad 8y-7 = 8+7y$

$\qquad y-7 = 8$

$\qquad y = 15$

7. a. The sector for business represents 17% of the circle.

  b. These sectors comprise a total of 17% + 4% or 21% of the circle.

  c. 17% of 253 = 0.17(253) = 43.01
About 43 of the 253 American travelers would be traveling solely for business.

8. a. $\dfrac{4(-3)-(-6)}{-8+4} = \dfrac{-12+6}{-4} = \dfrac{-6}{-4} = \dfrac{3}{2}$

  b. $\dfrac{3+(-3)(-2)^3}{-1-(-4)} = \dfrac{3+(-3)(-8)}{-1+4}$

$\qquad = \dfrac{3+24}{3}$

$\qquad = \dfrac{27}{3}$

$\qquad = 9$

9. a. $1.02\times 10^5 = 102{,}000$

  b. $7.358\times 10^{-3} = 0.007358$

  c. $8.4\times 10^7 = 84{,}000{,}000$

  d. $3.007\times 10^{-5} = 0.00003007$

10. a. $7{,}200{,}000 = 7.2\times 10^6$

  b. $0.000308 = 3.08\times 10^{-4}$

11. $(3x+2)(2x-5)$

$= (3x)(2x)-(3x)(5)+2(2x)-2(5)$

$= 6x^2-15x+4x-10$

$= 6x^2-11x-10$

12. $(7x+1)^2 = (7x)^2+2(7x)(1)+(1)^2$

$\qquad = 49x^2+14x+1$

13. $xy+2x+3y+6 = x(y+2)+3(y+2)$

$\qquad = (y+2)(x+3)$

14. $xy^2+5x-y^2-5 = x(y^2+5)-(y^2+5)$

$\qquad = (y^2+5)(x-1)$

15. $3x^2+11x+6 = (3x+2)(x+3)$

16. $3x^2+15x+18 = 3(x^2+5x+6)$

$\qquad = 3(x+2)(x+3)$

17. a. The expression is undefined for $x = 3$, since the denominator, $x-3$, is 0 when $x = 3$.

  b. The expression is undefined for $x = 2$ and $x = 1$, since the denominator, $x^2-3x+2 = (x-2)(x-1)$, is 0 when $x = 2$ or $x = 1$.

  c. There are no values for which the expression is undefined, since the denominator, 3, is never 0.

18. $\dfrac{2x^2+7x+3}{x^2-9} = \dfrac{(2x+1)(x+3)}{(x+3)(x-3)} = \dfrac{2x+1}{x-3}$

19. $\dfrac{x^2+4x+4}{x^2+2x}=\dfrac{(x+2)(x+2)}{x(x+2)}=\dfrac{x+2}{x}$

20. $\begin{aligned}\dfrac{12x^2y^3}{5}\div\dfrac{3y^3}{x}&=\dfrac{12x^2y^3}{5}\cdot\dfrac{x}{3y^3}\\[2mm]&=\dfrac{4x^2}{5}\cdot\dfrac{x}{1}\\[2mm]&=\dfrac{4x^3}{5}\end{aligned}$

21. a. $\dfrac{a}{4}-\dfrac{2a}{8}=\dfrac{a}{4}-\dfrac{a}{4}=0$

    b. $\dfrac{3}{10x^2}+\dfrac{7}{25x}=\dfrac{15}{50x^2}+\dfrac{14x}{50x^2}=\dfrac{15+14x}{50x^2}$

22. $y=mx+b$
    $y=-2x+4$

23.
$$\dfrac{4x}{x^2+x-30}+\dfrac{2}{x-5}=\dfrac{1}{x+6}$$
$$(x-5)(x+6)\left(\dfrac{4x}{(x-5)(x+6)}+\dfrac{2}{x-5}\right)=(x-5)(x+6)\left(\dfrac{1}{x+6}\right)$$
$$4x+2(x+6)=x-5$$
$$4x+2x+12=x-5$$
$$6x+12=x-5$$
$$5x+12=-5$$
$$5x=-17$$
$$x=-\dfrac{17}{5}$$

24. $\begin{aligned}4a^2+3a-2a^2+7a-5&=(4-2)a^2+(3+7)a-5\\&=2a^2+10a-5\end{aligned}$

25.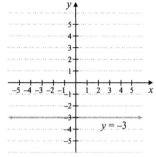

26. For $x = 0$:
$$2x + y = 6$$
$$2(0) + y = 6$$
$$y = 6$$

For $y = -2$:
$$2x + y = 6$$
$$2x - 2 = 6$$
$$2x = 8$$
$$x = 4$$

For $x = 3$:
$$2x + y = 6$$
$$2(3) + y = 6$$
$$6 + y = 6$$
$$y = 0$$

x	y
0	6
4	-2
3	0

27. $y = mx + b$
$$y = \frac{1}{4}x - 3$$

28. The slope of a line perpendicular to $y = 2x + 4$ has slope $-\frac{1}{2}$.
$$y - y_1 = m(x - x_1)$$
$$y - 5 = -\frac{1}{2}(x - 1)$$
$$y = -\frac{1}{2}x + \frac{1}{2} + 5$$
$$y = -\frac{1}{2}x + \frac{11}{2}$$

29. $\begin{cases} 3x + 4y = 13 \\ 5x - 9y = 6 \end{cases}$

$$\begin{array}{r} 15x + 20y = 65 \\ -15x + 27y = -18 \\ \hline 47y = 47 \\ y = 1 \end{array}$$

$$3x + 4y = 13$$
$$3x + 4(1) = 13$$
$$3x = 9$$
$$x = 3$$
The solution is (3, 1).

30. $\dfrac{x}{2} + y = \dfrac{5}{6}$

$$\begin{array}{r} 2x - y = \dfrac{5}{6} \\ \hline \dfrac{5}{2}x \quad = \dfrac{10}{6} \\ x = \dfrac{2}{3} \end{array}$$

$$\frac{x}{2} + y = \frac{5}{6}$$

$$\frac{1}{2}\left(\frac{2}{3}\right) + y = \frac{5}{6}$$

$$\frac{1}{3} + y = \frac{5}{6}$$

$$y = \frac{5}{6} - \frac{1}{3}$$

$$y = \frac{1}{2}$$

The solution is $\left(\frac{2}{3}, \frac{1}{2}\right)$.

31. Let x be Alfredo's walking speed. Then $x + 1$ is Louisa's walking speed.

$$2x + 2(x+1) = 15$$
$$2x + 2x + 2 = 15$$
$$4x + 2 = 15$$
$$4x = 13$$
$$x = \frac{13}{4} \text{ or } 3.25$$

$x + 1 = 3.25 + 1 = 4.25$
Alfredo's walking speed is 3.25 miles per hour, and Louisa's is 4.25 miles per hour.

32. Let x be the speed of the slower streetcar. Then $x + 15$ is the speed of the faster streetcar. Twelve minutes is $\frac{1}{5}$ hour.

$$\frac{1}{5}x + \frac{1}{5}(x+15) = 11$$
$$\frac{1}{5}x + \frac{1}{5}x + 3 = 11$$
$$\frac{2}{5}x + 3 = 11$$
$$\frac{2}{5}x = 8$$
$$x = 20$$

$x + 15 = 20 + 15 = 35$
The streetcars are traveling at 20 miles per hour and 35 miles per hour.

33. $\sqrt[3]{1} = 1$, because $1^3 = 1$.

34. $\sqrt{121} = 11$, because $11^2 = 121$ and 11 is positive.

35. $\sqrt[3]{-27} = -3$, because $(-3)^3 = -27$.

36. $\sqrt{\frac{1}{4}} = \frac{1}{2}$, because $\left(\frac{1}{2}\right)^2 = \frac{1}{4}$.

37. $\sqrt[3]{\frac{1}{125}} = \frac{1}{5}$, because $\left(\frac{1}{5}\right)^3 = \frac{1}{125}$.

38. $\sqrt{\frac{25}{144}} = \frac{5}{12}$, because $\left(\frac{5}{12}\right)^2 = \frac{25}{144}$.

39. $\sqrt{54} = \sqrt{9 \cdot 6} = \sqrt{9}\sqrt{6} = 3\sqrt{6}$

40. $\sqrt{63} = \sqrt{9 \cdot 7} = \sqrt{9}\sqrt{7} = 3\sqrt{7}$

41. $\sqrt{200} = \sqrt{100 \cdot 2} = \sqrt{100}\sqrt{2} = 10\sqrt{2}$

42. $\sqrt{500} = \sqrt{100 \cdot 5} = \sqrt{100}\sqrt{5} = 10\sqrt{5}$

43. $7\sqrt{12} - 2\sqrt{75} = 7\sqrt{4 \cdot 3} - 2\sqrt{25 \cdot 3}$
$$= 7\sqrt{4}\sqrt{3} - 2\sqrt{25}\sqrt{3}$$
$$= 7 \cdot 2\sqrt{3} - 2 \cdot 5\sqrt{3}$$
$$= 14\sqrt{3} - 10\sqrt{3}$$
$$= (14 - 10)\sqrt{3}$$
$$= 4\sqrt{3}$$

44. $\left(\sqrt{x} + 5\right)\left(\sqrt{x} - 5\right) = \left(\sqrt{x}\right)^2 - 5^2 = x - 25$

45. $2\sqrt{x^2} - \sqrt{25x^5} + \sqrt{x^5}$
$$= 2x - \sqrt{25x^4 \cdot x} + \sqrt{x^4 \cdot x}$$
$$= 2x - \sqrt{25x^4}\sqrt{x} + \sqrt{x^4}\sqrt{x}$$
$$= 2x - 5x^2\sqrt{x} + x^2\sqrt{x}$$
$$= 2x + (-5x^2 + x^2)\sqrt{x}$$
$$= 2x - 4x^2\sqrt{x}$$

46.
$$\left(\sqrt{6}+2\right)^2 = \left(\sqrt{6}\right)^2 + 2\left(\sqrt{6}\right)(2) + (2)^2$$
$$= 6 + 4\sqrt{6} + 4$$
$$= 10 + 4\sqrt{6}$$

47.
$$\frac{2}{\sqrt{7}} = \frac{2}{\sqrt{7}} \cdot \frac{\sqrt{7}}{\sqrt{7}} = \frac{2\sqrt{7}}{7}$$

48.
$$\frac{x+3}{\frac{1}{x}+\frac{1}{3}} = \frac{x+3}{\frac{3}{3x}+\frac{x}{3x}} = \frac{x+3}{\frac{3+x}{3x}} = (x+3) \cdot \frac{3x}{3+x} = 3x$$

49.
$$\sqrt{x} = \sqrt{5x-2}$$
$$\left(\sqrt{x}\right)^2 = \left(\sqrt{5x-2}\right)^2$$
$$x = 5x - 2$$
$$-4x = -2$$
$$x = \frac{-2}{-4}$$
$$x = \frac{1}{2}$$

The solution is $\frac{1}{2}$.

50.
$$\sqrt{x+4} = \sqrt{3x-1}$$
$$\left(\sqrt{x+4}\right)^2 = \left(\sqrt{3x-1}\right)^2$$
$$x + 4 = 3x - 1$$
$$-2x + 4 = -1$$
$$-2x = -5$$
$$x = \frac{-5}{-2}$$
$$x = \frac{5}{2}$$

The solution is $\frac{5}{2}$.

Chapter 9

Section 9.1 Practice Problems

1.
$$x^2 - 25 = 0$$
$$(x+5)(x-5) = 0$$
$$x+5 = 0 \quad \text{or} \quad x-5 = 0$$
$$x = -5 \qquad\qquad x = 5$$
The solutions are −5 and 5.

2.
$$2x^2 - 3x = 9$$
$$2x^2 - 3x - 9 = 0$$
$$(2x+3)(x-3) = 0$$
$$2x+3 = 0 \quad \text{or} \quad x-3 = 0$$
$$2x = -3 \qquad\qquad x = 3$$
$$x = -\frac{3}{2}$$
The solutions are $-\frac{3}{2}$ and 3.

3. $x^2 - 16 = 0$
$$x^2 = 16$$
$$x = \sqrt{16} \quad \text{or} \quad x = -\sqrt{16}$$
$$x = 4 \qquad\qquad x = -4$$
The solutions are 4 and −4.

4. $3x^2 = 11$
$$x^2 = \frac{11}{3}$$
$$x = \sqrt{\frac{11}{3}} \quad \text{or} \quad x = -\sqrt{\frac{11}{3}}$$
$$x = \frac{\sqrt{11} \cdot \sqrt{3}}{\sqrt{3} \cdot \sqrt{3}} \qquad x = -\frac{\sqrt{11} \cdot \sqrt{3}}{\sqrt{3} \cdot \sqrt{3}}$$
$$x = \frac{\sqrt{33}}{3} \qquad\qquad x = -\frac{\sqrt{33}}{3}$$
The solutions are $\frac{\sqrt{33}}{3}$ and $-\frac{\sqrt{33}}{3}$.

5. $(x-4)^2 = 49$
$$x-4 = \sqrt{49} \quad \text{or} \quad x-4 = -\sqrt{49}$$
$$x-4 = 7 \qquad\qquad x-4 = -7$$
$$x = 11 \qquad\qquad x = -3$$
The solutions are 11 and −3.

6. $(x-5)^2 = 18$
$$x-5 = \sqrt{18} \quad \text{or} \quad x-5 = -\sqrt{18}$$
$$x-5 = 3\sqrt{2} \qquad\qquad x-5 = -3\sqrt{2}$$
$$x = 5+3\sqrt{2} \qquad\qquad x = 5-3\sqrt{2}$$
$$x = 5 \pm 3\sqrt{2}$$
The solutions are $5 \pm 3\sqrt{2}$.

7. $(x+3)^2 = -5$
This equation has no real solution because the square root of −5 is not a real number.

8. $(4x+1)^2 = 15$
$$4x+1 = \sqrt{15} \quad \text{or} \quad 4x+1 = -\sqrt{15}$$
$$4x = -1+\sqrt{15} \qquad\qquad 4x = -1-\sqrt{15}$$
$$x = \frac{-1+\sqrt{15}}{4} \qquad\qquad x = \frac{-1-\sqrt{15}}{4}$$
$$x = \frac{-1 \pm \sqrt{15}}{4}$$
The solutions are $\dfrac{-1 \pm \sqrt{15}}{4}$.

9.
$$h = 16t^2$$
$$650 = 16t^2$$
$$40.625 = t^2$$
$$6.4 = t \text{ or } -6.4 = t$$
−6.4 is rejected because time cannot be negative.
It takes the object 6.4 seconds to fall 650 feet.

Exercise Set 9.1

1. $k^2 - 49 = 0$
$(k + 7)(k - 7) = 0$
$k + 7 = 0$ or $k - 7 = 0$
$\quad k = -7 \qquad k = 7$
The solutions are $k = -7$ and $k = 7$.

3. $m^2 + 2m = 15$
$m^2 + 2m - 15 = 0$
$(m + 5)(m - 3) = 0$
$m + 5 = 0$ or $m - 3 = 0$
$\quad m = -5 \qquad m = 3$
The solutions are $m = -5$ and $m = 3$.

5. $2x^2 - 32 = 0$
$2(x^2 - 16) = 0$
$2(x + 4)(x - 4) = 0$
$x + 4 = 0$ or $x - 4 = 0$
$\quad x = -4 \qquad x = 4$
The solutions are $x = -4$ and $x = 4$.

7. $4a^2 - 36 = 0$
$4(a^2 - 9) = 0$
$4(a - 3)(a + 3) = 0$
$a - 3 = 0$ or $a + 3 = 0$
$\quad a = 3 \qquad a = -3$
The solutions are $a = 3$ and $a = -3$.

9. $x^2 + 7x = -10$
$x^2 + 7x + 10 = 0$
$(x + 2)(x + 5) = 0$
$x + 2 = 0$ or $x + 5 = 0$
$\quad x = -2 \qquad x = -5$
The solutions are $x = -2$ and $x = -5$.

11. $x^2 = 64$
$x = \sqrt{64}$ or $x = -\sqrt{64}$
$x = 8 \qquad x = -8$
The solutions are $x = \pm 8$.

13. $x^2 = 21$
$x = \sqrt{21}$ or $x = -\sqrt{21}$
The solutions are $x = \pm\sqrt{21}$.

15. $x^2 = \dfrac{1}{25}$

$x = \sqrt{\dfrac{1}{25}}$ or $x = -\sqrt{\dfrac{1}{25}}$

$x = \dfrac{1}{5} \qquad x = -\dfrac{1}{5}$

The solutions are $x = \pm\dfrac{1}{5}$.

17. $x^2 = -4$ has no real solution because the square root of -4 is not a real number.

19. $3x^2 = 13$

$x^2 = \dfrac{13}{3}$

$x = \sqrt{\dfrac{13}{3}}$ or $x = -\sqrt{\dfrac{13}{3}}$

$x = \dfrac{\sqrt{13}}{\sqrt{3}} \cdot \dfrac{\sqrt{3}}{\sqrt{3}} \qquad x = -\dfrac{\sqrt{13}}{\sqrt{3}} \cdot \dfrac{\sqrt{3}}{\sqrt{3}}$

$x = \dfrac{\sqrt{39}}{3} \qquad x = -\dfrac{\sqrt{39}}{3}$

The solutions are $x = \pm\dfrac{\sqrt{39}}{3}$.

21. $7x^2 = 4$

$x^2 = \dfrac{4}{7}$

$x = \sqrt{\dfrac{4}{7}}$ or $x = -\sqrt{\dfrac{4}{7}}$

$x = \dfrac{\sqrt{4}}{\sqrt{7}} \cdot \dfrac{\sqrt{7}}{\sqrt{7}} \qquad x = -\dfrac{\sqrt{4}}{\sqrt{7}} \cdot \dfrac{\sqrt{7}}{\sqrt{7}}$

$x = \dfrac{2\sqrt{7}}{7} \qquad x = -\dfrac{2\sqrt{7}}{7}$

The solutions are $x = \pm\dfrac{2\sqrt{7}}{7}$.

23. $x^2 - 2 = 0$

$x^2 = 2$

$x = \sqrt{2}$ or $x = -\sqrt{2}$

The solutions are $x = \pm\sqrt{2}$.

25. $(x-5)^2 = 49$

$x - 5 = \sqrt{49}$ or $x - 5 = -\sqrt{49}$

$x - 5 = 7$ $x - 5 = -7$

$x = 12$ $x = -2$

The solutions are $x = -2$ and $x = 12$.

27. $(x+2)^2 = 7$

$x + 2 = \sqrt{7}$ or $x + 2 = -\sqrt{7}$

$x = -2 + \sqrt{7}$ $x = -2 - \sqrt{7}$

The solutions are $x = -2 \pm \sqrt{7}$.

29. $\left(m - \dfrac{1}{2}\right)^2 = \dfrac{1}{4}$

$m - \dfrac{1}{2} = \sqrt{\dfrac{1}{4}}$ or $m - \dfrac{1}{2} = -\sqrt{\dfrac{1}{4}}$

$m - \dfrac{1}{2} = \dfrac{1}{2}$ $m - \dfrac{1}{2} = -\dfrac{1}{2}$

$m = 1$ $m = 0$

The solutions are $m = 0$ and $m = 1$.

31. $(p+2)^2 = 10$

$p + 2 = \sqrt{10}$ or $p + 2 = -\sqrt{10}$

$p = -2 + \sqrt{10}$ $p = -2 - \sqrt{10}$

The solutions are $p = -2 \pm \sqrt{10}$.

33. $(3y+2)^2 = 100$

$3y + 2 = \sqrt{100}$ or $3y + 2 = -\sqrt{100}$

$3y + 2 = 10$ $3y + 2 = -10$

$3y = 8$ $3y = -12$

$y = \dfrac{8}{3}$ $y = -4$

The solutions are $y = -4$ and $y = \dfrac{8}{3}$.

35. $(z-4)^2 = -9$ has no real solution because the square root of -9 is not a real number.

37. $(2x-11)^2 = 50$

$2x - 11 = \sqrt{50}$ or $2x - 11 = -\sqrt{50}$

$2x - 11 = 5\sqrt{2}$ $2x - 11 = -5\sqrt{2}$

$2x = 11 + 5\sqrt{2}$ $2x = 11 - 5\sqrt{2}$

$x = \dfrac{11 + 5\sqrt{2}}{2}$ $x = \dfrac{11 - 5\sqrt{2}}{2}$

The solutions are $x = \dfrac{11 \pm 5\sqrt{2}}{2}$.

39. $(3x-7)^2 = 32$

$3x - 7 = \sqrt{32}$ or $3x - 7 = -\sqrt{32}$

$3x - 7 = 4\sqrt{2}$ $3x - 7 = -4\sqrt{2}$

$3x = 7 + 4\sqrt{2}$ $3x = 7 - 4\sqrt{2}$

$x = \dfrac{7 + 4\sqrt{2}}{3}$ $x = \dfrac{7 - 4\sqrt{2}}{3}$

The solutions are $x = \dfrac{7 \pm 4\sqrt{2}}{3}$.

41. $x^2 - 2 = 0$

$x^2 = 2$

$x = \sqrt{2}$ or $x = -\sqrt{2}$

The solutions are $x = \pm\sqrt{2}$.

43. $(x+6)^2 = 24$

$x + 6 = \sqrt{24}$ or $x + 6 = -\sqrt{24}$

$x + 6 = 2\sqrt{6}$ $x + 6 = -2\sqrt{6}$

$x = -6 + 2\sqrt{6}$ $x = -6 - 2\sqrt{6}$

The solutions are $x = -6 \pm 2\sqrt{6}$.

45. $\dfrac{1}{2}n^2 = 5$

$n^2 = 10$

$n = \sqrt{10}$ or $n = -\sqrt{10}$

The solutions are $n = \pm\sqrt{10}$.

47. $(4x-1)^2 = 5$

$4x-1 = \sqrt{5}$ or $4x-1 = -\sqrt{5}$

$4x = 1+\sqrt{5}$ $4x = 1-\sqrt{5}$

$x = \dfrac{1+\sqrt{5}}{4}$ $x = \dfrac{1-\sqrt{5}}{4}$

The solutions are $x = \dfrac{1\pm\sqrt{5}}{4}$.

49. $3z^2 = 36$

$z^2 = 12$

$z = \sqrt{12}$ or $z = -\sqrt{12}$

$z = 2\sqrt{3}$ $z = -2\sqrt{3}$

The solutions are $z = \pm 2\sqrt{3}$.

51. $(8-3x)^2 - 45 = 0$

$(8-3x)^2 = 45$

$8-3x = \sqrt{45}$ or $8-3x = -\sqrt{45}$

$8-3x = 3\sqrt{5}$ $8-3x = -3\sqrt{5}$

$-3x = -8+3\sqrt{5}$ $-3x = -8-3\sqrt{5}$

$x = \dfrac{-8+3\sqrt{5}}{-3}$ $x = \dfrac{-8-3\sqrt{5}}{-3}$

The solutions are $x = \dfrac{-8\pm 3\sqrt{5}}{-3}$.

53. $h = 16t^2$

$87.6 = 16t^2$

$\dfrac{87.6}{16} = t^2$

$5.475 = t^2$

$\sqrt{5.475} = t$ or $-\sqrt{5.475} = t$

$2.3 \approx t$ $-2.3 \approx t$

Since the time of a dive is not a negative number, reject the solution -2.3. The dive lasted approximately 2.3 seconds.

55. $h = 16t^2$

$343 = 16t^2$

$21.44 = t^2$

$\sqrt{21.44} = t$ or $-\sqrt{21.44} = t$

$4.6 \approx t$ $-4.6 \approx t$

Since the time of a jump is not a negative number, reject the solution -4.6. The fall lasted approximately 4.6 seconds.

57. $A = s^2$

$20 = s^2$

$\sqrt{20} = s$ or $-\sqrt{20} = s$

$2\sqrt{5} = s$ $-2\sqrt{5} = s$

Since the length of a side is not a negative number, reject the solution $-2\sqrt{5}$. The sides have length $2\sqrt{5} \approx 4.47$ inches.

59. $A = s^2$

$31,329 = s^2$

$\sqrt{31,329} = s$ or $-\sqrt{31,329} = s$

$177 = s$ $-177 = s$

Since the length of a side is not a negative number, reject the solution -177. The sides have length 177 meters.

61. $x^2 + 6x + 9 = (x)^2 + 2(x)(3) + (3)^2 = (x+3)^2$

63. $x^2 - 4x + 4 = x^2 - 2(x)(2) + (2)^2 = (x-2)^2$

65. answers may vary

67. $x^2 + 4x + 4 = 16$

$(x+2)^2 = 16$

$x+2 = \sqrt{16}$ or $x+2 = -\sqrt{16}$

$x+2 = 4$ $x+2 = -4$

$x = 2$ $x = -6$

The solutions are $x = 2$ and $x = -6$.

69. $A = \pi r^2$

$36\pi = \pi r^2$

$36 = r^2$

$\sqrt{36} = r$ or $-\sqrt{36} = r$

$\quad 6 = r \qquad\qquad -6 = r$

Since the radius is not a negative number, reject the solution -6. The radius of the circle is
6 inches.

71. $x^2 = 1.78$

$x = \sqrt{1.78}$ or $x = -\sqrt{1.78}$

$x \approx \pm 1.33$

The solutions are $x = \pm 1.33$.

73. $y = -0.07(x - 192.5)^2 + 3135$

$727 = -0.07(x - 192.5)^2 + 3135$

$-2408 = -0.07(x - 192.5)^2$

$\dfrac{-2408}{-0.07} = (x - 192.5)^2$

$34,400 = (x - 192.5)^2$

$\sqrt{34,400} = x - 192.5$

$20\sqrt{86} = x - 192.5$

$192.5 + 20\sqrt{86} = x$

$\qquad 378 \approx x$

or $\quad -\sqrt{34,400} = x - 192.5$

$\qquad -20\sqrt{86} = x - 192.5$

$192.5 - 20\sqrt{86} = x$

$\qquad\qquad 7 \approx x$

Since $7 < 378$, the solution is $x = 7$. There will be 727 Barnes and Noble Booksellers open for business when $x = 7$, which is 7 years after 1999, or 2006.

Section 9.2 Practice Problems

1. $x^2 + 8x + 1 = 0$

$x^2 + 8x = -1$

$x^2 + 8x + 16 = -1 + 16$

$(x + 4)^2 = 15$

$x + 4 = \sqrt{15}$ or $x + 4 = -\sqrt{15}$

$\quad x = -4 + \sqrt{15} \qquad x = -4 - \sqrt{15}$

The solutions are $-4 \pm \sqrt{15}$.

2. $x^2 - 14x = -32$

$x^2 - 14x + 49 = -32 + 49$

$(x - 7)^2 = 17$

$x - 7 = \sqrt{17}$ or $x - 7 = -\sqrt{17}$

$\quad x = 7 + \sqrt{17} \qquad x = 7 - \sqrt{17}$

The solutions are $7 \pm \sqrt{17}$.

3. $4x^2 - 16x - 9 = 0$

$x^2 - 4x - \dfrac{9}{4} = 0$

$x^2 - 4x = \dfrac{9}{4}$

$x^2 - 4x + 4 = \dfrac{9}{4} + 4$

$(x - 2)^2 = \dfrac{25}{4}$

$x - 2 = \sqrt{\dfrac{25}{4}}$ or $x - 2 = -\sqrt{\dfrac{25}{4}}$

$x - 2 = \dfrac{5}{2} \qquad\qquad x - 2 = -\dfrac{5}{2}$

$x = 2 + \dfrac{5}{2} \qquad\qquad x = 2 - \dfrac{5}{2}$

$x = \dfrac{9}{2} \qquad\qquad\quad x = -\dfrac{1}{2}$

The solutions are $\dfrac{9}{2}$ and $-\dfrac{1}{2}$.

4. $2x^2 + 10x = -13$

$x^2 + 5x = -\dfrac{13}{2}$

$x^2 + 5x + \dfrac{25}{4} = -\dfrac{13}{2} + \dfrac{25}{4}$

$\left(x + \dfrac{5}{2}\right)^2 = -\dfrac{1}{4}$

There is no real solution to this equation since the square root of a negative number is not a real number.

5. $2x^2 = -6x + 5$

$x^2 = -3x + \dfrac{5}{2}$

$x^2 + 3x = \dfrac{5}{2}$

$x^2 + 3x + \dfrac{9}{4} = \dfrac{5}{2} + \dfrac{9}{4}$

$\left(x + \dfrac{3}{2}\right)^2 = \dfrac{19}{4}$

$x + \dfrac{3}{2} = \sqrt{\dfrac{19}{4}}$ or $x + \dfrac{3}{2} = -\sqrt{\dfrac{19}{4}}$

$x + \dfrac{3}{2} = \dfrac{\sqrt{19}}{2}$ $x + \dfrac{3}{2} = -\dfrac{\sqrt{19}}{2}$

$x = -\dfrac{3}{2} + \dfrac{\sqrt{19}}{2}$ $x = -\dfrac{3}{2} - \dfrac{\sqrt{19}}{2}$

The solutions are $\dfrac{-3 \pm \sqrt{19}}{2}$.

Mental Math

1. $p^2 + 8p \Rightarrow \left(\dfrac{8}{2}\right)^2 = 4^2 = 16$

2. $p^2 + 6p \Rightarrow \left(\dfrac{6}{2}\right)^2 = 3^2 = 9$

3. $x^2 + 20x \Rightarrow \left(\dfrac{20}{2}\right)^2 = 10^2 = 100$

4. $x^2 + 18x \Rightarrow \left(\dfrac{18}{2}\right)^2 = 9^2 = 81$

5. $y^2 + 14y \Rightarrow \left(\dfrac{14}{2}\right)^2 = 7^2 = 49$

6. $y^2 + 2y \Rightarrow \left(\dfrac{2}{2}\right)^2 = 1^2 = 1$

Exercise Set 9.2

1. $x^2 + 8x = -12$

$x^2 + 8x + \left(\dfrac{8}{2}\right)^2 = -12 + \left(\dfrac{8}{2}\right)^2$

$x^2 + 8x + 4^2 = -12 + 4^2$

$x^2 + 8x + 16 = -12 + 16$

$(x + 4)^2 = 4$

$x + 4 = \sqrt{4}$ or $x + 4 = -\sqrt{4}$

$x = -4 + 2$ $x = -4 - 2$

$x = -2$ $x = -6$

The solutions are $x = -6$ and $x = -2$.

3. $x^2 + 2x - 7 = 0$

$x^2 + 2x = 7$

$x^2 + 2x + \left(\dfrac{2}{2}\right)^2 = 7 + \left(\dfrac{2}{2}\right)^2$

$x^2 + 2x + 1 = 7 + 1$

$(x + 1)^2 = 8$

$x + 1 = \sqrt{8}$ or $x + 1 = -\sqrt{8}$

$x + 1 = 2\sqrt{2}$ $x + 1 = -2\sqrt{2}$

$x = -1 + 2\sqrt{2}$ $x = -1 - 2\sqrt{2}$

The solutions are $x = -1 \pm 2\sqrt{2}$.

5.
$$x^2 - 6x = 0$$
$$x^2 - 6x + \left(\frac{-6}{2}\right)^2 = 0 + \left(\frac{-6}{2}\right)^2$$
$$x^2 - 6x + (-3)^2 = 0 + (-3)^2$$
$$x^2 - 6x + 9 = 9$$
$$(x-3)^2 = 9$$
$$x - 3 = \sqrt{9} \quad \text{or} \quad x - 3 = -\sqrt{9}$$
$$x = 3 + 3 \qquad\qquad x = 3 - 3$$
$$x = 6 \qquad\qquad\quad x = 0$$
The solutions are $x = 0$ and $x = 6$.

7.
$$z^2 + 5z = 7$$
$$z^2 + 5z + \left(\frac{5}{2}\right)^2 = 7 + \left(\frac{5}{2}\right)^2$$
$$z^2 + 5z + \frac{25}{4} = 7 + \frac{25}{4}$$
$$\left(z + \frac{5}{2}\right)^2 = \frac{53}{4}$$
$$z + \frac{5}{2} = \sqrt{\frac{53}{4}} \quad \text{or} \quad z + \frac{5}{2} = -\sqrt{\frac{53}{4}}$$
$$z = -\frac{5}{2} + \frac{\sqrt{53}}{2} \qquad\qquad z = -\frac{5}{2} - \frac{\sqrt{53}}{2}$$
The solutions are $z = \dfrac{-5 \pm \sqrt{53}}{2}$.

9.
$$x^2 - 2x - 1 = 0$$
$$x^2 - 2x = 1$$
$$x^2 - 2x + \left(\frac{-2}{2}\right)^2 = 1 + \left(\frac{-2}{2}\right)^2$$
$$x^2 - 2x + (-1)^2 = 1 + (-1)^2$$
$$x^2 - 2x + 1 = 1 + 1$$
$$(x-1)^2 = 2$$
$$x - 1 = \sqrt{2} \quad \text{or} \quad x - 1 = -\sqrt{2}$$
$$x = 1 + \sqrt{2} \qquad\qquad x = 1 - \sqrt{2}$$
The solutions are $x = 1 \pm \sqrt{2}$.

11.
$$y^2 + 5y + 4 = 0$$
$$y^2 + 5y = -4$$
$$y^2 + 5y + \left(\frac{5}{2}\right)^2 = -4 + \left(\frac{5}{2}\right)^2$$
$$\left(y + \frac{5}{2}\right)^2 = \frac{9}{4}$$
$$y + \frac{5}{2} = \sqrt{\frac{9}{4}} \quad \text{or} \quad y + \frac{5}{2} = -\sqrt{\frac{9}{4}}$$
$$y = -\frac{5}{2} + \frac{3}{2} \qquad\qquad y = -\frac{5}{2} - \frac{3}{2}$$
$$y = -\frac{2}{2} = -1 \qquad\qquad y = -\frac{8}{2} = -4$$
The solutions are $y = -1$ and $y = -4$.

13.
$$3x^2 - 6x = 24$$
$$x^2 - 2x = 8$$
$$x^2 - 2x + \left(\frac{-2}{2}\right)^2 = 8 + \left(\frac{-2}{2}\right)^2$$
$$x^2 - 2x + (-1)^2 = 8 + (-1)^2$$
$$x^2 - 2x + 1 = 8 + 1$$
$$(x-1)^2 = 9$$
$$x - 1 = \sqrt{9} \quad \text{or} \quad x - 1 = -\sqrt{9}$$
$$x = 1 + 3 \qquad\qquad x = 1 - 3$$
$$x = 4 \qquad\qquad\quad x = -2$$
The solutions are $x = -2$ and $x = 4$.

15.
$$5x^2 + 10x + 6 = 0$$
$$x^2 + 2x + \frac{6}{5} = 0$$
$$x^2 + 2x = -\frac{6}{5}$$
$$x^2 + 2x + \left(\frac{2}{2}\right)^2 = -\frac{6}{5} + \left(\frac{2}{2}\right)^2$$
$$(x+1)^2 = -\frac{1}{5}$$

This has no real solutions because $\sqrt{-\dfrac{1}{5}}$ does not have a real solution.

17.
$$2x^2 = 6x + 5$$
$$2x^2 - 6x = 5$$
$$x^2 - 3x = \frac{5}{2}$$
$$x^2 - 3x + \left(\frac{-3}{2}\right)^2 = \frac{5}{2} + \left(\frac{-3}{2}\right)^2$$
$$x^2 - 3x + \frac{9}{4} = \frac{5}{2} + \frac{9}{4}$$
$$\left(x - \frac{3}{2}\right)^2 = \frac{10}{4} + \frac{9}{4}$$
$$\left(x - \frac{3}{2}\right)^2 = \frac{19}{4}$$
$$x - \frac{3}{2} = \sqrt{\frac{19}{4}} \quad \text{or} \quad x - \frac{3}{2} = -\sqrt{\frac{19}{4}}$$
$$x = \frac{3}{2} + \frac{\sqrt{19}}{2} \qquad x = \frac{3}{2} - \frac{\sqrt{19}}{2}$$
$$x = \frac{3 + \sqrt{19}}{2} \qquad x = \frac{3 - \sqrt{19}}{2}$$
The solutions are $x = \frac{3 \pm \sqrt{19}}{2}$.

19.
$$2y^2 + 8y + 5 = 0$$
$$y^2 + 4y + \frac{5}{2} = 0$$
$$y^2 + 4y = -\frac{5}{2}$$
$$y^2 + 4y + \left(\frac{4}{2}\right)^2 = -\frac{5}{2} + \left(\frac{4}{2}\right)^2$$
$$(y + 2)^2 = \frac{3}{2}$$
$$y + 2 = \sqrt{\frac{3}{2}} \quad \text{or} \quad y + 2 = -\sqrt{\frac{3}{2}}$$
$$y = -2 + \frac{\sqrt{6}}{2} \qquad y = -2 - \frac{\sqrt{6}}{2}$$
The solutions are $y = -2 \pm \frac{\sqrt{6}}{2}$.

21.
$$x^2 + 6x - 25 = 0$$
$$x^2 + 6x = 25$$
$$x^2 + 6x + \left(\frac{6}{2}\right)^2 = 25 + \left(\frac{6}{2}\right)^2$$
$$x^2 + 6x + 3^2 = 25 + 3^2$$
$$x^2 + 6x + 9 = 25 + 9$$
$$(x + 3)^2 = 34$$
$$x + 3 = \sqrt{34} \quad \text{or} \quad x + 3 = -\sqrt{34}$$
$$x = -3 + \sqrt{34} \qquad x = -3 - \sqrt{34}$$
The solutions are $x = -3 \pm \sqrt{34}$.

23.
$$x^2 - 3x - 3 = 0$$
$$x^2 - 3x = 3$$
$$x^2 - 3x + \left(-\frac{3}{2}\right)^2 = 3 + \left(-\frac{3}{2}\right)^2$$
$$\left(x - \frac{3}{2}\right)^2 = \frac{21}{4}$$
$$x - \frac{3}{2} = \sqrt{\frac{21}{4}} \quad \text{or} \quad x - \frac{3}{2} = -\sqrt{\frac{21}{4}}$$
$$x = \frac{3}{2} + \frac{\sqrt{21}}{2} \qquad x = \frac{3}{2} - \frac{\sqrt{21}}{2}$$
The solutions are $x = \frac{3 \pm \sqrt{21}}{2}$.

25.
$$2y^2 - 3y + 1 = 0$$
$$2y^2 - 3y = -1$$
$$y^2 - \frac{3}{2}y = -\frac{1}{2}$$
$$y^2 - \frac{3}{2}y + \left(\frac{-\frac{3}{2}}{2}\right)^2 = -\frac{1}{2} + \left(\frac{-\frac{3}{2}}{2}\right)^2$$
$$y^2 - \frac{3}{2}y + \left(-\frac{3}{4}\right)^2 = -\frac{1}{2} + \left(-\frac{3}{4}\right)^2$$
$$y^2 - \frac{3}{2}y + \frac{9}{16} = -\frac{1}{2} + \frac{9}{16}$$
$$\left(y - \frac{3}{4}\right)^2 = \frac{1}{16}$$

$$y - \frac{3}{4} = \sqrt{\frac{1}{16}} \quad \text{or} \quad y - \frac{3}{4} = -\sqrt{\frac{1}{16}}$$

$$y = \frac{3}{4} + \frac{1}{4} \qquad\qquad y = \frac{3}{4} - \frac{1}{4}$$

$$y = 1 \qquad\qquad\qquad y = \frac{1}{2}$$

The solutions are $y = \frac{1}{2}$ and $y = 1$.

27.
$$x(x+3) = 18$$
$$x^2 + 3x = 18$$
$$x^2 + 3x + \left(\frac{3}{2}\right)^2 = 18 + \left(\frac{3}{2}\right)^2$$
$$\left(x + \frac{3}{2}\right)^2 = \frac{81}{4}$$
$$x + \frac{3}{2} = \sqrt{\frac{81}{4}} \quad \text{or} \quad x + \frac{3}{2} = -\sqrt{\frac{81}{4}}$$
$$x = -\frac{3}{2} + \frac{9}{2} \qquad\qquad x = -\frac{3}{2} - \frac{9}{2}$$
$$x = \frac{6}{2} = 3 \qquad\qquad x = -\frac{12}{2} = -6$$

The solutions are $x = 3$ and $x = -6$.

29.
$$3z^2 + 6z + 4 = 0$$
$$3z^2 + 6z = -4$$
$$z^2 + 2z = -\frac{4}{3}$$
$$z^2 + 2z + \left(\frac{2}{2}\right)^2 = -\frac{4}{3} + \left(\frac{2}{2}\right)^2$$
$$z^2 + 2z + 1^2 = -\frac{4}{3} + 1^2$$
$$z^2 + 2z + 1 = -\frac{4}{3} + 1$$
$$(z + 1)^2 = -\frac{1}{3}$$

The equation has no solution, since the square root of $-\frac{1}{3}$ is not a real number.

31.
$$4x^2 + 16x = 48$$
$$x^2 + 4x = 12$$
$$x^2 + 4x + \left(\frac{4}{2}\right)^2 = 12 + \left(\frac{4}{2}\right)^2$$
$$(x + 2)^2 = 16$$
$$x + 2 = \sqrt{16} \quad \text{and} \quad x + 2 = -\sqrt{16}$$
$$x = -2 + 4 \qquad\qquad x = -2 - 4$$
$$x = 2 \qquad\qquad\qquad x = -6$$

The solutions are $x = 2$ and $x = -6$.

33. $\dfrac{3}{4} - \sqrt{\dfrac{25}{16}} = \dfrac{3}{4} - \dfrac{\sqrt{25}}{\sqrt{16}}$

$$= \frac{3}{4} - \frac{5}{4}$$
$$= \frac{3 - 5}{4}$$
$$= \frac{-2}{4}$$
$$= -\frac{1}{2}$$

35. $\dfrac{1}{2} - \sqrt{\dfrac{9}{4}} = \dfrac{1}{2} - \dfrac{3}{2} = -\dfrac{2}{2} = -1$

37. $\dfrac{6 + 4\sqrt{5}}{2} = \dfrac{6}{2} + \dfrac{4\sqrt{5}}{2} = 3 + 2\sqrt{5}$

39. $\dfrac{3 - 9\sqrt{2}}{6} = \dfrac{3}{6} - \dfrac{9\sqrt{2}}{6} = \dfrac{1}{2} - \dfrac{3\sqrt{2}}{2} = \dfrac{1 - 3\sqrt{2}}{2}$

41. answers may vary

43. a. $\quad x^2 + 6x + 9 = 11$
$$(x + 3)^2 = 11$$
$$x + 3 = \sqrt{11} \quad \text{or} \quad x + 3 = -\sqrt{11}$$
$$x = -3 + \sqrt{11} \qquad\qquad x = -3 - \sqrt{11}$$

The solutions are $x = -3 \pm \sqrt{11}$.

b. answers may vary

45. $x^2 + kx + \left(\dfrac{k}{2}\right)^2$ is a perfect square trinomial.

If $x^2 + kx + 16$ is a perfect square trinomial,

then $\left(\dfrac{k}{2}\right)^2 = 16$

$\dfrac{k}{2} = \sqrt{16}$ or $\dfrac{k}{2} = -\sqrt{16}$

$k = 2 \cdot 4$ $k = 2(-4)$

$k = 8$ $k = -8$

$x^2 + kx + 16$ is a perfect square trinomial
when $k = 8$ or $k = -8$.

47.
$$y = 10x^2 + 513x + 15{,}743$$
$$20{,}487 = 10x^2 + 513x + 15{,}743$$
$$4744 = 10x^2 + 513x$$
$$474.4 = x^2 + 51.3x$$
$$\left(\dfrac{51.3}{2}\right)^2 + 474.4 = x^2 + 51.3x + \left(\dfrac{51.3}{2}\right)^2$$
$$1132.3 = (x + 25.65)^2$$

$\sqrt{1132.3} = x + 25.65$ or

$33.65 = x + 25.65$

$8 = x$

$-\sqrt{1132.3} = x + 25.65$

$-33.65 = x + 25.65$

$-59.3 = x$

Since time is not a negative number, reject
the solution -59.3. The retail sales for U.S.
bookstores will be \$20,487 million in
8 years or 2009.

49. The solutions are $x = -6$ and $x = -2$.

51. The solutions are $x = -0.68$ and $x = 3.68$.

Section 9.3 Practice Problems

1. $2x^2 - x - 5 = 0$

$a = 2, b = -1, c = -5$

$x = \dfrac{-b \pm \sqrt{b^2 - 4ac}}{2a}$

$x = \dfrac{-(-1) \pm \sqrt{(-1)^2 - 4(2)(-5)}}{2(2)}$

$= \dfrac{1 \pm \sqrt{1 + 40}}{4}$

$= \dfrac{1 \pm \sqrt{41}}{4}$

The solutions are $\dfrac{1 \pm \sqrt{41}}{4}$.

2. $3x^2 + 8x = 3$

$3x^2 + 8x - 3 = 0$

$a = 3, b = 8, c = -3$

$x = \dfrac{-b \pm \sqrt{b^2 - 4ac}}{2a}$

$x = \dfrac{-8 \pm \sqrt{8^2 - 4(3)(-3)}}{2(3)}$

$= \dfrac{-8 \pm \sqrt{64 + 36}}{6}$

$= \dfrac{-8 \pm \sqrt{100}}{6}$

$= \dfrac{-8 \pm 10}{6}$

$x = \dfrac{-8 + 10}{6} = \dfrac{1}{3}$ or $\dfrac{-8 - 10}{6} = -3$

The solutions are $\dfrac{1}{3}$ and -3.

3.

$$5x^2 = 2$$
$$5x^2 - 2 = 0$$
$$a = 5, b = 0, c = -2$$
$$x = \frac{-b \pm \sqrt{b^2 - 4ac}}{2a}$$
$$x = \frac{-0 \pm \sqrt{0^2 - 4(5)(-2)}}{2(5)}$$
$$= \pm \frac{\sqrt{40}}{10}$$
$$= \pm \frac{2\sqrt{10}}{10}$$
$$= \pm \frac{\sqrt{10}}{5}$$

The solutions are $\pm \dfrac{\sqrt{10}}{5}$.

4.

$$x^2 = -2x - 3$$
$$x^2 + 2x + 3 = 0$$
$$a = 1, b = 2, c = 3$$
$$x = \frac{-b \pm \sqrt{b^2 - 4ac}}{2a}$$
$$x = \frac{-2 \pm \sqrt{2^2 - 4(1)(3)}}{2(1)} = \frac{-2 \pm \sqrt{-8}}{2}$$

There is no real number solution because $\sqrt{-8}$ is not a real number.

5.

$$\frac{1}{3}x^2 - x = 1$$
$$\frac{1}{3}x^2 - x - 1 = 0$$
$$a = \frac{1}{3}, \ b = -1, c = -1$$
$$x = \frac{-b \pm \sqrt{b^2 - 4ac}}{2a}$$

$$x = \frac{-(-1) \pm \sqrt{(-1)^2 - 4\left(\frac{1}{3}\right)(-1)}}{2\left(\frac{1}{3}\right)}$$
$$= \frac{1 \pm \sqrt{\frac{7}{3}}}{\frac{2}{3}}$$
$$= \frac{1 \pm \frac{\sqrt{21}}{3}}{\frac{2}{3}}$$
$$= \frac{3 \pm \sqrt{21}}{2}$$

The solutions are $\dfrac{3 \pm \sqrt{21}}{2}$.

6. $\dfrac{1 + \sqrt{41}}{4} \approx 1.9$

$$1 - \frac{\sqrt{41}}{4} \approx -1.4$$

Mental Math

1. $2x^2 + 5x + 3 = 0 \Rightarrow a = 2, b = 5, c = 3$

2. $5x^2 - 7x + 1 = 0 \Rightarrow a = 5, b = -7, c = 1$

3. $10x^2 - 13x - 2 = 0 \Rightarrow a = 10, b = -13, c = -2$

4. $x^2 + 3x - 7 = 0 \Rightarrow a = 1, b = 3, c = -7$

5. $x^2 - 6 = 0 \Rightarrow a = 1, b = 0, c = -6$

6. $9x^2 - 4 = 0 \Rightarrow a = 9, b = 0, c = -4$

Exercise Set 9.3

1. $x^2 - 3x + 2 = 0$

    $a = 1, b = -3, c = 2$

    $x = \dfrac{-b \pm \sqrt{b^2 - 4ac}}{2a}$

    $x = \dfrac{-(-3) \pm \sqrt{(-3)^2 - 4(1)(2)}}{2(1)}$

    $= \dfrac{3 \pm \sqrt{9 - 8}}{2}$

    $= \dfrac{3 \pm \sqrt{1}}{2}$

    $= \dfrac{3 \pm 1}{2}$

    $x = \dfrac{3 + 1}{2} = 2$ or $x = \dfrac{3 - 1}{2} = 1$

    The solutions are $x = 2$ and $x = 1$.

3. $3k^2 + 7k + 1 = 0$

    $a = 3, b = 7, c = 1$

    $k = \dfrac{-b \pm \sqrt{b^2 - 4ac}}{2a}$

    $k = \dfrac{-7 \pm \sqrt{7^2 - 4(3)(1)}}{2(3)}$

    $= \dfrac{-7 \pm \sqrt{49 - 12}}{6}$

    $= \dfrac{-7 \pm \sqrt{37}}{6}$

    The solutions are $k = \dfrac{-7 \pm \sqrt{37}}{6}$.

5. $4x^2 - 3 = 0$

    $4x^2 + 0x - 3 = 0$

    $a = 4, b = 0, c = -3$

    $x = \dfrac{-b \pm \sqrt{b^2 - 4ac}}{2a}$

    $x = \dfrac{-0 \pm \sqrt{0^2 - 4(4)(-3)}}{2(4)}$

    $= \dfrac{0 \pm \sqrt{0 + 48}}{8}$

    $= \dfrac{\pm \sqrt{48}}{8}$

    $= \pm \dfrac{4\sqrt{3}}{8}$

    $= \pm \dfrac{\sqrt{3}}{2}$

    The solutions are $x = \pm \dfrac{\sqrt{3}}{2}$.

7. $5z^2 - 4z + 3 = 0$

    $a = 5, b = -4, c = 3$

    $z = \dfrac{-b \pm \sqrt{b^2 - 4ac}}{2a}$

    $z = \dfrac{-(-4) \pm \sqrt{(-4)^2 - 4(5)(3)}}{2(5)}$

    $= \dfrac{4 \pm \sqrt{16 - 60}}{10}$

    $= \dfrac{4 \pm \sqrt{-44}}{10}$

    The equation has no solution, since the square root of -44 is not a real number.

9. $\qquad\quad y^2 = 7y + 30$

    $y^2 - 7y - 30 = 0$

    $a = 1, b = -7, c = -30$

    $y = \dfrac{-b \pm \sqrt{b^2 - 4ac}}{2a}$

    $y = \dfrac{-(-7) \pm \sqrt{(-7)^2 - 4(1)(-30)}}{2(1)}$

    $= \dfrac{7 \pm \sqrt{49 + 120}}{2}$

    $= \dfrac{7 \pm \sqrt{169}}{2}$

    $= \dfrac{7 \pm 13}{2}$

$y = \dfrac{7+13}{2} = 10$ or $y = \dfrac{7-13}{2} = -3$

The solutions are $y = 10$ and $y = -3$.

11. $2x^2 = 10$

$2x^2 - 10 = 0$

$a = 2, b = 0, c = -10$

$x = \dfrac{-b \pm \sqrt{b^2 - 4ac}}{2a}$

$x = \dfrac{-0 \pm \sqrt{0^2 - 4(2)(-10)}}{2(2)}$

$= \dfrac{\pm\sqrt{80}}{4}$

$= \dfrac{\pm 4\sqrt{5}}{4}$

$= \pm\sqrt{5}$

The solutions are $x = \pm\sqrt{5}$.

13. $m^2 - 12 = m$

$m^2 - m - 12 = 0$

$a = 1, b = -1, c = -12$

$m = \dfrac{-b \pm \sqrt{b^2 - 4ac}}{2a}$

$m = \dfrac{-(-1) \pm \sqrt{(-1)^2 - 4(1)(-12)}}{2(1)}$

$= \dfrac{1 \pm \sqrt{1 + 48}}{2}$

$= \dfrac{1 \pm \sqrt{49}}{2}$

$= \dfrac{1 \pm 7}{2}$

$m = \dfrac{1+7}{2} = 4$ or $m = \dfrac{1-7}{2} = -3$

The solutions are $m = 4$ and $m = -3$.

15. $3 - x^2 = 4x$

$0 = x^2 + 4x - 3$

$a = 1, b = 4, c = -3$

$x = \dfrac{-b \pm \sqrt{b^2 - 4ac}}{2a}$

$x = \dfrac{-4 \pm \sqrt{4^2 - 4(1)(-3)}}{2(1)}$

$= \dfrac{-4 \pm \sqrt{16 + 12}}{2}$

$= \dfrac{-4 \pm \sqrt{28}}{2}$

$= \dfrac{-4 \pm 2\sqrt{7}}{2}$

$= -2 \pm \sqrt{7}$

The solutions are $x = -2 \pm \sqrt{7}$.

17. $6x^2 + 9x = 2$

$6x^2 + 9x - 2 = 0$

$a = 6, b = 9, c = -2$

$x = \dfrac{-b \pm \sqrt{b^2 - 4ac}}{2a}$

$x = \dfrac{-9 \pm \sqrt{9^2 - 4(6)(-2)}}{2(6)}$

$= \dfrac{-9 \pm \sqrt{81 + 48}}{12}$

$= \dfrac{-9 \pm \sqrt{129}}{12}$

The solutions are $x = \dfrac{-9 \pm \sqrt{129}}{12}$.

19. $7p^2 + 2 = 8p$

$7p^2 - 8p + 2 = 0$

$a = 7, b = -8, c = 2$

$p = \dfrac{-b \pm \sqrt{b^2 - 4ac}}{2a}$

$$p = \frac{-(-8) \pm \sqrt{(-8)^2 - 4(7)(2)}}{2(7)}$$

$$= \frac{8 \pm \sqrt{64 - 56}}{14}$$

$$= \frac{8 \pm \sqrt{8}}{14}$$

$$= \frac{8 \pm 2\sqrt{2}}{14}$$

$$= \frac{4 \pm \sqrt{2}}{7}$$

The solutions are $p = \dfrac{4 \pm \sqrt{2}}{7}$.

21. $x^2 - 6x + 2 = 0$

$a = 1, b = -6, c = 2$

$$x = \frac{-b \pm \sqrt{b^2 - 4ac}}{2a}$$

$$x = \frac{-(-6) \pm \sqrt{(-6)^2 - 4(1)(2)}}{2(1)}$$

$$= \frac{6 \pm \sqrt{36 - 8}}{2}$$

$$= \frac{6 \pm \sqrt{28}}{2}$$

$$= \frac{6 \pm 2\sqrt{7}}{2}$$

$$= \frac{2\left(3 \pm \sqrt{7}\right)}{2}$$

$$= 3 \pm \sqrt{7}$$

The solutions are $x = 3 \pm \sqrt{7}$.

23. $2x^2 - 6x + 3 = 0$

$a = 2, b = -6, c = 3$

$$x = \frac{-b \pm \sqrt{b^2 - 4ac}}{2a}$$

$$x = \frac{-(-6) \pm \sqrt{(-6)^2 - 4(2)(3)}}{2(2)}$$

$$= \frac{6 \pm \sqrt{36 - 24}}{4}$$

$$= \frac{6 \pm \sqrt{12}}{4}$$

$$= \frac{6 \pm 2\sqrt{3}}{4}$$

$$= \frac{3 \pm \sqrt{3}}{2}$$

The solutions are $x = \dfrac{3 \pm \sqrt{3}}{2}$.

25. $\qquad 3x^2 = 1 - 2x$

$3x^2 + 2x - 1 = 0$

$a = 3, b = 2, c = -1$

$$x = \frac{-b \pm \sqrt{b^2 - 4ac}}{2a}$$

$$x = \frac{-2 \pm \sqrt{2^2 - 4(3)(-1)}}{2(3)}$$

$$= \frac{-2 \pm \sqrt{4 + 12}}{6}$$

$$= \frac{-2 \pm \sqrt{16}}{6}$$

$$= \frac{-2 \pm 4}{6}$$

$$x = \frac{-2 + 4}{6} = \frac{1}{3} \text{ or } x = \frac{-2 - 4}{6} = -1$$

The solutions are $x = \dfrac{1}{3}$ and $x = -1$.

27. $\qquad 4y^2 = 6y + 1$

$4y^2 - 6y - 1 = 0$

$a = 4, b = -6, c = -1$

$$y = \frac{-b \pm \sqrt{b^2 - 4ac}}{2a}$$

$$y = \frac{-(-6) \pm \sqrt{(-6)^2 - 4(4)(-1)}}{2(4)}$$

$$= \frac{6 \pm \sqrt{36 + 16}}{8}$$

$$= \frac{6 \pm \sqrt{52}}{8}$$

$$= \frac{6 \pm 2\sqrt{13}}{8}$$

$$= \frac{3 \pm \sqrt{13}}{4}$$

The solutions are $y = \dfrac{3 \pm \sqrt{13}}{4}$.

29.
$$20y^2 = 3 - 11y$$
$$20y^2 + 11y - 3 = 0$$
$$a = 20, \, b = 11, \, c = -3$$
$$y = \frac{-b \pm \sqrt{b^2 - 4ac}}{2a}$$
$$y = \frac{-11 \pm \sqrt{11^2 - 4(20)(-3)}}{2(20)}$$
$$= \frac{-11 \pm \sqrt{121 + 240}}{40}$$
$$= \frac{-11 \pm \sqrt{361}}{40}$$
$$= \frac{-11 \pm 19}{40}$$
$$y = \frac{-11 + 19}{40} = \frac{1}{5} \text{ or } y = \frac{-11 - 19}{40} = -\frac{3}{4}$$

The solutions are $y = \dfrac{1}{5}$ and $y = -\dfrac{3}{4}$.

31. $x^2 + x + 2 = 0$
$$a = 1, \, b = 1, \, c = 2$$
$$x = \frac{-b \pm \sqrt{b^2 - 4ac}}{2a}$$

$$x = \frac{-1 \pm \sqrt{1^2 - 4(1)(2)}}{2(1)}$$

$$= \frac{-1 \pm \sqrt{1 - 8}}{2}$$

$$= \frac{-1 \pm \sqrt{-7}}{2}$$

The equation has no solution, since the square root of -7 is not a real number.

33.
$$\frac{m^2}{2} = m + \frac{1}{2}$$
$$\frac{m^2}{2} - m - \frac{1}{2} = 0$$
$$m^2 - 2m - 1 = 0$$
$$a = 1, \, b = -2, \, c = -1$$
$$m = \frac{-b \pm \sqrt{b^2 - 4ac}}{2a}$$
$$m = \frac{-(-2) \pm \sqrt{(-2)^2 - 4(1)(-1)}}{2(1)}$$
$$= \frac{2 \pm \sqrt{4 + 4}}{2}$$
$$= \frac{2 \pm \sqrt{8}}{2}$$
$$= \frac{2 \pm 2\sqrt{2}}{2}$$
$$= \frac{2\left(1 \pm \sqrt{2}\right)}{2}$$
$$= 1 \pm \sqrt{2}$$

The solutions are $m = 1 \pm \sqrt{2}$.

35. $3p^2 - \dfrac{2}{3}p + 1 = 0$
$$9p^2 - 2p + 3 = 0$$
$$a = 9, \, b = -2, \, c = 3$$
$$p = \frac{-b \pm \sqrt{b^2 - 4ac}}{2a}$$

$$p = \frac{-(-2) \pm \sqrt{(-2)^2 - 4(9)(3)}}{2(9)}$$

$$= \frac{2 \pm \sqrt{4 - 108}}{18}$$

$$= \frac{2 \pm \sqrt{-104}}{18}$$

The equation has no solution, since the square root of -104 is not a real number.

37.
$$4p^2 + \frac{3}{2} = -5p$$

$$4p^2 + 5p + \frac{3}{2} = 0$$

$$8p^2 + 10p + 3 = 0$$

$$a = 8, \, b = 10, \, c = 3$$

$$p = \frac{-b \pm \sqrt{b^2 - 4ac}}{2a}$$

$$p = \frac{-10 \pm \sqrt{10^2 - 4(8)(3)}}{2(8)}$$

$$= \frac{-10 \pm \sqrt{100 - 96}}{16}$$

$$= \frac{-10 \pm \sqrt{4}}{16}$$

$$= \frac{-10 \pm 2}{16}$$

$$p = \frac{-10 + 2}{16} = -\frac{1}{2} \text{ or } p = \frac{-10 - 2}{16} = -\frac{3}{4}$$

The solutions are $p = -\frac{1}{2}$ and $p = -\frac{3}{4}$.

39.
$$5x^2 = \frac{7}{2}x + 1$$

$$5x^2 - \frac{7}{2}x - 1 = 0$$

$$10x^2 - 7x - 2 = 0$$

$$a = 10, \, b = -7, \, c = -2$$

$$x = \frac{-b \pm \sqrt{b^2 - 4ac}}{2a}$$

$$x = \frac{-(-7) \pm \sqrt{(-7)^2 - 4(10)(-2)}}{2(10)}$$

$$= \frac{7 \pm \sqrt{49 + 80}}{20}$$

$$= \frac{7 \pm \sqrt{129}}{20}$$

The solutions are $x = \frac{7 \pm \sqrt{129}}{20}$.

41. $x^2 - \frac{11}{2}x - \frac{1}{2} = 0$

$$2x^2 - 11x - 1 = 0$$

$$a = 2, \, b = -11, \, c = -1$$

$$x = \frac{-b \pm \sqrt{b^2 - 4ac}}{2a}$$

$$x = \frac{-(-11) \pm \sqrt{(-11)^2 - 4(2)(-1)}}{2(2)}$$

$$= \frac{11 \pm \sqrt{121 + 8}}{4}$$

$$= \frac{11 \pm \sqrt{129}}{4}$$

The solutions are $x = \frac{11 \pm \sqrt{129}}{4}$.

43.
$$5z^2 - 2z = \frac{1}{5}$$

$$5z^2 - 2z - \frac{1}{5} = 0$$

$$25z^2 - 10z - 1 = 0$$

$$a = 25, \, b = -10, \, c = -1$$

$$z = \frac{-b \pm \sqrt{b^2 - 4ac}}{2a}$$

$$z = \frac{-(-10) \pm \sqrt{(-10)^2 - 4(25)(-1)}}{2(25)}$$

$$= \frac{10 \pm \sqrt{100 + 100}}{50}$$

$$= \frac{10 \pm \sqrt{200}}{50}$$

$$= \frac{10 \pm 10\sqrt{2}}{50}$$

$$= \frac{1 \pm \sqrt{2}}{5}$$

The solutions are $z = \dfrac{1 \pm \sqrt{2}}{5}$.

45.
$$3x^2 = 21$$
$$3x^2 - 21 = 0$$
$$3x^2 + 0x - 21 = 0$$
$$a = 3, \, b = 0, \, c = -21$$
$$x = \frac{-b \pm \sqrt{b^2 - 4ac}}{2a}$$
$$x = \frac{-0 \pm \sqrt{0^2 - 4(3)(-21)}}{2(3)}$$
$$= \frac{\pm\sqrt{0 + 252}}{6}$$
$$= \frac{\pm\sqrt{252}}{6}$$
$$= \frac{\pm 6\sqrt{7}}{6}$$
$$= \pm\sqrt{7}$$

The solutions are $x = \pm\sqrt{7}$ or $x \approx -2.6$ and $x \approx 2.6$.

47. $x^2 + 6x + 1 = 0$
$$a = 1, \, b = 6, \, c = 1$$
$$x = \frac{-b \pm \sqrt{b^2 - 4ac}}{2a}$$
$$x = \frac{-6 \pm \sqrt{6^2 - 4(1)(1)}}{2(1)}$$
$$= \frac{-6 \pm \sqrt{36 - 4}}{2}$$
$$= \frac{-6 \pm \sqrt{32}}{2}$$
$$= \frac{-6 \pm 4\sqrt{2}}{2}$$
$$= -3 \pm 2\sqrt{2}$$

The solutions are $x = -3 \pm 2\sqrt{2}$ or $x \approx -5.8$ and $x \approx -0.2$.

49.
$$x^2 = 9x + 4$$
$$x^2 - 9x - 4 = 0$$
$$a = 1, \, b = -9, \, c = -4$$
$$x = \frac{-b \pm \sqrt{b^2 - 4ac}}{2a}$$
$$x = \frac{-(-9) \pm \sqrt{(-9)^2 - 4(1)(-4)}}{2(1)}$$
$$= \frac{9 \pm \sqrt{81 + 16}}{2}$$
$$= \frac{9 \pm \sqrt{97}}{2}$$

The solutions are $x = \dfrac{9 \pm \sqrt{97}}{2}$ or $x \approx 9.4$ and $x \approx -0.4$.

51. $3x^2 - 2x - 2 = 0$
$$a = 3, \, b = -2, \, c = -2$$
$$x = \frac{-b \pm \sqrt{b^2 - 4ac}}{2a}$$

$$x = \frac{-(-2) \pm \sqrt{(-2)^2 - 4(3)(-2)}}{2(3)}$$

$$= \frac{2 \pm \sqrt{4 + 24}}{6}$$

$$= \frac{2 \pm \sqrt{28}}{6}$$

$$= \frac{2 \pm 2\sqrt{7}}{6}$$

$$= \frac{1 \pm \sqrt{7}}{3}$$

The solutions are $x = \dfrac{1 \pm \sqrt{7}}{3}$ or $x \approx 1.2$ and $x \approx -0.5$.

53. $y = -3$ is a horizontal line with y-intercept $(0, -3)$.

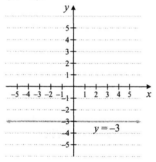

55. $y = 3x - 2$ is a line with slope of 3 and y-intercept $(0, -2)$.

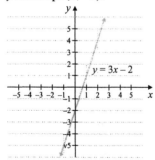

57. $5x^2 + 2 = x$

$5x^2 - x + 2 = 0$

$b = -1$, which is choice c.

59. $7y^2 = 3y$

$0 = -7y^2 + 3y + 0$

$a = -7$, which is choice b.

61. Let x be the width of the chocolate bar. Then the length is $2x + 0.5$.

Area = (length)(width)\

$50.8 = (2x + 0.5)(x)$

$50.8 = 2x^2 + 0.5x$

$0 = 2x^2 + 0.5x - 50.8$

$0 = 20x^2 + 5x - 508$

$a = 20, b = 5, c = -508$

$$x = \frac{-b \pm \sqrt{b^2 - 4ac}}{2a}$$

$$x = \frac{-5 \pm \sqrt{5^2 - 4(20)(-508)}}{2(20)}$$

$$= \frac{-5 \pm \sqrt{25 + 40,640}}{40}$$

$$= \frac{-5 \pm \sqrt{40,665}}{40}$$

$$x = \frac{-5 + \sqrt{40,665}}{40} \approx 4.9 \text{ or}$$

$$x = \frac{-5 - \sqrt{40,665}}{40} \approx -5.2$$

Since the width is not a negative number, discard the solution $x \approx -5.2$.

$2x + 0.5 = 2(4.9) + 0.5 = 9.8 + 0.5 = 10.3$

The width was approximately 4.9 feet and the length was approximately 10.3 feet.

63. $x^2 + 3\sqrt{2}x - 5 = 0$

$a = 1, \ b = 3\sqrt{2}, \ c = -5$

$$x = \frac{-b \pm \sqrt{b^2 - 4ac}}{2a}$$

$$x = \frac{-3\sqrt{2} \pm \sqrt{\left(3\sqrt{2}\right)^2 - 4(1)(-5)}}{2(1)}$$

$$= \frac{-3\sqrt{2} \pm \sqrt{18 + 20}}{2}$$

$$= \frac{-3\sqrt{2} \pm \sqrt{38}}{2}$$

The solutions are $x = \dfrac{-3\sqrt{2} \pm \sqrt{38}}{2}$.

65. answers may vary

67. $7.3^2 + 5.4z - 1.1 = 0$

$a = 7.3,\ b = 5.4,\ c = -1.1$

$$z = \frac{-b \pm \sqrt{b^2 - 4ac}}{2a}$$

$$z = \frac{-5.4 \pm \sqrt{(5.4)^2 - 4(7.3)(-1.1)}}{2(7.3)}$$

$$= \frac{-5.4 \pm \sqrt{29.16 + 32.12}}{14.6}$$

$$= \frac{-5.4 \pm \sqrt{61.28}}{14.6}$$

$$z = \frac{-5.4 + \sqrt{61.28}}{14.6} \approx 0.2 \ \text{ or } \ z = \frac{-5.4 - \sqrt{61.28}}{14.6} \approx -0.9$$

The solutions are $z \approx 0.2$ and $z = -0.9$.

69. $h = -16t^2 + 120t + 80$

$0 = -16t^2 + 120t + 80$

$a = -16,\ b = 120,\ c = 80$

$$t = \frac{-b \pm \sqrt{b^2 - 4ac}}{2a}$$

$$t = \frac{-120 \pm \sqrt{120^2 - 4(-16)(80)}}{2(-16)}$$

$$= \frac{-120 \pm \sqrt{14,400 + 5120}}{-32}$$

$$= \frac{-120 \pm \sqrt{19,520}}{-32}$$

$$t = \frac{-120 + \sqrt{19,520}}{-32} \approx -0.6 \text{ or } t = \frac{-120 - \sqrt{19,520}}{-32} \approx 8.1$$

Since the time of the flight is not a negative number, discard the solution −0.6. The rocket will strike the ground approximately 8.1 seconds after it is launched.

71.
$$y = 21,400x^2 - 16,100x + 1,111,000$$
$$2,351,800 = 21,400x^2 - 16,100x + 1,111,000$$
$$0 = 21,400x^2 - 16,100x - 1,240,800$$
$$a = 21,400, \ b = -16,100, \ c = -1,240,800$$
$$x = \frac{-b \pm \sqrt{b^2 - 4ac}}{2a}$$
$$x = \frac{-(-16,100) \pm \sqrt{(-16,100)^2 - 4(21,400)(-1,240,800)}}{2(21,400)}$$
$$= \frac{16,100 \pm \sqrt{259,210,000 + 106,212,480,000}}{42,800}$$
$$= \frac{16,100 \pm \sqrt{106,471,690,000}}{42,800}$$
$$= \frac{16,100 \pm 326,300}{42,800}$$
$$x = \frac{16,100 + 326,300}{42,800} = 8 \text{ or } x = \frac{16,100 - 326,300}{42,800} = -7.2$$

Since time is not negative, discard $x = -7.2$.
The average NFL salary will be approximately \$2,351,800 when $x = 8$, which is 8 years after 2000, or 2008.

The Bigger Picture

1. $7.9 - 9.7 = -1.8$

2. $5 + (-3) + (-7) = -5$

3. $(-4)^2 - 5^2 = 16 - 25 = -9$

4. $7x - 2 + \frac{1}{3}(9x - 3) + 5 = 7x - 2 + 3x - 1 + 5$
$$= 10x + 2$$

5. $\left(\frac{1}{2}x + 5\right)\left(\frac{1}{2}x - 5\right) = \left(\frac{1}{2}x\right)^2 - 5^2 = \frac{1}{4}x^2 - 25$

6. $\dfrac{9x^2y + 3xy - 12y}{3xy} = \dfrac{9x^2y}{3xy} + \dfrac{3xy}{3xy} - \dfrac{12y}{3xy}$

$\qquad\qquad\qquad\quad = 3x + 1 - \dfrac{4}{x}$

7. $\dfrac{x^2}{(x-5)(x-4)} - \dfrac{3x+10}{(x-5)(x-4)}$

$\quad = \dfrac{x^2 - 3x - 10}{(x-5)(x-4)}$

$\quad = \dfrac{(x-5)(x+2)}{(x-5)(x-4)}$

$\quad = \dfrac{x+2}{x-4}$

8. $\dfrac{x}{x-10} + \dfrac{5}{x+3}$

$\quad = \dfrac{x(x+3)}{(x-10)(x+3)} + \dfrac{5(x-10)}{(x+3)(x-10)}$

$\quad = \dfrac{x^2 + 3x + 5x - 50}{(x-10)(x+3)}$

$\quad = \dfrac{x^2 + 8x - 50}{(x-10)(x+3)}$

9. $\sqrt{50} = \sqrt{25 \cdot 2} = \sqrt{25} \cdot \sqrt{2} = 5\sqrt{2}$

10. $\dfrac{\sqrt{30a^2b^3}}{\sqrt{3ab}} = \sqrt{\dfrac{30a^2b^3}{3ab}} = \sqrt{10ab^2} = b\sqrt{10a}$

11. $\sqrt{\dfrac{2}{3}} = \dfrac{\sqrt{2}}{\sqrt{3}} = \dfrac{\sqrt{2}}{\sqrt{3}} \cdot \dfrac{\sqrt{3}}{\sqrt{3}} = \dfrac{\sqrt{6}}{\sqrt{9}} = \dfrac{\sqrt{6}}{3}$

12. $\dfrac{7x - 14}{x^2 - 4} \cdot \dfrac{x^2 + 5x + 6}{49}$

$\quad = \dfrac{7(x-2)}{(x-2)(x+2)} \cdot \dfrac{(x+3)(x+2)}{49}$

$\quad = \dfrac{x+3}{7}$

13. $x^2 + 3x - 5 = 0$

$\quad a = 1,\, b = 3,\, c = -5$

$\quad x = \dfrac{-b \pm \sqrt{b^2 - 4ac}}{2a}$

$\quad x = \dfrac{-3 \pm \sqrt{3^2 - 4(1)(-5)}}{2(1)}$

$\qquad = \dfrac{-3 \pm \sqrt{9 + 20}}{2}$

$\qquad = \dfrac{-3 \pm \sqrt{29}}{2}$

The solutions are $x = \dfrac{-3 \pm \sqrt{29}}{2}$.

14. $x^2 + x = x^2 + 6$

$\qquad x = 6$

The solution is $x = 6$.

15. $-2x \le 5.6$

$\quad\ x \ge -2.8$

The solution set is $\{x | x \ge -2.8\}$.

16. $\qquad 2x^2 + 15x = 8$

$\qquad 2x^2 + 15x - 8 = 0$

$\qquad (2x - 1)(x + 8) = 0$

$\qquad 2x - 1 = 0 \quad \text{or} \quad x + 8 = 0$

$\qquad\quad 2x = 1 \qquad\qquad x = -8$

$\qquad\quad x = \dfrac{1}{2}$

The solutions are $x = \dfrac{1}{2}$ and $x = -8$.

17. $\sqrt{x+2} + 4 = x$

$\qquad \sqrt{x+2} = x - 4$

$\qquad \left(\sqrt{x+2}\right)^2 = (x-4)^2$

$\qquad\qquad x + 2 = x^2 - 8x + 16$

$\qquad\qquad\quad 0 = x^2 - 9x + 14$

$\quad a = 1,\, b = -9,\, c = 14$

$\quad x = \dfrac{-b \pm \sqrt{b^2 - 4ac}}{2a}$

$$x = \frac{-(-9) \pm \sqrt{(-9)^2 - 4(1)(14)}}{2(1)}$$

$$= \frac{9 \pm \sqrt{81 - 56}}{2}$$

$$= \frac{9 \pm \sqrt{25}}{2}$$

$$= \frac{9 \pm 5}{2}$$

$$x = \frac{9+5}{2} = \frac{14}{2} = 7 \text{ or } x = \frac{9-5}{2} = \frac{4}{2} = 2$$

Check: $x = 7$

$\sqrt{x+2} + 4 = x$

$\sqrt{7+2} + 4 \stackrel{?}{=} 7$

$\sqrt{9} + 4 \stackrel{?}{=} 7$

$3 + 4 \stackrel{?}{=} 7$

$7 = 7$ True

Check: $x = 2$

$\sqrt{x+2} + 4 = x$

$\sqrt{2+2} + 4 \stackrel{?}{=} 2$

$\sqrt{4} + 4 \stackrel{?}{=} 2$

$2 + 4 \stackrel{?}{=} 2$

$6 \neq 2$ False

Therefore the solution for this equation is $x = 7$.

18.
$$\frac{5}{x} - \frac{3}{x-4} = \frac{7+x}{x(x-4)}$$

$$x(x-4)\left(\frac{5}{x} - \frac{3}{x-4}\right) = x(x-4)\left(\frac{7+x}{x(x-4)}\right)$$

$$5(x-4) - 3x = 7 + x$$

$$5x - 20 - 3x = 7 + x$$

$$2x - 20 = 7 + x$$

$$x = 27$$

The solution is $x = 27$.

Integrated Review

1. $5x^2 - 11x + 2 = 0$
$(5x - 1)(x - 2) = 0$
$5x - 1 = 0$ or $x - 2 = 0$
$\quad 5x = 1 \qquad\qquad x = 2$
$\quad x = \frac{1}{5}$

The solutions are $x = \frac{1}{5}$ and $x = 2$.

2. $5x^2 + 13x - 6 = 0$
$(5x - 2)(x + 3) = 0$
$5x - 2 = 0$ or $x + 3 = 0$
$\quad 5x = 2 \qquad\qquad x = -3$
$\quad x = \frac{2}{5}$

The solutions are $x = \frac{2}{5}$ and $x = -3$.

3. $\quad x^2 - 1 = 2x$
$x^2 - 2x - 1 = 0$
$a = 1, b = -2, c = -1$

$$x = \frac{-b \pm \sqrt{b^2 - 4ac}}{2a}$$

$$x = \frac{-(-2) \pm \sqrt{(-2)^2 - 4(1)(-1)}}{2(1)}$$

$$= \frac{2 \pm \sqrt{4+4}}{2}$$

$$= \frac{2 \pm \sqrt{8}}{2}$$

$$= \frac{2 \pm 2\sqrt{2}}{2}$$

$$= 1 \pm \sqrt{2}$$

The solutions are $x = 1 \pm \sqrt{2}$.

4.
$$x^2 + 7 = 6x$$
$$x^2 - 6x + 7 = 0$$
$$a = 1, b = -6, c = 7$$
$$x = \frac{-b \pm \sqrt{b^2 - 4ac}}{2a}$$
$$x = \frac{-(-6) \pm \sqrt{(-6)^2 - 4(1)(7)}}{2(1)}$$
$$= \frac{6 \pm \sqrt{36 - 28}}{2}$$
$$= \frac{6 \pm \sqrt{8}}{2}$$
$$= \frac{6 \pm 2\sqrt{2}}{2}$$
$$= 3 \pm \sqrt{2}$$
The solutions are $x = 3 \pm \sqrt{2}$.

5. $a^2 = 20$
$$a = \pm\sqrt{20}$$
$$a = \pm 2\sqrt{5}$$
The solutions are $a = \pm 2\sqrt{5}$.

6. $a^2 = 72$
$$a = \pm\sqrt{72}$$
$$a = \pm 6\sqrt{2}$$
The solutions are $a = \pm 6\sqrt{2}$.

7. $x^2 - x + 4 = 0$
$$a = 1, b = -1, c = 4$$
$$x = \frac{-b \pm \sqrt{b^2 - 4ac}}{2a}$$
$$x = \frac{-(-1) \pm \sqrt{(-1)^2 - 4(1)(4)}}{2(1)}$$
$$= \frac{1 \pm \sqrt{1 - 16}}{2}$$
$$= \frac{1 \pm \sqrt{-15}}{2}$$
The equation has no solution, since the square root of −15 is not a real number.

8. $x^2 - 2x + 7 = 0$
$$a = 1, b = -2, c = 7$$
$$x = \frac{-b \pm \sqrt{b^2 - 4ac}}{2a}$$
$$x = \frac{-(-2) \pm \sqrt{(-2)^2 - 4(1)(7)}}{2(1)}$$
$$= \frac{2 \pm \sqrt{4 - 28}}{2}$$
$$= \frac{2 \pm \sqrt{-24}}{2}$$
The equation has no solution, since the square root of −24 is not a real number.

9. $3x^2 - 12x + 12 = 0$
$$3(x^2 - 4x + 4) = 0$$
$$3(x - 2)(x - 2) = 0$$
$$x - 2 = 0 \quad \text{or} \quad x - 2 = 0$$
$$x = 2 \qquad\qquad x = 2$$
The solution is $x = 2$.

10. $5x^2 - 30x + 45 = 0$
$$5(x^2 - 6x + 9) = 0$$
$$5(x - 3)(x - 3) = 0$$
$$x - 3 = 0 \quad \text{or} \quad x - 3 = 0$$
$$x = 3 \qquad\qquad x = 3$$
The solution is $x = 3$.

11.
$$9 - 6p + p^2 = 0$$
$$(3 - p)(3 - p) = 0$$
$$3 - p = 0 \quad \text{or} \quad 3 - p = 0$$
$$3 = p \qquad\qquad 3 = p$$
The solution is $p = 3$.

12.
$$49 - 28p + 4p^2 = 0$$
$$(7 - 2p)(7 - 2p) = 0$$
$$7 - 2p = 0 \quad \text{or} \quad 7 - 2p = 0$$
$$7 = 2p \qquad\qquad 7 = 2p$$
$$\frac{7}{2} = p \qquad\qquad \frac{7}{2} = p$$
The solution is $p = \dfrac{7}{2}$.

13. $4y^2 - 16 = 0$

$4(y^2 - 4) = 0$

$4(y - 2)(y + 2) = 0$

$y - 2 = 0$ or $y + 2 = 0$

$y = 2$ $y = -2$

The solutions are $y = \pm 2$.

14. $3y^2 - 27 = 0$

$3(y^2 - 9) = 0$

$3(y - 3)(y + 3) = 0$

$y - 3 = 0$ or $y + 3 = 0$

$y = 3$ $y = -3$

The solutions are $y = \pm 3$.

15. $x^2 - 3x + 2 = 0$

$(x - 2)(x - 1) = 0$

$x - 2 = 0$ or $x - 1 = 0$

$x = 2$ $x = 1$

The solutions are $x = 1$ and $x = 2$.

16. $x^2 + 7x + 12 = 0$

$(x + 3)(x + 4) = 0$

$x + 3 = 0$ or $x + 4 = 0$

$x = -3$ $x = -4$

The solutions are $x = -3$ and $x = -4$.

17. $(2z + 5)^2 = 25$

$2z + 5 = \sqrt{25}$ or $2z + 5 = -\sqrt{25}$

$2z + 5 = 5$ $2z + 5 = -5$

$2z = 0$ $2z = -10$

$z = 0$ $z = -5$

The solutions are $z = 0$ and $z = -5$.

18. $(3z - 4)^2 = 16$

$3z - 4 = \sqrt{16}$ or $3z - 4 = -\sqrt{16}$

$3z - 4 = 4$ $3z - 4 = -4$

$3z = 8$ $3z = 0$

$z = \dfrac{8}{3}$ $z = 0$

The solutions are $z = \dfrac{8}{3}$ and $z = 0$.

19. $30x = 25x^2 + 2$

$0 = 25x^2 - 30x + 2$

$a = 25, b = -30, c = 2$

$x = \dfrac{-b \pm \sqrt{b^2 - 4ac}}{2a}$

$x = \dfrac{-(-30) \pm \sqrt{(-30)^2 - 4(25)(2)}}{2(25)}$

$= \dfrac{30 \pm \sqrt{900 - 200}}{50}$

$= \dfrac{30 \pm \sqrt{700}}{50}$

$= \dfrac{30 \pm 10\sqrt{7}}{50}$

$= \dfrac{3 \pm \sqrt{7}}{5}$

The solutions are $x = \dfrac{3 \pm \sqrt{7}}{5}$.

20. $12x = 4x^2 + 4$

$0 = 4x^2 - 12x + 4$

$0 = 4(x^2 - 3x + 1)$

$a = 1, b = -3, c = 1$

$x = \dfrac{-b \pm \sqrt{b^2 - 4ac}}{2a}$

$x = \dfrac{-(-3) \pm \sqrt{(-3)^2 - 4(1)(1)}}{2(1)}$

$= \dfrac{3 \pm \sqrt{9 - 4}}{2}$

$= \dfrac{3 \pm \sqrt{5}}{2}$

The solutions are $x = \dfrac{3 \pm \sqrt{5}}{2}$.

21. $\frac{2}{3}m^2 - \frac{1}{3}m - 1 = 0$

$2m^2 - m - 3 = 0$

$(2m - 3)(m + 1) = 0$

$2m - 3 = 0$ or $m + 1 = 0$

$2m = 3$ $\qquad\qquad m = -1$

$m = \frac{3}{2}$

The solutions are $m = \frac{3}{2}$ and $m = -1$.

22. $\frac{5}{8}m^2 + m - \frac{1}{2} = 0$

$5m^2 + 8m - 4 = 0$

$(5m - 2)(m + 2) = 0$

$5m - 2 = 0$ or $m + 2 = 0$

$5m = 2$ $\qquad\qquad m = -2$

$m = \frac{2}{5}$

The solutions are $m = \frac{2}{5}$ and $m = -2$.

23. $x^2 - \frac{1}{2}x - \frac{1}{5} = 0$

$10x^2 - 5x - 2 = 0$

$a = 10, b = -5, c = -2$

$x = \frac{-b \pm \sqrt{b^2 - 4ac}}{2a}$

$x = \frac{-(-5) \pm \sqrt{(-5)^2 - 4(10)(-2)}}{2(10)}$

$= \frac{5 \pm \sqrt{25 + 80}}{20}$

$= \frac{5 \pm \sqrt{105}}{20}$

The solutions are $x = \frac{5 \pm \sqrt{105}}{20}$.

24. $x^2 + \frac{1}{2}x - \frac{1}{8} = 0$

$8x^2 + 4x - 1 = 0$

$a = 8, b = 4, c = -1$

$x = \frac{-b \pm \sqrt{b^2 - 4ac}}{2a}$

$x = \frac{-4 \pm \sqrt{4^2 - 4(8)(-1)}}{2(8)}$

$= \frac{-4 \pm \sqrt{16 + 32}}{16}$

$= \frac{-4 \pm \sqrt{48}}{16}$

$= \frac{-4 \pm 4\sqrt{3}}{16}$

$= \frac{-1 \pm \sqrt{3}}{4}$

The solutions are $\frac{-1 \pm \sqrt{3}}{4}$.

25. $4x^2 - 27x + 35 = 0$

$(4x - 7)(x - 5) = 0$

$4x - 7 = 0$ or $x - 5 = 0$

$4x = 7$ $\qquad\qquad x = 5$

$x = \frac{7}{4}$

The solutions are $x = \frac{7}{4}$ and $x = 5$.

26. $9x^2 - 16x + 7 = 0$

$(9x - 7)(x - 1) = 0$

$9x - 7 = 0$ or $x - 1 = 0$

$9x = 7$ $\qquad\qquad x = 1$

$x = \frac{7}{9}$

The solutions are $x = \frac{7}{9}$ and $x = 1$.

27. $(7 - 5x)^2 = 18$

$7 - 5x = \sqrt{18}$ or $7 - 5x = -\sqrt{18}$

$-5x = -7 + 3\sqrt{2}$ $\qquad -5x = -7 - 3\sqrt{2}$

$x = \frac{7 - 3\sqrt{2}}{5}$ $\qquad x = \frac{7 + 3\sqrt{2}}{5}$

The solutions are $x = \frac{7 \pm 3\sqrt{2}}{5}$.

28. $(5-4x)^2 = 75$

$$5-4x = \sqrt{75} \quad \text{or} \quad 5-4x = -\sqrt{75}$$
$$-4x = -5+5\sqrt{3} \qquad -4x = -5-5\sqrt{3}$$
$$x = \frac{5-5\sqrt{3}}{4} \qquad\qquad x = \frac{5+5\sqrt{3}}{4}$$

The solutions are $x = \dfrac{5\pm5\sqrt{3}}{4}$.

29. $3z^2 - 7z = 12$

$$3z^2 - 7z - 12 = 0$$
$$a = 3, b = -7, c = -12$$
$$z = \frac{-b\pm\sqrt{b^2-4ac}}{2a}$$
$$z = \frac{-(-7)\pm\sqrt{(-7)^2-4(3)(-12)}}{2(3)}$$
$$= \frac{7\pm\sqrt{49+144}}{6}$$
$$= \frac{7\pm\sqrt{193}}{6}$$

The solutions are $z = \dfrac{7\pm\sqrt{193}}{6}$.

30. $6z^2 + 7z = 6$

$$6z^2 + 7z - 6 = 0$$
$$a = 6, b = 7, c = -6$$
$$z = \frac{-b\pm\sqrt{b^2-4ac}}{2a}$$
$$z = \frac{-7\pm\sqrt{(7)^2-4(6)(-6)}}{2(6)}$$
$$= \frac{-7\pm\sqrt{49+144}}{12}$$
$$= \frac{-7\pm\sqrt{193}}{12}$$

The solutions are $z = \dfrac{-7\pm\sqrt{193}}{12}$.

31. $x = x^2 - 110$

$$0 = x^2 - x - 110$$
$$0 = (x-11)(x+10)$$
$$x-11 = 0 \quad \text{or} \quad x+10 = 0$$
$$x = 11 \qquad\qquad x = -10$$

The solutions are $x = 11$ and $x = -10$.

32. $x = 56 - x^2$

$$0 = 56 - x - x^2$$
$$0 = (8+x)(7-x)$$
$$8+x = 0 \quad \text{or} \quad 7-x = 0$$
$$x = -8 \qquad\qquad x = 7$$

The solutions are $x = 7$ and $x = -8$.

33. $\dfrac{3}{4}x^2 - \dfrac{5}{2}x - 2 = 0$

$$3x^2 - 10x - 8 = 0$$
$$(3x+2)(x-4) = 0$$
$$3x+2 = 0 \quad \text{or} \quad x-4 = 0$$
$$3x = -2 \qquad\qquad x = 4$$
$$x = -\frac{2}{3}$$

The solutions are $x = -\dfrac{2}{3}$ and $x = 4$.

34. $x^2 - \dfrac{6}{5}x - \dfrac{8}{5} = 0$

$$5x^2 - 6x - 8 = 0$$
$$(5x+4)(x-2) = 0$$
$$5x+4 = 0 \quad \text{or} \quad x-2 = 0$$
$$5x = -4 \qquad\qquad x = 2$$
$$x = -\frac{4}{5}$$

The solutions are $x = -\dfrac{4}{5}$ and $x = 2$.

35. $x^2 - 0.6x + 0.05 = 0$

$$(x-0.5)(x-0.1) = 0$$
$$x-0.5 = 0 \quad \text{or} \quad x-0.1 = 0$$
$$x = 0.5 \qquad\qquad x = 0.1$$

The solutions are $x = 0.5$ and $x = 0.1$.

36. $x^2 - 0.1x - 0.06 = 0$

$(x - 0.3)(x + 0.2) = 0$

$x - 0.3 = 0$ or $x + 0.2 = 0$

     $x = 0.3$ $x = -0.2$

The solutions are $x = 0.3$ and $x = -0.2$.

37. $10x^2 - 11x + 2 = 0$

$a = 10, b = -11, c = 2$

$$x = \frac{-b \pm \sqrt{b^2 - 4ac}}{2a}$$

$$x = \frac{-(-11) \pm \sqrt{(-11)^2 - 4(10)(2)}}{2(10)}$$

$$= \frac{11 \pm \sqrt{121 - 80}}{20}$$

$$= \frac{11 \pm \sqrt{41}}{20}$$

The solutions are $x = \dfrac{11 \pm \sqrt{41}}{20}$.

38. $20x^2 - 11x + 1 = 0$

$a = 20, b = -11, c = 1$

$$x = \frac{-b \pm \sqrt{b^2 - 4ac}}{2a}$$

$$x = \frac{-(-11) \pm \sqrt{(-11)^2 - 4(20)(1)}}{2(20)}$$

$$= \frac{11 \pm \sqrt{121 - 80}}{40}$$

$$= \frac{11 \pm \sqrt{41}}{40}$$

The solutions are $x = \dfrac{11 \pm \sqrt{41}}{40}$.

39. $\dfrac{1}{2}z^2 - 2z + \dfrac{3}{4} = 0$

$2z^2 - 8z + 3 = 0$

$a = 2, b = -8, c = 3$

$$z = \frac{-b \pm \sqrt{b^2 - 4ac}}{2a}$$

$$z = \frac{-(-8) \pm \sqrt{(-8)^2 - 4(2)(3)}}{2(2)}$$

$$= \frac{8 \pm \sqrt{64 - 24}}{4}$$

$$= \frac{8 \pm \sqrt{40}}{4}$$

$$= \frac{8 \pm 2\sqrt{10}}{4}$$

$$= \frac{4 \pm \sqrt{10}}{2}$$

The solutions are $z = \dfrac{4 \pm \sqrt{10}}{2}$.

40. $\dfrac{1}{5}z^2 - \dfrac{1}{2}z - 2 = 0$

$2z^2 - 5z - 20 = 0$

$a = 2, b = -5, c = -20$

$$z = \frac{-b \pm \sqrt{b^2 - 4ac}}{2a}$$

$$z = \frac{-(-5) \pm \sqrt{(-5)^2 - 4(2)(-20)}}{2(2)}$$

$$= \frac{5 \pm \sqrt{25 + 160}}{4}$$

$$= \frac{5 \pm \sqrt{185}}{4}$$

The solutions are $z = \dfrac{5 \pm \sqrt{185}}{4}$.

41. answers may vary

Section 9.4 Practice Problems

1. $y = -3x^2$

x	y
-2	-12
-1	-3
0	0
1	-3
2	-12

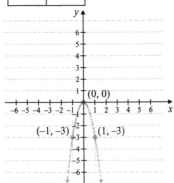

2. $y = x^2 - 9$

x	y
-3	0
-2	-5
-1	-8
0	-9
1	-8
2	-5
3	0

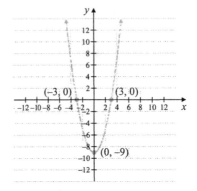

3. $y = x^2 - 2x - 3$

x	y
-1	0
0	-3
1	-4
2	-3
3	0

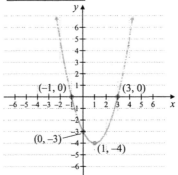

4. $y = x^2 - 4x + 1$

$a = 1,\ b = -4,\ c = 1$

$x = \dfrac{-b}{2a} = \dfrac{-(-4)}{2(1)} = \dfrac{4}{2} = 2$

$y = 2^2 - 4(2) + 1 = 4 - 8 + 1 = -3$

vertex $= (2, -3)$

$x = \dfrac{-b \pm \sqrt{b^2 - 4ac}}{2a}$

$$x = \frac{-(-4) \pm \sqrt{(-4)^2 - 4(1)(1)}}{2(1)}$$

$$= \frac{4 \pm \sqrt{16 - 4}}{2}$$

$$= \frac{4 \pm \sqrt{12}}{2}$$

$$= \frac{4 \pm 2\sqrt{3}}{2}$$

$$= 2 \pm \sqrt{3}$$

$$= 3.7 \text{ or } 0.3$$

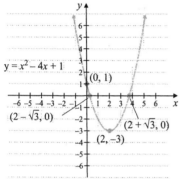

$y = x^2 - 4x + 1$

$(0, 1)$

$(2 - \sqrt{3}, 0)$

$(2 + \sqrt{3}, 0)$

$(2, -3)$

x	y
2	−3
$2 + \sqrt{3} \approx 3.7$	0
$2 - \sqrt{3} \approx 0.3$	0
0	1
4	1

Calculator Explorations

1. $x^2 - 7x - 3 = 0$ $x \approx -0.41, 7.41$

2. $2x^2 - 11x - 1 = 0$ $x \approx -0.09, 5.59$

3. $-1.7x^2 + 5.6x - 3.7 = 0$ $x \approx 0.91, 2.38$

4. $-5.8x^2 + 2.3x - 3.9 = 0$ No real solutions

5. $5.8x^2 - 2.6x - 1.9 = 0$ $x \approx -0.39, 0.84$

6. $7.5x^2 - 3.7x - 1.1 = 0$ $x \approx -0.21, 0.70$

Exercise Set 9.4

1.

x	$y = 2x^2$
−2	$2(-2)^2 = 2(4) = 8$
−1	$2(-1)^2 = 2(1) = 2$
0	$2(0)^2 = 2(0) = 0$
1	$2(1)^2 = 2(1) = 2$
2	$2(2)^2 = 2(4) = 8$

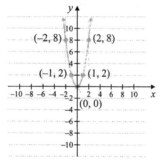

$(-2, 8)$ $(2, 8)$

$(-1, 2)$ $(1, 2)$

$(0, 0)$

3.

x	$y = -x^2$
−2	$-(-2)^2 = -(4) = -4$
−1	$-(-1)^2 = -(1) = -1$
0	$-(0)^2 = -(0) = 0$
1	$-(1)^2 = -(1) = -1$
2	$-(2)^2 = -(4) = -4$

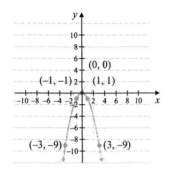

5. $y = x^2 - 1$

$y = 0$:　　$0 = x^2 - 1$

$1 = x^2$

$\pm\sqrt{1} = x$

$\pm 1 = x$

x-intercepts: $(-1, 0)$, $(1, 0)$

$x = 0$: $y = 0^2 - 1 = -1$

y-intercept: $(0, -1)$

$y = x^2 + 0x - 1$

$a = 1$, $b = 0$, $c = 1$

$\dfrac{-b}{2a} = \dfrac{-0}{2(1)} = \dfrac{0}{2} = 0$

$x = 0$: $y = 0^2 - 1 = -1$

The vertex is $(0, -1)$.

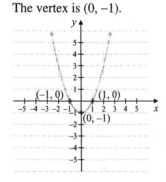

7. $y = x^2 + 4$

$y = 0$:　　$0 = x^2 + 4$

$-4 = x^2$

$\pm\sqrt{-4} = x$

x-intercepts: none

$x = 0$: $y = 0^2 + 4 = 4$

y-intercept: $(0, 4)$

$y = x^2 + 0x + 4$

$a = 1$, $b = 0$, $c = 4$

$\dfrac{-b}{2a} = \dfrac{-0}{2(1)} = \dfrac{0}{2} = 0$

$x = 0$: $y = 0^2 + 4 = 4$

The vertex is $(0, 4)$.

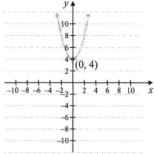

9. $y = -x^2 + 4x - 4$

$y = 0$: $0 = -x^2 + 4x - 4$

$0 = x^2 - 4x + 4$

$0 = (x - 2)^2$

$0 = x - 2$

$2 = x$

The x-intercept is $(2, 0)$.

$x = 0$: $y = -0^2 + 4(0) - 4 = -4$

The y-intercept is $(0, -4)$.

$y = -x^2 + 4x - 4$

$a = -1$, $b = 4$, $c = -4$

$\dfrac{-b}{2a} = \dfrac{-4}{2(-1)} = \dfrac{-4}{-2} = 2$

$x = 2$: $y = -(2)^2 + 4(2) - 4 = -4 + 8 - 4 = 0$

The vertex is $(2, 0)$.

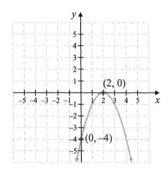

11. $y = x^2 + 5x + 4$

$y = 0$: $0 = x^2 + 5x + 4$

$\ 0 = (x+4)(x+1)$

$\ 0 = x + 4 \quad$ or $\quad 0 = x + 1$

$\ {-4} = x \phantom{+4 \quad \text{or}} \quad {-1} = x$

x-intercepts: $(-4, 0)$ and $(-1, 0)$

$x = 0$: $y = 0^2 + 5(0) + 4 = 4$

y-intercept: $(0, 4)$

$y = x^2 + 5x + 4$

$a = 1, b = 5, c = 4$

$\dfrac{-b}{2a} = \dfrac{-5}{2(1)} = \dfrac{-5}{2}$

$x = \dfrac{-5}{2}:\ y = \left(-\dfrac{5}{2}\right)^2 + 5\left(-\dfrac{5}{2}\right) + 4$

$\phantom{x = \dfrac{-5}{2}:\ y} = \dfrac{25}{4} - \dfrac{25}{2} + 4$

$\phantom{x = \dfrac{-5}{2}:\ y} = \dfrac{25}{4} - \dfrac{50}{4} + \dfrac{16}{4}$

$\phantom{x = \dfrac{-5}{2}:\ y} = -\dfrac{9}{4}$

The vertex is $\left(-\dfrac{5}{2},\ -\dfrac{9}{4}\right)$ or $\left(-2\dfrac{1}{2},\ -2\dfrac{1}{4}\right)$.

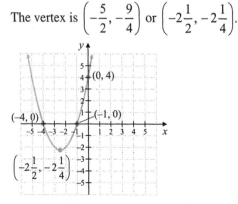

13. $y = x^2 - 4x + 5$

$y = 0$: $0 = x^2 - 4x + 5$

$\ a = 1, b = -4, c = 5$

$\ y = \dfrac{-b \pm \sqrt{b^2 - 4ac}}{2a}$

$\ y = \dfrac{-(-4) \pm \sqrt{(-4)^2 - 4(1)(5)}}{2(1)}$

$\ = \dfrac{4 \pm \sqrt{16 - 20}}{2}$

$\ = \dfrac{4 \pm \sqrt{-4}}{2}$

There are no x-intercepts, since $\sqrt{-4}$ is not a real number.

$x = 0$: $y = 0^2 - 4(0) + 5 = 5$

y-intercept: $(0, 5)$

$y = x^2 - 4x + 5$

$a = 1, b = -4, c = 5$

$\dfrac{-b}{2a} = \dfrac{-(-4)}{2(1)} = \dfrac{4}{2} = 2$

$x = 2$: $y = 2^2 - 4(2) + 5 = 4 - 8 + 5 = 1$

vertex: $(2, 1)$

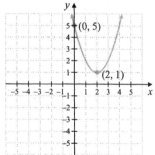

15. $y = 2 - x^2$

$y = 0$: $\quad 0 = 2 - x^2$

$\ x^2 = 2$

$\ x = \pm\sqrt{2}$

x-intercepts: $\left(\sqrt{2}, 0\right)$ and $\left(-\sqrt{2}, 0\right)$

$x = 0$: $y = 2 - 0^2 = 2$

y-intercept: $(0, 2)$

$y = 2 - x^2$

$a = -1, b = 0, c = 2$

$$\frac{-b}{2a} = \frac{-0}{2(-1)} = \frac{0}{-2} = 0$$

$x = 0: \ y = 2 - 0^2 = 2$

vertex: $(0, 2)$

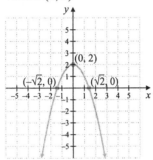

17. $y = \dfrac{1}{3}x^2$

$y = 0: \ 0 = \dfrac{1}{3}x^2$

$\qquad\qquad 0 = x^2$

$\qquad\qquad 0 = x$

x-intercept: $(0, 0)$

The y-intercept is also $(0, 0)$.

$y = \dfrac{1}{3}x^2 + 0x + 0$

$a = \dfrac{1}{3}, \ b = 0, \ c = 0$

$$\frac{-b}{2a} = \frac{-0}{2\left(\frac{1}{3}\right)} = \frac{0}{\frac{2}{3}} = 0$$

$x = 0: \ y = \dfrac{1}{3}(0)^2 = \dfrac{1}{3}(0) = 0$

vertex: $(0, 0)$

For additional points, use $x = \pm 3$.

$x = -3: \ y = \dfrac{1}{3}(-3)^2 = \dfrac{1}{3}(9) = 3$

$x = 3: \ y = \dfrac{1}{3}(3)^2 = \dfrac{1}{3}(9) = 3$

$(-3, 3)$ and $(3, 3)$ are also on the graph.

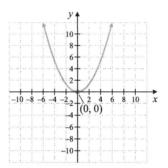

19. $y = x^2 + 6x$

$y = 0: \ 0 = x^2 + 6x$

$\qquad\qquad 0 = x(x + 6)$

$\qquad\qquad x = 0 \quad \text{or} \quad x + 6 = 0$

$\qquad\qquad\qquad\qquad\qquad\qquad x = -6$

x-intercepts: $(0, 0)$ and $(-6, 0)$

$x = 0: \ y = 0^2 + 6(0) = 0$

y-intercept: $(0, 0)$

$y = x^2 + 6x + 0$

$a = 1, b = 6, c = 0$

$$\frac{-b}{2a} = \frac{-6}{2(1)} = \frac{-6}{2} = -3$$

$x = -3: \ y = (-3)^2 + 6(-3) = 9 - 18 = -9$

vertex: $(-3, -9)$

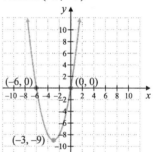

21. $y = x^2 + 2x - 8$

$y = 0: \ 0 = x^2 + 2x - 8$

$\qquad\qquad 0 = (x + 4)(x - 2)$

$\qquad\qquad 0 = x + 4 \quad \text{or} \quad 0 = x - 2$

$\qquad\qquad -4 = x \qquad\qquad\qquad 2 = x$

x-intercepts: $(-4, 0), (2, 0)$

$x = 0$: $y = 0^2 + 2(0) - 8 = -8$

y-intercept: $(0, -8)$

$y = x^2 + 2x - 8$

$a = 1$, $b = 2$, $c = -8$

$\dfrac{-b}{2a} = \dfrac{-2}{2(1)} = \dfrac{-2}{2} = -1$

$x = -1$: $y = (-1)^2 + 2(-1) - 8 = 1 - 2 - 8 = -9$

vertex: $(-1, -9)$

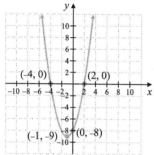

23. $y = -\dfrac{1}{2}x^2$

$y = 0$: $0 = -\dfrac{1}{2}x^2$

$\qquad\quad 0 = x^2$

$\qquad\quad 0 = x$

x-intercept: $(0, 0)$

The y-intercept is also $(0, 0)$.

$y = -\dfrac{1}{2}x^2 + 0x + 0$

$a = -\dfrac{1}{2}$, $b = 0$, $c = 0$

$\dfrac{-b}{2a} = \dfrac{-0}{2\left(-\frac{1}{2}\right)} = \dfrac{0}{-1} = 0$

$x = 0$: $y = -\dfrac{1}{2}(0)^2 = -\dfrac{1}{2}(0) = 0$

vertex: $(0, 0)$

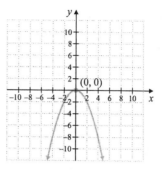

25. $y = 2x^2 - 11x + 5$

$y = 0$: $0 = 2x^2 - 11x + 5$

$\qquad\quad 0 = (2x - 1)(x - 5)$

$\qquad\quad 0 = 2x - 1$ or $0 = x - 5$

$\qquad\quad 1 = 2x$ $5 = x$

$\qquad\quad \dfrac{1}{2} = x$

x-intercepts: $\left(\dfrac{1}{2}, 0\right)$, $(5, 0)$

$x = 0$: $y = 2(0)^2 - 11(0) + 5 = 5$

y-intercept: $(0, 5)$

$y = 2x^2 - 11x + 5$

$a = 2$, $b = -11$, $c = 5$

$\dfrac{-b}{2a} = \dfrac{-(-11)}{2(2)} = \dfrac{11}{4}$

$x = \dfrac{11}{4}$: $y = 2\left(\dfrac{11}{4}\right)^2 - 11\left(\dfrac{11}{4}\right) + 5$

$\qquad\qquad = 2\left(\dfrac{121}{16}\right) - \dfrac{121}{4} + 5$

$\qquad\qquad = \dfrac{121}{8} - \dfrac{121}{4} + 5$

$\qquad\qquad = \dfrac{121}{8} - \dfrac{242}{8} + \dfrac{40}{8}$

$\qquad\qquad = -\dfrac{81}{8}$

vertex: $\left(\dfrac{11}{4}, -\dfrac{81}{8}\right)$

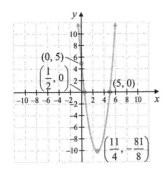

27. $y = -x^2 + 4x - 3$

$y = 0:\ 0 = -x^2 + 4x - 3$

$\qquad 0 = x^2 - 4x + 3$

$\qquad 0 = (x - 3)(x - 1)$

$\qquad x - 3 = 0 \quad \text{or} \quad x - 1 = 0$

$\qquad\quad x = 3 \qquad\qquad x = 1$

x-intercepts: $(3, 0)$ and $(1, 0)$

$x = 0:\ y = -0^2 + 4(0) - 3 = -3$

y-intercept: $(0, -3)$

$y = -x^2 + 4x - 3$

$a = -1,\ b = 4,\ c = -3$

$\dfrac{-b}{2a} = \dfrac{-4}{2(-1)} = \dfrac{-4}{-2} = 2$

$x = 2:\ y = -(2)^2 + 4(2) - 3 = -4 + 8 - 3 = 1$

vertex: $(2, 1)$

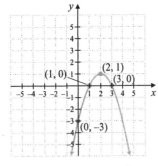

29. $\dfrac{\frac{1}{7}}{\frac{2}{5}} = \dfrac{1}{7} \div \dfrac{2}{5} = \dfrac{1}{7} \cdot \dfrac{5}{2} = \dfrac{5}{14}$

31. $\dfrac{\frac{1}{x}}{\frac{2}{x^2}} = \dfrac{1}{x} \div \dfrac{2}{x^2} = \dfrac{1}{x} \cdot \dfrac{x^2}{2} = \dfrac{x}{2}$

33. $\dfrac{2x}{1 - \frac{1}{x}} = \dfrac{x(2x)}{x\left(1 - \frac{1}{x}\right)} = \dfrac{2x^2}{x - 1}$

35. $\dfrac{\frac{a-b}{2b}}{\frac{b-a}{8b^2}} = \dfrac{a-b}{2b} \div \dfrac{b-a}{8b^2} = \dfrac{a-b}{2b} \cdot \dfrac{8b^2}{b-a} = -4b$

37. a. The maximum height appears to be about 256 feet.

    b. The fireball appears to reach its maximum height when $t = 4$ seconds.

    c. The fireball appears to return to the ground when $t = 8$ seconds.

39. With $a > 0$, the parabola opens upward. An upward-opening parabola that crosses the x-axis twice is graph A.

41. With $a < 0$, the parabola opens downward. A downward-opening parabola that does not touch or cross the x-axis (no x-intercept) is graph D.

43. With $a > 0$, the parabola opens upward. An upward-opening parabola that crosses the x-axis once is graph F.

Chapter 9 Vocabulary Check

1. If $x^2 = a$, then $x = \sqrt{a}$ or $x = -\sqrt{a}$. This property is called the <u>square root</u> property.

2. The formula $\dfrac{-b}{2a}$ where $y = ax^2 + bx + c$ is called the <u>vertex</u> formula.

3. The process of solving a quadratic equation by writing it in the form $(x + a)^2 = c$ is called <u>completing the square</u>.

4. The formula $x = \dfrac{-b \pm \sqrt{b^2 - 4ac}}{2a}$ is called

the <u>quadratic</u> formula.

Chapter 9 Review

1. $x^2 - 121 = 0$

$$x^2 = 121$$
$$x = \pm\sqrt{121}$$
$$x = \pm 11$$

The solutions are $x = \pm 11$.

2. $y^2 - 100 = 0$

$$y^2 = 100$$
$$y = \pm\sqrt{100}$$
$$y = \pm 10$$

The solutions are $y = \pm 10$.

3. $3m^2 - 5m = 2$

$$3m^2 - 5m - 2 = 0$$
$$(3m + 1)(m - 2) = 0$$
$$3m + 1 = 0 \quad \text{or} \quad m - 2 = 0$$
$$3m = -1 \qquad\qquad m = 2$$
$$m = -\frac{1}{3}$$

The solutions are $m = -\dfrac{1}{3}$ and $m = 2$.

4. $7m^2 + 2m = 5$

$$7m^2 + 2m - 5 = 0$$
$$(7m - 5)(m + 1) = 0$$
$$7m - 5 = 0 \quad \text{or} \quad m + 1 = 0$$
$$7m = 5 \qquad\qquad m = -1$$
$$m = \frac{5}{7}$$

The solutions are $m = \dfrac{5}{7}$ and $m = -1$.

5. $x^2 = 36$

$$x = \pm\sqrt{36}$$
$$x = \pm 6$$

The solutions are $x = \pm 6$.

6. $x^2 = 81$

$$x = \pm\sqrt{81}$$
$$x = \pm 9$$

The solutions are $x = \pm 9$.

7. $k^2 = 50$

$$k = \pm\sqrt{50}$$
$$k = \pm 5\sqrt{2}$$

The solutions are $k = \pm 5\sqrt{2}$.

8. $k^2 = 45$

$$k = \pm\sqrt{45}$$
$$k = \pm 3\sqrt{5}$$

The solutions are $k = \pm 3\sqrt{5}$.

9. $(x - 11)^2 = 49$

$$x - 11 = \pm\sqrt{49}$$
$$x - 11 = \pm 7$$
$$x = 11 + 7 \quad \text{or} \quad x = 11 - 7$$
$$x = 18 \qquad\qquad x = 4$$

The solutions are $x = 18$ and $x = 4$.

10. $(x + 3)^2 = 100$

$$x + 3 = \pm\sqrt{100}$$
$$x + 3 = \pm 10$$
$$x = -3 + 10 \quad \text{or} \quad x = -3 - 10$$
$$x = 7 \qquad\qquad x = -13$$

The solutions are $x = 7$ and $x = -13$.

11. $(4p + 5)^2 = 41$

$$4p + 5 = \pm\sqrt{41}$$
$$4p = -5 \pm \sqrt{41}$$
$$p = \frac{-5 \pm \sqrt{41}}{4}$$

The solutions are $p = \dfrac{-5 \pm \sqrt{41}}{4}$.

12. $(3p+7)^2 = 37$

$3p+7 = \pm\sqrt{37}$

$3p = -7 \pm \sqrt{37}$

$p = \dfrac{-7 \pm \sqrt{37}}{3}$

The solutions are $p = \dfrac{-7 \pm \sqrt{37}}{3}$.

13. $h = 16t^2$

$100 = 16t^2$

$6.25 = t^2$

$\pm 2.5 = t$

Reject -2.5 because time is not negative. It will take Kara 2.5 seconds to hit the water.

14. $5 \text{ miles} \cdot \dfrac{5280 \text{ ft}}{1 \text{ mile}} = 26,400 \text{ feet}$

$y = 16t^2$

$26,400 = 16t^2$

$1650 = t^2$

$\pm 40.6 = t$

Reject -40.6 because time is not negative. A 5-mile freefall will take 40.6 seconds.

15. $x^2 - 9x = -8$

$x^2 - 9x + \left(-\dfrac{9}{2}\right)^2 = -8 + \left(-\dfrac{9}{2}\right)^2$

$\left(x - \dfrac{9}{2}\right)^2 = -8 + \dfrac{81}{4}$

$\left(x - \dfrac{9}{2}\right)^2 = \dfrac{49}{4}$

$x - \dfrac{9}{2} = \pm\sqrt{\dfrac{49}{4}}$

$x - \dfrac{9}{2} = \pm\dfrac{7}{2}$

$x = \dfrac{9}{2} + \dfrac{7}{2} = \dfrac{16}{2} = 8$ or $x = \dfrac{9}{2} - \dfrac{7}{2} = \dfrac{2}{2} = 1$

The solutions are $x = 8$ and $x = 1$.

16. $x^2 + 8x = 20$

$x^2 + 8x + \left(\dfrac{8}{2}\right)^2 = 20 + \left(\dfrac{8}{2}\right)^2$

$(x+4)^2 = 36$

$x + 4 = \pm\sqrt{36}$

$x + 4 = \pm 6$

$x = -4 + 6 = 2$ or $x = -4 - 6 = -10$

The solutions are $x = 7$ and $x = -10$.

17. $x^2 + 4x = 1$

$x^2 + 4x + \left(\dfrac{4}{2}\right)^2 = 1 + \left(\dfrac{4}{2}\right)^2$

$(x+2)^2 = 5$

$x + 2 = \pm\sqrt{5}$

$x = -2 \pm \sqrt{5}$

The solutions are $x = -2 \pm \sqrt{5}$.

18. $x^2 - 8x = 3$

$x^2 - 8x + \left(-\dfrac{8}{2}\right)^2 = 3 + \left(-\dfrac{8}{2}\right)^2$

$(x-4)^2 = 19$

$x - 4 = \pm\sqrt{19}$

$x = 4 \pm \sqrt{19}$

The solutions are $x = 4 \pm \sqrt{19}$.

19. $x^2 - 6x + 7 = 0$

$x^2 - 6x = -7$

$x^2 - 6x + \left(-\dfrac{6}{2}\right)^2 = -7 + \left(-\dfrac{6}{2}\right)^2$

$(x-3)^2 = 2$

$x - 3 = \pm\sqrt{2}$

$x = 3 \pm \sqrt{2}$

The solutions are $3 \pm \sqrt{2}$.

20.
$$x^2 + 6x + 7 = 0$$
$$x^2 + 6x = -7$$
$$x^2 + 6x + \left(\frac{6}{2}\right)^2 = -7 + \left(\frac{6}{2}\right)^2$$
$$(x+3)^2 = 2$$
$$x + 3 = \pm\sqrt{2}$$
$$x = -3 \pm \sqrt{2}$$
The solutions are $-3 \pm \sqrt{2}$.

21.
$$2y^2 + y - 1 = 0$$
$$2y^2 + y = 1$$
$$y^2 + \frac{1}{2}y = \frac{1}{2}$$
$$y^2 + \frac{1}{2}y + \left(\frac{1}{4}\right)^2 = \frac{1}{2} + \left(\frac{1}{4}\right)^2$$
$$\left(y + \frac{1}{4}\right)^2 = \frac{9}{16}$$
$$y + \frac{1}{4} = \pm\sqrt{\frac{9}{16}}$$
$$y + \frac{1}{4} = \pm\frac{3}{4}$$
$$y = -\frac{1}{4} + \frac{3}{4} = \frac{2}{4} = \frac{1}{2} \quad \text{or}$$
$$y = -\frac{1}{4} - \frac{3}{4} = -\frac{4}{4} = -1$$

The solutions are $y = \frac{1}{2}$ and $y = -1$.

22.
$$y^2 + 3y - 1 = 0$$
$$y^2 + 3y = 1$$
$$y^2 + 3y + \left(\frac{3}{2}\right)^2 = 1 + \left(\frac{3}{2}\right)^2$$
$$\left(y + \frac{3}{2}\right)^2 = \frac{13}{4}$$
$$y + \frac{3}{2} = \pm\sqrt{\frac{13}{4}}$$
$$y + \frac{3}{2} = \pm\frac{\sqrt{13}}{2}$$

$$y = -\frac{3}{2} \pm \frac{\sqrt{13}}{2} \quad \text{or} \quad \frac{-3 \pm \sqrt{13}}{2}$$
The solutions are $y = \dfrac{-3 \pm \sqrt{13}}{2}$.

23. $9x^2 + 30x + 25 = 0$
$a = 9$, $b = 30$, $c = 25$
$$x = \frac{-b \pm \sqrt{b^2 - 4ac}}{2a}$$
$$x = \frac{-30 \pm \sqrt{30^2 - 4(9)(25)}}{2(9)}$$
$$= \frac{-30 \pm \sqrt{900 - 900}}{18}$$
$$= \frac{-30 \pm \sqrt{0}}{18}$$
$$= \frac{-30}{18}$$
$$= -\frac{5}{3}$$

The solution is $x = -\dfrac{5}{3}$.

24. $16x^2 - 72x + 81 = 0$
$a = 16$, $b = -72$, $c = 81$
$$x = \frac{-b \pm \sqrt{b^2 - 4ac}}{2a}$$
$$x = \frac{-(-72) \pm \sqrt{(-72)^2 - 4(16)(81)}}{2(16)}$$
$$= \frac{72 \pm \sqrt{5184 - 5184}}{32}$$
$$= \frac{72 \pm \sqrt{0}}{32}$$
$$= \frac{72}{32}$$
$$= \frac{9}{4}$$
The solution is $x = \dfrac{9}{4}$.

25. $7x^2 = 35$

$7x^2 - 35 = 0$

$a = 7, b = 0, c = -35$

$x = \dfrac{-b \pm \sqrt{b^2 - 4ac}}{2a}$

$x = \dfrac{-0 \pm \sqrt{0^2 - 4(7)(-35)}}{2(7)}$

$= \dfrac{\pm\sqrt{980}}{14}$

$= \dfrac{\pm 14\sqrt{5}}{14}$

$= \pm\sqrt{5}$

The solutions are $x = \pm\sqrt{5}$.

26. $11x^2 = 33$

$11x^2 - 33 = 0$

$a = 11, b = 0, c = -33$

$x = \dfrac{-b \pm \sqrt{b^2 - 4ac}}{2a}$

$x = \dfrac{-0 \pm \sqrt{0^2 - 4(11)(-33)}}{2(11)}$

$= \dfrac{\pm\sqrt{1452}}{22}$

$= \dfrac{\pm 22\sqrt{3}}{22}$

$= \pm\sqrt{3}$

The solutions are $x = \pm\sqrt{3}$.

27. $x^2 - 10x + 7 = 0$

$a = 1, b = -10, c = 7$

$x = \dfrac{-b \pm \sqrt{b^2 - 4ac}}{2a}$

$x = \dfrac{-(-10) \pm \sqrt{(-10)^2 - 4(1)(7)}}{2(1)}$

$= \dfrac{10 \pm \sqrt{100 - 28}}{2}$

$= \dfrac{10 \pm \sqrt{72}}{2}$

$= \dfrac{10 \pm 6\sqrt{2}}{2}$

$= 5 \pm 3\sqrt{2}$

The solutions are $x = 5 \pm 3\sqrt{2}$.

28. $x^2 + 4x - 7 = 0$

$a = 1, b = 4, c = -7$

$x = \dfrac{-b \pm \sqrt{b^2 - 4ac}}{2a}$

$x = \dfrac{-4 \pm \sqrt{4^2 - 4(1)(-7)}}{2(1)}$

$= \dfrac{-4 \pm \sqrt{16 + 28}}{2}$

$= \dfrac{-4 \pm \sqrt{44}}{2}$

$= \dfrac{-4 \pm 2\sqrt{11}}{2}$

$= -2 \pm \sqrt{11}$

The solutions are $x = -2 \pm \sqrt{11}$.

29. $3x^2 + x - 1 = 0$

$a = 3, b = 1, c = -1$

$x = \dfrac{-b \pm \sqrt{b^2 - 4ac}}{2a}$

$x = \dfrac{-1 \pm \sqrt{(1)^2 - 4(3)(-1)}}{2(3)}$

$= \dfrac{-1 \pm \sqrt{1 + 12}}{6}$

$= \dfrac{-1 \pm \sqrt{13}}{6}$

The solutions are $x = \dfrac{-1 \pm \sqrt{13}}{6}$.

30. $x^2 + 3x - 1 = 0$

$a = 1, b = 3, c = -1$

$x = \dfrac{-b \pm \sqrt{b^2 - 4ac}}{2a}$

$x = \dfrac{-3 \pm \sqrt{3^2 - 4(1)(-1)}}{2(1)}$

$= \dfrac{-3 \pm \sqrt{9 + 4}}{2}$

$= \dfrac{-3 \pm \sqrt{13}}{2}$

The solutions are $x = \dfrac{-3 \pm \sqrt{13}}{2}$.

31. $2x^2 + x + 5 = 0$

$a = 2, b = 1, c = 5$

$x = \dfrac{-b \pm \sqrt{b^2 - 4ac}}{2a}$

$x = \dfrac{-1 \pm \sqrt{1^2 - 4(2)(5)}}{2(2)}$

$= \dfrac{-1 \pm \sqrt{1 - 40}}{4}$

$= \dfrac{-1 \pm \sqrt{-39}}{4}$

The equation has no solution, since the square root of -39 is not a real number.

32. $7x^2 - 3x + 1 = 0$

$a = 7, b = -3, c = 1$

$x = \dfrac{-b \pm \sqrt{b^2 - 4ac}}{2a}$

$x = \dfrac{-(-3) \pm \sqrt{(-3)^2 - 4(7)(1)}}{2(7)}$

$= \dfrac{3 \pm \sqrt{9 - 28}}{14}$

$= \dfrac{3 \pm \sqrt{-19}}{14}$

The equation has no solution since the square root of -19 is not a real number.

33. $x = \dfrac{-1 + \sqrt{13}}{6} \approx 0.4$ or $x = \dfrac{-1 - \sqrt{13}}{6} \approx -0.8$

34. $x = \dfrac{-3 + \sqrt{13}}{2} \approx 0.3$ or $x = \dfrac{-3 - \sqrt{13}}{2} \approx -3.3$

35. $y = 38x^2 - 43x + 446$

$1556 = 38x^2 - 43x + 446$

$0 = 38x^2 - 43x - 1110$

$a = 38, b = -43, c = -1110$

$x = \dfrac{-b \pm \sqrt{b^2 - 4ac}}{2a}$

$x = \dfrac{-(-43) \pm \sqrt{(-43)^2 - 4(38)(-1110)}}{2(38)}$

$= \dfrac{43 \pm \sqrt{1849 + 168,720}}{76}$

$= \dfrac{43 \pm \sqrt{170,569}}{76}$

$= \dfrac{43 \pm 413}{76}$

$x = \dfrac{43 + 413}{76} = 6$ or $x = \dfrac{43 - 413}{76} \approx -4.9$

Discard $x = -49$ since the number of years is not negative.

The price of silver will be 1556 cents per ounce 6 years after 2001 or 2007.

36. $y = 64x^2 - 87x + 545$

$2327 = 64x^2 - 87x + 545$

$0 = 64x^2 - 87x - 1782$

$a = 64, b = -87, c = -1782$

$x = \dfrac{-b \pm \sqrt{b^2 - 4ac}}{2a}$

$$x = \frac{-(-87) \pm \sqrt{(-87)^2 - 4(64)(-1782)}}{2(64)}$$

$$= \frac{87 \pm \sqrt{7569 + 456,192}}{128}$$

$$= \frac{87 \pm \sqrt{463,761}}{128}$$

$$= \frac{87 \pm 681}{128}$$

$$x = \frac{78 + 681}{128} = 6 \quad \text{or} \quad x = \frac{87 - 681}{128} \approx -4.6$$

Discard $x = -4.6$ since the number of years is not negative.
The price of platinum will be 2327 dollars per ounce 6 years after 2001 or 2007.

37. $y = 5x^2$

$y = 0$: $\quad 0 = 5x^2$
$\qquad\qquad 0 = x^2$
$\qquad\qquad 0 = x$
x-intercept: $(0, 0)$
The y-intercept is also $(0, 0)$.

x	$y = 5x^2$
-2	$5(-2)^2 = 5(4) = 20$
-1	$5(-1)^2 = 5(1) = 5$
0	$5(0)^2 = 5(0) = 0$
1	$5(1)^2 = 5(1) = 5$
2	$5(2)^2 = 5(4) = 20$

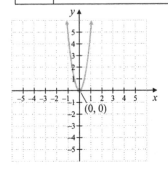

38. $y = -\frac{1}{2}x^2$

$y = 0$: $\quad 0 = -\frac{1}{2}x^2$
$\qquad\qquad 0 = x^2$
$\qquad\qquad 0 = x$
x-intercept: $(0, 0)$
The y-intercept is also $(0, 0)$.

x	$y = -\frac{1}{2}x^2$
-2	$-\frac{1}{2}(-2)^2 = -\frac{1}{2}(4) = -2$
-1	$-\frac{1}{2}(-1)^2 = -\frac{1}{2}(1) = -\frac{1}{2}$
0	$-\frac{1}{2}(0)^2 = -\frac{1}{2}(0) = 0$
1	$-\frac{1}{2}(1)^2 = -\frac{1}{2}(1) = -\frac{1}{2}$
2	$-\frac{1}{2}(2)^2 = -\frac{1}{2}(4) = -2$

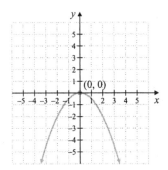

39. $y = x^2 - 25$

$y = 0$: $\qquad 0 = x^2 - 25$
$\qquad\qquad 25 = x^2$
$\qquad\qquad \pm\sqrt{25} = x$
$\qquad\qquad \pm 5 = x$
x-intercepts: $(5, 0)$ and $(-5, 0)$
$x = 0$: $\ y = 0^2 - 25 = -25$
y-intercept: $(0, -25)$
$y = x^2 + 0x - 25$

$a = 1, b = 0, c = -25$

$$\frac{-b}{2a} = \frac{-0}{2(1)} = \frac{0}{2} = 0$$

$x = 0$: $y = 0^2 - 25 = -25$

vertex: $(0, -25)$

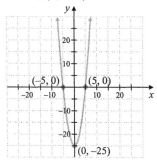

40. $y = x^2 - 36$

$y = 0$: $\qquad 0 = x^2 - 36$

$$36 = x^2$$
$$\pm\sqrt{36} = x$$
$$\pm 6 = x$$

x-intercepts: $(6, 0)$ and $(-6, 0)$

$x = 0$: $y = 0^2 - 36 = -36$

y-intercept: $(0, -36)$

$y = x^2 + 0x - 36$

$a = 1, b = 0, c = -36$

$$\frac{-b}{2a} = \frac{-0}{2(1)} = \frac{0}{2} = 0$$

$x = 0$: $y = 0^2 - 36 = -36$

vertex: $(0, -36)$

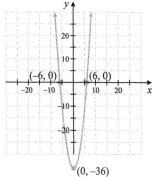

41. $y = x^2 + 3$

$y = 0$: $\qquad 0 = x^2 + 3$

$$-3 = x^2$$
$$\pm\sqrt{-3} = x$$

x-intercepts: none

$x = 0$: $y = 0^2 + 3 = 3$

y-intercept: $(0, 3)$

$y = x^2 + 0x + 3$

$a = 1, b = 0, c = 3$

$$\frac{-b}{2a} = \frac{-0}{2(1)} = \frac{0}{2} = 0$$

$x = 0$: $y = 0^2 + 3 = 3$

vertex: $(0, 3)$

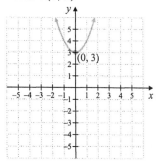

42. $y = x^2 + 8$

$y = 0$: $\qquad 0 = x^2 + 8$

$$-8 = x^2$$
$$\pm\sqrt{-8} = x$$

x-intercepts: none

$x = 0$: $y = 0^2 + 8 = 8$

y-intercept: $(0, 8)$

$y = x^2 + 0x + 8$

$a = 1, b = 0, c = 8$

$$\frac{-b}{2a} = \frac{0}{2(1)} = \frac{0}{2} = 0$$

$x = 0$: $y = 0^2 + 8 = 8$

vertex: $(0, 8)$

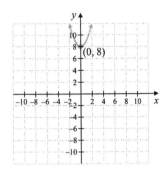

43. $y = -4x^2 + 8$

$\quad y = 0: \qquad 0 = -4x^2 + 8$

$\qquad\qquad\qquad 0 = -4(x^2 - 2)$

$\qquad\qquad\qquad 0 = x^2 - 2$

$\qquad\qquad\qquad 2 = x^2$

$\qquad\qquad\qquad \pm\sqrt{2} = x$

$\quad$ *x*-intercepts: $\left(\sqrt{2}, 0\right)$ and $\left(-\sqrt{2}, 0\right)$

$\quad x = 0: \ y = -4(0)^2 + 8 = 8$

$\quad$ *y*-intercept: $(0, 8)$

$\quad y = -4x^2 + 0x + 8$

$\quad a = -4, b = 0, c = 8$

$\quad \dfrac{-b}{2a} = \dfrac{-0}{2(-4)} = \dfrac{0}{-8} = 0$

$\quad x = 0: \ y = -4(0)^2 + 8 = 8$

$\quad$ vertex: $(0, 8)$

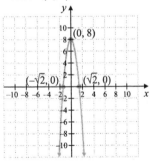

44. $y = -3x^2 + 9$

$\quad y = 0: \qquad 0 = -3x^2 + 9$

$\qquad\qquad\qquad 0 = -3(x^2 - 3)$

$\qquad\qquad\qquad 0 = x^2 - 3$

$\qquad\qquad\qquad 3 = x^2$

$\qquad\qquad\qquad \pm\sqrt{3} = x$

$\quad$ *x*-intercepts: $\left(\sqrt{3}, 0\right)$ and $\left(-\sqrt{3}, 0\right)$

$\quad x = 0: \ y = -3(0)^2 + 9 = 9$

$\quad$ *y*-intercept: $(0, 9)$

$\quad y = -3x^2 + 0x + 9$

$\quad a = -3, b = 0, c = 8$

$\quad \dfrac{-b}{2a} = \dfrac{-0}{2(-3)} = \dfrac{0}{-6} = 0$

$\quad x = 0: \ y = -3(0)^2 + 9 = 9$

$\quad$ vertex: $(0, 9)$

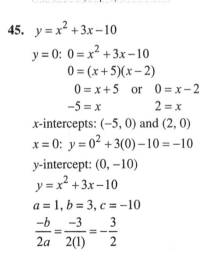

45. $y = x^2 + 3x - 10$

$\quad y = 0: \ 0 = x^2 + 3x - 10$

$\qquad\qquad 0 = (x + 5)(x - 2)$

$\qquad\qquad 0 = x + 5 \quad$ or $\quad 0 = x - 2$

$\qquad\qquad -5 = x \qquad\qquad\quad 2 = x$

$\quad$ *x*-intercepts: $(-5, 0)$ and $(2, 0)$

$\quad x = 0: \ y = 0^2 + 3(0) - 10 = -10$

$\quad$ *y*-intercept: $(0, -10)$

$\quad y = x^2 + 3x - 10$

$\quad a = 1, b = 3, c = -10$

$\quad \dfrac{-b}{2a} = \dfrac{-3}{2(1)} = -\dfrac{3}{2}$

$$x = -\frac{3}{2}: \quad y = \left(-\frac{3}{2}\right)^2 + 3\left(-\frac{3}{2}\right) - 10$$

$$= \frac{9}{4} - \frac{9}{2} - 10$$

$$= \frac{9}{4} - \frac{18}{4} - \frac{40}{4}$$

$$= -\frac{49}{4}$$

vertex: $\left(-\dfrac{3}{2}, -\dfrac{49}{4}\right)$

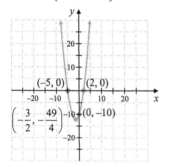

46. $y = x^2 + 3x - 4$

$$y = 0: \quad 0 = x^2 + 3x - 4$$
$$0 = (x+4)(x-1)$$
$$0 = x+4 \quad \text{or} \quad 0 = x-1$$
$$-4 = x \qquad\qquad 1 = x$$

x-intercepts: $(-4, 0)$ and $(1, 0)$

$x = 0: \quad y = 0^2 + 3(0) - 4 = -4$

y-intercept: $(0, -4)$

$y = x^2 + 3x - 4$

$a = 1, b = 3, c = -4$

$$\frac{-b}{2a} = \frac{-3}{2(1)} = \frac{-3}{2}$$

$$x = -\frac{3}{2}: \quad y = \left(-\frac{3}{2}\right)^2 + 3\left(-\frac{3}{2}\right) - 4$$

$$= \frac{9}{4} - \frac{9}{2} - 4$$

$$= \frac{9}{4} - \frac{18}{4} - \frac{16}{4}$$

$$= -\frac{25}{4}$$

vertex: $\left(-\dfrac{3}{2}, -\dfrac{25}{4}\right)$

47. $y = -x^2 - 5x - 6$

$$y = 0: \quad 0 = -x^2 - 5x - 6$$
$$0 = x^2 + 5x + 6$$
$$0 = (x+3)(x+2)$$
$$0 = x+3 \quad \text{or} \quad 0 = x+2$$
$$-3 = x \qquad\qquad -2 = x$$

x-intercepts: $(-3, 0)$ and $(-2, 0)$

$x = 0: \quad y = -0^2 - 5(0) - 6 = -6$

y-intercept: $(0, -6)$

$y = -x^2 - 5x - 6$

$a = -1, b = -5, c = -6$

$$\frac{-b}{2a} = \frac{-(-5)}{2(-1)} = \frac{5}{-2} = -\frac{5}{2}$$

$$x = -\frac{5}{2}: \quad y = -\left(-\frac{5}{2}\right)^2 - 5\left(-\frac{5}{2}\right) - 6$$

$$y = -\frac{25}{4} + \frac{25}{2} - 6$$

$$y = -\frac{25}{4} + \frac{50}{4} - \frac{24}{4}$$

$$y = \frac{1}{4}$$

vertex: $\left(-\dfrac{5}{2}, \dfrac{1}{4}\right)$

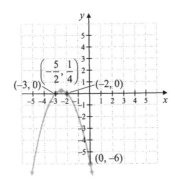

48. $y = 3x^2 - x - 2$

$y = 0$: $0 = 3x^2 - x - 2$

$0 = (3x + 2)(x - 1)$

$0 = 3x + 2$ or $0 = x - 1$

$-2 = 3x$ $1 = x$

$-\dfrac{2}{3} = x$

x-intercepts: $\left(-\dfrac{2}{3}, 0\right)$ and $(1, 0)$

$x = 0$: $y = 3(0)^2 - 0 - 2 = -2$

y-intercept: $(0, -2)$

$y = 3x^2 - x - 2$

$a = 3$, $b = -1$, $c = -2$

$\dfrac{-b}{2a} = \dfrac{-(-1)}{2(3)} = \dfrac{1}{6}$

$x = \dfrac{1}{6}$: $y = 3\left(\dfrac{1}{6}\right)^2 - \dfrac{1}{6} - 2$

$y = \dfrac{3}{36} - \dfrac{1}{6} - 2$

$y = \dfrac{3}{36} - \dfrac{6}{36} - \dfrac{72}{36}$

$y = -\dfrac{75}{36} = -\dfrac{25}{12}$

vertex: $\left(\dfrac{1}{6}, -\dfrac{25}{12}\right)$

49. $y = 2x^2 - 11x - 6$

$y = 0$: $0 = 2x^2 - 11x - 6$

$0 = (2x + 1)(x - 6)$

$0 = 2x + 1$ or $0 = x - 6$

$-1 = 2x$ $6 = x$

$-\dfrac{1}{2} = x$

x-intercepts: $\left(-\dfrac{1}{2}, 0\right)$ and $(6, 0)$

$x = 0$: $y = 2(0)^2 - 11(0) - 6 = -6$

y-intercept: $(0, -6)$

$y = 2x^2 - 11x - 6$

$a = 2$, $b = -11$, $c = -6$

$\dfrac{-b}{2a} = \dfrac{-(-11)}{2(2)} = \dfrac{11}{4}$

$x = \dfrac{11}{4}$: $y = 2\left(\dfrac{11}{4}\right)^2 - 11\left(\dfrac{11}{4}\right) - 6$

$y = \dfrac{242}{16} - \dfrac{121}{4} - 6$

$y = \dfrac{242}{16} - \dfrac{484}{16} - \dfrac{96}{16}$

$y = -\dfrac{338}{16} = -\dfrac{169}{8}$

vertex: $\left(\dfrac{11}{4}, -\dfrac{169}{8}\right)$

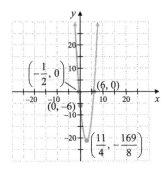

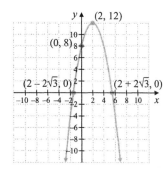

50. $y = -x^2 + 4x + 8$

$y = 0: \; 0 = -x^2 + 4x + 8$

$a = -1, b = 4, c = 8$

$x = \dfrac{-b \pm \sqrt{b^2 - 4ac}}{2a}$

$x = \dfrac{-4 \pm \sqrt{4^2 - 4(-1)(8)}}{2(-1)}$

$= \dfrac{-4 \pm \sqrt{16 + 32}}{2}$

$= \dfrac{-4 \pm \sqrt{48}}{-2}$

$= \dfrac{-4 \pm 4\sqrt{3}}{-2}$

$= 2 \pm 2\sqrt{3}$

x-intercepts: $\left(2 + 2\sqrt{3},\, 0\right)$ and $\left(2 - 2\sqrt{3},\, 0\right)$

$x = 0: \; y = -0^2 + 4(0) + 8 = 8$

y-intercept: (0, 8)

$y = -x^2 + 4x + 8$

$a = -1, b = 4, c = 8$

$\dfrac{-b}{2a} = \dfrac{-4}{2(-1)} = \dfrac{-4}{-2} = 2$

$x = 2: \; y = -(2)^2 + 4(2) + 8 = -4 + 8 + 8 = 12$

vertex: (2, 12)

51. $y = 2x^2$ matches with graph A.

52. $y = -x^2$ matches with graph D.

53. $y = x^2 + 4x + 4$ matches with graph B.

54. $y = x^2 + 5x + 4$ matches with graph C.

55. The graph crosses the *x*-axis once so there is one real solution.

56. The graph crosses the *x*-axis twice so there are two real solutions.

57. The graph doesn't cross the *x*-axis so there are no real solutions.

58. The graph crosses the *x*-axis twice so there are two real solutions.

59. $x^2 = 49$

$x = \pm\sqrt{49}$

$x = \pm 7$

The solutions are $x = \pm 7$.

60. $y^2 = 75$

$y = \pm\sqrt{75}$

$y = \pm 5\sqrt{3}$

The solutions are $y = \pm 5\sqrt{3}$.

61. $(x-7)^2 = 64$

$x - 7 = \pm\sqrt{64}$

$x - 7 = \pm 8$

$x - 7 = 8$ or $x - 7 = -8$

$x = 15$ $x = -1$

The solutions are $x = 15$ and $x = -1$.

62. $x^2 + 4x = 6$

$x^2 + 4x + \left(\dfrac{4}{2}\right)^2 = 6 + \left(\dfrac{4}{2}\right)^2$

$(x+2)^2 = 10$

$x + 2 = \pm\sqrt{10}$

$x = -2 \pm \sqrt{10}$

The solutions are $x = -2 \pm \sqrt{10}$.

63. $3x^2 + x = 2$

$x^2 + \dfrac{1}{3}x = \dfrac{2}{3}$

$x^2 + \dfrac{1}{3}x + \left(\dfrac{1}{6}\right)^2 = \dfrac{2}{3} + \left(\dfrac{1}{6}\right)^2$

$\left(x + \dfrac{1}{6}\right)^2 = \dfrac{25}{36}$

$x + \dfrac{1}{6} = \pm\sqrt{\dfrac{25}{36}}$

$x + \dfrac{1}{6} = \pm\dfrac{5}{6}$

$x = -\dfrac{1}{6} \pm \dfrac{5}{6}$

$x = -\dfrac{1}{6} + \dfrac{5}{6} = \dfrac{4}{6} = \dfrac{2}{3}$ or

$x = -\dfrac{1}{6} - \dfrac{5}{6} = -\dfrac{6}{6} = -1$

The solutions are $x = \dfrac{2}{3}$ and $x = -1$.

64. $4x^2 - x - 2 = 0$

$x^2 - \dfrac{1}{4}x - \dfrac{1}{2} = 0$

$x^2 - \dfrac{1}{4}x = \dfrac{1}{2}$

$x^2 - \dfrac{1}{4}x + \left(-\dfrac{1}{8}\right)^2 = \dfrac{1}{2} + \left(-\dfrac{1}{8}\right)^2$

$\left(x - \dfrac{1}{8}\right)^2 = \dfrac{33}{64}$

$x - \dfrac{1}{8} = \pm\sqrt{\dfrac{33}{64}}$

$x - \dfrac{1}{8} = \pm\dfrac{\sqrt{33}}{8}$

$x = \dfrac{1}{8} \pm \dfrac{\sqrt{33}}{8}$

$x = \dfrac{1 \pm \sqrt{33}}{8}$

The solutions are $x = \dfrac{1 \pm \sqrt{33}}{8}$.

65. $4x^2 - 3x - 2 = 0$

$a = 4,\ b = -3,\ c = -2$

$x = \dfrac{-b \pm \sqrt{b^2 - 4ac}}{2a}$

$x = \dfrac{-(-3) \pm \sqrt{(-3)^2 - 4(4)(-2)}}{2(4)}$

$= \dfrac{3 \pm \sqrt{9 + 32}}{8}$

$= \dfrac{3 \pm \sqrt{41}}{8}$

The solutions are $x = \dfrac{3 \pm \sqrt{41}}{8}$.

66. $5x^2 + x - 2 = 0$

$a = 5,\ b = 1,\ c = -2$

$x = \dfrac{-b \pm \sqrt{b^2 - 4ac}}{2a}$

$$x = \frac{-1 \pm \sqrt{1^2 - 4(5)(-2)}}{2(5)}$$

$$= \frac{-1 \pm \sqrt{1 + 40}}{10}$$

$$= \frac{-1 \pm \sqrt{41}}{10}$$

The solutions are $x = \dfrac{-1 \pm \sqrt{41}}{10}$.

67. $4x^2 + 12x + 9 = 0$

$a = 4,\ b = 12,\ c = 9$

$$x = \frac{-b \pm \sqrt{b^2 - 4ac}}{2a}$$

$$x = \frac{-12 \pm \sqrt{12^2 - 4(4)(9)}}{2(4)}$$

$$= \frac{-12 \pm \sqrt{144 - 144}}{8}$$

$$= \frac{-12 \pm \sqrt{0}}{8}$$

$$= -\frac{12}{8}$$

$$= -\frac{3}{2}$$

The solution is $x = -\dfrac{3}{2}$.

68. $2x^2 + x + 4 = 0$

$a = 2,\ b = 1,\ c = 4$

$$x = \frac{-b \pm \sqrt{b^2 - 4ac}}{2a}$$

$$x = \frac{-1 \pm \sqrt{1^2 - 4(2)(4)}}{2(2)}$$

$$= \frac{-1 \pm \sqrt{1 - 32}}{4}$$

$$= \frac{-1 \pm \sqrt{-31}}{4}$$

The equation has no solution, since the square root of -31 is not a real number.

69. $y = 4 - x^2$

$y = 0: \quad 0 = 4 - x^2$

$\qquad\qquad x^2 = 4$

$\qquad\qquad x = \pm\sqrt{4}$

$\qquad\qquad x = \pm 2$

x-intercepts: $(2, 0)$ and $(-2, 0)$

$x = 0:\ y = 4 - 0^2 = 4$

y-intercept: $(0, 4)$

$y = 4 + 0x - x^2$

$a = -1,\ b = 0,\ c = 4$

$$\frac{-b}{2a} = \frac{-0}{2(-1)} = \frac{0}{-2} = 0$$

$x = 0:\ y = 4 - 0^2 = 4$

vertex: $(0, 4)$

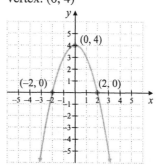

70. $y = x^2 + 4$

$y = 0: \qquad 0 = x^2 + 4$

$\qquad\qquad -4 = x^2$

$\qquad\quad \pm\sqrt{-4} = x$

x-intercepts: none

$x = 0:\ y = 0^2 + 4 = 4$

y-intercept: $(0, 4)$

$y = x^2 + 0x + 4$

$a = 1,\ b = 0,\ c = 4$

$$\frac{-b}{2a} = \frac{-0}{2(1)} = \frac{0}{2} = 0$$

$x = 0:\ y = 0^2 + 4 = 4$

vertex: $(0, 4)$

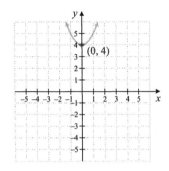

71. $y = x^2 + 6x + 8$

$y = 0$: $\quad 0 = x^2 + 6x + 8$

$\qquad 0 = (x + 4)(x + 2)$

$\qquad x + 4 = 0 \quad$ or $\quad x + 2 = 0$

$\qquad\quad x = -4 \qquad\qquad x = -2$

x-intercepts: $(-4, 0)$ and $(-2, 0)$

$x = 0$: $\quad y = 0^2 + 6(0) + 8 = 8$

y-intercept: $(0, 8)$

$y = x^2 + 6x + 8$

$a = 1, b = 6, c = 8$

$\dfrac{-b}{2a} = \dfrac{-6}{2(1)} = \dfrac{-6}{2} = -3$

$x = -3$:

$\quad y = (-3)^2 + 6(-3) + 8 = 9 - 18 + 8 = -1$

vertex: $(-3, -1)$

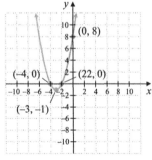

72. $y = x^2 - 2x - 4$

$y = 0$: $\quad 0 = x^2 - 2x - 4$

$a = 1, b = -2, c = -4$

$x = \dfrac{-b \pm \sqrt{b^2 - 4ac}}{2a}$

$x = \dfrac{-(-2) \pm \sqrt{(-2)^2 - 4(1)(-4)}}{2(1)}$

$\quad = \dfrac{2 \pm \sqrt{4 + 16}}{2}$

$\quad = \dfrac{2 \pm \sqrt{20}}{2}$

$\quad = \dfrac{2 \pm 2\sqrt{5}}{2}$

$\quad = 1 \pm \sqrt{5}$

x-intercepts: $\left(1 + \sqrt{5},\, 0\right)$ and $\left(1 - \sqrt{5},\, 0\right)$

$x = 0$: $\quad y = 0^2 - 2(0) - 4 = -4$

x-intercept: $(0, -4)$

$y = x^2 - 2x - 4$

$a = 1, b = -2, c = -4$

$\dfrac{-b}{2a} = \dfrac{-(-2)}{2(1)} = \dfrac{2}{2} = 1$

$x = 1$: $\quad y = 1^2 - 2(1) - 4 = 1 - 2 - 4 = -5$

vertex: $(1, -5)$

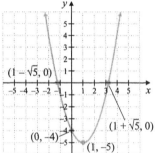

Chapter 9 Test

1. $\qquad x^2 - 400 = 0$

$\quad (x + 20)(x - 20) = 0$

$\quad x + 20 = 0 \quad$ or $\quad x - 20 = 0$

$\qquad\quad x = -20 \qquad\qquad x = 20$

The solutions are $x = \pm 20$.

2.
$$2x^2 - 11x = 21$$
$$2x^2 - 11x - 21 = 0$$
$$(2x+3)(x-7) = 0$$
$$2x+3 = 0 \quad \text{or} \quad x-7 = 0$$
$$2x = -3 \qquad\qquad x = 7$$
$$x = -\frac{3}{2}$$

The solutions are $x = -\dfrac{3}{2}$ and $x = 7$.

3. $5k^2 = 80$
$$k^2 = 16$$
$$k = \pm\sqrt{16}$$
$$k = \pm 4$$
The solutions are $k = \pm 4$.

4. $(3m-5)^2 = 8$
$$3m - 5 = \pm\sqrt{8}$$
$$3m - 5 = \pm 2\sqrt{2}$$
$$3m = 5 \pm 2\sqrt{2}$$
$$m = \frac{5 \pm 2\sqrt{2}}{3}$$

The solutions are $m = \dfrac{5 \pm 2\sqrt{2}}{3}$.

5.
$$x^2 - 26x + 160 = 0$$
$$x^2 - 26x = -160$$
$$x^2 - 26x + \left(\frac{-26}{2}\right)^2 = -160 + \left(\frac{-26}{2}\right)^2$$
$$x^2 - 26x + (-13)^2 = -160 + (-13)^2$$
$$x^2 - 26x + 169 = -160 + 169$$
$$(x-13)^2 = 9$$
$$x - 13 = \sqrt{9} \ \text{ or } \ x - 13 = -\sqrt{9}$$
$$x - 13 = 3 \qquad\quad x - 13 = -3$$
$$x = 16 \qquad\qquad x = 10$$
The solutions are $x = 10$ and $x = 16$.

6.
$$3x^2 + 12x - 4 = 0$$
$$x^2 + 4x - \frac{4}{3} = 0$$
$$x^2 + 4x = \frac{4}{3}$$
$$x^2 + 4x + \left(\frac{4}{2}\right)^2 = \frac{4}{3} + \left(\frac{4}{2}\right)^2$$
$$x^2 + 4x + (2)^2 = \frac{4}{3} + 2^2$$
$$(x+2)^2 = \frac{16}{3}$$
$$x + 2 = \pm\sqrt{\frac{16}{3}}$$
$$x + 2 = \pm\sqrt{\frac{16}{3}} \cdot \sqrt{\frac{3}{3}}$$
$$x + 2 = \pm\frac{4\sqrt{3}}{3}$$
$$x = -2 \pm \frac{4\sqrt{3}}{3}$$
$$x = \frac{-6 \pm 4\sqrt{3}}{3}$$

The solutions are $x = \dfrac{-6 \pm 4\sqrt{3}}{3}$.

7. $x^2 - 3x - 10 = 0$
$$a = 1, \, b = -3, \, c = -10$$
$$x = \frac{-b \pm \sqrt{b^2 - 4ac}}{2a}$$
$$x = \frac{-(-3) \pm \sqrt{(-3)^2 - 4(1)(-10)}}{2(1)}$$
$$= \frac{3 \pm \sqrt{9 + 40}}{2}$$
$$= \frac{3 \pm \sqrt{49}}{2}$$
$$= \frac{3 \pm 7}{2}$$
$$x = \frac{3+7}{2} = \frac{10}{2} = 5 \ \text{ or } \ x = \frac{3-7}{2} = \frac{-4}{2} = -2$$
The solutions are $x = 5$ and $x = -2$.

8. $p^2 - \dfrac{5}{3}p - \dfrac{1}{3} = 0$

$a = 1,\ b = -\dfrac{5}{3},\ c = -\dfrac{1}{3}$

$p = \dfrac{-b \pm \sqrt{b^2 - 4ac}}{2a}$

$p = \dfrac{-\left(-\frac{5}{3}\right) \pm \sqrt{\left(-\frac{5}{3}\right)^2 - 4(1)\left(-\frac{1}{3}\right)}}{2(1)}$

$= \dfrac{\frac{5}{3} \pm \sqrt{\frac{25}{9} + \frac{4}{3}}}{2}$

$= \dfrac{\frac{5}{3} \pm \sqrt{\frac{25}{9} + \frac{12}{9}}}{2}$

$= \dfrac{\frac{5}{3} \pm \sqrt{\frac{37}{9}}}{2}$

$= \dfrac{\frac{5}{3} \pm \frac{\sqrt{37}}{3}}{2}$

$= \dfrac{5 \pm \sqrt{37}}{6}$

The solutions are $p = \dfrac{5 \pm \sqrt{37}}{6}$.

9. $(3x - 5)(x + 2) = -6$

$3x^2 + 6x - 5x - 10 = -6$

$3x^2 + x - 10 = -6$

$3x^2 + x - 4 = 0$

$a = 3,\ b = 1,\ c = -4$

$x = \dfrac{-b \pm \sqrt{b^2 - 4ac}}{2a}$

$x = \dfrac{-1 \pm \sqrt{1^2 - 4(3)(-4)}}{2(3)}$

$= \dfrac{-1 \pm \sqrt{1 + 48}}{6}$

$= \dfrac{-1 \pm \sqrt{49}}{6}$

$= \dfrac{-1 \pm 7}{6}$

$x = \dfrac{-1 + 7}{6} = 1 \ \text{ or } \ x = \dfrac{-1 - 7}{6} = -\dfrac{4}{3}$

The solutions are $x = 1$ and $x = -\dfrac{4}{3}$.

10. $(3x - 1)^2 = 16$

$3x - 1 = \pm\sqrt{16}$

$3x - 1 = \pm 4$

$3x = 1 \pm 4$

$x = \dfrac{1 \pm 4}{3}$

$x = \dfrac{1 + 4}{3} = \dfrac{5}{3} \ \text{ or } \ x = \dfrac{1 - 4}{3} = \dfrac{-3}{3} = -1$

The solutions are $x = \dfrac{5}{3}$ and $x = -1$.

11. $3x^2 - 7x - 2 = 0$

$a = 3,\ b = -7,\ c = -2$

$x = \dfrac{-b \pm \sqrt{b^2 - 4ac}}{2a}$

$x = \dfrac{-(-7) \pm \sqrt{(-7)^2 - 4(3)(-2)}}{2(3)}$

$= \dfrac{7 \pm \sqrt{49 + 24}}{6}$

$= \dfrac{7 \pm \sqrt{73}}{6}$

The solutions are $x = \dfrac{7 \pm \sqrt{73}}{6}$.

12. $x^2 - 4x - 5 = 0$

$(x - 5)(x + 1) = 0$

$x - 5 = 0 \ \text{ or } \ x + 1 = 0$

$x = 5 \qquad\qquad x = -1$

The solutions are $x = 5$ and $x = -1$.

13. $3x^2 - 7x + 2 = 0$

$a = 3, b = -7, c = 2$

$x = \dfrac{-b \pm \sqrt{b^2 - 4ac}}{2a}$

$x = \dfrac{-(-7) \pm \sqrt{(-7)^2 - 4(3)(2)}}{2(3)}$

$= \dfrac{7 \pm \sqrt{49 - 24}}{6}$

$= \dfrac{7 \pm \sqrt{25}}{6}$

$= \dfrac{7 \pm 5}{6}$

$x = \dfrac{7 + 5}{6} = 2$ or $x = \dfrac{7 - 5}{6} = \dfrac{1}{3}$

The solutions are $x = \dfrac{1}{3}$ and $x = 2$.

14. $2x^2 - 6x + 1 = 0$

$a = 2, b = -6, c = 1$

$x = \dfrac{-b \pm \sqrt{b^2 - 4ac}}{2a}$

$x = \dfrac{-(-6) \pm \sqrt{(-6)^2 - 4(2)(1)}}{2(2)}$

$= \dfrac{6 \pm \sqrt{36 - 8}}{4}$

$= \dfrac{6 \pm \sqrt{28}}{4}$

$= \dfrac{6 \pm 2\sqrt{7}}{4}$

$= \dfrac{3 \pm \sqrt{7}}{2}$

The solutions are $x = \dfrac{3 \pm \sqrt{7}}{2}$.

15. Let x = the length of the base.

$4x$ = height

$A = \dfrac{1}{2}bh$

$18 = \dfrac{1}{2}(x)(4x)$

$36 = 4x^2$

$9 = x^2$

$\pm\sqrt{9} = x$

$\pm 3 = x$

$4x = 4(3) = 12$

The base is 3 feet. The height is 12 feet.

16. $y = -5x^2$

$y = 0:\ 0 = -5x^2$

$\qquad 0 = x^2$

$\qquad 0 = x$

x-intercept: $(0, 0)$

The y-intercept is also $(0, 0)$.

$y = -5x^2 + 0x + 0$

$a = -5, b = 0, c = 0$

$\dfrac{-b}{2a} = \dfrac{-0}{2(-5)} = \dfrac{0}{-10} = 0$

$x = 0:\ y = -5(0)^2 = 0$

vertex: $(0, 0)$

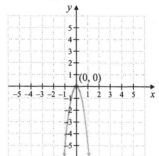

17. $y = x^2 - 4$

$y = 0:\ 0 = x^2 - 4$

$\qquad 0 = (x + 2)(x - 2)$

$\qquad\quad 0 = x + 2$ or $0 = x - 2$

$\qquad\quad -2 = x \qquad\qquad 2 = x$

x-intercepts: $(-2, 0), (2, 0)$

$x = 0$: $y = 0^2 - 4 = -4$

y-intercept: $(0, -4)$

$y = x^2 + 0x - 4$

$a = 1$, $b = 0$, $c = -4$

$$\frac{-b}{2a} = \frac{-0}{2(1)} = \frac{0}{2} = 0$$

$x = 0$: $y = 0^2 - 4 = -4$

vertex: $(0, -4)$

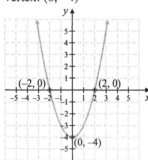

18. $y = x^2 - 7x + 10$

$y = 0$: $0 = x^2 - 7x + 10$

$\qquad 0 = (x - 5)(x - 2)$

$\qquad x - 5 = 0 \quad$ or $\quad x - 2 = 0$

$\qquad\qquad x = 5 \qquad\qquad x = 2$

x-intercepts: $(5, 0)$ and $(2, 0)$

$x = 0$: $y = 0^2 - 7(0) + 10 = 10$

y-intercept: $(0, 10)$

$y = x^2 - 7x + 10$

$a = 1$, $b = -7$, $c = 10$

$$\frac{-b}{2a} = \frac{-(-7)}{2(1)} = \frac{7}{2}$$

$x = \dfrac{7}{2}$: $\quad y = \left(\dfrac{7}{2}\right)^2 - 7\left(\dfrac{7}{2}\right) + 10$

$$= \frac{49}{4} - \frac{49}{2} + 10$$

$$= \frac{49}{4} - \frac{98}{4} + \frac{40}{4}$$

$$= -\frac{9}{4}$$

vertex: $\left(\dfrac{7}{2}, -\dfrac{9}{4}\right)$

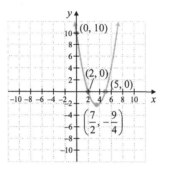

19. $y = 2x^2 + 4x - 1$

$y = 0$: $0 = 2x^2 + 4x - 1$

$a = 2$, $b = 4$, $c = -1$

$$x = \frac{-b \pm \sqrt{b^2 - 4ac}}{2a}$$

$$x = \frac{-4 \pm \sqrt{4^2 - 4(2)(-1)}}{2(2)}$$

$$= \frac{-4 \pm \sqrt{16 + 8}}{4}$$

$$= \frac{-4 \pm \sqrt{24}}{4}$$

$$= \frac{-4 \pm 2\sqrt{6}}{4}$$

$$= \frac{-2 \pm \sqrt{6}}{2}$$

x-intercepts: $\left(\dfrac{-2 - \sqrt{6}}{2}, 0\right)$ and

$\left(\dfrac{-2 + \sqrt{6}}{2}, 0\right)$

$x = 0$: $y = 2(0)^2 + 4(0) - 1 = -1$

y-intercept: $(0, -1)$

$y = 2x^2 + 4x - 1$

$a = 2$, $b = 4$, $c = -1$

$$\frac{-b}{2a} = \frac{-4}{2(2)} = \frac{-4}{4} = -1$$

$x = -1$:

$y = 2(-1)^2 + 4(-1) - 1 = 2 - 4 - 1 = -3$

vertex: $(-1, -3)$

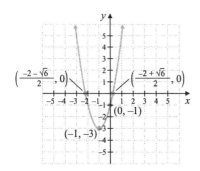

20. $d = \dfrac{n^2 - 3n}{2}$

$9 = \dfrac{n^2 - 3n}{2}$

$18 = n^2 - 3n$

$0 = n^2 - 3n - 18$

$0 = (n - 6)(n + 3)$

$n - 6 = 0$ or $n + 3 = 0$

    $n = 6$ $n = -3$

A 6-sided polygon has 9 diagonals.

21. $h = 16t^2$

       $120.75 = 16t^2$

       $\dfrac{120.75}{16} = t^2$

       $7.546875 = t^2$

  $\sqrt{7.546875} = t$ or $-\sqrt{7.546875} = t$

      $2.7 \approx t$ $-2.7 \approx t$

Since the time of the dive is not a negative number, discard the solution $t \approx -2.7$. The dive took approximately 2.7 seconds.

22. $y = 26x^2 - 136x + 607$

   $727 = 26x^2 - 136x + 607$

     $0 = 26x^2 - 136x - 120$

     $0 = 13x^2 - 68x - 60$

$a = 13, b = -68, c = -60$

$x = \dfrac{-b \pm \sqrt{b^2 - 4ac}}{2a}$

$x = \dfrac{-(-68) \pm \sqrt{(-68)^2 - 4(13)(-60)}}{2(13)}$

$= \dfrac{68 \pm \sqrt{4624 + 3120}}{26}$

$= \dfrac{68 \pm \sqrt{7744}}{26}$

$= \dfrac{68 \pm 88}{26}$

$x = \dfrac{68 + 88}{26} = 6$ or $x = \dfrac{68 - 88}{26} \approx -0.78$

Discard $x \approx -0.78$ since the number of years cannot be negative.
The value of mineral production is $727 million in 6 years after 2000 or 2006.

Cumulative Review Chapters 1–9

1. $y + 0.6 = -1.0$

       $y = -1.6$

2. $8x - 14 = 6x - 20$

       $2x = -6$

        $x = -3$

3. $8(2 - t) = -5t$

   $16 - 8t = -5t$

       $16 = 3t$

      $\dfrac{16}{3} = t$

4. $2(x + 7) = 5(2x - 3)$

   $2x + 14 = 10x - 15$

       $29 = 8x$

      $\dfrac{29}{8} = x$

5. Let x be the number of Democratic Representatives. Let $x + 15$ be the number of Republican Representatives.

   $x + x + 15 = 431$

      $2x + 15 = 431$

          $2x = 416$

           $x = 208$

$x + 15 = 223$

There are 208 Democratic Representatives and 223 Republican Representatives.

6. Let n be the first integer. Then $n + 1$ and $n + 2$ are the next two consecutive integers.
$$n + (n+1) + (n+2) = 438$$
$$3n + 3 = 438$$
$$3n = 435$$
$$n = 145$$
$n + 1 = 146$
$n + 2 = 147$
The three consecutive integers are 145, 146, and 147.

7. $3^0 = 1$

8. $\left(\dfrac{-6x}{y^3}\right)^3 = \dfrac{(-6x)^3}{(y^3)^3} = \dfrac{(-6)^3 x^3}{y^{3\cdot 3}} = -\dfrac{216x^3}{y^9}$

9. $(5x^3 y^2)^0 = 1$

10. $\dfrac{a^2 b^7}{(2b^2)^5} = \dfrac{a^2 b^7}{32 b^{10}} = \dfrac{a^2}{32 b^3}$

11. $-4^0 = -1$

12. $\dfrac{(3y)^2}{y^2} = \dfrac{9y^2}{y^2} = 9$

13. $(3y+2)^2 = (3y)^2 + 2(2)(3y) + 2^2$
$$= 9y^2 + 12y + 4$$

14. $(x^2 + 5)(y - 1) = x^2 y - x^2 + 5y - 5$

15. $\dfrac{x^2 + 7x + 12}{x + 3} = x + 4$

$$
\begin{array}{r}
x+4 \\
x+3{\overline{\smash{\big)}\,x^2 + 7x + 12}} \\
\underline{x^2 + 3x} \\
4x + 12 \\
\underline{4x + 12} \\
0
\end{array}
$$

16. $2 + 8.1a + a - 6 = 8.1a + a + 2 - 6$
$$= (8.1 + 1) + (2 - 6)$$
$$= 9.1a - 4$$

17. $r^2 - r - 42 = (r - 7)(r + 6)$

18. a. $\dfrac{x - y}{7 - x} = \dfrac{-4 - 7}{7 - (-4)} = \dfrac{-11}{11} = -1$

   b. $x^2 + 2y = (-4)^2 + 2(7) = 16 + 14 = 30$

19. $10x^2 - 13xy - 3y^2 = (2x - 3y)(5x + y)$

20. $\dfrac{1}{x+2} + \dfrac{7}{x-1} = \dfrac{1(x-1)}{(x+2)(x-1)} + \dfrac{7(x+2)}{(x-1)(x+2)}$
$$= \dfrac{x - 1 + 7x + 14}{(x+2)(x-1)}$$
$$= \dfrac{8x + 13}{(x+2)(x-1)}$$

21. $8x^2 - 14x + 5 = (2x - 1)(4x - 5)$

22. $\dfrac{x^2 + 7x}{5x} \cdot \dfrac{10x + 25}{x^2 - 49} = \dfrac{x(x+7)}{5x} \cdot \dfrac{5(2x+5)}{(x-7)(x+7)}$
$$= \dfrac{2x + 5}{x - 7}$$

23. a. $4x^3 - 49x = x(4x^2 - 49)$
$$= x(2x - 7)(2x + 7)$$

   b. $162x^4 - 2 = 2(81x^4 - 1)$
$$= 2(9x^2 - 1)(9x^2 + 1)$$
$$= 2(9x^2 + 1)(3x - 1)(3x + 1)$$

24. $\dfrac{2x + 7}{3} = \dfrac{x - 6}{2}$
$$6\left(\dfrac{2x+7}{3}\right) = 6\left(\dfrac{x-6}{2}\right)$$
$$2(2x+7) = 3(x-6)$$
$$4x + 14 = 3x - 18$$
$$x = -32$$

25. $(5x-1)(2x^2+15x+18)=0$

$(5x-1)(2x+3)(x+6)=0$

$5x-1=0$ or $2x+3=0$ or

$5x=1$ $2x=-3$

$x=\dfrac{1}{5}$ $x=-\dfrac{3}{2}$

$x+6=0$

$x=-6$

The solutions are $x=\dfrac{1}{5}$, $x=-\dfrac{3}{2}$, and

$x=-6$.

26. a. $4x-3+7-5x=-x+4$

   b. $-6y+3y-8+8y=5y-8$

   c. $2+8.1a+a-6=9.1a-4$

   d. $2x^2-2x=2x^2-2x$

27. $\dfrac{x^2+8x+7}{x^2-4x-5}=\dfrac{(x+7)(x+1)}{(x-5)(x+1)}=\dfrac{x+7}{x-5}$

28. $\quad 2x^2+5x=7$

$2x^2+5x-7=0$

$(2x+7)(x-1)=0$

$2x+7=0$ or $x-1=0$

$2x=-7$ $x=1$

$x=-\dfrac{7}{2}$

The solutions are $x=-\dfrac{7}{2}$ and $x=1$.

29. $\quad \dfrac{n}{6}-\dfrac{5}{3}=\dfrac{n}{2}$

$6\left(\dfrac{n}{6}-\dfrac{5}{3}\right)=6\left(\dfrac{n}{2}\right)$

$n-10=3n$

$-10=2n$

$-5=n$

The number is -5.

30. $d=\sqrt{(x_2-x_1)^2+(y_2-y_1)^2}$

$d=\sqrt{(2-(-7))^2+(5-4)^2}$

$d=\sqrt{(9)^2+(1)^2}$

$d=\sqrt{81+1}$

$d=\sqrt{82}$

The distance is $\sqrt{82}$ units.

31.

	x	$y=3x$
a.	-1	-3
b.	0	0
c.	-3	-9

32. a. x-intercept: $(4, 0)$

       y-intercept: $(0, 1)$

   b. x-intercepts: $(-2, 0)$, $(0, 0)$, $(3, 0)$

       y-intercept: $(0, 0)$

33. a. $\qquad y=-\dfrac{1}{5}x+1$

$2x+10y=3$

$10y=-2x+3$

$y=\dfrac{-2}{10}x+\dfrac{3}{10}$

$y=-\dfrac{1}{5}x+\dfrac{3}{10}$

These two equations have the same slope, therefore they are parallel.

   b. $\quad x+y=3 \Rightarrow y=-x+3$

$-x+y=4 \Rightarrow y=x+4$

These two equations have slopes whose product is -1, therefore they are perpendicular.

   c. $3x+y=5 \Rightarrow y=-3x+5$

$2x+3y=6 \Rightarrow 3y=-2x+6$

$\Rightarrow y=-\dfrac{2}{3}x+2$

These two equations have different slopes (and their product is $\neq -1$), so they are neither parallel nor perpendicular.

34. $y = 3x + 7$

$x + 3y = -15 \Rightarrow 3y = -x - 15$

$$y = -\frac{x}{3} - 5$$

These two equations have slopes whose product is -1; therefore these lines are perpendicular.

35. a. $\{(-1, 1), (2, 3), (7, 3), (8, 6)\}$
This is a function because each x-value is assigned to only one y-value.

b. $\{(0, -2), (1, 5), (0, 3), (7, 7)\}$
This is not a function because the x-value 0 is paired with two different y-values.

36. a. $\sqrt{80} + \sqrt{20} = 4\sqrt{5} + 2\sqrt{5} = 6\sqrt{5}$

b. $2\sqrt{98} - 2\sqrt{18} = 2 \cdot 7\sqrt{2} - 2 \cdot 3\sqrt{2}$
$$= 14\sqrt{2} - 6\sqrt{2}$$
$$= 8\sqrt{2}$$

c. $\sqrt{32} + \sqrt{121} - \sqrt{12} = 4\sqrt{2} + 11 - 2\sqrt{3}$

37. $\begin{cases} 2x + y = 10 \\ x = y + 2 \end{cases}$

Substitute $y + 2$ for x in the first equation.
$2(y + 2) + y = 10$
$2y + 4 + y = 10$
$3y = 6$
$y = 2$
Solve for x.
$x = y + 2 = 2 + 2 = 4$
The solution for this system is (4, 2).

38. $\begin{cases} 5x + y = 3 \\ y = -5x \end{cases}$

Substitute $-5x$ for y is the first equation.
Solve for x.
$5x - 5x = 3$
$0 = 3$ False
There are no solutions for this system.

39. $\begin{cases} 2x - y = 7 \\ 8x - 4y = 1 \end{cases}$

Solve the first equation for y.
$2x - y = 7$
$2x - 7 = y$
Substitute $2x - 7$ for y in the second equation and solve for x.
$8x - 4(2x - 7) = 1$
$8x - 8x + 28 = 1$
$28 = 1$ False
There are no solution for this system.

40. $\begin{cases} -2x + y = 7 \\ 6x - 3y = -21 \Rightarrow -2x + y = 7 \end{cases}$

The two equations are the same. Therefore, there are infinite number of solutions.

41. $\sqrt{36} = 6$ because $6^2 = 36$.

42. $\sqrt{\dfrac{4}{25}} = \dfrac{2}{5}$ because $\left(\dfrac{2}{5}\right)^2 = \dfrac{4}{25}$.

43. $\sqrt{\dfrac{9}{100}} = \dfrac{3}{10}$ because $\left(\dfrac{3}{10}\right)^2 = \dfrac{9}{100}$.

44. $\sqrt{\dfrac{16}{121}} = \dfrac{4}{11}$ because $\left(\dfrac{4}{11}\right)^2 = \dfrac{16}{121}$.

45. $\dfrac{2}{1+\sqrt{3}} \cdot \dfrac{1-\sqrt{3}}{1-\sqrt{3}} = \dfrac{2 - 2\sqrt{3}}{1 - 3}$
$$= \dfrac{2 - 2\sqrt{3}}{-2}$$
$$= -1 + \sqrt{3}$$

46. $\dfrac{5}{\sqrt{8}} = \dfrac{5}{2\sqrt{2}} \cdot \dfrac{\sqrt{2}}{\sqrt{2}} = \dfrac{5\sqrt{2}}{4}$

47. $(x-3)^2 = 16$

$\quad x - 3 = \pm\sqrt{16}$

$\quad x - 3 = \pm 4$

$\quad\quad x = 3 \pm 4$

$x = 3 + 4 = 7$ or $x = 3 - 4 = -1$

The solutions are $x = 7$ and $x = -1$.

48. $3(x-4)^2 = 9$

$\quad (x-4)^2 = 3$

$\quad\quad x - 4 = \pm\sqrt{3}$

$\quad\quad\quad x = 4 \pm \sqrt{3}$

The solutions are $x = 4 \pm \sqrt{3}$.

49. $\dfrac{1}{2}x^2 - x = 2$

$\dfrac{1}{2}x^2 - x - 2 = 0$

$a = \dfrac{1}{2},\ b = -1, c = -2$

$x = \dfrac{-b \pm \sqrt{b^2 - 4ac}}{2a}$

$x = \dfrac{-(-1) \pm \sqrt{(-1)^2 - 4\left(\frac{1}{2}\right)(-2)}}{2\left(\frac{1}{2}\right)}$

$\quad = \dfrac{1 \pm \sqrt{1+4}}{1}$

$\quad = 1 \pm \sqrt{5}$

The solutions are $x = 1 \pm \sqrt{5}$.

50. $x^2 + 4x = 8$

$\quad x^2 + 4x - 8 = 0$

$\quad a = 1, b = 4, c = -8$

$x = \dfrac{-b \pm \sqrt{b^2 - 4ac}}{2a}$

$x = \dfrac{-4 \pm \sqrt{4^2 - 4(1)(-8)}}{2(1)}$

$\quad = \dfrac{-4 \pm \sqrt{16 + 32}}{2}$

$\quad = \dfrac{-4 \pm \sqrt{48}}{2}$

$\quad = \dfrac{-4 \pm 4\sqrt{3}}{2}$

$\quad = -2 \pm 2\sqrt{3}$

The solutions are $x = -2 \pm 2\sqrt{3}$.

Appendix

1. $a^3 + 27 = a^3 + 3^3$
$$= (a+3)[a^2 - (a)(3) + 3^2]$$
$$= (a+3)(a^2 - 3a + 9)$$

3. $8a^3 + 1 = (2a)^3 + 1^3$
$$= (2a+1)[(2a)^2 - (2a)(1) + 1^2]$$
$$= (2a+1)(4a^2 - 2a + 1)$$

5. $5k^3 + 40 = 5(k^3 + 8)$
$$= 5(k^3 + 2^3)$$
$$= 5(k+2)[k^2 - (k)(2) + 2^2]$$
$$= 5(k+2)(k^2 - 2k + 4)$$

7. $x^3 y^3 - 64 = (xy)^3 - 4^3$
$$= (xy-4)[(xy)^2 + (xy)(4) + 4^2]$$
$$= (xy-4)(x^2 y^2 + 4xy + 16)$$

9. $x^3 + 125 = x^3 + 5^3$
$$= (x+5)[x^2 - (x)(5) + 5^2]$$
$$= (x+5)(x^2 - 5x + 25)$$

11. $24x^4 - 81xy^3 = 3x(8x^3 - 27y^3)$
$$= 3x[(2x)^3 - (3y)^3]$$
$$= 3x(2x-3y)[(2x)^2 + (2x)(3y) + (3y)^2]$$
$$= 3x(2x-3y)(4x^2 + 6xy + 9y^2)$$

13. $27 - t^3 = 3^3 - t^3$
$$= (3-t)[3^2 + (3)(t) + t^2]$$
$$= (3-t)(9 + 3t + t^2)$$

15. $8r^3 - 64 = 8(r^3 - 8)$
$$= 8(r^3 - 2^3)$$
$$= 8(r - 2)[r^2 + (r)(2) + 2^2]$$
$$= 8(r - 2)(r^2 + 2r + 4)$$

17. $t^3 - 343 = t^3 - 7^3$
$$= (t - 7)[t^2 + (t)(7) + 7^2]$$
$$= (t - 7)(t^2 + 7t + 49)$$

19. $s^3 - 64t^3 = s^3 - (4t)^3$
$$= (s - 4t)[s^2 + (s)(4t) + (4t)^2]$$
$$= (s - 4t)(s^2 + 4st + 16t^2)$$

Appendix C Exercise Set

1. mean $= \dfrac{21 + 28 + 16 + 42 + 38}{5} = \dfrac{145}{5} = 29$

16, 21, $\underline{28}$, 38, 42
median $= 28$
no mode

3. mean $= \dfrac{7.6 + 8.2 + 8.2 + 9.6 + 5.7 + 9.1}{6}$

$= \dfrac{48.4}{6}$

≈ 8.1

5.7, 7.6, $\underline{8.2, 8.2}$, 9.1, 9.6

median $= \dfrac{8.2 + 8.2}{2} = 8.2$

mode: 8.2

5. mean $= \dfrac{0.2 + 0.3 + 0.5 + 0.6 + 0.6 + 0.9 + 0.2 + 0.7 + 1.1}{9}$

$= \dfrac{5.1}{9}$

≈ 0.6

0.2, 0.2, 0.3, 0.5, $\underline{0.6}$, 0.6, 0.7, 0.9, 1.1
median $= 0.6$
mode: 0.2 and 0.6

7. $\text{mean} = \dfrac{231 + 543 + 601 + 293 + 588 + 109 + 334 + 268}{8}$

$= \dfrac{2967}{8}$

≈ 370.9

109, 231, 268, <u>293, 334</u>, 543, 588, 601

$\text{median} = \dfrac{293 + 334}{2} = 313.5$

no mode

9. $\dfrac{1454 + 1250 + 1136 + 1127 + 1107}{5} = \dfrac{6074}{5}$

$= 1214.8$

The mean height of the five tallest buildings is 1214.8 feet.

11. 1002, 1023, 1046, <u>1107, 1127</u>, 1136, 1250, 1454

$\text{median} = \dfrac{1107 + 1127}{2} = 1117$

The median height of the eight tallest buildings is 1117 feet.

13. $\dfrac{7.8 + 6.9 + 7.5 + 4.7 + 6.9 + 7.0}{6} = \dfrac{40.8}{6} = 6.8$

The mean time was 6.8 seconds.

15. The mode time is 6.9 seconds.

17. 74, 77, <u>85, 86</u>, 91, 95

$\text{median} = \dfrac{85 + 86}{2} = 85.5$

The median test score was 85.5.

19. $\dfrac{78 + 80 + 66 + \cdots + 72}{15} = \dfrac{1095}{15} = 73$

The mean pulse rate is 73.

21. The values 70 and 71 both occur twice while other values only occur once, so the mode of the pulse rates is 70 and 71.

23. There were 9 rates lower than the mean of 73.

25. Since the mode is 21, the value 21 must occur at least twice in the set of numbers. Since there are an odd number of numbers in the set, the median is the middle number. That is, the median of 20 is one of the numbers in the set.
The missing numbers are 21, 21, and 20.

Appendix D Exercise Set

1. The set of negative integers from −10 to −5 in roster form is {−9, −8, −7, −6}.

3. The set of the days of the week starting with the letter *T* in roster form is {Tuesday, Thursday}.

5. The set of whole numbers in roster form is {0, 1, 2, 3, 4, ...}.

7. There are no integers between 1 and 2, so the answer is { } or ∅.

9. Since 3 is a listed element of the set {1, 3, 5, 7, 9}, the statement 3 ∈ {1, 3, 5, 7, 9} is true.

11. {3} ⊆ {1, 3, 5, 7, 9} since every element of the left-hand set is also an element of the right-hand set.

13. {*a*, *e*, *i*, *o*, *u*} ⊆ {*a*, *e*, *i*, *o*, *u*} since every element of the left-hand set is also an element of the right-hand set.

15. {May} ⊆ the set of the days of the week is false since the element in the left-hand set is not an element of the right-hand set. Note that the days of the week are Sunday, Monday, Tuesday, Wednesday, Thursday, Friday, and Saturday.

17. Since 9 is not an even number, it is not an element of the set {*x*|*x* is an even number}. Thus the statement 9 ∉ {*x*|*x* is an even number} is true.

19. {*a*} ⊄ the set of vowels is false since the element in the left-hand set is a vowel.

21. $A \cup B = \{1, 2, 3, 4, 5, 6\} \cup \{2, 4, 6\}$
$= \{1, 2, 3, 4, 5, 6\}$

23. $A \cap B = \{1, 2, 3, 4, 5, 6\} \cap \{2, 4, 6\}$
$= \{2, 4, 6\}$

25. $C \cup D = \{1, 3, 5\} \cup \{7\} = \{1, 3, 5, 7\}$

27. $B \cap D = \{2, 4, 6\} \cap \{7\} = \varnothing$

29. Since every element of $B = \{2, 4, 6\}$ is also an element of $A = \{1, 2, 3, 4, 5, 6\}$, the statement $B \subseteq A$ is true.

31. Since the element of $D = \{7\}$ is not an element of $C = \{1, 3, 5\}$, the statement $D \subseteq C$ is false.

33. Since 2 is an element of $A = \{1, 2, 3, 4, 5, 6\}$, but not of $C = \{1, 3, 5\}$, the statement $A \subseteq C$ is false.

35. Since not every element of $B = \{2, 4, 6\}$ is also an element of $C = \{1, 3, 5\}$, the statement $B \not\subseteq C$ is true.

37. Since the empty set is a subset of every set, the statement $\varnothing \subseteq A$ is true.

39. $A \cup D = \{1, 2, 3, 4, 5, 6\} \cup \{7\}$
$= \{1, 2, 3, 4, 5, 6, 7\}$
The statement is true.

41. Since the union of a set with the empty set is the original set, the statement that {*a*, *b*, *c*} ∪ { } is {*a*, *b*, *c*} is true.

Appendix E Exercise Set

1. 90° − 19° = 71°
The complement of a 19° angle is a 71° angle.

3. 90° − 70.8° = 19.2°
The complement of a 70.8° angle is a 19.2° angle.

5. $90° - 11\frac{1}{4}° = 78\frac{3}{4}°$

The complement of an $11\frac{1}{4}°$ angle is a

$78\frac{3}{4}°$ angle.

7. $180° - 150° = 30°$
The supplement of a 150° angle is a
30° angle.

9. $180° - 30.2° = 149.8°$
The supplement of a 30.2° angle is a
149.8° angle.

11. $180° - 79\frac{1}{2}° = 100\frac{1}{2}°$

The supplement of a $79\frac{1}{2}°$ angle is a

$100\frac{1}{2}°$ angle.

13. $\angle 1$ and the angle marked 110° are vertical
angles, so $m\angle 1 = 110°$.
$\angle 2$ and the angle marked 110° are
supplementary angles, so
$m\angle 2 = 180° - 110° = 70°$.
$\angle 3$ and the angle marked 110° are
supplementary angles, so
$m\angle 3 = 180° - 110° = 70°$.
$\angle 4$ and $\angle 3$ are alternate interior angles, so
$m\angle 4 = m\angle 3 = 70°$.
$\angle 5$ and the angle marked 110° are alternate
interior angles, so $m\angle 5 = 110°$.
$\angle 6$ and $\angle 5$ are supplementary angles, so
$m\angle 6 = 180° - m\angle 5 = 180° - 110° = 70°$.
$\angle 7$ and the angle marked 110° are
corresponding angles, so $m\angle 7 = 110°$.

15. $180° - 11° - 79° = 90°$
The third angle measures 90°.

17. $180° - 25° - 65° = 90°$
The third angle measures 90°.

19. $180° - 30° - 60° = 90°$
The third angle measures 90°.

21. Since the triangle is a right triangle, one
angle measures 90°.
$180° - 45° - 90° = 45°$
The other two angles of the triangle measure
45° and 90°.

23. Since the triangle is a right triangle, one
angle measures 90°.
$180° - 17° - 90° = 73°$
The other two angles of the triangle measure
90° and 73°.

25. Since the triangle is a right triangle, one
angle measures 90°.
$$180° - 39\frac{3}{4}° - 90° = 50\frac{1}{4}°$$
The other two angles of the triangle measure
$50\frac{1}{4}°$ and 90°.

27. $\dfrac{12}{4} = \dfrac{18}{x}$
$12x = 72$
$x = \dfrac{72}{12}$
$x = 6$

29. $\dfrac{6}{9} = \dfrac{3}{x}$
$6x = 27$
$x = \dfrac{27}{6}$
$x = 4.5$

31. $a^2 + b^2 = c^2$
$6^2 + 8^2 = c^2$
$36 + 64 = c^2$
$100 = c^2$
Since c represents a length, we assume that
c is positive. Since $c^2 = 100$, $c = 10$. The
hypotenuse of the right triangle has length
10.

33. $a^2 + b^2 = c^2$
$a^2 + 5^2 = 13^2$
$a^2 + 25 = 169$
$a^2 = 144$
Since a represents a length, we assume that
a is positive. Since $a^2 = 144$, $a = 12$. The
other leg of the right triangle has length 12.